FOOD SAFETY AND HUMAN HEALTH

FOOD SAFETY AND HUMAN HEALTH

Edited by

RAM LAKHAN SINGH

Department of Biochemistry, Dr. Rammanohar Lohia Avadh University, Faizabad, India

SUKANTA MONDAL

Animal Physiology Division, ICAR-National Institute of Animal Nutrition and Physiology, Adugodi, India

ELSEVIER

ACADEMIC PRESS

An imprint of Elsevier

Academic Press is an imprint of Elsevier
125 London Wall, London EC2Y 5AS, United Kingdom
525 B Street, Suite 1650, San Diego, CA 92101, United States
50 Hampshire Street, 5th Floor, Cambridge, MA 02139, United States
The Boulevard, Langford Lane, Kidlington, Oxford OX5 1GB, United Kingdom

Notices

Knowledge and best practice in this field are constantly changing. As new research and experience broaden our understanding, changes in research methods, professional practices, or medical treatment may become necessary.

Practitioners and researchers must always rely on their own experience and knowledge in evaluating and using any information, methods, compounds, or experiments described herein. In using such information or methods they should be mindful of their own safety and the safety of others, including parties for whom they have a professional responsibility.

To the fullest extent of the law, neither the Publisher nor the authors, contributors, or editors, assume any liability for any injury and/or damage to persons or property as a matter of products liability, negligence or otherwise, or from any use or operation of any methods, products, instructions, or ideas contained in the material herein.

British Library Cataloguing-in-Publication Data
A catalogue record for this book is available from the British Library

Library of Congress Cataloging-in-Publication Data
A catalog record for this book is available from the Library of Congress

ISBN: 978-0-12-816333-7

For Information on all Academic Press publications
visit our website at https://www.elsevier.com/books-and-journals

Publisher: Charlotte Cockle
Acquisition Editor: Patricia Osborn
Editorial Project Manager: Tasha Frank
Production Project Manager: Omer Mukthar
Cover Designer: Mark Rogers

Typeset by MPS Limited, Chennai, India

Dedication

Dedicated to

Our beloved daughters

Suneha (Daughter of Dr. (Mrs.) Uma Lakhan Singh and Dr. Ram Lakhan Singh), who always functions as our lifeline and remains concerned for every single bit related to us

&

Anoushka Mondal (Daughter of Mrs. Shrabanti Mondal and Dr. Sukanta Mondal), for making everything worthwhile and cheerfully tolerating many hours of absence involved in writing the book

Contents

12. International Laws and Food-Borne Illness

TEK CHAND BHALLA, MONIKA, SHEETAL AND SAVITRI

List of Contributors

Vinod R. Bhagwat Department of Biochemistry, S.B.H. Government Medical College, Dhule, India

Tek Chand Bhalla Department of Biotechnology, Himachal Pradesh University, Shimla, India

S. Biswas Department of Livestock Products Technology, West Bengal University of Animal and Fishery Sciences, Kolkata, India

Pushpa Choudhary ICAR-Central Institute of Freshwater Aquaculture, Bhabaneswar, India

Annada Das Department of Livestock Products Technology, West Bengal University of Animal and Fishery Sciences, Kolkata, India

Arun K. Das Eastern Regional Station, ICAR-Indian Veterinary Research Institute, Kolkata, India

Rajeeva Gaur Department of Microbiology (Centre of Excellence), Dr. Rammanohar Lohia Avadh University, Faizabad, India

Ankit Gupta Department of Biochemistry, All India Institute of Medical Sciences (AIIMS), Raibareli, India

Charu Gupta Amity Institute for Herbal Research & Studies, Amity University, Noida, India

Rasna Gupta Department of Biochemistry, Dr. Rammanohar Lohia Avadh University, Faizabad, India

Mozammel Hoque ICAR-Indian Veterinary Research Institute, Bareilly, India

Kamal Kishore Department of Pharmacy, M.J.P. Rohilkhand University, Bareilly, India

Neeraj Kumar Dr. R. M. L. Institute of Pharmacy, Kunwarpur Badagaon, Powayan, India

Sukanta Mondal Animal Physiology Division, ICAR-National Institute of Animal Nutrition and Physiology, Adugodi, India

Monika Department of Biotechnology, Himachal Pradesh University, Shimla, India

P.K. Nanda Eastern Regional Station, ICAR-Indian Veterinary Research Institute, Kolkata, India

Dhan Prakash Amity Institute for Herbal Research & Studies, Amity University, Noida, India

Mrinal Samanta ICAR-Central Institute of Freshwater Aquaculture, Bhabaneswar, India

Savitri Department of Biotechnology, Himachal Pradesh University, Shimla, India

Dinesh Kumar Sharma Amrapali Institute of Pharmacy and Sciences, Lamachaur, Haldwani, Uttarakhand, India

Sheetal Department of Biotechnology, Himachal Pradesh University, Shimla, India

Anita Singh Department of Pharmacy, Kumaun University, Bhimtal, India

Anurag Singh Department of Microbiology (Centre of Excellence), Dr. Rammanohar Lohia Avadh University, Faizabad, India

Pankaj Singh Department of Biochemistry, Jhunjhunwala P. G. College, Faizabad, India

Pradeep Kumar Singh Department of Biochemistry, Dr. Rammanohar Lohia Avadh University, Faizabad, India

Rajat Pratap Singh Department of Biotechnology, Dr. Rammanohar Lohia Avadh University, Faizabad, India

Ram Lakhan Singh Department of Biochemistry, Dr. Rammanohar Lohia Avadh University, Faizabad, India

Ravindra Pratap Singh Department of Biotechnology, Indira Gandhi National Tribal University, Amarkantak, India

Ashutosh Tripathi Department of Microbiology (Centre of Excellence), Dr. Rammanohar Lohia Avadh University, Faizabad, India

Preface

Food safety is an important public health issue that relates to human health and economic development. Safe food contributes to health and productivity and provides an effective mean for development and poverty alleviation. People are becoming increasingly concerned about the health risks posed by microbial pathogens and potentially hazardous chemicals in food. Outbreaks of food-borne illnesses continue to be a major problem worldwide while international trade in food products is increasing. According to the World Health Organization (WHO) estimates, 1.8 million deaths related to contaminated food or water occur every year. Food-borne diseases not only significantly affect people's health and well-being, but they also have economic consequences for individuals, families, communities, businesses, and countries. Developing countries in particular are experiencing rapid changes in their health and social environments, and the strains on their limited resources are compounded by expanding urbanization, increasing dependence on stored foods, and insufficient access to safe water and facilities for safe food preparation. The globalization of the food trade offers many benefits to consumers, as it results in a wider variety of high-quality foods that are accessible, affordable, and safe, meeting consumer demand. A diversity of foods in a balanced diet improves nutritional status and health. Food safety programs are increasingly focusing on a farm-to-table approach as an effective means of reducing food-borne hazards. The introduction of new technologies, including genetic engineering and irradiation, in this climate of concern about food safety is posing a special challenge. Some new technologies will increase agricultural production and make food safer, but their usefulness and safety must be demonstrated if they are to be accepted by consumers. Furthermore, the evaluation must be participatory, transparent, and conducted using internationally agreed methods. New technologies, such as genetic engineering, irradiation of food, ohmic heating, and modified atmosphere packaging, can be used to increase agricultural production, extend shelf life, or make food safer.

Effective food safety systems are vital to maintain consumer confidence in the food system and to provide a sound regulatory foundation for domestic and international trade in food, which supports economic development. The primary objective of food safety is safeguarding the public health by ensuring safe and good quality foods that do not create an adverse effect on human health. The food safety can be controlled in food industry by developing in-house food safety and food quality programs. Food-borne diseases are a widespread and growing public health problem, both in developed as well as developing nations of the world. These diseases have a great impact on the economy of a country. The factors that contribute to food-borne disease include improper storage of temperature, poor personal hygiene, contaminated equipment, inadequate cooking, food from unsafe sources, etc. Food-borne microbial diseases account for 20 million cases annually, and their incidence is increasing. In developing countries, food-borne diseases cause an estimated 2.2 million deaths each year, of which 1.9 million are children. Food should be the source of nourishment for human beings and not an opportunity for potential pathogens, which can cause serious and life-threatening illness. Food safety is all about making food safe to eat and free of disease-causing agents such as infectious microbes, toxic chemicals, and foreign bodies. Many of the microbes of current concern survive in the environment, in water, on pastures, and in food, unless proper precautions are taken.

It gives us immense pleasure to present this book, *Food Safety and Human Health.* This book is an outstanding collection of current research that integrates basic and advanced concepts of food toxicology with future development prospects. The book provides the basic principles of food toxicology, its processing, and safety for human health in order to help the students and professionals in better understanding of real problems of toxic materials in foods. This will help them to address the day-to-day problems reported regarding food contamination and safety. This book will help in identifying potential new approaches to develop new microbiologically safe functional foods and significantly contribute to decrease the incidence of bacterial food-borne illness outbreaks and gastrointestinal diseases. This will also serve as a reference book for scientists, researchers, educators, and professionals in this area.

ORGANIZATION OF THE BOOK

The book consists of 12 chapters that focus on current issues in food safety and strategies for minimizing the food-borne illness.

Chapter 1, Current Issues in Food Safety With Reference to Human Health, deals with brief introduction of food toxicology, food hazards, and their impact on human health, the need for food toxicity studies, and various aspects of food safety and toxicology vis-à-vis human health.

Chapter 2, Food Hazards: Physical, Chemical, and Biological, includes the physical, chemical, and biological food toxicants and their effects on human health. It also deals with toxicants generated during food processing and genetically modified foods.

Chapter 3, Toxicity of Food Additives, discusses contaminants in foods, food additive safety regulations, toxicity of food additives, toxicological evaluation of food additives, in vitro genetic toxicity tests, in vivo genetic toxicity tests, need of advancements in toxicity testing, and new technologies in toxicity studies.

Chapter 4, Food Allergies, covers food allergy and associated complexity, emerging problems, and recent recommendations associated with food allergy, allergens and their types, immunology of food allergy, associated food allergy conditions, food allergy versus food intolerances, precautions for food allergy, diagnosis for food allergens, guidelines for management of food allergies, and food labeling.

Chapter 5, Safety of Milk and Dairy Products, highlights milk-borne diseases, lactose intolerance, sourcing of milk and its proteins, milk allergy, antibiotic residues in milk, heavy metals and other pollutants in milk, adulteration of milk, artificial cow's milk, plant-based milk, synthetic milk, organic milk, A1–A2 milk, etc.

Chapter 6, Hazards and Safety Issues of Meat and Meat Products, includes sources of contamination at different stages of production, hazards associated with meat and meat

products, management and control of hazards associated with meat and meat products, and meat safety and human health.
Chapter 7, Safety of Fish and Seafood, focuses on bacterial contamination and naturally occurring toxins in fish and seafood products, fish and seafood product–borne parasitic diseases, detection of pathogens and spoilage of fish and seafoods, preservation and packaging, antimicrobial compounds used in fish and seafood products, and control of pathogens in these products.
Chapter 8, Microbial Environment of Food, describes the microbial food environment, extrinsic factors affecting microbial growth, intrinsic food factors and microbial physiology, and growth kinetics factors.
Chapter 9, Safety of Water Used in Food Production, illustrates resources of water used in food production, storage and preservation, water-borne food contaminants, health hazards in water and food contamination, water treatment, and soil and water quality.
Chapter 10, Safety of Fresh Fruits and Vegetables, describes sources of contamination in fruits and vegetables, deterioration, spoilage and post-harvest losses, causes of spoilage in fruits and vegetables, development of biopreservatives for fresh-harvest produce, strategies for safety of fruits and vegetables, and problems in natural products applications.
Chapter 11, Utility of Nanomaterials in Food Safety, discusses use of nanomaterials in food processing and packaging, nanomaterials in smart/active/intelligent food, its utility in polymer nanocomposites, nanomaterials as antimicrobial agents, nanomaterials in food pathogens detection, nanomaterials for protection from food allergens, nanomaterials in heavy metals detection in food, nanomaterials vis-a-vis food safety issues, as well as challenges, perspectives, and health risks.
Chapter 12, International Laws and Food-Borne Illness, describes food-borne illness, agents for nonbacterial food-borne illness, food-borne illness due to agricultural pesticides and insecticides, food-borne viral infections, fungi and mycotoxins, helminths, protozoa, and international laws related to food safety.

Ram Lakhan Singh and Sukanta Mondal

Acknowledgments

It is a pleasure to acknowledge our enormous debt to contributors who assisted materially in the preparation of this book. We express our gratitude to all those who helped directly or indirectly in the accomplishment of this work with their support, valuable guidance, and innumerable suggestions. We are grateful to both of our families, who cheerfully tolerated and supported many hours of our absence in finishing the book project. We wish to express special appreciation to the editorial and production staffs of Elsevier for their excellent work. The team of Elsevier publishing group has played a great role throughout, always helpful and supportive. Special thanks are due to Ms. Nancy Maragioglio, who has been instrumental in getting this proposal approved.

Last but not the least, we are grateful to all the research students who have contributed in so many ways for production of this text.

Ram Lakhan Singh and Sukanta Mondal

May 2019

CHAPTER

1

Current Issues in Food Safety With Reference to Human Health

Ram Lakhan Singh[1] *and Sukanta Mondal*[2]

[1]Department of Biochemistry, Dr. Rammanohar Lohia Avadh University, Faizabad, India

[2]Animal Physiology Division, ICAR-National Institute of Animal Nutrition and Physiology, Adugodi, India

OUTLINE

INTRODUCTION

Food safety is an increasingly global public health issue as humans suffer from a plethora of foodborne diseases. The diseases caused by foodborne pathogens constitute a worldwide public health concern. Ensuring food safety to protect public health remains a significant challenge in both developing and developed nations. Effective food safety systems are vital to maintain consumer confidence in the food system and to provide a sound

DOI: https://doi.org/10.1016/B978-0-12-816333-7.00001-1

regulatory foundation for domestic and international trade in food, which supports economic development. Food safety has become an increasingly important international concern, as food contamination creates a huge economic burden on the communities and their health (Fig. 1.1). Governments worldwide are intensifying their efforts to improve food safety so that no consumer experiences any infection or disease following the consumption of food. Developing nations are encountering quick changes in their well-being and social situations, and the strains on their restricted assets are aggravated by growing urbanization, expanding reliance on stored foods, inadequate access to safe water, and facilities for secure food production. Food security programs are progressively concentrating on a farm-to-table methodology as a successful method for decreasing foodborne risks. Among various factors, foodborne microbial diseases account for 20 million cases annually, and their incidence is increasing. In developing countries, foodborne diseases cause an estimated 2.2 million deaths each year, of which 1.9 million are children. Food should be the source of nourishment for human beings and not an opportunity for potential pathogens, which can cause serious and life-threatening illness. Globalization of the food supply has created conditions favorable for the emergence, reemergence, and spread of foodborne pathogens.

Many foodborne pathogens have reemerged due to factors related to lifestyle and political, economic, and ecological changes. In industrialized nations, one out of every three persons has a foodborne illness event every year. The globalization of food markets has increased the challenge to manage the microbial risks. The recent technologies, for example, genetic engineering, food irradiation, ohmic heating, and modified packaging, may be utilized to raise the agricultural production, broaden shelf life, or make food more secure.

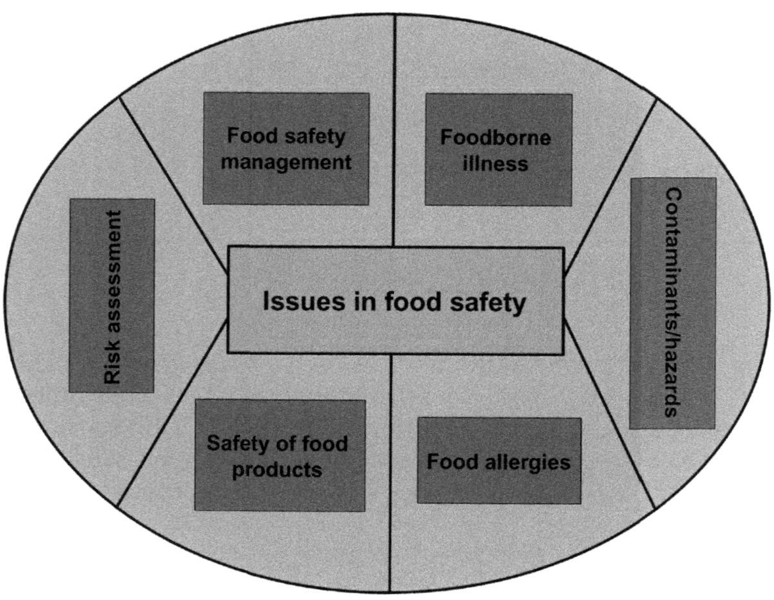

FIGURE 1.1 Issues in food safety.

FOODBORNE DISEASES

Microorganisms are defined as microscopic living organisms that occur ubiquitously in the environment. Although some are harmless, some are virulent and could cause infections. The infections are due to several factors including violation of certain basic food hygienic practices. For example, foodborne diseases (FBDs) are caused by microorganisms or toxins transmitted through person-to-person, animal-human, or human-animal contacts and through contact with the environment such as through human-to-surface or through equipment. Transmissions of infections are recorded to either be directly or indirectly from food and/or water, which, in most cases, act as vehicles for infection. Contamination by food poisoning agents can occur at various stages during the food chain in raw products, prior to harvesting, during slaughter, or in processing or as cross-contamination in the kitchen by the food handlers. The Centers for Disease Control and Prevention (CDC) further recorded that eating food containing pathogens or their toxins (poisons) is the leading cause of foodborne illness. The CDC further stated that there are four types of microorganisms that could contaminate food and cause FBDs: bacteria, viruses, parasites, and fungi. The National Restaurant Association has recorded the nutrients that supported these microorganisms to grow including food, acidity, temperature, time, oxygen, and moisture. Almost all of these are derived from foods that humans consume. According to the World Health Organization (WHO), foodborne illnesses are reported daily the worldwide in both developed and developing countries. Further reports indicated that illnesses caused by contaminated food constitute one of the most extensive issues and is a primary driver of decreased monetary profitability. The prevalence rate of FBDs has raised much concern, since the magnitude of the problem was previously unknown due to lack of reliable data. According to WHO, FBDs have increased, and even more challenging is the reemergence of drug-resistant microorganisms, which are viewed as a big threat to the hospitality industry. According to the CDC, *Escherichia coli* 0157, H. *Listeria monocytogenes*, and *Salmonella enterica* are three of the eight known pathogens that account for the vast majority of reported FBD, hospitalizations, and deaths each year (Sonnier et al., 2018). For example, outbreaks of Salmonellosis have been reported for decades, but within the past 25 years the disease has increased in many continents.

Salmonella

Infections are usually traceable to various food products derived from meat, eggs, milk, and poultry. Although *Salmonella* traditionally is thought of as being associated with animal products in the past, fresh produce also has been the source of major outbreaks, particularly recently. *Salmonella* infections can originate from household pets containing the bacteria; improperly prepared meats and seafood; or the surfaces of raw eggs, fruits, or vegetables that have not been adequately disinfected. In Europe and other Western countries, *Salmonella* serotype *enteritidis* (SE) had become the predominant strain (Boyce et al., 1996). In Australia, Canada, the United States, European countries, and South Africa, infections with *E. coli* serotype 0157:H7 are reported to be a major cause of bloody diarrhea and acute renal failure (Motarjemi and Kaferstein, 1998). Outbreaks of these infections are associated with beef consumption and were said to be fatal particularly in children. Many developed nations have encountered flare-ups of illnesses because of recently perceived foodborne pathogens, for example, *E. coli* 0157:H7, *Campylobacter jejuni*, and *L. monocytogenes*. *Campylobacter* and *Vibrio parahaemolyticus*, which are the most common pathogens in fish, are most probably transmitted in the market or while cooking, according to Adams and Moss (2000). Other contaminants like pesticide residues or environmental chemicals are also reported in fruits and vegetables. Examples included *Bacillus*, *Clostridium*, and *L. monocytogenes*, which are said to be introduced from the soil, as well as viruses such as rotavirus and bacteria including *Shigella*, *Salmonella*, and *E. coli*. The problem is not limited to developing countries alone. Studies in industrialized nations evaluated that every year, 5%–10% of the populace experienced FBD. In many countries, *Salmonella enteritidis* is cited as the predominant pathogen, and poultry, eggs, and egg products are identified as the major source. Globally, about 60%–80% of poultry are accounted to be contaminated with *S. enteritidis* which was mostly associated with dairy products beef, poultry, and eggs. In developing countries, in Latin America, Asia, and Africa, the rate of infection has been less documented, yet these countries have borne the brunt of the problem due to the presence of a wide range of FBDs. *Salmonella* and *Shigella* are considered as main pathogens responsible for most of the foodborne diseases and had been largely associated with diarrhea, abdominal pains, nausea, and vomiting (Motarjemi and Kaferstein, 1998). Although viruses did not grow on food, raw fruits and vegetables have been cited to provide avenues for infection.

Campylobacteriosis

The genus *Campylobacter* includes a number of species, but *C. jejuni* and *C. coli* are mainly responsible for human enterocolitis (Pal, 2007). *Campylobacter* is recognized as one of the principal causes of human acute gastroenteritis worldwide (Allos, 2001; Pal, 2005). *C. jejuni* represents 80%–90% of enteric sicknesses. *Campylobacter* species are part of typical intestinal microflora of wild animals, livestock, and birds. Animals utilized in production of food including poultry, swine, cattle, and sheep are the primary reservoir (Blaser, 1997). The family pets, for example, dogs, cats, and birds, are additional animal reservoirs. In spite of the fact that discharge of *Campylobacter* is not related to manifestations in poultry, diarrheal sicknesses are depicted in mammalian pets and domesticated animals, and this adds to the contamination of surface water. Ingestion of contaminated water, interaction with colonized pets, especially puppies and kittens, and consumption of unpasteurized milk or undercooked poultry or meat are all associated with human disease (Pal, 2007). *Campylobacter* is responsible for about 5%–14% of all diarrheal sicknesses worldwide (Anon, 2000a,b).

Brucellosis and Other Pathogens

Brucellosis is another FBD that has raised concern. It occurs worldwide; however, in North America and Western Europe, incidents have been reported to have decreased due to strict surveillance and the application of the hazard analysis and critical control point (HACCP) system. However, the disease still remains an important health problem in the

Mediterranean countries (Egypt, Greece, Italy, Morocco, and Tunisia), the Middle East (Iraq, Iran, and Saudi Arabia) Mexico, Peru, and some regions of China and India. *V. parahaemolyticus* is listed as another pathogen that causes acute gastroenteritis. In developing countries, especially in Africa, the microorganism constitutes a group of pathogens that caused persistent diarrhea. This group includes *Giardia lamblia, Cryptosporidium* sp., and *Entamoeba histolytica*. These pathogens mainly affected children and people with impaired immunity. Raw and uncooked meat and vegetables are major routes for their transmission. Other parasites transmitted through raw meat included *Trichinella spiralis, Taenia solium,* and *Taenia saginata* (Mensah, 2002). The most recent FBD is the avian influenza. According to WHO (2005), the virus, which attacks birds, is highly pathogenic. The flu started in Asia, moved to Europe, and more recently moved to Africa. It was reported that the virus infected human beings through contact with infected live or dead poultry. Exposure may have also occurred when the virus was inhaled through dust and possibly through contact with contaminated surfaces. The avian flu virus was found in the respiratory and gastrointestinal tracts of infected birds and not in the meat. Regardless, accessible information demonstrated that profoundly pathogenic viruses, for example, H5N1 strain, may have been spread through marketing and distribution channels, since low temperature was conducive for viruses. Available data further indicated that the virus survived in poultry droppings for at least 35 days at low temperature of 40°C, while at 37°C, it could survive for 6 days (WHO, 2005). It was also noted that H5N1 was not killed by refrigeration. Further reports indicated that the eggs could contain H5N1 virus both on the shell and on the inside (white and yolk). As a result, it is recommended that eggs from areas with H5N1 outbreak are not supposed to be consumed raw or partially cooked (runny yolk for breakfast). In line with this, cooked poultry is supposed to be served "piping hot." Although there is no evidence indicating that individuals have been contaminated with H5N1 infection after consumption of properly cooked poultry or eggs, previous studies showed that the major risk of exposure to the virus infection was through handling or slaughtering of live infected poultry. Good sterile practices are, therefore, essential during and postslaughter handling to check cross-contamination from poultry to other food and from food preparation surfaces and equipment.

Escherichia coli

According to WHO (2005), enterohemorrhagic *E. coli* is another serious pathogen with regard to food safety. Enterohemorrhagic *E. coli* (EHEC) 0157:H7 serotype was first identified as a human pathogen in 1892 in the United States. This was after two main outbreaks of hemorrhagic colitis (bloody diarrhea). Since then, it is reported that flare-ups of this pathogen have turned into a significant health issue throughout the different areas of the world (Schlundt, 2001; Clarke et al., 2002; Ram et al., 2009). Further examinations on the thermal sensitivity of *E. coli* 0157:H7 in ground beef uncovered that heating kills *E. coli* strains as well as *Salmonella* spp. The optimum temperature for growth of *E. coli* 0157:H7 is 37°C, whereas below 10°C, growth is not observed (Yanamala et al., 2011). On the other hand, *E. coli* 0157:H7 survived freezing. The study also reported that *E. coli* 0157:H7 was more acid resistant than other *E. coli* strains. Podolak et al. (2009) observed that monitoring of *Salmonella typhimurium* was of great importance because of its high survivability. The commonly involved food vehicles were raw or insufficiently cooked foods of bovine origin, particularly half-cooked, ground or minced beef, and unpasteurized milk. Likewise, various flare-ups were additionally connected with the utilization of raw or insignificantly processed fruits and green leafy vegetables. It was reported that beef was the main source of 46% of US foodborne outbreaks between the years 1993 and 1999. Other products include improperly pasteurized cow's milk. It was also realized that pasteurization eliminated pathogens from milk, including *E. coli* 0157:H7. Further research additionally affirmed that the fruits and vegetables contaminated with *E. coli* 0157:H7 brought about various flare-ups (Ackers et al., 1998). Green leafy vegetables were referred as the wellspring of 26% of FBD in the United States between 1998 and 1999. Although contamination of vegetables could occur in different ways, the utilization of manure or water contaminated with fecal matter is suspected to be the most possible route of infection (Solomon et al., 2002). In addition, suspected fertilizer from nearby cattle and poor sewage treatment is considered as another source of *E. coli* 0157:H7 strains detected in cabbage plants. The ingestion of *E. coli* 0157:H7 infection extended from asymptomatic to death. Incubation period is usually ranged from 1 to 8 days. Symptoms begin with abdominal cramps and nonbloody diarrhea, which further progresses to bloody diarrhea inside 2–3 days (Mead and Griffin, 1998). More severe manifestation of *E. coli* 0157:H7 infections included hemorrhagic colitis (bloody diarrhea), in which the most vulnerable groups are children and the elderly. Green leafy vegetables are said to grow low to the ground and are, therefore, recognized as another source of *E. coli* 0157:H7 infection.

Pseudomonas

Pseudomonas are described as rod-shaped Gram-negative aerobic, nonspore-forming type of bacteria commonly found in water, or some type of plant seeds. They are widely found in the environment such as soil and water plants. They thrive in moist areas and are found in hospital setups. The infection acquired in hospitals is referred to as nosocomial infection. Free bacteria found in wet areas such as sinks, antiseptic solutions, and urine receptacles cause pseudo infections. Healthy persons are usually not at risk of infection. *Pseudomonas* infections were considered opportunistic infections, that is, they only cause disease when a person's immune system is already impaired. These include patients in burns units, cancer patients undergoing chemotherapy, HIV patients, and cystic fibrosis or presence of a foreign body such as catheter. Infection in the blood is called bacteremia. Symptoms include fever, chills, fatigue, and muscle and joint pains. Infection of the lungs, pneumonia, was indicated by symptoms, which include chills, fever, productive cough, and difficulty in breathing. Others include skin infection called folliculitis—itchy rash, bleeding ulcers, and headache. The isolation of pseudomonas could be attributed to wet surfaces in the kitchen.

Chemical Hazards

Synthetic compounds are a critical wellspring of foodborne ailments, despite the fact that impacts are generally hard to connect with a specific food. Chemical compounds can enter in the food chain, either due to their existence in the environment through unexpected contamination of food, or to their purposeful use along the food production chain. For the most part, industrial pollutants are unexpected contaminants of foods; hence, even if regulated, it might be difficult to control. Chemicals used in agriculture are intentionally applied to land or crops during production, so their utilization may be controlled. Some harmful chemical compounds can present naturally in foods and in nature. Naturally occurring chemicals are generally incorporated in plants by chemical toxicants in food; for example, glycoalkaloids; natural contaminants, for example, mycotoxins and marine toxins; and environmental contaminants, such as mercury, lead, radionuclides, and dioxins. The nourishment supply is increased by incorporation of food additives and supplements (vitamins and essential minerals, pesticides, and veterinary medication residues), but it should be ensured that every single such use is safe. Food contamination by chemicals can influence well-being after a single exposure or, more frequently, after long-term exposure. However, the health consequences of chemical compounds exposure in food are not clearly understood.

Factors Contributing to Foodborne Diseases

Foodborne diseases occur for a number of reasons, including, among others, increase in international travel and trade, microbial adaptation, changes in the food production, and globalization of food supply. In North America, for example, from 1996 to 1997, an outbreak of cyclosporiosis was linked to contaminated raspberries imported from South America. According to WHO, introduction of pathogens into new geographical areas is also viewed as a contributing factor to the emergence of FBDs. For example, *Vibrio cholera* was introduced in the waters off the coast of the southern United States when a cargo sheep discharged contaminated ballast into the water in 1991. It is assumed that a similar mechanism led to the introduction of cholera in South America that same year. Further reports indicated that there are many risks associated with food safety due to industrialization and mass food production. Other factors such as the emergence of longer complex food chains, fast-food consumption, street-vended foods, and growing habits of eating foods out are cited as the major causes of food safety problems (Panisello, Quintic, and Knowles, 1999). In addition, travelers, refugees, and immigrants exposed to unfamiliar foodborne hazards while abroad are also contributing factors to FBDs. In Sweden, for example, it is estimated that about 90% of all cases of Salmonellosis are imported. Other factors experienced by many countries include changes in the microbial population, which lead to evolution of new pathogens and development of new variant strains in old pathogens. This transformation results in antibiotic resistant organisms making a disease more difficult to treat. This has been observed when a microorganism isolated in one country exhibits different characteristics in another country, making it difficult to be identified and controlled.

Increases in the global population of highly susceptible persons are a warning trend to various nations with respect to FBDs. The upward trend is attributed to aging, malnutrition, an HIV/AIDS pandemic, and other underlying medical conditions. Elderly individuals are likely to be infected more because they have low immunity to infection. In fact, people with weak immune system are even infected with pathogens at lower doses. For example, persons suffering from cancer or HIV and AIDS are more likely to succumb to infections with *Salmonella*, *Campylobacter*, listeria, toxoplasma, cryptosporidium, and other foodborne pathogens. In addition, in developing countries, poor nutritional status lead to reduced immunity, particularly in children and the elderly, who consequently become more susceptible to foodborne infections. A change in lifestyle contributes significantly to the spread of food-related infections. Behaviors and practices such as eating in restaurants, canteens, fast-food joints, and informal dining outlets increases chances of consumption of contaminated foods. In many countries, effective food safety education and control do not match the boom in food service establishments. As a result, unhygienic preparation of food provides good opportunities for contamination and growth and survival of foodborne pathogens. Processing factors that contribute significantly to FBDs are related to how food handlers manage various stages of food purchase and preparation, up to the point of service. Major documented factors associated with FBDs in homes and institutional settings, especially in the restaurants, include improper holding temperature, inadequate cooking, contaminated equipment, and poor personal hygiene. According to Cody and Keith (2001) and WHO (2005), these factors have to be controlled to keep the foods safe. Boyce et al. (1996) reported that undercooking contributes to outbreaks due to *Clostridium botulinum* (91%), *V. parahaemolyticus* (92%), *Clostridium perfringens* (65%), *Salmonella* species (67%), and *T. spiralis* (100%). Raw seafoods are sources of *V. parahaemolyticus* and enteric viruses. Raw pork meat is potentially infected with *Trichinella* larvae. All foods must be heated to the time, temperature, and values required to kill pathogens (Cody and Keith, 2001; WHO, 2005). Though FBDs are a burden to both developed and developing countries, the challenges in developing countries are aggravated by socioeconomic factors such as widespread poverty, large-scale migration to already crowded cities, and rapid growth of population, among others (Martin, 2001). Moreover, climatic factors exacerbated food hygiene problems because high ambient temperatures are conducive to the growth of mesospheric bacterial pathogens. The high levels of humidity are also cited as favorable for the growth of other microorganisms (Martin, 2001). Sanitary conditions are also considered as a big challenge. WHO (2006) reported that about 2.6 billion people in the developing world lack toilets, and about 1.1 billion have no access to potable water. Poverty at the individual level is also cited to increase food safety problems due to lack of facilities for the hygienic preparation and storage of food.

FOOD SAFETY MANAGEMENT SYSTEMS

Food Safety Management System is defined as a group of programs, procedures, and measures for preventing FBDs by actively controlling risks and hazards throughout the flow of food. Food safety systems address issues related to basic sanitation and operation conditions, which include personal hygiene programs, supplier selection, and food specification programs. This is in relation to the Codex Alimentarius Commission, an international body set up by WHO with the aim of ensuring the safety of the consumer and fair practices in the food trade. The European Union (EU) was among the organizations using the program. Codex Alimentarius Commission and Food Hygiene Committees made efforts to clarify the principles of food hygiene by elucidating the rationale behind these principles and providing examples as to how the principles are to be applied (Mitcham et al., 2007). Food safety management systems also entail active managerial control (AMC), which manages food safety risk and focuses on the five most common risk factors responsible for FBDs as identified by the CDC. The points include purchasing food from unsafe sources, failing to cook food adequately, holding food at improper temperatures, using contaminated equipment, and generally practicing poor personal hygiene (Mitcham et al., 2007).

Several systems addressing food safety are put in place in different countries. Such systems include HACCP, which ensures quality service and product delivery in the entire food flow, and the International Organization for Standardization (ISO), the world's largest developer of voluntary international standards. The ISO standards are related to different fields as follows:

ISO 9000: Quality Management Standard
ISO 14000: Environmental Management Standard
ISO 22000: Food Safety Management Standard
ISO 17025: Laboratory Management Standard

The most relevant standard to this study is ISO 22000, related to Food Safety Management Standard. The food hygiene standards, which have been in operation in many institutions, include The General Food Safety Hygiene regulations and Food and Agriculture Organization (FAO)/WHO Codex Alimentarius Commission Standards. Although HACCP has been introduced as one of the best food safety measures, the system is viewed as expensive and difficult to implement due to lack of capacity, and long processes, which many institutions feel, are tedious and out of reach for them due to financial constraints. Recently, the ISO certification came up with certification of institutions, which complied with the set standards for management of various activities in an organization. A brief look at ISO 22000 gave a clearer understanding of the management systems.

ISO 22000, an international standardization also known as generic food safety management system, is specifically designed to be used for certification or registry purposes, mainly because an accredited auditor formally registered an institution if it was compliant with the requirements of the system. The system comprehensively describes a set of general food safety requirements that applies to all organizations in the food chain. In this context, food chain is explained as a complete outline involved in the creation to consumption of food products. This includes every step from initial production to final consumption, and involves production, processing, distribution, storage, and handling of all food ingredients. However, since the food chain also includes the organizations that do not directly handle raw materials used in food production, a number of categories of organizations are included in the food flow:

- Primary producers: Farms, fisheries, ranches, dairies
- Processors: Fish, meat, poultry, feeds
- Manufacturer: Bread, soup, snack, cereal, canned food
- Food service providers: Restaurants, cafeteria, hospitals, airlines, cruise ships
- Nursing homes, senior lodging, grocery stores
- Other service providers including storage service provider, catering service provider, transportation, sanitation service provider, cleaning service provider, among others

ISO 22000 uses policies and structures defined by HACCP. The system, therefore, is involved in identification, prevention, and control of food safety hazards. This implies that ISO comprehensively deals with how to conduct a food safety hazard, identifies critical control points (CCPs), establishes limits for each CCPs, develops procedures to monitor CCPs, designs a corrective action to handle critical limit violations, creates food safety record keeping, and validates and verifies the system.

Critical control limits define a set of values that separate acceptability from unacceptability. These parameters, according to the CDC, if maintained within permissible limits confirm the safety of food products. In other words, critical control points provide a tolerance level in the food flow. For instance, critical control point analyses checked on optimal temperature for each step in a food flow and the duration required for any step in food production procedures. Questions asked in such analyses include how, where, when, and who. "How" defines the methodology used to monitor the critical limit, "where" defines the location for undertaking the activity, "when" defines the time or frequency of the activity, and "who" defines the responsibility for undertaking the monitor (Mitcham et al., 2007).

STRATEGIES FOR FOOD SAFETY

Strengthening Surveillance Systems of Foodborne Diseases

The worldwide burden of FBDs in humans is extensive, and the causative pathogens are generally zoonoses. There are an expected 76 million instances of FBDs in the United States annually (Mead et al., 1999). There were approximately 1.3 million instances of FBDs in England and Wales in 2000 (Adak et al., 2002). Surveillance of FBDs is mainly concerned with the public health issues in many nations. The main objectives are to evaluate the burden of FBDs, to survey its relative effect on well-being and financial matters, and to assess disease prevention and control programs. It is also essential for conducting risk assessment, and more comprehensively for risk management and communication. The principle targets of FBD surveillance are to build up the degree to which food acts as a route of transmission for specific pathogens and to recognize high-risk foods, practices, and populaces (Borgdorff and Motarjemi, 1997). Surveillance of FBDs should be coupled with food monitoring information along the whole feed-food chain. The detailing of foodborne sickness ought to be coordinated into the modification of the international health regulations.

Improving Risk Assessments

Risk analysis is widely recognized as the fundamental methodology underlying the development of food safety standards. As per Codex, risk management ought to pursue an organized methodology including the components of hazard assessment, risk management option assessment, usage of the executive's choice, monitoring, and audit. Risk assessment of hazards includes identification of a food security issue, establishment of a risk profile, ranking of hazards for risk assessment and risk management priority, establishment of policy for conduct of risk assessment, commissioning of the risk assessment, and review of the outcomes. Identification of the food security issue is the passage point for preliminary risk management exercises and may go to the consideration of the risk manager through disease surveillance information and request from a trading accomplice or consumer concern. The essential objective of food-related risk management is to secure general well-being of the public by managing such risks as essentially as conceivable through the choice and execution of suitable measures. The development of a globally accepted system for analysis of risk by Codex, which serves as a basis for setting food norms at universal dimensions, has focused attention on the acceptability of risk assessments.

In association with FAO, WHO gives the scientific guidance as the proof for Codex norms, and in additional standards, recommendations, and policy options. WHO has the authority and assembling power in the field of worldwide public health to attempt this essential function. Scientific guidance has been given for decades through long-standing and settled systems, in particular, the Joint FAO/WHO Expert Committee on Food Additives (JECFA), the Joint FAO/WHO Meeting on Pesticide Residues (JMPR), and the Joint FAO/WHO Expert Meeting on Microbiological Risk Assessment (JEMRA). Emerging and emergency problems and in addition complex evaluations should be addressed through targeted ad hoc expert gatherings.

Developing Methods for Assessing the Safety of the Products of New Technologies

The usage of biotechnological tools in production of food presents consumers with new difficulties and challenges. The greatest risk brought about by genetically engineered (GM) foods is that they may have negative impacts on the human body. It is a general presumption that utilization of such foods can result in diseases resistant to antibiotics. Additionally, as GM foods are new innovations, very little is known about their long-term impacts on humans.

Resolution WHA 53.15 perceived genetic engineering of food as a critical public health problem, and to settle that, WHO needs to fortify its ability to give a scientific premise for decisions on the impacts of genetically modified foods on human health. The uniformity approach was designed by the Joint FAO/WHO Expert Consultation on foods in June 2000 as a starting step in evaluating security and risks associated with genetically modified food. The evaluation of food safety itself requires a consolidated, reliable, case-by-case way to deal with the assessment of such foods. Major emphasis was laid on the allergenic capability of genetically modified foods by the experts. However, the allergenic capability of genetically engineered foods needs to be assessed. These discussions address the inception of a series of expert gatherings examining genetically modified foods. WHO continues to participate in discussions regarding this matter by giving expert advice on the health risks of these new advancements and by adding to a better comprehension of new improvements in order to address the concerns of consumers. WHO will continue to give a logical structure to the security and dietary evaluation of genetically modified foods, and also for the incorporation of other logical parts of such foods. WHO will support the broadening of assessment, with the goal that environmental, money-saving advantage, socioeconomic, and other concerns can be coordinated in a progressively rational framework.

Nanotechnological Approaches

Nanotechnology is one of the most advanced innovations in this century. The food industry has applied this innovation in the majority of its divisions. Nanotechnology has a huge prospective in food and agriculture, including enhancing agricultural and food production, improving food flavor and nutrition, and progressive food packaging and preservation. However, the novel properties of nanoscale materials that permit useful applications are additionally accompanied with vulnerabilities, even obscure dangers.

The uptake and targeted delivery has been improved by incorporation of nanoparticles in nanoceuticals and nutritional supplements (colloids of zinc nanoparticles and other nanosized minerals, and nano-encapsulates (Bouwmeester et al., 2009). Nanochips or nanosensors have been found to be useful in the detection of storage conditions accommodating for spoilage (e.g., temperature or moisture problems) (Bouwmeester et al., 2009). The detection of *E. coli* O157, *Campylobacter*, and *Salmonella* in food has been carried out by using nonnanotechnology biosensors (Patel, 2002). Even nano-based inserts are being developed to detect "Category B" agents (CDC list) such as *Salmonella*, *E. coli*, and other pathogens. Nanosensors can also be developed to detect "Category A" agents such as anthrax and botulism pathogen and also other toxic contaminants, for example, heavy metals (e.g., lead, mercury and arsenic) and chemicals (e.g., furans, dioxins, polychlorinated biphenyls [PCBs] and harmful pesticide residues).

International Regulatory Frameworks

The need of coordination and integration for management of food safety and plant and animal health from the farm to the table are frequently not given by the present international policy and regulatory system. The Codex Alimentarius Commission, WHO, and International Plant Protection Convention developed the international public guidelines. These guidelines and related sanitary and phytosanitary (SPS) standards are executed and implemented to a more noteworthy or lesser degree, depending upon accessible assets, through a variety of uncoordinated national activities regulated by different ministries in various nations. In spite of the fact that the majority of agricultural products are not traded internationally, national agricultural planning and agricultural knowledge, science and technology (AKST) speculation is progressively oriented toward export markets and intended to follow international trade rules. In principle, trade-related SPS guidelines and control measures may likewise be connected promptly to domestic SPS programs. In practice, developing nations enact few international measures domestically since they lack the assets and specialized limit for execution. The expense of meeting private international guidelines, for example, those administered by the Global Food Safety Initiative, is borne by primary producers. There are some studies that evaluate infrastructural and consistence expenses of international public and private standards execution and usage.

References

Ackers, M.L., Mahon, B.E., Leahy, E., Goode, B., Damrow, T., Hayes, P.S., et al., 1998. An outbreak of *Escherichia coli* O157:H7 infections associated with leaf lettuce consumption. J. Infect. Dis. 177, 1588–1593.

Adak, G.K., Long, S.M., O'Brien, S.J., 2002. Intestinal infection: trends in indigenous foodborne disease and deaths, England and Wales: 1992 to 2000. Gut 51, 832–841.

Adams, M., Moss, M.O., 2000. Food Microbiology, second ed. Royal Society of Chemistry, Cambridge.

Allos, B.M., 2001. *Campylobacter jejuni* infections: update of emerging issues and trends. Clin. Infect. Dis. 32, 202–226.

Anon, 2000a. Microbial pathogen data sheets. New Zealand Food Safety Authority, http://www.nzfsa.govt.nz/science-technology/data-sheets/ (accessed 28.10.04).

Anon, 2000b. Notifiable diseases on-line. Public Health Agency of Canada. http://dsol-smed.hc-sc.gc.ca/dsol-smed/ndis/index_e.html (accessed 2.11.04).

Blaser, M.J., 1997. Epidemiologic and clinical features of *Campylobacter jejuni* infections. J. Infect. Dis. 176, 103–105.

Borgdorff, M.W., Motarjemi, Y., 1997. Surveillance of foodborne diseases: what are the options? WHO/FSF/97 (Geneva, Switzerland).

Bouwmeester, H., Dekkers, S., Noordam, M.Y., Hagens, W.I., Bulder, A.S., Heer, C.D., et al., 2009. Review of health safety aspects of nanotechnologies in food production. Regul. Toxicol. Pharmacol. 53, 52–62.

Boyce, T.G., Koo, D., Swerdlow, D.L., et al., 1996. Recurrent outbreaks of *Salmonella enteritidis* infections in a Texas restaurant: phage type 4 arrives in the United States. Epidemiol. Infect. 117, 29–34.

Clarke, S.C., Haigh, R.D., Freestone, P.P., Williams, P.H., 2002. Enteropathogenic *Escherichia coli* infection: history and clinical aspects. Br. J. Biomed. Sci. 59 (2), 123–127.

Cody, M.M., Keith, M., 2001. Food Safety for Professionals: A Reference and Study Guide, second ed. American Dietetic Association, Chicago.

Martin, S., 2001. Forced migration and professionalism. Int. Migr. Rev. 35 (1), 226–243.

Mead, P., Griffin, P., 1998. *Escherichia coli* O157:H7. Lancet 352, 1207–1212.

Mead, P.S., Slutsker, L., Dietz, V., McCaig, L.F., Bresee, J.S., Shapiro, C., et al., 1999. Food-related illness and death in the United States. Emerg. Infect. Dis. 5 (5), 607–625.

Mensah, P., 2002. Street foods in Ghana: how safe are they? Bull. WHO 3–9.

Mitcham, E.J., Crisosto, C.H., Kader, A.A., 2007. Bushberries: blackberry, blueberry, cranberry, raspberry. Recommendations for Maintaining Postharvest Quality. Davis, Calif., Dept. of Pomology, Univ. of California. Available at: http://postharvest.ucdavis.edu.

Motarjemi, Y., Kaferstein, F., 1998. Food safety, hazard analysis and critical control point and the increase in foodborne diseases: a paradox? Food Control 10, 325–333.

Pal, M., 2005. Importance of zoonoses in public health. Indian J. Anim. Sci. 75, 586–591.

Pal, M., 2007. Zoonoses, second ed. Satyam Publishers, Jaipur, India.

Panisello, P.J., Quintic, P.C., Knowles, M.J., 1999. Towards the implantation of HACCP: results of a U.K. regional survey. Food Control 10 (2), 87–98.

Patel, P.D., 2002. (Bio)sensors for measurement of analytes implicated in food safety: a review. Trends Anal. Chem. 21 (2), 96–115.

Podolak, R., Enache, E., Stone, W., Black, D.G., Elliott, P.H., 2009. Sources and risk factors for contamination, survival, persistence, and heat resistance of *Salmonella* in low-moisture foods. J. Food. Prot. 73 (10), 1919–1936.

Ram, S., Bajpai, P., Singh, R.L., Shanker, R., 2009. Surface water of a perennial river exhibits multi-antimicrobial resistant shiga toxin and enterotoxin producing *Escherichia coli*. Ecotoxicol. Environ. Safety 72 (2), 490–495.

Schlundt, J., 2001. Emerging food-borne pathogens. Biomed. Environ. Sci. 14 (1–2), 44–52.

Solomon, E.B., Potenski, C.J., Matthews, K.R., 2002. Effect of irrigation method on transmission to and persistence of *Escherichia coli* O157:H7 on lettuce. J. Food. Prot. 65 (4), 673–676.

Sonnier, J.L., Karns, J.S., Lombard, J.E., Kopral, C.A., Haley, B.J., Kim, S.W., et al., 2018. Prevalence of *Salmonella enterica, Listeria monocytogenes*, and pathogenic *Escherichia coli* in bulk tank milk and milk filters from US dairy operations in the National Animal Health Monitoring System Dairy 2014 study. J. Dairy Sci. 101 (3), 1943–1956.

WHO, 2005. Working together for safe food. Weekly update GEMS/FOOD WHO. Geneva, Switzerland.

WHO, 2006. Five keys to safer food manual. www.who.int/entity/foodsafety/publication/consumermanual_keys.pdf.

Yanamala, S., Miller, M.F., Loneragan, G.H., Gragg, S.E., Brashears, M.M., 2011. Potential for microbial contamination of spinach through feed yard air/dust growing in close proximity to cattle feed yard operations. J. Food Saf. 31 (4), 525–529.

Further Reading

Jeremy, T., Branen, L., 2005. Nanotechnology and Food Safety. Presentation at the "Nanotech and Food Safety." September 21. Seminar to Nanobio Convergence, Bay Area, California.

Lee, J.B., Roh, Y.H., Um, S.H., Funabashi, H., Cheng, W., Cha, J.J., et al., 2009. Multifunctional nanoarchitectures from DNA-based ABC monomers. Nat. Nanotechnol. 4, 430–436.

CHAPTER

2

Food Hazards: Physical, Chemical, and Biological

Pradeep Kumar Singh[1], *Rajat Pratap Singh*[2], *Pankaj Singh*[3] *and Ram Lakhan Singh*[1]

[1]Department of Biochemistry, Dr. Rammanohar Lohia Avadh University, Faizabad, India
[2]Department of Biotechnology, Dr. Rammanohar Lohia Avadh University, Faizabad, India
[3]Department of Biochemistry, Jhunjhunwala P. G. College, Faizabad, India

OUTLINE

INTRODUCTION

Food is one of the most fundamental materials for the survival of living beings. Generally, the term *food* has been utilized for those substances that are necessary and valuable for human body. Food has been characterized as eatable materials comprising

Food Safety and Human Health
DOI: https://doi.org/10.1016/B978-0-12-816333-7.00002-3

elementary parts that maintain life and vital processes (David et al., 2012). Our ancestors probably endeavored to eat foods derived from plants and animals in their journey for sustenance and may have perceived that there were both beneficial and harmful effects linked with the utilization of such nourishments. Food include a variety of natural chemicals, and every one of these are not supplements, yet in reality some natural chemicals may diminish nutritional value or may be toxic (e.g., naturally occurring toxicants). Likewise, during preparation and processing of foods, synthetic compounds can be added, either deliberately or inadvertently. Food hazards can be of different inceptions as metabolic products of plants, animals, and microbes; as chemical and biological hazards from the environment; as purposefully added food additives; and as those generated during the food processing. Although food is necessary for our body, if it is contaminated with pathogenic microbes or their toxins or environmental contaminants, it can play a role in transmission/onset of diseases. The overall number of diseases caused by foodborne microbial pathogens makes microbiological standard the most imperative food safety factor (Gorham and Zarek, 2006; Scallan et al., 2011). Contamination of food by pathogenic microbes or chemical hazards is a significant problem since it can prompt an extensive variety of health issues. Food contamination is accountable for more than 200 diseases such as typhoid, diarrhea, and other foodborne diseases and can lead to the deaths all over the world (WHO, 2005). It is estimated that obscure pathogenic agents represent 81% of sicknesses and hospitalizations and 64% of deaths because of foodborne disease (Mead et al., 1999). Despite the foodborne diseases, contamination of foods with microbial toxins, pesticides, and drug residues and industrial chemicals is a major issue that influences the public health. This chapter is an attempt to provide an overview of various physical, chemical, and biological food hazards.

FOOD TOXICANTS AND HUMAN HEALTH

Naturally produced toxic components can be found in food of animal and plant origin, as well as in higher fungi, which is used as a food source. Dangerous food components fall into different chemical classes, including simple amines and amino acids, fatty acids, organic acids, phenolic compounds, and encompass also more complex alkaloids, cyanogenic glycosides, and very complex proteins. Being diverse in their chemical structure, the mode of toxic action varies considerably among naturally occurring food toxicants. The toxins may impair specific organs and systems, such as the skin or cardiovascular system, or may have systematic effects by binding to hormone receptors or affecting the nervous system. A food safety hazard refers to any agent present in the food that causes adverse health consequences for consumers. Food safety hazards occur when food is exposed to hazardous agents (Fig. 2.1). Hazard Analysis and Critical Control Point (HACCP) is a systematic approach to be used in food production as a means to ensure food safety. According to the National Advisory Committee on Microbiological Criteria for Foods (NACMCF), any biological, chemical, or physical properties that impose an undesirable consumer health risk are considered as hazards. Thus by definition one must be concerned with three classes of hazards: physical, chemical, and biological.

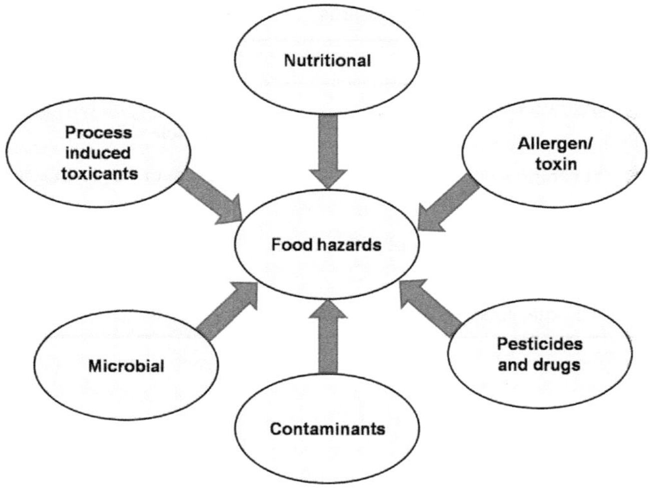

FIGURE 2.1 Different types of food hazards.

Hard or sharp substances such as metal, plastic, stones, glass, pits, wood, or even bone are well-known examples of physical hazards. Physical hazards can lead to injuries such as choking, cuts, or broken teeth. Some foreign material in food products such as hair, insects, or sand that are not likely to cause injuries may also come under the category of physical hazards. Chemical hazards are the substances that are used in processing at various levels but can lead to illness or injury if consumed at too high concentrations. Biological hazards include microorganisms such as bacteria, viruses, yeasts, molds, and parasites. Some of these are potent pathogens or associated with toxins production. A pathogenic microorganism causes disease, but degree of disease severity is highly variable. Examples of biological hazards include *Salmonella*, *Escherichia coli*, and *Clostridium botulinum*.

Physical Toxicants

Physical hazards are generally harmful extraneous matter that is not commonly part of food. When these materials reach the body, they lead to a number of injurious health effects. The physical hazards are easy to identify as they immediately can cause injury (Table 2.1).

Extraneous material includes all materials (except bacteria and their toxins, viruses, and parasites) that may be present in a food and are foreign for a particular food. These materials are normally nonhazardous but are associated with unsanitary conditions such as processing, production, storage, and distribution of food. Extraneous material can be considered hazardous due to its hardness, sharpness, size, or shape. It may cause lacerations, perforations, and wounds or may become a choking hazard. Extraneous materials can be differentiated into two categories: unavoidable and avoidable.

The food contains several extraneous materials that arise during processing as the byproduct or that may be inherent in nature; thus they are considered as unavoidable

TABLE 2.1 Injury Risk of Physical Hazards

Injury Risk	Commodity	Size of Physical Hazard
High	Infant foods	Any size (including small particle < 2 mm)
	Beverage	2 mm or larger in size in any one dimension
Moderate	All other foods (except infant food and beverage)	2 mm or larger in size in any one dimension
Low	All other foods (except infant food)	2 mm in size in all dimensions

TABLE 2.2 Common Physical Hazards and Sources

Material	Sources
Glass	Bottles, jars, utensils, light fixtures
Plastic	Fields, pellets, plant packing materials
Stones	Field, buildings
Metal	Machinery, fields, wire
Insects and other filth	Fields, plant postprocess entry
Bone	Fields, improper plant processing
Wood	Fields, building pallets, boxes
Bullet/needles	Animals shot in the field, hypodermic needles used for infections
Insulation	Building materials

materials. Materials like dirt on potatoes, remnants of insect fragment in figs, stems in blueberries, etc., are all common unavoidable extraneous matters. Some extraneous materials are avoidable as they can be prevented by using proper methods. Small glass fragments, pieces of jewelry, animal debris, pieces of plastics, etc., are the different forms of avoidable physical hazards that are present in food (Table 2.2). Sometimes in certain food products, a crystal-like structure appears as in tuna (struvite), processed cheese, soya sauce and fish sauce, etc. These are not glass, but they are mineral crystals. This can be verified by dissolving the crystals in heated vinegar or lemon juice.

Temperature

Leaving food out for too long time at room temperature can cause bacteria such as *Staphylococcus aureus*, *Salmonella enteritidis*, *E. coli*, and *Campylobacter* to grow at dangerous levels that can cause illness. Bacteria grow very fast in the temperature range between 5°C and 60°C. This range of temperatures is often called the danger zone. Food safety agencies, such as the US Food Safety and Inspection Service (FSIS), define the danger zone as roughly 5°C–60°C. The FSIS suggests that potentially hazardous food should not be stored at this range, to prevent contamination of food from bacteria. The food stored at this range for more than 2 hours should not be consumed. Foodborne microorganisms grow rapidly in the middle of the zone, between 21°C and 47°C. Control of time and temperature is very crucial for food safety. To reduce the damage, the time for which food is stored in the danger zone must be minimized. Bacteria need both time and the right temperature to multiply to dangerous levels. A logarithmic relationship exists between microbial cell death and temperature (Forsythe, 2010). Never leave food out of refrigeration over 2 hours. If the temperature is above 32°C, food should not be left out more than 1 hour.

By heat processing food, we can reduce the risk of microbial spoilage and can extend the product shelf life, but during such heat processing, there are a number of harmful chemicals formed, such as acrylamide, benzo (a) pyrene, and heterocyclic aromatic amines. Acrylamide has been shown to be a carcinogen and genotoxic in animal studies (Besaratinia and Pfeifer, 2007). It is also neurotoxic in humans (Erkekoglu and Baydar, 2014). On the basis of animal studies, the International Agency for Research on Cancer has classified acrylamide as Group 2A, that is, probably carcinogenic to humans. Thermal decomposition of lipid and amino acid produces furans, although there is evidence that furan levels can be reduced in some foods through volatilization during cooking.

Irradiation

Food irradiation is the processing of food products by ionizing radiation (gamma rays, electrons, or X-rays) in order to control foodborne pathogens. It reduces microbial load or destroys bacteria and fungi and other parasites that cause human disease or cause food to get spoiled. Irradiation destroys harmful bacteria such as *E. coli* O157:H7, *Salmonella, Listeria, Campylobacter,* and *Vibrio* that are major contributors of foodborne illnesses worldwide. Exposure of food to ionizing radiation may lead to a series of chemical reactions by primary and secondary radiolysis effects. Hydrocarbons and 2-alkylcyclobutanones are major reported radiolytic products that are produced from the major fatty acids present in food. Several other radiolytic products are also formed from food components that have been subjected to processing treatments other than irradiation. Currently, 40 countries have approved irradiation methods for 50 food products. A billion pounds of food products and ingredients are irradiated annually worldwide. In the United States alone, approximately 170 million pounds of spices are irradiated every year.

Irradiation extends shelf life of food in several ways. First, it reduces or inhibits spoilage bacteria and molds that are able to grow under cooling storage. Radiation splits the DNA or damages other vital molecules that promote killing or inhibit the process of reproduction in bacteria. Another way that irradiation is used is in delaying the ripening process of fruits and vegetables to expand their shelf life. Electron beams or gamma rays or X-rays are used to irradiate food.

Irradiation is a physical treatment in which foodstuff is exposed to a definite dose of ionizing radiation. Ionizing radiation at low doses may not be appropriate for all food, as it can produce undesirable odors and flavors in some foods. Radiolytic products are produced when food and water absorb energy during their exposure to ionizing radiation. These radiolytic products are short-lived, unstable, and reactive molecules causing damage to biological cells or even contaminating microorganisms or insects. Radiolytic products are not unique to irradiated food, however, as identical products can be found in food that has been cooked, frozen, or pasteurized and even in unprocessed food.

Free radicals are also formed as by-products of normal and vital metabolic processes in human physiology (e.g., oxidation and respiration), as well as in pathological processes leading to diseases (Pryor, 1984). In foods, identical free radicals are also formed by irradiation, heat pasteurization, and by cooking (infrared, microwaves), boiling, baking, broiling, or frying foods. Therefore free radicals formed in the irradiation process are not unique or different in nature than those formed in biological or other common cooking processes.

The Codex General Standard on Irradiated Food was revised in 2003 (CAC, 2003) which states that "For the irradiation of any food, the minimum absorbed dose should be sufficient to obtain the technological purpose and the maximum absorbed dose should be less than that which does not affect the consumer safety. The maximum absorbed dose should not be greater than 10 kGy, except in some cases."

Retail food products are required to display the following Radura symbol in green color.

Treated with irradiation

The degree of irradiation-based induction of chemical reaction in food components largely depends on various parameters such as absorbed dose, dose rate, presence of oxygen, and temperature and facility type. The physical state (solid, liquid, or powder, frozen or fresh) and composition of food also control the radiation-based induced reaction and formed products. Foods, dosage, and purpose for irradiation are listed in Table 2.3. Chemical reaction and the product produced from main food components such as carbohydrates, fat, proteins, and vitamins are described in the following sections.

EFFECTS OF IRRADIATION ON FOOD CONSTITUENTS

PROTEINS Irradiation-based chemical reactions of proteins largely depend on various factors, including amino acid composition, types of protein structure (globular, fibrous), physical status, state (native or denatured), and other substances in food. The important changes including oxidation, aggregation, dissociation, and cross-linking occur during treatment. For instance, the denaturation and aggregation of proteins take place at a dose of 10 kGy gamma irradiation incidents on hazelnuts (Dogan et al., 2007). Variety of low-molecular-weight radiolytic products such as keto acids, ammonia, diamino acids, and

TABLE 2.3 Foods, Dosage, Purpose, and Date Approved for Irradiation by the US Food and Drug Administration (IFT, 1998)

Product	Dose (kGy)	Purpose	Date Approved
Wheat, wheat flour	0.2–0.5	Insect disinfestations	1963
White potatoes	0.05–0.15	Sprout inhibition	1964
Pork	0.3–1.0	Control *Trichinella spiralis*	7/22/85
Enzymes (dehydrated)	10 max	Microbial control	4/18/86
Fruit	1 max	Disinfestation, delay ripening	4/18/86
Vegetables, fresh	1 max	Disinfestation	4/18/86
Herbs	30 max	Microbial control	4/18/86
Spices	30 max	Microbial control	4/18/86
Vegetable seasonings	30 max	Microbial control	4/18/86
Poultry, fresh, or frozen	3 max	Microbial control	5/2/90
Meat, packaged, and frozen	44 or greater	Sterilization	3/8/95
Animal feed and pet food	2–25	*Salmonella* control	9/28/95
Meat, uncooked, and chilled	4.5 max	Microbial control	12/2/97
Meat, uncooked, and frozen	7.0 max	Microbial control	12/2/97

amide-like intermediates are formed after irradiation of peptide (Delincee, 1983a). Modification of amino acid by treatment of radiation is very common. Out of all common amino acids, the aromatic and sulfur-containing amino acids are more prone to modification by irradiation. Duliu et al. (2004) studied the radiation effect of gamma rays or e-beam irradiation of dose ranges from 1 to 20 kGy on four enzymes (i.e., fungal α-amylase, microbial α-amylase, glucoamylase, and pectinase) and inferred that enzyme activity decreased up to half of the original activity.

LIPIDS Irradiation-based chemical reactions of lipids depend on physical status (liquid or solid), lipid concentration, unsaturation profile, presence of antioxidants, environmental conditions (oxygen, moisture, pH, light, heat), storage conditions, and type of storage (Delincee, 1983b). Irradiation accelerates the lipid peroxidation (O'Bryan et al., 2008), and it is mostly found in foods with high unsaturated fatty acids and larger fat content by generating free radicals. Lipid peroxidation can be minimized by using low temperature and reducing availability of oxygen (Stefanova et al., 2010). The antioxidants are also used in retardation of lipid peroxidation. Natural antioxidants such as oregano and rosemary extracts in beef burger are useful in reducing lipid oxidation (da Trindade et al., 2009). Sterols and stanols are the naturally occurring phytosterols present in cereals, fruit, nut, vegetables, and seeds, which have structural homology with cholesterol, and they can be oxidized by irradiation and heating.

CARBOHYDRATES Monosaccharides and polysaccharides are usually distorted by ionizing radiation (Adam, 1983). The major radiolytic products, for example, formic acid, hydrogen peroxide, and aldehydes (formaldehyde, acetaldehyde, and malonaldehyde) are generated by the application of 6.2 kGy/h dose of gamma radiation on starches from various foods such as wheat, rice, maize, or potatoes (Raffi et al., 1981a,b).

VITAMINS Irradiation has a direct effect on vitamins by decreasing their activity. Vitamins A, C, and E are more susceptible to higher doses of irradiation, similar to postthermal processing losses. Vitamin E is the most sensitive of fat-soluble vitamins to irradiation. Oxygen has a direct effect on postirradiation loss of vitamin E (Josephson et al., 1975). Pork liver treated with 5 kGy at 0°C has 4% less vitamin A as compared to untreated pork after 1 week and 13% reduction in vitamin A attributed to 4-week storage (Diehl, 1995). Diehl (1991) reported 2%–7% loss of β-carotene in 1 kGy gamma irradiated fresh-milled wheat flour. Loss of vitamin D is much less during irradiation. Vitamin losses due to irradiation are generally smaller in food matrix compared to pure solutions (Zegota, 1988). The factors responsible to minimize the radiation-mediated loss of vitamins are freezing temperatures and absence of oxygen (Diehl, 1991; WHO, 1999).

Thiamine (vitamin B1) has been found to be the most vulnerable against radiation and is, therefore, used to demonstrate worst-case results (WHO, 1994). The significant losses usually can occur in irradiated meat products (Fox et al., 1995; Graham et al., 1998; Thayer, 1987). However, the extent of these losses is attributed to processing conditions (temperature and dose) and can be overcome by using packaging techniques (Fox et al., 1997).

FOOD ADDITIVES Some food additives generate harmful radiolytic products on irradiation, for example, an antimicrobial agent potassium benzoate used at 0.1% concentration in turkey ham. When this ham is treated with 2 kGy radiation and stored under refrigeration for 6 weeks, it produces the volatile compound benzene by decarboxylation of potassium benzoate (Zhu et al., 2005). It is very similar to the production of volatile benzene in acidic beverages blended with benzoic acid and ascorbic acid (McNeal et al., 1993).

Chemical Toxicants

Food Additives

The human diet consists of thousands of structurally diverse chemical substances. The US Food and Drug Administration (FDA) defines a food additive as any substance that directly or indirectly becomes a component or otherwise affects the characteristics of any food. They may be either of natural origin or intentionally/unintentionally added, such as nutrients, colorants, sweeteners, herbicides, pesticides, and flavor-imparting substances. These substances may be added to food during processing and preparation that causes some chemical changes in raw agricultural product. Basically, these substances are added to attain definite technical effects, such as color, sweetening, flavoring, preservation, and other physicochemical effects. Human diet may also contain some other impurities from natural sources, for example, microorganisms and their metabolites and plant-derived substances. The FDCA recognizes three categories of food additives (Roberts, 1981):

1. Intentionally added substances
2. Substances that are natural components of food
3. Substances that may contaminate food

Additives are used to maintain or improve freshness, nutritional value, safety, taste, texture, and appearance. The use of food additives has become more popular in recent years due to the increased production of prepared, processed, and convenient foods.

COLOR

Food color is a visible sensory characteristic that positively influences flavor perception (Hutchings, 2003). Food color is considered one of the most remarkable and pleasant features of foodstuffs, which directly control the selection and eating desires of consumers (Shim et al., 2011). Food coloring agents have the property to absorb light due to the presence of conjugated double bonds (Schoefs, 2002). Most manufacturers use a range of colors in the manufacturing of soft drinks, beverages, cosmetics, toffees, bakery products, ice creams, jams and jellies, etc. FDA regulations outline a color additive as any dye, pigment, or other substance that gives color to a food, cosmetic, pharmaceutical, or drink. The Pure Food and Drug Act of 1906 permitted seven synthetic colors (Amaranth, Erythrosine, Ponceau 3R, Indigotine, Light Green SF, Naphthol Yellow 1, and Orange 1) as food additives (Young, 1989). Certified colors are artificially synthesized and used as they provide uniform color, low cost, and combine more simply to generate a variety of shades. These colors are highly resistant to exposure to pH, heat, and light. Colors from natural sources such as pigment derived from vegetables and animals are exempt from certification, but they are unstable in a broad range of pH, heat, and light and also have lower shelf life. Several colors such as annatto extract (yellow) frequently used in butter, caramel (yellow to tan), and dehydrated beets (bluish red to brown) used in soft drinks are the examples of exempt color. These additives have more cost than certified colors and are used to improve flavors of foods. These natural color additives are commonly used in the United States to color foods, drugs, and cosmetics. Natural food colorants are more effective than synthetic ones as they offer some advantages such as being safer, providing health benefits, exerting two or more benefits as food ingredients, and also contributing functional properties to food products (Carocho et al., 2014; Rodriguez-Amaya, 2016). Synthetic food colorants are extensively used to improve the aesthetic value of numerous foodstuffs and may impart blue, red, orange, yellow, green, and white color. Currently, the FDA and the European Food Safety Authority (EFSA) allow their application in food products with already established acceptable daily intake (ADI) doses. Several naturally occurring food colorants, for example, anthocyanins, beet colorants, carotenoids, and phenolic compounds are also used in various eatables. Anthocyanins are the most widely used natural food colorants, derived from leaves, flowers, fruits, and even whole plants. Commercial anthocyanins, such as cyanidin 3-glucoside, peonidin 3-glucoside, and pelargonidin 3-glucoside have also been utilized and their beneficial properties have been assessed. Stability of anthocyanin pigment as colorants is largely depending on pH, temperature, stress conditions, humidity, salinity, and even storage conditions. Thus the anthocyanin colors may vary from red to purple and blue color (Jimenez-Aguilar et al., 2011; Nontasan et al., 2012). Synthetic colors are reliable and inexpensive for restoring the original shade of the foods, whereas natural colors are less stable, costly, and possess lower tinctorial power. The National Institute of Nutrition, India (NIN, 1994) suggested that consumption of some synthetic food colorants could sometimes lead to detrimental effects. Many unpacked foods such as snacks, chewing gum, colored ice balls, colored candies, and cold drinks prepared by prohibited colors and toxic chemicals in quantities higher than the toxic levels, without following regulatory parameters on such products, are cheaper compared to the products prepared with permitted colors and are mostly consumed by low-income populations. Titanium dioxide, synthetic colorants currently used in food products, causes allergic reactions and also is reported to affect the children's behavior (Gostner et al., 2015). Council Regulation (EC 1333/2008) has identified attention deficit hyperactivity disorder (ADHD) promoted by use of six synthetic food colorants: Tartrazine, Quinolone Yellow, Carmoisine, Sunset Yellow FCF, Ponceau 4R, and Allura Red AC in consumers. Food colorants are normally used as mixtures of two or more dyes to form different shades (Sharma et al., 2008). Metanil yellow, an azo dye, has been proved to be hepatotoxic in albino rats (Singh et al., 1987, 1988). The metabolic disposition of Metanil Yellow and Orange II has also been studied using rats and guinea pigs as model systems (Singh, 1989; Singh et al., 1991a,b). Two azo dye Tartrazine (lemon yellow color) and Carmoisine (red color) are used as food colorants in several foodstuffs and some of nonfood products like chips, soft drinks, cakes, soaps, shampoos, ice cream, medicinal capsules, and drugs. On ingestion, intestinal microflora and mammalian azo reductase in the liver can reduce these azo dyes to aromatic amines (Chequer et al., 2011; Singh et al., 1991b). Azo dyes and their derivative aromatic amines are carcinogenic to humans and can accumulate in food chains by interaction with secretions such as saliva and gastric secretions. Long-term exposure of azo dyes can cause some diseases such as pathological lesions in the brain, spleen, liver, and kidney; tumors; growth deficiency; mental disorders; anemia; indigestion; and hypersensitive response (Sayed et al., 2012). Sunset Yellow, Carmoisine, Quinoline Yellow, Allura Red, Tartrazine, and Ponceau 4R are azo compounds that are proved to be harmful to children when used as additives in food and drinks (Singh and Singh, 2017). The presence of dyes imparts an intense color to effluents, which leads to environmental as well as aesthetic problems (Singh et al., 2015).

SWEETENERS

By the end of the 20th century, obesity became a major concern worldwide (Disse et al., 2010), as it promotes other life-threatening diseases, for example, cardiac problems, diabetes, and hypertension (Fujiwara et al., 2012; Ribeiro and Santos, 2013). To cure and control the obesity and its related diseases, there is a need to isolate some naturally occurring sweeteners and also development of synthetic sweeteners. These sweeteners have potential to replace sugar partially or totally. One of the drawbacks of these sweeteners is that they are unable to trigger physiological satiety mechanisms as compared to sugar (Raben et al., 2002; Swithers et al., 2010).

High-intensive sweeteners are normally used in food products, beverages, and some oral pharmaceuticals as sugar substitutes or sugar alternatives that offer zero calories and are much sweeter than sucrose. The general philosophy of Directive 94/35/EC on sweeteners used in foodstuffs is that they replace sugar and are involved in production of noncariogenic foods, foods without added sugars and energy reduced foods, etc.

Aspartame, acesulfame K, saccharin, cyclamate, and sucralose are the most popular noncaloric sweeteners (Butchko et al., 2001). Most of them have a high degree of sweetness and replace the sweetness of a much larger amount of sugar by their small amount (Gardner et al., 2012) and are known as high-potency sweeteners. Chemically, sweeteners are very diverse compounds and are categorized under natural/synthetic sweeteners, sugar alcohols, natural sugars, and sweet-tasting proteins. These compounds are of great interest as they are used in production of low-calorie food products (Klug and Lipinski, 2011).

Sweeteners are either natural or synthetic in origin (Tables 2.4 and 2.5). Several artificial sweeteners are available but their use is limited because they have some harmful side effects (Tandel, 2011). As a consequence, there is need to explore some natural sugar sources. Various plants contain plenty of sugars and/or polyols or other sweet ingredients. Presently, the available sweeteners in the global market are of synthetic nature, but their use needs approval by legislative authority (Grenby, 1991). Although synthetic sweeteners cover huge global markets, some concerns also arise regarding their stability, safety, cost, and/or quality of taste.

Unlike sucrose, increased concentration of most sweeteners leads to bad tastes, changing from a sweet to a bitter or metallic taste (Riera et al., 2007). Long-term use of these sweeteners may cause several side effects such as heart failure, mental disorders, psychological problems, brain tumors, and bladder cancer (Sun et al., 2006).

TABLE 2.4 Natural Sweeteners and Their Food Applications

S. No.	Sweeteners	Food Applications (References)
1.	Brazzein	Low-calorie sweetener (Kant, 2005)
2.	Curculin	Low-calorie sweetener (Kant, 2005)
3.	Erythritol	Beverages (Priya et al., 2011)
4.	Glycyrrhizic acid	Natural flavor and flavor enhancer (Kroger et al., 2006)
5.	Miraculin	Sour beverages (Rodrigues et al., 2016)
6.	Mogrol glycosides	Beverages and foods (DuBois and Prakash, 2012)
7.	Steviol glycosides	Beverages, chewing gum, and dairy products (Priya et al., 2011), fruit drinks (Khattab et al., 2017)
8.	Thaumatin	Flavor enhancer (Kant, 2005)

TABLE 2.5 Synthetic Sweeteners and Their Food Applications

S. No.	Sweeteners	Food Applications (References)
1.	Acesulfame K	Baked goods, fruit-flavored dairy products (Klug and Lipinski, 2011), carbonated and uncarbonated beverages (Kroger et al., 2006)
2.	Alitame	Carbonated beverages (Nelson et al., 2001)
3.	Aspartame	Pasteurized and sterilized flavored milk (Kumari et al., 2016), sweeteners, yogurt (Butchko et al., 2001)
4.	Neohesperidin	Sweeteners, artificial flavor (DuBois and Prakash, 2012)
5.	Neotame	Beverages (Kroger et al., 2006), sweeteners (DuBois and Prakash, 2012)
6.	Saccharin	Sweetener blends, cooked food, beverages (DuBois and Prakash, 2012)
7.	Sucralose	Cooked and baked food, sweeteners, beverages (Kroger et al., 2006)

FLAVORS

Flavors of foodstuffs are usually the result of many volatile and nonvolatile compounds having different chemical and physical properties. Nonvolatile components give mainly the taste and volatile ones offer both taste and aroma (Table 2.6). A number of compounds may be responsible for the food-derived aroma such as alcohols, esters, dicarbonyls, aldehydes, short chain free fatty acids, lactones, phenolic compounds, methyl ketones, and sulfur compounds (Urbach, 1997). Fruit and vegetable flavoring compounds are mainly generated during the ripening process and are produced as secondary metabolites by the catabolism of small quantities of carbohydrates, lipids, and amino acids. Earlier, range of flavor compounds were extracted naturally from plant sources. Later, after revelation of their structure, synthetic flavors were synthesized by chemical synthesis. Currently, flavors cover almost one-fourth of the market of food additives, and their origin is either chemical or natural. Although flavors are also formed by chemical alteration of natural compounds, these products cannot legally be labeled as natural. Moreover, chemical synthesis is not an eco-friendly process and has no substrate selectivity, which increases the purification costs.

PRESERVATIVES

Addition of preservatives in food is done with the aim that it prevents degradation, or to restore nutritional value and flavor for an extended period (Abdulmumeen et al., 2012). Preservatives are used to eradicate microorganisms from the food and avert their growth. Furthermore, they are used to extend the shelf life of certain products and ensure their safety for longer periods. Preservatives impede bacteria-mediated degradation of food, which results in the production of various toxins and causes food poisoning (Lee, 2012). Several methods such as high concentration of salt or reduced water content in food

TABLE 2.6 Flavoring Substances Derived From Essential Oils

S. No.	Flavoring Substance	Aroma/Odor	Origin
1.	Anethol	Anisic	Anise (*Pimpinella anisum*), star anise (*Illicium verum*)
2.	Allyl isothiocyanate	Pungent	Black mustard (*Brassica nigra*)
3.	Benzaldehyde	Bitter almond	Bitter almond (*Prunus amygdalus* var. *amara*)
4.	1,8-Cineole	Fresh, cool	Eucalyptus (*Eucalyptus globulus*)
5.	Cinnamic aldehyde	Warm, spicy	Cinnamon (*Cinnamomum zeylanicum*)
6.	Citral	Lemon	Lemongrass (*Cymbopogon citrates*)
7.	Decanal	Orange	Orange (*Citrus sinensis*)
8.	Dimethyl sulfide	Sharp, green radish	Commini (*Mentha arvensis*)
9.	Eugenol	Spicy	Clove (*Syzygium aromaticum*)
10.	Geraniol	Floral	Palmarosa (*Cymbopogon martini*)
11.	Geranyl acetate	Fruity, sweet	Lemon grass (*Cymbopogon citratus*)
12.	Linalool	Woody	Basil (*Ocimum basilicum*), camphor tree (*Cinnamomum camphora*)
13.	Linalyl acetate	Floral fruity	Bergamot mint (*Mentha citrata*)
14.	Massoia lactone	Coconut	Massoia tree (*Cryptocaria massoia*)
15.	Methyl chavicol	Sweet	Basil (*Ocimum basilicum*)
16.	Methyl anthranilate	Sweet	Mandarin (*Citrus reticulata*)
17.	Terpinenol-4	Warm peppery	Tea tree (*Melaleuca alternifolia*)
18.	Thymol	Sweet medicinal	Thyme (*Thymus vulgaris*)

inhibit spoilage by microbial growth. Traditional methods of preservation are known to exclude air, moisture, and microorganisms or to provide environments that are not suitable for the survival of spoilage microorganisms (Daniel, 2007). The removal of air may be achieved by sealing the foods inside containers, or the layering of food surfaces with hot paraffin. Food preservation is commonly achieved for three purposes, including preservation of appearance, preservation of nutritional characteristics, and for extended shelf life. Antimicrobial preservatives are used to hamper the growth of fungi, bacteria, and mold, whereas antioxidant preservatives are used to prevent the oxidation of food constituents. Calcium propionate, sodium nitrite, sodium bisulfate, sodium nitrate, sulfites, potassium hydrogen sulfite, and disodium EDTA are the commonly used chemical preservatives (Dalton, 2002). On the other hand, vinegar, alcohol, salt, sugar, and diatomaceous earth are natural substances commonly used as traditional preservatives. Processes including freezing and pickling are also meant to preserve food.

Some chemical preservatives may pose harmful effects, for example, sulfur dioxide (wines preservative) is irritating to the bronchial tubes of asthma patient, and nitrate, nitrites (meat preservative) are considered as potent carcinogens when converted into nitrous acid (Sanchez-Echaniz et al., 2001; Dusdieker et al., 1994). Use of sulfites as chemical fruit preservative may lead to several harmful effects such as headaches, palpitations, allergies, and even cancer. The antimicrobial preservatives such as benzoates have been reported to cause allergies, asthma, and skin rashes, whereas sorbic acid have been reported to cause urticaria and contact dermatitis (Kinderlerer and Hatton, 1990).

Nuclear radiation when used for preservation does not make foods radioactive but may cause alteration in food color or texture (John, 2003). The ionizing radiation utilized to irradiate foods products are high-intensity X-rays, which disrupt bacterial DNA and thus prevent the microbial growth.

ANTIOXIDANTS

Antioxidants are substances that significantly inhibit or delay oxidative processes. The use of antioxidants has become popular in the food industry over the last few decades to avoid the lipids from oxidative degradation (Vaya and Aviram, 1999). Antioxidants defend cells against the effects of harmful free radicals, which are generated during various metabolic processes. Several group of scientists proved that production of a large amount of free radicals in living beings may lead to processes such as aging (Calabrese and Maines, 2006) and medical conditions including cancer (Johnson, 2001), rheumatoid arthritis (Firuzi et al., 2006), atherosclerosis (Siekmeier et al., 2007), stroke (Spence, 2006), and diabetes (Rahimi et al., 2005). Several studies also have revealed that antioxidants may prevent diseases caused by free radicals. According to the Prevention of Food Adulteration Act (PFA), antioxidants are a substance that delays or prevents oxidative deterioration of food when used as food additives (PFA, 2008). The major antioxidants used as food additives are monohydroxy or polyhydroxy phenol compounds with various ring substitutions. Antioxidants are used in food that may be either natural or synthetic in origin.

The search for natural antioxidants with beneficial activity to be added to foods for specific population groups (hypertensive, diabetics, etc.) is necessary. It is believed that natural antioxidants derived from fruit and vegetables (Table 2.7) are safer than synthetic antioxidants. The well-known naturally occurring antioxidants such as lycopenes, nordihydroguaretic acid (NDGA), avonoids, sesamol, gossypol, phytochemicals, minerals (Zinc, Selenium), enzymes (catalase, glutathione peroxidase, super oxide dismutase), lecithin (Cuppett, 2001), and vitamin E are commonly used as food additives (McCarthy et al., 2001). Synthetic antioxidants are also extensively used in food products to inhibit the mechanism of lipid peroxidation. However, addition of these to food products is illegal in some countries. Currently, butylated hydroxy anisole (BHA), *tert*-butyl hydroquinone or *t*-butylhydroquinone (TBHQ) and esters of gallic acid are the common synthetic antioxidants used in many countries (Yanishlieva-Maslarova, 2001). Several natural antioxidants prevent the generation of free radicals and propagation of reactive oxygen species (ROS), while others may scavenge free radicals (Ozsoy et al., 2009). Only permitted antioxidants are allowed for use in food that has been rigorously tested for safety and should be safe even at excessive doses.

TABLE 2.7 Natural Antioxidant Compounds Present in Fruits and Vegetables

S. No.	Source	Natural Antioxidant (References)
1.	Apple	Anthocyanins, Flavonols (Heinonen et al., 1989), Carotenoids (β-Carotene) (Markowski and Plocharski, 2006)
2.	Bilberries	Flavonols, Hydroxycinnamates, Carotenoids (β-Carotene) (Hakkinen et al., 2000)
3.	Broccoli	Quercetin (Plumb et al.,1997), Vitamin C (Van den Berg et al., 2000)
4.	Brussels sprouts	Carotenoids (β-Carotene) (Sorensen et al., 2001)
5.	Grapes	Hydroxycinnamates (Santos-Buelga and Scalbert, 2000), Resveratrol (Gehm et al., 1997)
6.	Carrots	Carotenoids (β-Carotene) (Van den Berg et al., 2000)
7.	Orange	Hydroxycinnamates, Carotenoids (β-Carotene) (Van den Berg et al., 2000)
8.	Raspberries (red)	Flavonols, Hydroxycinnamates, Carotenoids (β-Carotene) (Kahkonen et al., 2001)
9.	Pea	Flavonols (quercetin), Carotenoids (β-Carotene) (Heinonen et al., 1989)
10.	Potatoes	Hydroxycinnamates, Carotenoids (β-Carotene) (Heinonen et al., 1989)
11.	Spinach	Carotenoids (β-Carotene) (Van den Berg et al., 2000)
12.	Tomatoes	Quercetin, Carotenoids (β-Carotene) (Rao and Kiran, 2011), Lycopene (Yaping et al., 2002)
13.	Sweet red pepper	Carotenoids (β-Carotene) (Van den Berg et al., 2000)
14.	Peach	Flavonols, Hydroxycinnamates, Carotenoids (β-Carotene) (Tomas-Barberan et al., 2001)
15.	Strawberries	Flavonols, Hydroxycinnamates, Carotenoids (β-Carotene) (Kahkonen et al., 2001)

Antioxidants offer many positive health effects, but they may also be associated with some harmful side effects, for example, vitamin C is used to prevent the common cold (McClain and Jochen Bausch, 2003) but its use can lead to certain cancers and cardiovascular disease (Goodman, 1980). Similarly, polyphenolic antioxidants used to cure arthrosclerosis (Shahidi and Ho, 2005) also possibly prevent oxidative DNA damage, which leads to cancer (Yao et al., 2004). The normal recommended intake of synthetic antioxidants BHT and BHA are about 0.1 mg/kg, but at higher dose they have enzyme or lipid alteration or carcinogenic effects (Goulds, 1995).

Agricultural Residues

PESTICIDES

Pesticides are chemicals used to kill insects and pests from agricultural crops and are assumed to be an economic, labor-saving, and efficient means for pest management (Damalas and Eleftherohorinos, 2011). Pesticide use is believed to be essential for maintaining current production and yield levels as well as to maintain a high-quality standard of life (Delcour et al., 2015). Modern agricultural technologies have led to widespread use of pesticides along with other modern inputs in growing economies (Pingali and Rola, 1995). Extensive use of pesticides reduces agricultural losses to insect and pests and enhances availability of food at a reasonable price to increasing populations (Cooper and Dobson, 2007). Therefore pesticides are considered as an essential part of modern life that prevent growth of unwanted species (Bolognesi, 2003).

According to the calculation by Popp et al. (2013), the expected 30% increase of world population will be 9.2 billion by 2050 with projected demand to increase food production by 70%. Worldwide use of agricultural pesticide has raised agricultural production and thereby contributes to food security (Fisher et al., 2012). Oerke (2006) reported that worldwide, an average of 35% of crop yield is lost due to preharvest pests.

Based on the chemical structure, the most popular pesticides are divided into the following groups:

1. Organochlorines
2. Organophosphates
3. Fluorine-containing compounds
4. Carbamic and thiocarbamic derivatives
5. Phenol and nitrophenol derivatives
6. Metal organic and inorganic compounds
7. Heterocyclic compounds such as benzimidazole and triazole derivatives
8. Hydrocarbons, ketones, aldehydes, and their derivatives
9. Urea derivatives
10. Natural and synthetic pyrethroids

Pesticide use is involved in damage to noncrop vegetation and nontarget organisms (birds, fish, beneficial insects) and polluting the air, water, and soil (Andreu and Pico, 2004). Toxic pesticides present in food have severe adverse effects on human health. Pesticidal ingestion, inhalation, or dermal absorption leads to toxicity. In addition, there is a high risk of direct exposures of pesticides for agricultural workers and pesticide factory workers (Verger and Boobis, 2013). It was estimated that in developing countries, 1–5 million farmworkers face pesticide poisoning each year, and also at least 20,000 deaths are recorded annually from its exposure (World Bank, 2006).

Extensive toxicological studies in animals reveal that several pesticides to which the population is commonly exposed are possible carcinogens, neurotoxins, reproductive toxins, and immunotoxins (Baker and Wilkenson, 1990). Pesticide poisoning also leads to development of neurodegenerative diseases in human (Franco et al., 2010). It is also evident that pesticides have a negative impact on biochemical parameters, especially on protein metabolism (Li et al., 2011), endocrine (Cooper et al., 2000), and reproductive systems (Abarikwu et al., 2009).

Because of toxic effects of pesticides against nontarget organisms, it is necessary to discover certain ideal pesticides that are effective but nontoxic to humans. Currently, pesticide manufacturers have successfully produced less-toxic and less-persistent pesticides while maintaining their efficacy. Pyrethroids and *Bacillus thuringiensis* (Bt)–based pesticides are safer to use and are nontoxic/less toxic to living organisms (Zhang et al., 2005).

These pesticides are less toxic to the environment and to human health compared with older available pesticides.

FUNGICIDES

Fungicides are used to suppress the growth of fungi or fungal spores. Fungicides have a role in protection of fruits, vegetables, and tubers during storage. They are also useful in saving standing crops, tress, ornamental plants, and turf grasses (Gupta and Aggarwal, 2007). Fungicides have vast applications in agriculture and in prevention of fungal infection in animals. Fungicides are grouped as contact, translaminar, or systemic in nature. Contact fungicides protect plant tissue topically, translaminar fungicides are redistributed from the upper sprayed leaf surface to the lower unsprayed surface, and systemic fungicides enter into plant tissue and are distributed by xylem vessels throughout the plant. About 90% sulfur is present in powdered fungicides, which have severe toxic effects. Fungicides are also prepared by blending some other active ingredients like jojoba oil, rosemary oil, neem oil, and the bacterium *Bacillus subtilis*.

Fungicides are classified on the basis of their mode of application, origin, and also according to the chemical structure. According to the origin, two major groups of fungicides are available: biological and chemical based. The bio-fungicides are composed of living microorganisms like bacteria and fungi as active ingredients and are effective against the pathogens that cause turf disease. The bio-fungicide ecoguard has *Bacillus licheniformis* and Bio-Trek 22G has *Trichoderma harzianum* that are frequently applied in agriculture. The chemical fungicides are prepared from organic and inorganic chemicals. Use of some fungicides are dangerous to humans, for example, vinclozolin, which has now been totally banned (Hrelia, 1996). Generally, fungicides have low to moderate mammalian toxicity, but it is believed that they are potent carcinogens as compared to other pesticides (Costa, 1997). It has been estimated that more than 80% of all oncogenic incidence from the use of pesticides originate from a few fungicides (NAS, 1987). According to an exposure report from Poison Control Centers, a small proportion of fungicides are related human deaths yearly worldwide (Blondell, 1997; Gray et al., 1999; Litovitz et al., 1994). Chemical fungicides may also be nonbiodegradable. Fungicide residues can deposit in the soil (Athiel et al., 1995) and may be transferred throughout the food chain. Worldwide, consumers are increasingly aware of the potential environmental and health threats (Draper et al., 2003) linked with the build-up of toxic residues, mainly in food products (Mukherjee et al., 2003). Every year, livestock are unintentionally poisoned by fungicides applied to grains, fodder, or other agricultural materials. Generally, newer classes of fungicides have low to moderate toxicity (Gupta and Aggarwal, 2007). The mode of action differs among fungicides but specific reproductive, teratogenic, mutagenic, and carcinogenic effects may persist in the population according to ingested fungicide (Hayes and Laws, 1990; US Environmental Protection Agency, 1999).

HERBICIDES

Weeds have considerable effect on the yield and quality of crops. Much energy is spent in arable farming with mechanical operations aimed at removing weeds and to provide a suitable environment for the growth of crop plants. Weeds are considered a biotic stress that has significant impact on the world crop yield available for human consumption.

In the last few decades, the world population has grown exponentially while adequate supplies of food have diminished due to weeds and other biotic and abiotic stresses. Crude chemicals such as rock salts, crushed arsenical ores, sulfuric acid, creosote, oil wastes, and copper salts have been used for total weed control, which removes all plants from railway tracks, timber yards, and car roads (Green et al., 1987). These chemicals that remove all plants function as total herbicides, and treated areas remain toxic to plants for years. For agricultural purposes, chemicals that selectively kill the weeds but do not harm the crop plants are preferred. Modern agriculture has developed a range of effective herbicides (weed killers) to overcome the effect of weeds on crop yield. These herbicides are either narrow or broad spectrum in nature. Classification of herbicides is based on their mode of application, chemical affinity, structure similarity, and mode of action (Rao, 2000).

About 20 mechanisms of action for herbicides are known, and out of these, some have common target sites with mammalian system. Acetyl-CoA carboxylase (ACCase), an enzyme commonly found in both plants and mammals, can be inhibited by the herbicides cyclohexanediones and the aryloxyphenoxypropionates which kill plants by suppressing de novo synthesis of fatty acids, but do not affect ACCase in mammals (Incledon and Hall, 1997; Shaner, 2003). Currently, several herbicides such as 2,4-D, glyphosate, atrazine, bialaphos, and bromoxynil are in use, which belong to different families with different modes of action. In the chemical industry where herbicides are produced and during applications of herbicides to vegetation and crops, there is the chance for contact, inhalation, and even ingestion of these toxic agents. Gonzalez et al. (2005) reported DNA damages in Chinese hamster ovary cells on exposure to 2,4-D. Exposure of human beings to herbicides may lead to moderate to severe toxic effects. In addition to accidental exposure, herbicides may enter into the food chain of humans in the form of residues by fruits, vegetable, and dairy products. Herbicide poisoning can lead to accidental deaths of exposed individuals. Dermal and respiratory exposure to herbicide during spraying operation in the fields has resulted in acute poisoning of farmers.

Heavy Metals

Food is a vital material required by all living organisms for growth, development, and maintenance of the body. The origin of most food materials are plants (fruits, vegetables, tuber, cereals, etc.) and animals. Heavy metals are naturally occurring elements with high atomic numbers and densities that are higher than the density of water (Tchounwou et al., 2012). Heavy metals are found in a variety of food materials including tea (Garba et al., 2015); fish (Ogamba et al., 2016); prepared foods such as beans, cake, pudding, burgers, bread, fried yams, egg rolls, hot dogs, fried bean cake, herbal drinks; fruits such as apples, watermelons, bananas, oranges, pineapples; and beverages (Al Othman, 2010). Deposits of these metals in various body parts of humans, including the liver, heart, kidney, and spleen, lead to various diseases (Woyessa et al., 2015). Heavy metals are considered a potential health hazard as they are usually toxic in very low amounts. Metals that are mainly toxic to humans include lead, mercury, copper, cadmium, arsenic, and molybdenum. Nitrate intake from water and food leads to the condition methemoglobinemia (blue baby syndrome) in infants and its reported implication in various types of cancer. Nitrosamines, generated from nitrate reduction, are known to be potent carcinogens.

Biological Toxicants

Biological toxicants are organisms, or substances produced by living beings, that represent a danger to human well-being. They are a noteworthy concern in food processing because they cause most foodborne illness outbreaks. Biological hazards can be introduced to food from the environment or from poor sanitation practices and cross-contamination during transportation, handling, processing, and storage of foods.

Microbial Toxicants

Pathogenic bacteria, fungi, parasites, and viruses cause microbial foodborne sickness. Foodborne ailments are any disease caused by utilization of contaminated food. Microbial-related foodborne diseases can lead to either infections or intoxications. Foodborne illness caused by a pathogen itself is known as an infection, whereas foodborne illness caused by toxic products (toxin, toxic metabolites) of pathogens is known as intoxication. The establishment of pathogenic microbes in the host's body through contaminated food is responsible for foodborne infection. These pathogenic microbes can develop or colonize in the digestive tracts, frequently attacking the mucosa or other tissues and causing intrusive infections. These microorganisms produce toxins that affect the different organs or tissue functions. All classes of foodborne pathogens (viruses, bacteria, parasitic protozoa, and other parasites) are infectious agents.

BACTERIAL TOXICANTS

Bacteria are single-celled living organisms thought to be the most important causative agents of foodborne diseases. The common food items that support the growth of bacteria are milk, shell eggs, poultry, fish, meat, and shellfish. Bacterial toxicants are toxic compounds frequently associated with pathogenesis due to the harmful effect on host cells. Toxin-producing bacterial pathogens are a consistent element of the environment. The bacterial toxins are moderately high-molecular-weight substances such as proteins, peptides, or lipopolysaccharides. Bacterial toxins with diverse structure, origins, immunological identity, or mode of action can promptly be recognized from each other regarding their physical or chemical attributes, the organism of origin, as well as the illness symptoms that are the outcome from toxin activity in the susceptible host. The bacteria associated with foodborne diseases (either due to intoxication or infection) are listed in Table 2.8.

Bacillus cereus

Bacillus cereus, a gram-positive, spore-forming, rod-shaped bacteria produce two different toxins. One is a heat-sensitive, high-molecular-weight protein responsible for diarrheal illness (Jones, 1993; Beecher and Lee, 1994). The toxin produced by *B. cereus* (diarrheal type) in the intestinal tract is responsible for toxicoinfection with the symptoms of nausea, abdominal cramps, diarrhea, and some vomiting (Jones, 1993; Frazier and Westhoff, 1988). The other toxin is a heat-stable, low-molecular-weight peptide that delivers an emetic response joined by gastric pain. The symptoms occur between 1 and 6 hours after ingestion of food. An extensive assortment of foods, for example, meat, fish, vegetables, and dairy products, can be associated with *B. cereus* poisoning; however, rice products, meat dishes, and casseroles have been typically linked with the emetic response.

TABLE 2.8 Bacterial Food Toxicants and Their Associated Food Sources

Bacteria	Associated Foods	Symptoms
Staphylococcus aureus	Red meats of pork, beef, chicken, custard and cream-filled baked goods, potato	Food poisoning, nausea, vomiting, abdominal pain, retching, sometimes diarrhea and fever
Clostridium perfringens	Raw meats of pork, chicken, beef, incompletely cooked or reheated meat products and gravies, fish, dried foods such as spices and vegetables	Food poisoning, watery diarrhea, and intense abdominal pain
Bacillus cereus (diarrheal)	Meat, rice, and vegetables dishes, cereals, spices, vegetables, sauces, custards, puddings	Watery diarrhea and abdominal cramps, nausea
B. cereus (emetic)	Rice products and starchy foods such as potatoes, pasta, cheese, soups, and salads	Nausea, vomiting, occasionally diarrhea
Clostridium botulinum	Vacuum-packed foods, home-canned or bottled foods such as meat and vegetables, under-processed canned foods, inadequately processed smoked foods, improperly processed peppers, asparagus, soup, spinach	Botulism may include vomiting, diarrhea, abdominal distention, constipation, difficulty in breathing, difficulty in swallowing, muscle weakness, double or blurred vision, progressive nervous system involvement, and paralysis
Vibrio cholerae	Uncooked or undercooked seafood like shellfish, or by human fecal matter that contaminates food and water	Stomach cramps, sickness, emetic response, diarrhea, fever, and chills
Vibrio vulnificus and *Vibrio parahaemolyticus*	Undercooked or raw seafood, shellfish, crustaceans (oysters)	Gastroenteritis, vomiting, watery diarrhea, dehydration, abdominal pain, fever, blood-borne infection
Salmonella sp.	Raw meats, pork, seafood, shellfish, poultry, undercooked foods, eggs, raw milk and dairy products, unpasteurized juice, reheated food, raw seed sprouts, raw vegetables	Acute gastroenteritis, abdominal pain, diarrhea, vomiting, nausea, chills, fever, headache and body aches, loss of appetite
Yersinia enterocolytica	Raw milk, raw and cooked pork and beef, raw vegetables	Lymph node inflammation, fever, diarrhea, vomiting, abdominal pain, appendicitis-like symptom
Escherichia coli	Raw beef, poultry, pork, raw milk, vegetables and fruits, raw seed sprouts, fecal contamination of food or water	Hemorrhagic colitis, diarrhea, severe abdominal pain and vomiting
Shigella sp.	Raw vegetables and herbs, fecal contamination of food	Bacillary dysentery includes abdominal pain, diarrhea, fever, emetic response, mucus in feces
Campylobacter jejuni	Raw meats, poultry, eggs, shellfish, unpasteurized milk, mushroom	Nausea, watery or bloody diarrhea, vomiting, abdominal cramps, fever, headache, muscle pain
Listeria monocytogenes	Raw meat, meat products, poultry, eggs, milk and dairy products, vegetables and salads, smoked seafood	Fever, muscular pain, sickness and diarrhea; pregnant women may have sepsis; infection can prompt premature delivery or stillbirth, meningitis, encephalitis

Clostridium botulinum

The botulism toxins responsible for a severe paralytic sickness are produced by *C. botulinum*. Foodborne botulism is caused by eating foods that contain the botulinum toxin, which is formed during growth of *C. botulinum*. *C. botulinum* is gram-positive, motile, rod-shaped, spore-forming anaerobic bacterium present in the soil, and its spores can regularly be found on fruit products, vegetables, sausages, fish, and meat (Sakaguchi et al., 1988). The spore of *C. botulinum* can tolerate adverse environmental conditions, which makes it a critical contaminant. Foods prepared with inadequate heating are responsible for *C. botulinum* poisoning. The toxin produced by *C. botulinum* is a heat-sensitive high-molecular-weight protein that is extremely lethal. Only a few nanograms of the toxin can cause sickness. There are seven immunogenic types of the toxins assigned by the letters A to G produced by *C. botulinum* that have been identified and represent the most known intense toxins. These toxins have enterotoxic, hemotoxic, and neurotoxic properties, and toxicity of these toxins varies from one species to another. The *botulinum* neurotoxins block the release of acetylcholine at the synapse. A, B, and E types of *C. botulinum* toxins are regularly responsible for botulism in humans, whereas B, C, and D types cause botulism in cattle, and C and E types adversely affect birds. Type F has once in a while been engaged with botulism in humans (Wong and Carito, 1990; Sakaguchi et al., 1988). The toxin in foods can be destroyed by heating the foods to 80°C for about 30 minutes. Symptoms of botulism take place between 12 and 72 hours after consumption of contaminated food, which include nausea, emetic response, cerebral pains, tiredness, and muscular paralysis.

Clostridium perfringens

C. perfringens is a gram-positive, rod-shaped, nonmotile, spore-forming, anaerobic bacterium that is widespread in the soil and found in intestinal tracts of human and animals. *C. perfringens* spores can survive in soil and various foods, for example, raw meat, poultry, fish, and vegetables (Hobbs, 1983). *C. perfringens* spores can survive at high temperatures. During cooling and holding of food at warm temperatures, the spores germinate and the subsequent vegetative cells of the bacteria grow. The bacterium at that point delivers sufficient toxin in the intestines to cause disease. Few numbers of *C. perfringens* are regularly present in foods after cooking, and during cooling and storage of prepared foods they multiply at a level that is responsible for food poisoning. The foods usually associated with *C. perfringens* food poisoning are meats, meat products, and gravy. The symptoms of *C. perfringens* poisoning include serious abdominal pain, nausea, and acute diarrhea, which begin 8–22 hours after ingestion of extensive quantities of *C. perfringens*–contaminated foods. During sporulation of the ingested *C. perfringens*, an enterotoxin is discharged in the gut, which is responsible for fluid accumulation in the intestinal lumen (Frazier and Westhoff, 1988). The sickness is normally finished within 24 hours, yet less extreme indications can persist in a few people for 1 or 2 weeks. *C. perfringens* produces five different types of toxins (A, B, C, D, and E). The A type toxin is a cytolysin (phospholipase C), which hydrolyzes the membrane phospholipids.

Staphylococcus aureus

S. aureus is gram-positive, nonmotile, facultative anaerobic cocci responsible for producing toxins causing staphylococcal food poisoning. Foods including dairy products, meat, poultry, and salads are regularly connected with staphylococcal food poisoning. Staphylococcal food poisoning is an intoxication that is extremely offensive but usually not lethal. Ingestion of enterotoxins produced by *S. aureus* within different foods is responsible for food poisoning (Bae and Miller, 1992). Five distinctive enterotoxins (designated SEA, SEB, SEC, SED, and SEE) are produced by *S. aureus* (Harris et al., 1993). These toxins are cytolysins, which hydrolyzes the membrane phospholipids. The measure of enterotoxin required to create sickness symptoms is in the vicinity of 0.1 and 1.0 μg. The enterotoxins A and D are responsible for food poisoning, which restrains the absorption of water from the intestinal lumen and induces diarrhea. These enterotoxins are likewise responsible for vomiting by acting on the emetic receptor sites. Enterotoxin B causes hydrolysis of membrane phospholipids, damages the intestinal epithelium, and produces colitis. B-toxin was the first bacterial toxin that appeared to be an enzyme. It is Mg^{2+}-dependent phospholipase C with a substrate particularly restricted to sphingomyelin and lysophosphatidyl choline. Symptoms, including nausea, vomiting, cerebral pain, retching, weakness, abdominal pain, diarrhea, chills, and fever, may occur within a few hours (1–6 hours) after eating the *S. aureus*–contaminated food.

Salmonella

Salmonella species are facultative anaerobic, gram-negative, nonspore-forming bacilli, yet *S. typhimurium* has a capsule. Most of the *Salmonella* species are pathogenic to mankind and are conveyed by wild and domestic animals, reptiles, birds, and insects. The pathogenicity of *Salmonella* in humans is reliant on the strain and sensitivity of the individual. Salmonellosis is the foodborne illness related to *Salmonella* and may incorporate septicemia, typhoid fever, and enteric disease. Improperly cooked meat, undercooked eggs, dairy products, cheese, salads, cold sandwiches, and contaminated food can serve as a significant wellspring of human infection. The incubation period is 8–48 hours after consumption of contaminated food or water. The symptoms of salmonellosis include the sudden beginning of nausea, vomiting, diarrhea, headache, chills, fever, and stomach pains. The most dangerous form of *Salmonella* ailment is typhoid fever.

Campylobacter jejuni

Campylobacter jejuni is a gram-negative, curved, motile, and microaerophilic bacterium. *C. jejuni* infection causes foodborne enterotoxigenic-like sickness. Symptoms of illness fluctuate from irrelevant enteritis to enterocolitis, including stomach pain, diarrhea, fever, vomiting, headache, muscle pain, and, in extreme cases, bloody stools. The incubation period of sickness is generally 2–5 days. *Campylobacters* are found in all foods derived from animals. The implicated vehicles of disease transmission in humans are poultry, undercooked chicken, processed turkey, drinking water, and raw milk (Jones, 1993).

Escherichia coli

E. coli is found in the intestinal tract of humans and other warm-blooded animals. They are gram-negative, nonspore-forming rods; most are motile with flagella. *E. coli* is responsible for gastroenteritis in humans. Foodborne diarrheagenic *E. coli* are classified into four classes based on pathogenic properties, clinical syndromes, epidemiology, and distinct O: H serogroups. These are EPEC (enteropathogenic *E. coli*) causing gastroenteritis or infantile diarrhea, EIEC (enteroinvasive *E. coli*) causing bacillary dysentery, ETEC (enterotoxigenic *E. coli*) causing travelers' diarrhea, and EHEC (enterohemorrhagic *E. coli*) causing hemorrhagic colitis. EHEC strains, including *E. coli* O157:H7 and *E. coli* O26:H11 are the most severe, with potential for serious consequences, for example, hemolytic uremic syndrome, particularly in young children (Padhye and Doyle, 1992; Ram et al., 2009). Symptoms vary for the different forms of disease, including abdominal pain, diarrhea, vomiting, fever, chills, dehydration, electrolyte imbalance, high body fluid acidity, and general discomfort. A variety of foods, such as raw meats (beef, pork, chicken), milk, vegetables, and fruits, and fecal contamination of food or water are associated with *E. coli*.

Vibrio cholerae

The genus *Vibrio* comprises gram-negative-curved, nonencapsulated rod, facultative anaerobe and motile bacteria. *Vibrio cholerae* produces cholera toxin (choleragen), which is responsible for cholera disease (Williams, 1991). The disease is transmitted through consumption of uncooked or undercooked seafood, like shellfish, or by human fecal matter that contaminates food and water. Patients influenced with this pathogen can be asymptomatic, and may have mild or watery diarrhea, vomiting, and abdominal cramping (Madden et al., 1989).

Shigella

The shigellae are gram-negative, rod-shaped, nonsporulating, nonmotile, and facultative anaerobes. *Shigella dysenteriae, Shigella flexneri, Shigella boydii,* and *Shigella sonnei* are the four species of *Shigella*. It causes shigellosis or bacterial dysentery. Among the various toxins produced *by Shigella* sp., Shiga toxin is the best characterized. The Shiga toxin is enterotoxic, neurotoxic, and cytotoxic in nature. Transmission of shigellosis or bacterial dysentery is for the most part by contact with infected person, fecal-oral course, house flies, and contaminated food and water. The associated food sources include raw vegetables and herbs, salads, watermelon, raw meats, eggs, shellfish, and milk products. Persons infected with this pathogen may be asymptomatic, and it might shift from mild diarrhea to dysentery accompanied by abdominal cramps. Extreme circumstances may incorporate bloody stools, dehydration, fever, chills, and emetic response. The incubation period of shigellosis or bacterial dysentery is 1–7 days.

FUNGAL TOXICANTS

Fungi or molds are capable of producing an extensive range of chemicals (fungal metabolites) that are biologically active. Certain fungal metabolites are highly desired components in the production of foods such as cheese and medicines (antibiotics). Some fungi, especially filamentous fungi, can produce toxic metabolites known as mycotoxins

TABLE 2.9 Fungal Toxicants (Mycotoxins) and Their Associated Food Sources

Mycotoxins	Fungi	Associated Food Sources
Aflatoxins	*Aspergillus flavus*, *Aspergillus parasiticus*, and *Aspergillus nominus*	Rice, wheat, corn, barley, bran, peanuts, cottonseed, soybeans, chili peppers, figs, millet, green coffee beans, sorghum, and dried fruits
Alternariol	*Alternaria alternate*	Pecans, sorghum
Citrinin	*Penicillium citrinum*	Groundnuts, wheat, oats, barley, rye
Ergot alkaloids	*Claviceps purpurea*	Rye, wheat, barley, and other cereals
Fumonisins	*Fusarium moniliforme*	Corn, corn-based foods, and feeds
3-Nitropropionic acid	*Arthrinium* sp.	Sugarcane
Ochratoxins	*Aspergillus ochraceus*	Wheat, rice, barley, corn, flour, rye, oats, peas, beans, green coffee beans, dried fruits, and mixed feeds
Patulin	*Penicillum expansum*	Apples and their juice, pears, grape juice, bananas, pineapples, grapes, peaches, apricots
Rubratoxins	*Penicillum rubrum*	Corn
Trichothecenes	*Fusarium* sp.	Wheat, corn, barley and other cereals, bread, snack foods, cake
Zearalenone	*Fusarium graminearum*	Barley, corn, sorghum, sesame meal, feedstuffs

(Table 2.9). Mycotoxins are fungal secondary metabolites with various structures and toxicological properties. When foods contaminated with mycotoxins are consumed, it exerts adverse impacts on humans and animals, called mycotoxicosis (Peraica et al., 1999). These mycotoxins are potent acute or chronic toxins, carcinogens, mutagens, and teratogens (Bennett and Klich, 2003). The chemical structures of mycotoxins vary extensively; however, they are relatively low-molecular-weight organic compounds. The production of mycotoxins and their contamination with foods depend on environmental conditions such as weather and moisture.

AFLATOXIN

Aflatoxins comprise a class of highly toxic metabolites, carcinogens, and hapatotoxic compounds (Peraica et al., 1999). These mycotoxins are produced by the typical agriculturally important fungi *Aspergillus flavus*, *Aspergillus nominus*, and *Aspergillus parasiticus* (Wilson and Payne, 1994). These fungi are found everywhere. They can grow on a wide range of agricultural products under favorable conditions. *A. flavus* is widespread in various important food and feed products, including stored rice, wheat, corn, barley, bran, peanuts, cottonseed, soybeans, chili peppers, figs, millet, green coffee beans, sorghum, and dried fruits. Aflatoxins were first perceived in the 1960s in peanuts. On an overall premise, maize is the most important food contaminated with aflatoxin. The disease caused by aflatoxin is known as aflatoxicosis. Aflatoxins are a series of bisfuran polycyclic fungal metabolites. The aflatoxins are classified in four major types (B1, B2, G1, and G2) that fluoresce strongly in ultraviolet light. These are lipid soluble and heat stable and survive most forms of food processing. The acute and chronic effects of aflatoxins on animals are quite different depending on species, dosage level, age, etc. The organ mainly affected by toxicity and carcinogenicity of aflatoxins is the liver, but other organs may be affected also. Prominent acute toxicity may occur within 3 weeks of consumption of aflatoxin-contaminated foods. Clinical symptoms include hepatic lesions with edema, biliary proliferation, parenchymal cell necrosis, jaundice, vomiting, and anorexia.

OCHRATOXIN A

Ochratoxins are secondary metabolites produced by *Aspergillus ochraceus* and related species, and in addition certain *Penicillium* species (Scott, 1994). Ochratoxins have been found in grains, rice, wheat, barley, corn, rye, peas, beans, soybeans, peanuts, dried fruits, mixed feeds, cheese, and all kinds of food commodities of animal origin (Speijers and Van Egmond, 1993). The production of ochratoxin A by *A. ochraceus* is supported by moist conditions and moderate temperature. Ochratoxin A has been related to porcine nephropathy. The Ochratoxin A exhibits nephrotoxicity in mammals, birds, and fishes and is teratogenic to mice, rats, and chicken embryos. Symptoms developed after 24 hours of transitory epigastric tension, respiratory distress, and retrosternal burning. The human disease balkan endemic nephropathy (BEN) is associated with Ochratoxin A (Smith and Moss, 1985). BEN is a noninflammatory, chronic kidney disease that leads to kidney failure.

ERGOT ALKALOIDS AND ERGOTISM

Ergot is the common name of fungus *Claviceps purpurea*, which produces ergot alkaloids. *C. purpurea* is a common preharvest grain fungus that grows on rye and cereals. Ergot alkaloids are also produced by some strains of *Aspergillus*, *Rhizopus*, and *Penicillium* sp. (Flieger et al., 1997). The consumption of ergot alkaloids–contaminated food leads to human disease known as ergotism. Ergot alkaloids mainly contaminate rye, wheat, and barley. The ergot alkaloids include ergotamine, ergocristine, ergonovine (ergometrine), ergosine, ergocornine, and ergocryptine. All of these ergot alkaloids are pharmacologically active compounds. Ergonovine (ergometrine) is observed as a potent inducer of uterine contraction. Ergot alkaloids influence the smooth muscles, resulting in peripheral artery constriction and neurological disorders. The manifestations of ergotism are of two types. The first type of ergot toxicity is gangrenous type, which is characterized by severe pain, inflammation, and a burned appearance of the extremities, which may become blackened. The second type of ergot poisoning is convulsive ergotism, which is quite different from the gangrenous type. The symptoms are mainly neurological in nature and include writhing, tremors, numbness, muscle cramps, vomiting, headache, blindness, convulsions, hallucinations, and psychological disorders.

TRICHOTHECENES

Trichothecenes are the mycotoxins mainly produced by members of the *Fusarium* genus. It comprises a group of more than 80 sesquiterpene derivatives of 12, 13-epoxytrichothecene. The main trichothecenes are vomitoxin (deoxynivalenol), neosolaniol,

T-2 toxin, HT-2 toxin, and diacetoxyscirpenol (DAS). These toxins contaminate a variety of food items such as corn, moldy millet, oats, wheat, barley, rye, buckwheat, and other cereals and promptly persist into foodstuffs, for example, bread, snacks, and cake (Yoshizawa et al., 1981). The most widely recognized human mycotoxicosis is alimentary toxic aleukia (ATA), which is caused by trichothecene T-2 toxin. The common symptoms of trichothecene acute toxicity are generally neurological; however, the chronic toxicity is specified by cellular damage of the bone marrow, thymus, spleen, and GI tract, inflammation, hemorrhages, leucopenia, and sometimes nausea, vomiting, and diarrhea. The other mycotoxins (nontrichothecenes) produced by *Fusarium* include fumonisins, zearalenone, fusarochromanones, wortmannin, fusarins, and moniliformin.

VIRAL TOXICANTS

Viruses are very small particles that exist everywhere. They can be considered as obligatory intracellular parasites. Viruses can be present in foods without multiplying; hence, they do not require food, water, or air for their survival. An appropriate host is required for multiplication of viruses. Human beings are also considered appropriate hosts by a few viruses. All foodborne viruses generally contain single-stranded RNA and are coated with structural protein. They do not cause spoilage. They cause ailments by infecting living cells and multiply within the host cell by utilizing materials from it. Pathogenicity of viruses include killing of the cells, loss of unique capacity of cells, or multiplication within the cells. Poor hygienic practices are generally responsible for contamination of foods by viruses. Viruses are transmitted enterically, shed with human excrement, and contaminate by being ingested. Viruses can enter in the food supply through various ways, for example, by food handlers infected with viruses or by sewage contamination. Viruses can survive for a considerable length of time in the digestive tract of humans, in contaminated water, and in frozen foods (Table 2.10).

HEPATITIS A VIRUS

The hepatitis A virus is responsible for infectious hepatitis and is spread through the fecal-oral route. The potential source of hepatitis A virus is milk, fruits like strawberries and raspberries, shellfish, meat, poultry, contaminated egg products, and human fecal matter–contaminated vegetables. Flies and cockroaches may likewise be a vehicle of Hepatitis A virus. Fever, abdominal discomfort, and jaundice are common symptoms of the hepatitis A virus infection. In hepatic infection, the liver becomes inflamed and enlarged. The onset ranges from 15 to 20 days.

HEPATITIS E VIRUS

Hepatitis E is a calici-like virus spread by the fecal-oral route. It causes sickness between an incubation time of 2–9 weeks. Hepatitis E occurs in both epidemic and sporadic-endemic forms. The transmission of hepatitis E is related with contaminated drinking water, contaminated food products, and person-to-person contact. Symptoms caused by the hepatitis E virus may incorporate anorexia, fever, discomfort, nausea, and vomiting, followed by manifestations of liver damage, for example, passage of dark urine and jaundice.

TABLE 2.10 Viral Food Toxicants and Associated Food Sources

Viral Toxicants	Associated Food Sources	Symptoms
Hepatitis A	Raw foods, cooked and uncooked foods contaminated by infected person through fecal-oral course, shellfish, vegetables, milk	Hepatitis A including flu-like symptoms, fever, nausea, diarrhea, dark urine, anorexia, abdominal pain, jaundice, loss of appetite, tiredness
Hepatitis B	Contaminated food products	Hepatomas, liver cirrhosis, liver cancer, liver failure
Hepatitis E	Contaminated food products	Fever, anorexia, discomfort, nausea, vomiting, liver damage, dark urine, jaundice
Noroviruses	Raw foods, foods contaminated via fecal-oral route, raw or inadequately steamed shellfish, clams and oysters	Acute viral gastroenteritis, food poisoning, sickness, diarrhea, emetic response, malaise, stomach pain, fever, chills, cerebral pain, loss of appetite
Rotaviruses and Reoviruses	Foods contaminated by infected person via fecal-oral route	Nausea, vomiting, diarrhea, malaise, abdominal pain, headache, and fever
Norwalk-like virus	Salads, raw oysters, clams	Gastroenteritis
Caliciviruses	Contaminated foods and water via fecal-oral routes	Acute viral gastroenteritis, sickness, emetic response, diarrhea, malaise, stomach cramp, cerebral pain, and fever
Astroviruses	Contaminated foods and water via fecal-oral routes	Viral gastroenteritis, sickness, emetic response, diarrhea, malaise, stomach cramp, cerebral pain, and fever

ROTAVIRUSES

Rotaviruses have a place with family Reoviridae. They are related with gastroenteritis, particularly in newborn children and those under 5 years of age. The symptoms of rotavirus infection are fever, sickness, vomiting, and watery diarrhea. The mode of transmission of rotavirus is food contaminated by an infected person by means of fecal-oral route.

PARASITIC TOXICANTS

Parasites are organisms that need a host for their survival. A few parasites might be spread by food or water contaminated by fecal matter of infected hosts (Table 2.11). Parasitic contaminations are usually connected with raw or half-cooked foods, such as meat products, freshwater fish, and freshwater snails. There are two types of parasites that can infect people through food or water: parasitic protozoa and worms.

TABLE 2.11 Parasitic Food Toxicants and Associated Food Sources

Parasitic Food Toxicants	Associated Food Sources	Symptoms
Toxoplasma gondii	Raw or undercooked meat, especially pork, contaminated water	Fever, headache, muscle aches, swollen lymph nodes, sore throat, mental problems, for example, confusion and psychosis; complications for pregnant women and fetus followed by miscarriage, fever, aching muscles, rash
Entamoeba histolytica	Raw or undercooked meat, shellfish, eggs, salad, pealed fruits, sauces, ice cream, ice cubes, contaminated tap water	Amoebiasis characterized by high fever, stomach cramp, excessive gas, rectal pain, amoebic dysentery, loose stool (may be bloody), enlarged liver and unintentional weight loss
Cryptosporidium	Raw food, food contaminated by infected person, contaminated drinking water	Watery diarrhea, abdominal cramps, upset stomach, fever
Cyclospora cayetanensis	Different variety of fresh produce, imported berries, lettuce, basil	Watery diarrhea, constipation, loss of appetite, weight loss, stomach cramps, bloating
Giardia lamblia	Contaminated waters	Nausea, abdominal cramps, diarrhea, weakness and weight loss
Trichinella spiralis	Pork, bear, seal meat	Sickness, diarrhea, emetic response, fatigue, fever, stomach pain, constipation, cerebral pain, cough, eye swelling, joints and muscle pains, itchy skin
Anisakis simplex	Different fish such as cod, haddock, pacific salmon, herring, flounder, monkfish	Extreme abdominal pain
Tapeworms (cestodes) *Taenia solium, Diphyllobothrium latum, Taenia saginata*	Pork, beef, fish, contaminated food and water	Some gastrointestinal symptoms, abdominal discomfort, diarrhea, loss of appetite

PROTOZOA

Protozoa are single-cell animals. *Entamoeba histolytica, Giardia lamblia, Toxoplasma gondii,* and *Cryptosporidium parvum* are disease-causing protozoans. They are obligate intracellular parasite and produce a cyst, which is the infectious stage. The transmission of cyst is mainly by fecal-oral route. The important sources of cyst are contaminated water, vegetables, and foods such as improperly cooked pork or beef. Symptoms of disease caused by these protozoans may incorporate diarrhea (sometimes bloody diarrhea), abdominal cramps, abdominal distention, nausea, weakness, and sometimes fever. Onset time is highly variable.

WORMS

Parasitic worms are also known as helminths. They are large, multicellular organisms including roundworms (nematodes), tapeworms (cestodes), and flukes (trematodes).

ROUNDWORMS (NEMATODES)

Ascaris lumbricoides, *Trichinella spiralis*, and *Anisakis simplex* are pathogenic parasites. The disease might be spread through eggs or larval cysts of these worms. Contaminations of foods by these pathogenic worms are more typical where sanitation is poor. The potential sources of roundworm are half-cooked foods, raw fish, seafood, vegetables, and fruits. The disease symptoms may incorporate abdominal discomfort, intestinal ulcer, bloody sputum, fever, nauseous feeling, and muscle and joint aches.

TAPEWORMS (CESTODES)

Tapeworms live in human intestines, where they feed on the partially digested food there. The three common types of tapeworms are *Taenia solium*, found in pork; *Taenia saginata*, found in beef; and *Diphyllobothrium latum*, found in fish. Tapeworm eggs are for the most part ingested through food, water, or soil contaminated with human or animal host excrement. After ingestion, they develop into larvae, which can move out of the intestines and form cysts in different tissues, for example, lungs and liver. The infection of tapeworms in the intestine usually causes no symptoms. However, some people experience upper abdominal discomfort, diarrhea, loss of appetite, and sometimes anemia. Sickness is by and large perceived when the infected person passes segments of proglottids in the stool.

FLUKES (TREMATODES)

Flukes are a kind of parasitic flatworm under the class trematoda inside the phylum platyhelminthes. Most trematodes have an intricate life cycle with at least two hosts. The primary host is a vertebrate, where the flukes reproduce sexually. The intermediate host is typically a snail, where asexual reproduction occurs. Flukes can be found in any place where untreated human waste is utilized as manure. Few flukes (*Fasciola hepatica*) live on the gills, skin, or outside of their hosts, while others, like blood flukes (*Schistosoma*), live inside their hosts. Humans are infected by *Fasciola hepatica* when raw or improperly cooked food is ingested.

Animal Toxicants

NATURAL TOXINS IN MARINE FOODSTUFFS

Fishes are important wellsprings of food and income all over the world. On a worldwide scale, the vast majority of the general population depends on fish as an important source of animal protein. There are many species of toxic and poisonous marine organisms (Table 2.12). Many of these toxicities originate from toxins produced by blue green algae and bacteria that contaminate marine fish or shellfish.

TABLE 2.12 Animal Toxicants and Their Toxic Effects

Animal Toxicants	Animal Source	Toxic Effects
Tetrodotoxin	Marine and terrestrial animals such as puffer fish include blowfish, porcupine fish, toadfish, fugu and molas, frogs, newts, starfish, crabs, octopus, marine snails	Initial numbness, tingling of the lips, tongue, gastrointestinal disturbances, including vomiting, abdominal pain, and diarrhea, neurotoxic, results in paralysis of the central nervous system and peripheral nerves, general weakness, followed by paralysis of the limbs and chest muscles, drop in blood pressure; death can occur within 30 min
Ciguatoxin	Ciguatera fish	Gastrointestinal symptoms, for example, sickness, vomiting, stomach cramp, diarrhea, neurological symptoms such as headaches, muscle aches, inhibits cholinesterase, paresthesia, numbness, ataxia, and in some cases hallucinations
Saxitoxin	Shellfish	Paralytic shellfish poisoning, rapid onset of symptoms of paresthesia, increasing paralysis, and eventual death by respiratory failure
Brevetoxin	Shellfish	Neurotoxic shellfish poisoning; numbness; tingling in the mouth, arms, and legs; incoordination; and gastrointestinal upset
Domoic acid	Shellfish	Amnesic shellfish poisoning, gastrointestinal distress, dizziness, headache, and disorientation, permanent short-term memory loss

Plant Toxicants

Despite the common major (protein, fat, carbohydrate, and fiber) and minor (vitamins, minerals) supplements, our food contains a huge number of naturally present toxic plant compounds or antinutritional compounds (Table 2.13). Healthy people may endure naturally occurring toxicants. Nevertheless, there are various situations under which these toxicants can create health issues. Many plant species contain hazardous levels of natural toxic constituents (Dolan et al., 2010). The plant toxins might be present naturally in common food source plants, for example, fruits and vegetables. Distinctive kinds of natural toxins might be found in different plants and in various parts of a plant, for example, foliage, buds, stems, roots, fruits, and tubers. They are normally metabolic products or secondary metabolites produced by plants to provide a defense mechanism against various pathogenic bacteria, fungi, insects, predators, and adverse conditions (Wink, 1988). If these metabolites or toxins are consumed in adequate quantities, they can result in adverse effects on human or animal health. Toxicological impacts following ingestion of plant toxins may range from acute effects of gastroenteritis to more severe toxicities in the central nervous system leading to death.

TABLE 2.13 Plant Toxicants and Their Toxic Effects

Plant Toxicants	Plant Source	Toxic Effects
Glucosinolates (Goitrin)	*Brassica* species (cabbage, broccoli, turnip, rutabaga, mustard greens), soybeans, cassava, sweet potatoes, peaches, spinach, strawberries, pears, peanuts	Goitrogens, inhibition of iodine binding to thyroid gland
Canavanine	Alfalfa sprouts and legumes such as jack bean	Causing autoimmune disorders such as lupus erythematosus
Cyanogenic glycosides	Seeds from apples, apricots, cherries, peaches, pears, almonds, cashew nuts, sorghum, lima beans, cassava, corn, chickpeas	Acute life-threatening anoxia, birth defects, endemic goiter, cyanide poisoning
Allyl isothiocyanates	Mustard, horseradish, broccoli, cabbage, cassava, and other tropical staple foods	Toxic endemic goitrogens
Alkenylbenzenes (safrole, estragole, myristicin, asarone, piperine, and isosafrole)	Spices, essential oils, and herbs	Some of them induce tumorogenesis
Pyrrolizidine alkaloids	Species of flowering plants, from genera, for example, *Senecio*, *Crotalaria*, and *Cynoglossum*	Carcinogenic, mutagenic, teratogenic, and chronically hepatotoxic
Ptaquiloside (norsesqiterpenoid glucoside)	Bracken fern (*Pteridium aquilinum*, *Pteridium esculentum*)	Carcinogen
Glycoalkaloids	Members of the family Solanaceae	Gastric pain, muscle weakness, nausea, vomiting, disability of nervous system, acetylcholinesterase inhibitors, ataxia, convulsions, coma, fatal
Tannins (polyphenols)	Nearly every plant-derived food such as bananas, raisins, spinach, red wines, bracken fern, coffee, and tea	Cause a reduced weight gain and reduced efficiency of nutrient usage, inhibition of several enzymes, carcinogens, liver injury (necrosis and fatty liver)
Caffeic acid and chlorogenic acid	Fruits such as grapes, berries, and vegetables such as eggplant, tomatoes, lettuce, potatoes, radish	Carcinogens
Coumarin	Cabbage, radish, and spinach	High doses cause liver damage, carcinogens
Psoralen	Plant families such as Apiaceae, Rutaceae (e.g., bergamot, limes, cloves), and Moraceae	Carcinogens
Flavonoids (quercetin, ellagic acid, kaempferol, and rutin)	Plant-derived foods, including fruits and fruit juices, vegetables, buckwheat, tea, cocoa, red wine, dill, soybeans, bracken fern, and others	Some of them may be mutagenic
Lectins (phytohemagglutinins)	*Legumes*, *Pisum*, *Vicia*, *Lens*, and *Canavalia* spp., *Glycine* spp. *Arachis*, *Phaseolus vulgaris*	Association with specific blood groups, agglutination of tumor cells, mitogenesis, and toxicity to animals, growth depression, fatal

TOXICITY OF NUTRIENTS

Sufficient amount of vitamin and mineral intake from a balanced diet is necessary for maintaining health and preventing diseases (such as pellagra and night blindness) that are the result of dietary deficiencies. Nutrients are essential to maintain good health and pertain to a group of substances that includes carbohydrates, proteins, fats, vitamins, and minerals. Food stronghold refers to the addition of one or more supplements (vitamins, minerals, and amino acids) to a food product, which plays an imperative role in ensuring the health. But at too high a level, some nutrients can be toxic. Nutrients that do not supply calories to your diet can be dangerous if taken in too many quantities (Table 2.14). Fat-soluble vitamins that are stored in our adipose tissue are especially prone to cause adverse effects. Toxic levels of vitamin A lead to birth defects, while higher intake of vitamin D increases blood calcium level, which results in deposition of calcium in soft tissues. Minerals can also accumulate to toxic levels and are harmful to humans; for example, excess dietary calcium may cause kidney stones and it also hinders the absorption of other nutrients such as phosphorus, iron, zinc, and magnesium. High sodium intake has severe effects on the cardiovascular system. Similarly, excess copper, zinc, and manganese intake results in liver damage, suppression of the immune system, and neurotoxicity, respectively. Excess niacin results in flushing reactions, tingling, rashes, itching and reddening of the skin, and nausea. High doses can cause liver toxicity, with symptoms such as jaundice, glucose intolerance, and blurred vision (IOM, 1998).

Numerous case studies have shown the risks of excessive intake of vitamin A for infants, toddlers, and children. Among the more common signs of vitamin A toxicity are brittle nails, hair loss, fever, headaches, and weight loss (IOM, 2001). Excess retinoic acid, the active form of vitamin A, inhibits bone formation (Lind et al., 2013). Carbohydrates, proteins, and fats all contribute calories to the diet in which carbohydrate provides primary fuel source. However, once energy needs are fulfilled, excess carbohydrates are converted to fatty acids and stored in the adipose tissue, which leads to unwanted weight gain and increases the risk of nutritional toxicities.

The human liver has a limited capacity to metabolize proteins, so excessive protein intake ($>35\%$ of total calorie intake) may result in increased blood levels of amino acids (hyperaminoacidemia), ammonium (hyperammonemia), and insulin (hyperinsulinemia), and even death. Common health problems associated with increased protein excretion via urine include systemic infections, urinary tract infections (UTI), kidney disorders, heart disease, high blood pressure, diabetes mellitus (excessive urination, fatigue), rheumatoid arthritis, systemic lupus erythematosus (SLE), certain cancers, lithium, lead levels, and mercury intoxication.

TOXICANTS GENERATED DURING FOOD PROCESSING

During the irradiation of food, free radicals are produced. These free radicals are oxidizers (i.e., accept electrons) and react very strongly. As per the free-radical theory, free radicals can interact with cellular macromolecules and alter several cellular proteins, lipids,

TABLE 2.14 Nutrient Chart—Function, Deficiency and Toxicity Symptoms, and Major Food Sources (Maher and Escott-Stump, 2004)

Nutrient	Function	Deficiency Symptoms	Toxicity Symptoms	Major Food Sources
MACROMOLECULES				
Protein	Anabolism of tissue proteins; helps maintain fluid balance; energy source; formation of immunoglobulins; maintenance of acid-base balance; important part of enzymes and hormones	Kwashiorkor-edema; reddish pigmentation of hair and skin; fatty liver; retardation of growth in children; diarrhea; dermatosis; decreased T cell lymphocytes with increased secondary infections; marasmus—muscle and fat wasting; anemia	Azotemia; acidosis; hyperammonemia	Breast milk, infant formula, meat, fish, poultry, egg yolk, cheese, yogurt, legumes
Carbohydrate	Major energy source; protein sparing; necessary for normal fat metabolism; glucose is the sole source of energy for the brain; many sources also provide dietary fiber	Ketosis	–	Breast milk; infant formula; whole-grain breads, cereals, and other fortified or enriched grain products; potatoes; corn; legumes; fruits; vegetables
Fat	Concentrated energy source; protein sparing; insulation for temperature maintenance; supplies essential fatty acids; carries fat-soluble vitamins A, D, E, K	Eczema; low growth rate in infants; lowered resistance in infection; hair loss	–	Breast milk, infant formula, protein-rich foods (meats, dairy products, egg yolk, nuts), butter, margarine, cream, salad oils and dressings, cooking and meat fats
VITAMINS				
Ascorbic acid (Vitamin C)	Essential in the synthesis of collagen (thus strengthens tissues and improves wound healing and resistance to infection); iron absorption and transport; water soluble antioxidant; functions in folacin metabolism	Scurvy, pinpoint peripheral hemorrhages, bleeding gums, osmotic diarrhea	Nausea, abdominal cramps, diarrhea, possible formation of kidney stones	Breast milk, infant formula, fruits (especially citrus fruits, papaya, cantaloupe, strawberries), vegetables (potatoes, cabbage)
Biotin	Essential component of enzymes; important in reactions involving the lengthening of carbon chains; coenzyme carrier of carbon dioxide; plays an important role in the metabolism of fatty acids and amino acids	Seborrheic dermatitis; glossitis; nausea; insomnia	–	Breast milk, infant formula, liver, meat, egg yolk, yeast, bananas, most vegetables, strawberries, grapefruit, watermelon

(Continued)

TABLE 2.14 (Continued)

Nutrient	Function	Deficiency Symptoms	Toxicity Symptoms	Major Food Sources
Niacin	Part of the enzyme system for oxidation, energy release; necessary for synthesis of glycogen and the synthesis and breakdown of fatty acids	Pellegra: dermatitis, diarrhea, dementia	Transient due to the vasodilating effects of niacin (does not occur with niacinamide)—flushing, tingling, dizziness, nausea; liver abnormalities; hyperuricemia; decreased LDL and increased HDL cholesterol	Breast milk; infant formula; meat; poultry; fish; whole-grain breads, cereals, and fortified or enriched grain products; egg yolk
Pantothenic acid	Functions in the synthesis and breakdown of many vital body compounds; essential in the intermediary metabolism of carbohydrate, fat, and protein	Fatigue; sleep disturbances; nausea; muscle cramps; impair antibody production	Diarrhea; water retention	Breast milk; infant formula; meat; fish; poultry; liver; egg yolk; yeast; whole-grain breads, cereals, and other grain products; legumes; vegetables
Riboflavin (Vitamin B2)	Essential for growth; plays enzymatic role in tissue respiration and acts as a transporter of hydrogen ions; synthesis of FMN and FAD	Photophobia, cheilosis, glossitis, corneal vascularization, poor growth	–	Breast milk; infant formula, meat; dairy products; egg yolk; legumes; green vegetables; whole-grain breads, cereals, and fortified or enriched grain products
Vitamin A	Preserves integrity of epithelial cells; formation of rhodopsin for vision in dim light; necessary for wound healing, growth, and normal immune function	Night blindness, dry eyes, poor bone growth, impaired resistance to infection, papillary hyperkeratosis of the skin	Fatigue; night sweats; vertigo; headache; dry and fissured skin; lips; hyperpigmentation; retarded growth; bone pain; abdominal pain; vomiting; jaundice; hypercalcemia	Breast milk, infant formula, liver, egg yolk, dark green and deep yellow vegetables and fruits
Vitamin D	Necessary for the formation of normal bone; promotes the absorption of calcium and phosphorus in the intestines	Rickets (symptoms: costochondral beading, epiphyseal enlargement, cranial bossing, bowed legs, persistently open anterior fontanel)	Abnormally high blood calcium (hypercalcemia), retarded growth, vomiting, nephrocalcinosis	Infant formula, egg yolk, liver, fatty fish, sunlight (activation of 7-dehydrocholesterol in the skin)
Vitamin E	May function as an antioxidant in the tissues; may also have a role as a coenzyme; neuromuscular function	Hemolytic anemia in the premature and newborn; hyporeflexia, and spinocerebellar and retinal degeneration	May interfere with vitamin K activity leading to prolonged clotting and bleeding time; in anemia, suppresses the normal hematologic response to iron	Breast milk; infant formula; vegetable oils; liver; egg yolk; butter; green leafy vegetables; whole-grain breads, cereals, and other fortified or enriched grain products; wheat germ

Vitamin K	Catalyzes prothrombin synthesis; required in the synthesis of other blood clotting factors; synthesis by intestinal bacteria	Prolonged bleeding and prothrombin time; hemorrhagic manifestations (especially in newborns)	Possible hemolytic anemia; hyperbilirubinemia (jaundice)	Infant formula, vegetable oils, green leafy vegetables, pork, liver
MINERALS				
Calcium	Builds and maintains bones and teeth; essential in clotting of blood; influences transmission of ions across cell membranes; required in nerve transmission	Rickets—abnormal development of bones. Osteomalacia—failure to mineralize bone matrices; tetany; possibly hypertension	Excessive calcification of bone; calcification of soft tissue; hypercalcemia; vomiting; lethargy	Breast milk, infant formula, yogurt, cheese, fortified or enriched grain products, some green leafy vegetables (such as collards, kale mustard greens, and turnip greens), tofu (if made with calcium sulfate), sardines, salmon
Chloride	Helps regulate acid-base equilibrium and osmotic pressure of body fluids; component of gastric juices	Usually accompanied by sodium depletion; see Sodium	–	Breast milk, infant formula, sodium chloride (table salt)
Chromium	Required for normal glucose metabolism; insulin cofactor	Glucose intolerance; impaired growth; peripheral neuropathy; negative nitrogen balance; decreased respiratory quotient	–	Meat; whole-grain breads, cereals, and other fortified or enriched grain products; brewer's yeast; corn oil
Copper	Facilitates the function of many enzymes and iron; may be an integral part of RNA, DNA molecules	Pallor, retarded growth, edema, anorexia	Wilson's disease—copper deposits in the cornea; cirrhosis of liver; deterioration of neurological processes	Liver; kidney; poultry; shellfish; legumes; whole-grain breads, cereals, and other grain products
Fluoride	Helps protect teeth against tooth decay; may minimize bone loss	Increased dental caries	Mottled, discolored teeth; possible increase in bone density; calcified muscle insertions and exotosis	Fluoridated water
Iodine	Helps regulate thyroid hormones; important in regulation of cellular oxidation and growth	Endemic goiter; depressed thyroid function; cretinism	Possible thyroid enlargement	Breast milk, infant formula, seafood, iodized salt
Iron	Essential for the formation of hemoglobin and oxygen transport; increases resistance to infection; functions as part of enzymes involved in tissue respiration	Hypochromic microcytic anemia; malabsorption; irritability; anorexia; pallor, lethargy	Hemochromatosis; hemosiderosis	Breast milk; infant formula; meat; liver; legumes; whole-grain breads, cereals, or fortified or enriched grain products; and dark green vegetables

(Continued)

TABLE 2.14 (Continued)

Nutrient	Function	Deficiency Symptoms	Toxicity Symptoms	Major Food Sources
Magnesium	Required for many coenzyme oxidation phosphorylation reactions, nerve impulse transmissions, and for muscle contraction	Muscle tremors; convulsions; irritability; tetany; hyper- or hypoflexia	Diarrhea; transient hypocalcemia	Breast milk; infant formula; whole-grain breads, cereals, and other grain products; tofu; legumes; green vegetables
Manganese	Essential part of several enzyme systems involved in protein and energy metabolism	Impaired growth; skeletal abnormalities; neonatal ataxia	In extremely high exposure from contamination: severe psychiatric and neurologic disorders	Whole-grain breads, cereals, and other grain products; legumes; fruits; vegetables (leafy)
Molybdenum	Part of the enzymes xanthine oxidase and aldehyde oxidase, possibly helps reduce incidence of dental caries	–	Gout like syndrome	Organ meats; breads, cereals, and other grain products; dark green leafy vegetables; legumes
Phosphorus	Builds and maintains bones and teeth; component of nucleic acids, phospholipids; as coenzyme functions in energy metabolism; buffers intracellular fluid	Phosphate depletion unusual —affects renal, neuromuscular, skeletal systems as well as blood chemistries	Hypocalcemia (when parathyroid gland not fully functioning)	Breast milk; infant formula; cheese; egg yolk; meat; poultry; fish; whole-grain breads, cereals, and other grain products; legumes
Potassium	Helps regulate acid-base equilibrium and osmotic pressure of body fluids; influences muscle activity, especially cardiac muscle	Muscle weakness; decreased intestinal tone and distension; cardiac arrhythmias; respiratory failure	–	Breast milk; infant formula; fruits especially orange juice, bananas, and dried fruits; yogurt; potatoes; meat; fish; poultry; soy products; vegetables
Selenium	May be essential to tissue respiration; associated with fat metabolism and vitamin E; acts as an antioxidant	Myalgia; muscle tenderness; cardiac myopathy; increased fragility of red blood cells; degeneration of pancreas	–	Whole-grain breads, cereals, and other fortified or enriched grain products; onions; meats; seafood; dependent on soil content—vegetables
Sodium	Helps regulate acid-base equilibrium and osmotic pressure of body fluids; plays a role in normal muscle irritability and contractility; influences cell permeability	Nausea; cramps; vomiting; dizziness; apathy; exhaustion; possible respiratory failure	–	Sodium chloride (table salt), abundant in most foods except fruit
Zinc	Component of many enzyme systems and insulin	Decreased wound healing, hypogonadism, mild anemia, decreased taste acuity, hair loss, diarrhea growth failure, skin changes	Acute gastrointestinal upset; vomiting; sweating; dizziness; copper deficiency	Breast milk; infant formula; meat; liver; egg yolk; oysters and other seafood; whole-grain breads, cereals, and other fortified or enriched grain products; legumes

and DNA, which results in altered target cell functions. Oxidative stress occurs in a cell or tissue when the level of free radicals generation exceeds the antioxidant capability of that cell. Free radicals can be produced both endogenously and exogenously. Endogenous oxidative stress can be the result of normal cellular metabolism and oxidative phosphorylation. Exogenous sources of free radicals can also impact the overall oxidative status of a cell. Drugs, hormones, and other xenobiotic chemicals can produce ROS by either direct or indirect mechanisms. However, much of the free radicals generated in the body are destroyed in the metabolic process (Karthikeyan et al., 2011). Several compounds, most notably 2-alkylcyclobutanones, acrylamide, polycyclic aromatic hydrocarbons, and furan are generated during irradiation and heating of the food.

2-Alkylcyclobutanones

2-Alkycyclobutanones (2-ACBs) are produced from irradiation of fat-containing food due to radiation-induced breakage of triglycerides (LeTellier and Nawar, 1972). These are thought to be unique radiolytic products. The 2-ACBs has been found solely in fat-containing irradiated food and have never been identified in nonirradiated foods treated by other food processes (Crone et al., 1993; Ndiaye et al., 1999). Subsequently, these compounds were considered to be unique markers for food irradiation. In irradiated foods, the level of generated 2-ACBs is proportional to the fat content and absorbed dose (Hartwig et al., 2007). Depending on the dose absorbed, the concentration of 2-ACBs in irradiated food ranged from 0.2 to 2 μg/g of fat (Marchioni et al., 2004). The radiation doses needed

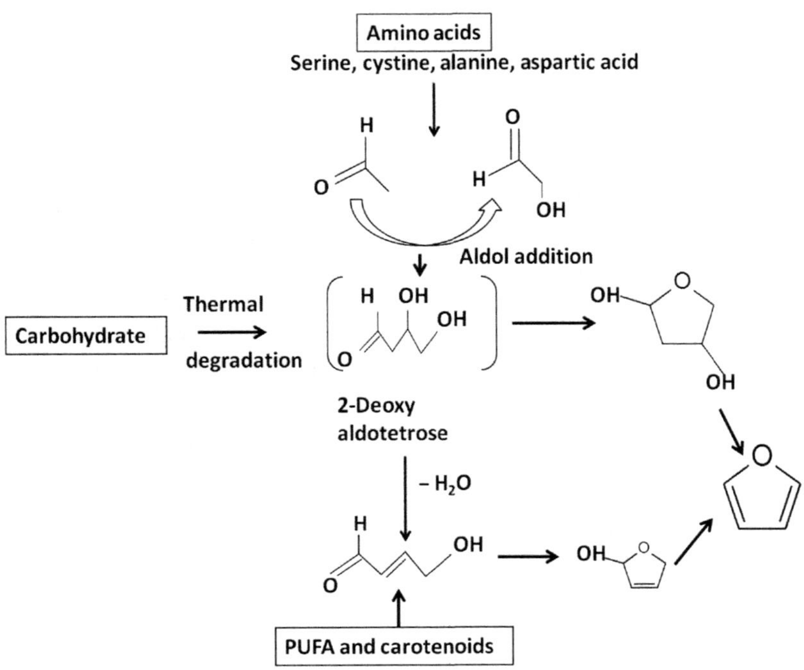

FIGURE 2.2 Three main precursors (amino acid, lipid, and carbohydrate) for the formation of furan (Yaylayan, 2006).

to cause toxic changes are much higher than the doses used during irradiation, taking into account the presence of 2-ACBs along with other free radicals.

Furan

Furan is considered a conceivable carcinogen by the International Agency for Research on Cancer (IARC, 1995). Amino acid, lipid, and carbohydrate are the three main precursors for the formation of furan (Fig. 2.2).

A number of studies have been conducted on the impact of gamma irradiation on the level of furan in foods. It was also reported that the level of furan increased linearly with increasing irradiation dose (Fan, 2005). It was found that fruits such as grapes and pineapples produced low levels of furan by irradiation due to the presence of a large amount of simple sugars and low pH (Fan and Sokorai, 2008). Nevertheless, the level of furan identified in irradiated foods acquired from a general store in the United States was much lower than those in some thermal-processed foods (Pauli, 2006).

Polycyclic Aromatic Hydrocarbons

Heterocyclic amines (HCAs) and polycyclic aromatic hydrocarbons (PAHs) are the toxic chemicals produced when muscle meat such as fish, beef, pork, and poultry is cooked by using high temperature such as during pan frying or grilling directly over an open flame (Cross and Sinha, 2004). HCAs and PAHs have been found to be mutagenic in laboratory experiments because they alter the DNA sequence, which may lead to increased risk of cancer. HCAs are produced at high temperatures when sugars, amino acids, and creatine or creatinine (present in muscles) react to each other. On the other hand, PAHs are produced with flames and smoke when meat is grilled directly over a heated surface. The PAHs present in smoke then stick to the surface of the meat.

PAHs can likewise be produced during incomplete burning of charcoal (Hamidi et al., 2016). It is reported that PAHs are toxic and one of the causing agent of cancer (Domingo and Nadal, 2016). Many studies showed that PAHs have been connected to an increased risk of breast and prostate malignancies (White et al., 2016; Mordukhovich et al., 2010). Grilled meats may increase the risk of kidney cancer due to high amounts of PAHs (Daniel et al., 2012). The most grounded relationship has been found between grilled meats and colon cancer (Diggs et al., 2011). This association with colon malignancy has just been found in red meats, for example, beef, pork, and sheep. Chicken meat seems to have either an impartial or defensive impact on the risk of colon malignancy (Cross et al., 2010). When a high concentration of calcium was added artificially to restore meat, markers of malignant growth-causing compounds diminished in both animals and human stools (Pierre et al., 2013).

During hydrogenation of unsaturated oils so as to transform them into solid fats, trans fats are generated, which can prompt various medical issues after consumption (Dorfman et al., 2009). Various studies have shown that utilization of trans fat is responsible for inflammation and negative impacts on the heart (Iwata et al., 2011). A study on 730 women showed that inflammatory markers were most noteworthy in the individuals who ate the trans fats, including 73% higher levels of C-reactive protein (CRP), which is a strong risk factor for coronary illness (Lopez-Garcia et al., 2005). Controlled investigations in humans have affirmed that trans fats are responsible for inflammation, which has significantly negative impacts on heart. This incorporates weakened capacity of arteries to appropriately dilate and keep blood circulating (Baer et al., 2004). In many experiments, rodents fed with diets containing HCAs developed a variety of tumors in organs and tissues such as breast, liver, skin, colon, lung, and prostate (Ito et al., 1991; Kato et al., 1988; Sugimura et al., 2004).

Acrylamide

Acrylamide is an odorless and colorless crystalline solid with a melting point of 84.5°C that is formed during thermal processing in carbohydrate-rich and protein-low plant foods at high temperatures and low-moisture conditions associated with frying, baking, and roasting (Tareke et al., 2000). Acrylamide forms in naturally occurring components in certain foods when cooked at sufficiently high temperatures (Fig. 2.3).

This is only achieved when the temperature during cooking is sufficiently high. Acrylamide is a toxic compound with mutagenic and carcinogenic properties in experimental mammalians that has been found in several carbohydrate foods processed at high temperatures. There is an urgent need to reduce the content of dietary acrylamide in order to prevent adverse in vivo effects in human beings.

GENETICALLY MODIFIED FOODS AND HUMAN HEALTH

Genetically modified (GM) foods are derived from genetically modified organisms (GMOs), particularly plants and animals of agricultural significance. GMOs are characterized as organisms whose genomes have been altered or modified using genetic

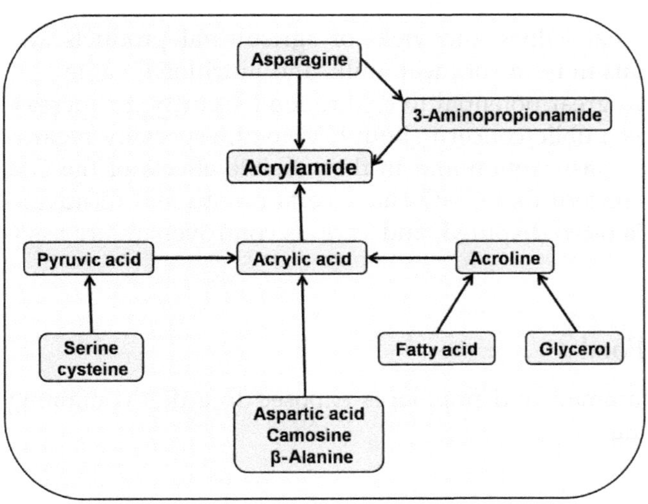

FIGURE 2.3 The basic route of formation of acrylamide in food.

TABLE 2.15 Some Genetically Modified (GM) Food Species

Food Species	Genetic Modifications	Properties
Cottonseed oil	Bt crystal protein gene transferred into plant genome	Pest-resistant cotton
Soya bean	Herbicide-resistant gene from bacteria inserted into soybean	Resistance to glyphosate herbicides
Tomato	An antisense copy of the gene responsible for the production of polygalacturonase (PG) enzyme added into plant genome	Production of PG enzyme is suppressed resulting in delayed ripening of tomato
Rice	Transfer of three genes: two from daffodils and one from a bacterium	Golden rice: genetically modified to contain beta carotene, which is a source of vitamin A
Canola	New genes added into plant genome	Resistance to glyphosate herbicides, high laurate canola
Alfalfa	New genes added into plant genome.	Resistant to glyphosate herbicides
Corn (maize)	New genes, some from the bacterium *Bacillus thuringiensis*, added into plant genome	Resistant to glyphosate herbicides; insect resistance through producing Bt proteins

engineering or recombinant DNA technologies. Genetic engineering technologies permit the transfer of one or more genes from one species to another and produce plants with the exact desired characteristic very rapidly and with incredible accuracy (Verma et al., 2011; Wieczorek, 2013). Various genetically engineered (GE) varieties that have been commercialized include soybeans, corn, sugar beets, cotton, canola, papaya, and squash (Table 2.15). The potential application of GM technology is a production of industrial products (amino acids, vitamins, hormones, enzymes, organic acids, alcohols, etc.) using genetic alteration of microorganisms, production of pest-resistant and herbicide-resistant plant varieties, enhancing the nutritional values and yield of agricultural products and increasing the productiveness of animals in terms of meat, milk, and nutritional value.

Despite the fact that GM foods have great potential to ensure food security for increasing population worldwide, there is yet public concern about GM food, especially regarding the safety of GM food for human consumption and to the possible effects of the GM technology on the environment. The concern that GMOs have certain risks and disadvantages in addition to their advantages is often disputed, and various controversies are associated with GM foods.

Hazards of Genetically Modified Food

Conceivable risks of GM food for animals and populaces exposed to a diet containing GM products may include the following.

Allergenicity

GM foods can possibly cause allergic responses by way of the proteins produced by newly inserted genes. Most of the genes utilized for the production of GM foods are novel and do not have a history of safe food use. The other potential hazard is the introduction of novel protein into foods that did not already exist in the food chain, which may evoke possible harmful immunological reactions, including allergic hypersensitivity (Conner et al., 2003). GM soybean variety producing methionine from Brazil nuts (Nordlee et al., 1996) and GE corn variety altered to express a Bt endotoxin, Cry9C, are examples.

Increase in Antinutrients

The addition of a novel gene for production of GM foods may sometimes lead to an increased level of antinutrients. A few of the antinutrients, for example, phytoestrogens, glucinins, and phytic acid, are heat-stable, which cannot be decreased with heat treatment. They have been found to cause infertility issues and allergenic reactions in sheep and cattle (Dona and Arvanitoyannis, 2009).

Gene Transfer

The other possible concern related to GM foods is the possibility that genes inserted into the GM foods can be taken up by cells of the human body or the microbial flora in the gut (Dona and Arvanitoyannis, 2009). DNA from consumed GM food is not totally degraded during digestion, and small pieces of DNA from GM foods have been found in various parts of the gastrointestinal tract. This might result in horizontal gene transfer because of absorption of DNA fragments by gut microflora or somatic cells lining intestinal cells. Theoretically, antibiotic-resistant genes introduced into GM plants could be transferred to humans in a similar way.

Pleitropic and Insertional Effects

The genes inserted for production of GM foods may cause the silencing of existing genes or changes in their level of expression, or may turn on the genes that were previously not expressed (Conner and Jacobs, 1999). The transgene may interact with the activity of existing genes and biochemical pathways of plants in unpredictable ways and lead to generation of toxic products. The possibility of the presence of an unidentified compound in GM food makes it crucial, and it is important that whole GM products rather than single proteins ought to be tested for toxicity (Dona and Arvanitoyannis, 2009).

RISK ASSESSMENT AND MANAGEMENT

Food safety and foodborne illness remains a critical challenge in both developed and developing nations. A physical, chemical, or biological agent that can possibly cause an adverse health effect is known as a hazard. Foodborne risks to human well-being can emerge from these hazards. Risk is a measure of the likelihood of an adverse health impact and the seriousness of that impact, significant to a food hazard. A systematic disciplined approach for diminishing foodborne sickness and fortifying food safety frameworks

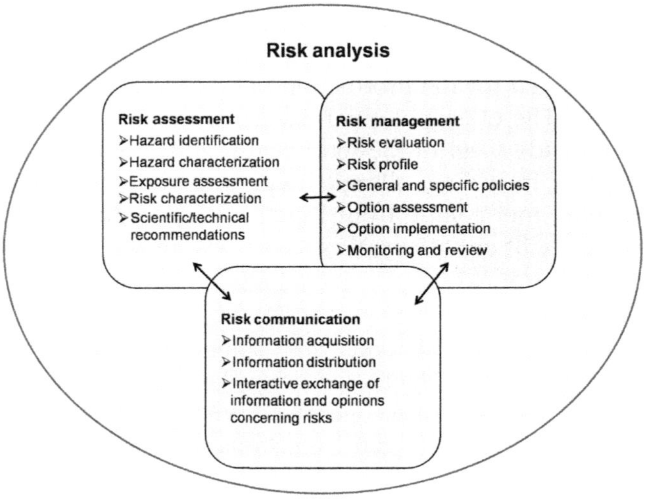

FIGURE 2.4 Components of risk analysis.

is known as risk analysis. Risk analysis approach has now gained wide acknowledgment as the favored method to assess possible connections between hazards in the food chain and actual risks to human well-being. It has the capacity to improve food safety decision-making processes and enhancements in public health. The Food and Agriculture Organization of the United Nations (FAO) and the World Health Organization (WHO) have assumed the main role in the advancement of food safety risk analysis. Risk analysis has been defined by Codex Alimentarius Commission (CAC) as "a procedure comprising of three components: risk assessment, risk management and risk communication" (FAO/WHO, 2008) (Fig. 2.4).

Risk Assessment

Risk assessment is the initial step in risk analysis. It encourages the facility to settle on the level of risk for each hazard. Risk assessment ought to give complete information to permit the risk management group to make the best possible decisions. Risk assessment is a scientifically based process that includes hazard identification, hazard characterization, and exposure assessment and risk characterization. Risk assessment is imperative in developing a HACCP system. The HACCP system has been introduced as a new quality assurance standard for the avoidance of health hazards.

Risk Management

Risk management is about choosing the most ideal approach to decrease or manage the risk. The principle objective of food safety risk management is to secure public health. This is done by controlling risks as much as could reasonably be expected. The connection between risk assessment and risk management is an interactive procedure to set up the extent of the analysis, especially during issue formulation (otherwise called risk profiling).

Risk Communication

Risk communication is an interactive exchange of information and opinions including the explanation of risk assessment findings and the premise of risk management decisions throughout the risk analysis process. Open communication concerning risk, risk-related factors, and risk perceptions among all stakeholders (from employees to consumers, the academic community, and other interested parties) will improve the overall risk management.

CONCLUSION

Food is the fundamental material that fulfill our various nutritional requirements. It consists of nutritive components that support life and important biological processes, supply energy, and offer growth, maintenance, and health of the body. Various food hazards (physical, chemical, and biological) are added to food either intentionally or unintentionally at the time of harvesting, processing, or storage. These food hazards can lead to several foodborne diseases in human beings, resulting in the loss of health. To remove or minimize the negative effect of food hazards, we must follow and apply health rules during harvesting, processing, and storage of food. Biological hazards are considered as the primary food safety concern as compared to physical and chemical hazards. The HACCP concept for food safety requires control over hazardous materials and other substances in foods that cause them to be injurious to public health. Risk analysis is a systematic approach for reducing foodborne illness and strengthening food safety. The primary focus of the strategy is the control and elimination of critical biological, chemical, and physical hazards from the food supply.

References

Abarikwu, S.O., Adesiyan, A.C., Oyeloja, T.O., Oyeyemi, M.O., Farombi, E.O., 2009. Changes in sperm characteristics and induction of oxidative stress in the testis and epididymis of experimental rats by a herbicide, atrazine. Arch. Environ. Contam. Toxicol. 58 (3), 874–882.

Abdulmumeen, H.A., Ahmed, N.R., Agboola, R.S., 2012. Food: its preservatives, additives and applications. Int. J. Chem. Biochem. Sci. 1, 36–47.

Adam, S., 1983. Recent developments in radiation chemistry of carbohydrates. In: Elias, P.S., Cohen, A.J. (Eds.), Recent Advances in Food Irradiation. Elsevier, pp. 149–170.

Al Othman, Z.A., 2010. Lead contamination in selected foods from Riyadh city market and estimation of the daily intake. Molecules 15, 7482–7497.

Andreu, V., Pico, Y., 2004. Determination of pesticides and their degradation products in soil: critical review and comparison of methods. Trends Anal. Chem. 23, 772–789.

Athiel, P., Alfizar Mercadier, C., Vega, D., Bastide, J., Davet, P., Brunel, B., et al., 1995. Degradation of iprodione by a soil *Arthrobacter*-like strain. App. Environ. Microbiol. 61, 3216–3220.

Bae, J.H., Miller, K.J., 1992. Identification of two proline transport systems in *Staphylococcus aureus* and their possible roles in osmoregulation. Appl. Environ. Microbiol. 58 (2), 471–475.

Baer, D.J., Judd, J.T., Clevidence, B.A., Tracy, R.P., 2004. Dietary fatty acids affect plasma markers of inflammation in healthy men fed controlled diets: a randomized crossover study. Am. J. Clin. Nutr. 79 (6), 969–973.

Baker, S.R., Wilkenson, C.F., 1990. The Effects of Pesticides on Human Health. Princeton Sci. Pub, Princeton.

Beecher, D.J., Lee, A.C., 1994. Identification and analysis of the antigens detected by two commercial *Bacillus cereus* diarrheal enterotoxin immunoassay kits. Appl. Environ. Microbiol. 60, 4614–4616.

Bennett, J.W., Klich, M., 2003. Mycotoxins. Clin. Microbiol. Rev. 16, 497–516.

Besaratinia, A., Pfeifer, G.P., 2007. A review of mechanisms of acrylamide carcinogenicity. Carcinogenesis 28 (3), 519–528.

Blondell, J., 1997. Epidemiology of pesticide poisoning in the United States, with special reference to occupational cases. Occup. Med., State Art Rev. 12, 209–220.

Bolognesi, C., 2003. Genotoxicity of pesticides: a review of human biomonitoring studies. Mutat. Res. 543 (3), 251–272.

Butchko, H.H., et al., 2001. Aspartame. In: Nabors, L.O. (Ed.), Alternative Sweeteners. Marcel Dekker, New York, pp. 41–48.

CAC (Codex Alimentarius Commission), 2003. Revised Codex General Standard for Irradiated Foods (CODEX STAN 106-1983, REV.1-2003).

Calabrese, V., Maines, M., 2006. Antiaging medicine: antioxidants and aging. Antioxid. Redox Signal. 8 (3–4), 362–364.

Carocho, M., Barreiro, M.F., Morales, P., Ferreira, I.C.F.R., 2014. Adding molecules to food, pros and cons: a review on synthetic and natural food additives. Compr. Rev. Food Sci. Food Safety 13 (4), 377–399.

Chequer, F.M.D., Lizier, T.M., de Felicio, R., Zanoni, M.V.B., Debonsi, H.M., Lopes, N.P., et al., 2011. Analyses of the genotoxic and mutagenic potential of the products formed after the biotransformation of the azo dye Disperse Red 1. Toxicol. In Vitro 25, 2054–2063.

Conner, A.J., Jacobs, J.M.E., 1999. Genetic engineering of crops as potential source of genetic hazard in the human diet. Mutat. Res. 443, 223–234.

Conner, A.J., Glare, T.R., Nap, J.P., 2003. The release of genetically modified crops into the environment: Part II. Overview of ecological risk assessment. Plant J. 33, 19–46.

Cooper, R.L., Stoker, T.E., Tyrey, L., Goldman, J.M., McElroy, W.K., 2000. Atrazine disrupts hypothalamic control of pituitary-ovarian function. Toxicol. Sci. 53 (2), 297–307.

Cooper, J., Dobson, H., 2007. The benefits of pesticides to mankind and the environment. Crop. Prot. 26 (9), 1337–1348.

Costa, L.G., 1997. Basic toxicology of pesticides. Occup. Med. State Art Rev. 12, 251–268.

Crone, A.V.J., Hand, M.V., Hamilton, J.T.G., Sharman, N.D., Boyd, D.R., Stevenson, M.H., 1993. Synthesis, characterisation and use of 2-tetradecylcyclobutanones together with other cyclobutanones as markers of irradiated liquid whole egg. J. Sci. Food Agric. 62, 361–367.

Cross, A.J., Ferrucci, L.M., Risch, A., Graubard, B.I., Ward, M.H., Park, Y., et al., 2010. A large prospective study of meat consumption and colorectal cancer risk: an investigation of potential mechanisms underlying this association. Cancer Res. 70 (6), 2406–2414.

Cross, A.J., Sinha, R., 2004. Meat-related mutagens/carcinogens in the etiology of colorectal cancer. Environ. Mol. Mutagen. 44 (1), 44–55.

Cuppett, S.L., 2001. The use of natural antioxidants in food products of animal origin. In: Pokorny, J., Yanishlieva, N., Gordon, M. (Eds.), Antioxidants in Food Practical Applications. Cambridge, CRC Press, Woodhead Publishing Ltd, pp. 284–310.

da Trindade, R.A., Mancini, J., Villavicencio, A., 2009. Effects of natural antioxidants on the lipid profile of electron beam-irradiated beef burgers. Europ. J. Lipid Sci. Technol. 111, 1161–1168.

Dalton, L., 2002. Food preservatives. Chem. Eng. News 50 (45), 40.

Damalas, C.A., Eleftherohorinos, I.G., 2011. Pesticide exposure, safety issues, and risk assessment indicators. Int. J. Environ. Res. Public Health 8, 1402–1419.

Daniel, M., 2007. Reactions to Food Additives and Preservatives.

Daniel, C.R., Cross, A.J., Graubard, B.I., Park, Y., Ward, M.H., Rothman, N., et al., 2012. Large prospective investigation of meat intake, related mutagens, and risk of renal cell carcinoma. Am. J. Clin. Nutr. 95 (1), 155–162.

David, R.J., Jaako, M., Katie, A.M., 2012. The importance of food. Arch. Pediatr. Adolesc. Med. 166 (2), 187–188.

Delcour, I., Spanoghe, P., Uyttendaele, M., 2015. Literature review: impact of climate change on pesticide use. Food Res. Int. 68, 7–15.

Delincee, H., 1983a. In: Elias, P.S., Cohen, A.J. (Eds.), Recent Advances in Food Irradiation. Elsevier Biomedical Press, Amsterdam, pp. 89–114.

Delincee, H., 1983b. In: Elias, P.S., Cohen, A.J. (Eds.), Recent Advances in Food Irradiation. Elsevier Biomedical Press, Amsterdam, pp. 129–147.

Diehl, J.F., 1991. Nutritional effects of combining irradiation with other treatments. Food Control. 2, 20–25.

Diehl, J.F., 1995. Safety of Irradiated Food, second ed. Marcel Dekker, New York.

Diggs, D.L., Huderson, A.C., Harris, K.L., Myers, J.N., Banks, L.D., Rekhadevi, P.V., et al., 2011. Polycyclic aromatic hydrocarbons and digestive tract cancers: a perspective. J. Environ. Sci. Health C. Environ. Carcinog. Ecotoxicol. Rev. 29 (4), 324–357.

Disse, E., Bussier, A.L., Veyrat-Durebex, C., Deblon, N., Pfluger, P.T., Tschop, M.H., et al., 2010. Peripheral ghrelin enhances sweet taste food consumption and preference, regardless of its caloric content. Physiol. Behav. 101 (2), 277–281.

Dogan, A., Siyakus, G., Severcan, F., 2007. FTIR spectroscopic characterization of irradiated hazelnut (*Corylus avellana* L). Food Chem. 100, 1106–1114.

Dolan, L.C., Matulka, R.A., Burdock, G.A., 2010. Naturally occurring food toxins. Toxins 2, 2289–2332.

Domingo, J.L., Nadal, M., 2016. Carcinogenicity of consumption of red and processed meat: what about environmental contaminants? Environ. Res. 145, 109–115.

Dona, A., Arvanitoyannis, I.S., 2009. Health risks of genetically modified foods. Crit. Rev. Food Sci. Nutr. 49, 164–175.

Dorfman, S.E., Laurent, D., Gounarides, J.S., Li, X., Mullarkey, T.L., Rocheford, E.C., et al., 2009. Metabolic implications of dietary trans-fatty acids. Obesity. (Silver Spring) 17 (6), 1200–1207.

Draper, A., Cullinan, P., Jones, M., Newman Taylor, A., 2003. Occupational asthma from fungicides fluazinam and chlorothalonil. Occup. Environ. Med. 60, 76–77.

DuBois, G.E., Prakash, I., 2012. Non-caloric sweeteners, sweeteners modulators and sweetener enhancers. Annu. Rev. Food Sci. Technol. 3 (1), 353–380.

Duliu, O.G., Ferdes, M., Ferdes, O.S., 2004. EPR study of some irradiated food enzymes. J. Radioanal. Nucl. Chem. 260, 273–277.

Dusdieker, L.B., Getchell, J.P., Liarakos, T.M., Hausler, W.J., Dungy, C.I., 1994. Nitrate in baby foods. Adding to the nitrate mosaic. Arch. Pediatr. Adolesc. Med. 148 (5), 490–494.

Erkekoglu, P., Baydar, T., 2014. Acrylamide neurotoxicity. Nutr. Neurosci. 17 (2), 49–57.

Fan, X.T., 2005. Formation of furan from carbohydrates and ascorbic acid following exposure to ionizing radiation and thermal processing. J. Agric. Food Chem. 53, 7826–7831.

Fan, X., Sokorai, K.J.B., 2008. Effect of ionizing radiation on furan formation in fresh-cut fruits and vegetable. J. Food Sci. 73, 79–83.

FAO/WHO, 2008. Codex Alimentarius Commission Procedural Manual, eighteenth ed. Food and Agriculture Organization of the United Nations, Codex Alimentarius Commission, Rome, <ftp://ftp.fao.org/codex/Publications/ProcManuals/Manual_18e.pdf>.

Firuzi, O., Fuksa, L., Spadaro, C., Bousova, I., Riccieri, V., Spadaro, A., et al., 2006. Oxidative stress parameters in different systemic rheumatic diseases. J. Pharm. Pharmacol. 58 (7), 951–957.

Fisher, M.C., Henk, D.A., Briggs, C.J., Brownstein, J.S., Madoff, L.C., McCraw, S.L., et al., 2012. Emerging fungal threats to animal, plant and ecosystem health. Nature 484, 186–194.

Flieger, M., Wurst, M., Shelby, R., 1997. Ergot alkaloids—sources, structures and analytical methods. Folia. Microbiol. (Praha). 42, 3–30.

Forsythe, S., 2010. The Microbiology of Safe Food. Blackwell Publishing Limited, p. 320.

Fox, J.B., Lakritz, L., Hampson, J., Richardson, R., Ward, K., Thayer, D.W., 1995. Gamma irradiation effects on thiamin and riboflavin in beef, lamb, pork, and turkey. J. Food Sci. 60, 596–598, 603.

Fox, J.B., Lakritz, L., Thayer, D.W., 1997. Thiamin, riboflavin, and a-tocopherol retention in processed and stored irradiated pork. J. Food Sci. 62, 1022–1025.

Franco, R., Lia, S., Rodriguez-Rocha, H., Burns, M., Panayiotidis, M.I., 2010. Molecular mechanisms of pesticide induced neurotoxicity: relevance to Parkinson's disease. Chem. Biol. Interact. 188 (2), 289–300.

Frazier, W.C., Westhoff, D.C., 1988. Food Microbiology, fourth ed. McGraw-Hill, New York.

Fujiwara, S., Omada, T., Nakagita, T., Dkada, S., Nammoku, T., Abe, K., et al., 2012. Sweeteners interacting with the transmembrane domain of the human sweet taste receptor induce sweet-taste synergism in binary mixtures. Food Chem. 130 (3), 561–568.

Garba, Z.N., Ubam, S., Babando, A.A., Galadima, A., 2015. Quantitative assessment of heavy metals from selected tea brands marketed in Zaria, Nigeria. J. Phys. Sci. 26, 43–51.

Gardner, C., Wylie-Rosett, J., Gidding, S.S., Steffen, L.M., Johnson, R.K., Reader, D., et al., 2012. Nonnutritive Sweeteners I: current use and health perspectives. Diabetes. Care 35 (8), 1798–1808.

Gehm, B.D., Mcandrews, J.M., Chien, P.-Y., Jameson, J.L., 1997. Resveratrol, a polyphenolic compound found in grapes and wine, is an agonist for the estrogen receptor. Proc. Natl. Acad. Sci. U.S.A. 94, 14138–14143.

Gonzalez, M., Soloneski, S., Reigosa, M.A., Larramendy, M.L., 2005. Genotoxicity of the herbicide 2,4-dichlorophenoxyacetic and a commercial formulation, 2,4-dichlorophenoxyacetic acid dimethylamine salt. I. Evaluation of DNA damage and cytogenetic endpoints in Chinese Hamster ovary (CHO) cells. Toxicol. In Vitro 19 (2), 289–297.

Goodman, D.W.S., 1980. Vitamin A metabolism. Federation Proceedings—Fed of Am Societies for Experimental Biology 39, 2716–2722.

Gorham, T.R., Zarek, L., 2006. Filth and other foreign objects in foods, a review of analytical methods and health significance. In: Hui, Y. (Ed.), Handwork of Food Science Tech and Engineering. Taylor and Francis, CRC Press, Boca Raton, FL.

Gostner, J.M., Becker, K., Ueberall, F., Fuchs, D., 2015. The good and bad of antioxidant foods: an immunological perspective. Food Chem. Toxicol. 80, 72–79.

Goulds, G.W., 1995. Biodeterioration of foods and an overview of preservation in the food and dairy industries. Int. Biodeterior. Biodegradation. 36, 267–277.

Graham, W.D., Stevenson, M.H., Stewart, E.M., 1998. Effect of irradiation dose and irradiation temperature on the thiamin content of raw and cooked chicken breast meat. J. Sci. Food Ag. 78 (4), 559–564.

Gray Jr, L.E., Wolf, C., Lambright, C., Mann, P., Price, M., Cooper, R.L., et al., 1999. Administration of potentially antiandrogenic pesticides (procymidone, linuron, iprodione, chlozolinate, p, p DDE, and ketoconazole) and toxic substances (dibuthyl- and diethylhexylphtalan PCB 169, and ethane dimethane sulphonate) during sexual differentiation produces diverse profiles of reproductive malformations in the male rat. Toxicol. Ind. Health 15, 94–118.

Green, M.B., Hartley, G.S., West, T.F., 1987. Chemicals for Crop Improvement and Pest Management, third ed. Pergamon Press, Oxford, p. 207.

Grenby, T.H., 1991. Ontense sweeteners for the food industryI: an overview. Trends Food Sci. Technol. 2, 2–6.

Gupta, P.K., Aggarwal, M., 2007. Toxicity of fungicides. In: Gupta, R.C. (Ed.), Veterinary Toxicology, first ed. Elsevier, New York, pp. 587–601.

Hakkinen, S., Karenlampi, S., Mykkanen, H., Torronen, A., 2000. Influence of domestic processing and storage on flavonol contents in berries. J. Agric. Food Chem. 48 (7), 2960–2965.

Hamidi, E.N., Hajeb, P., Selamat, J., Abdull Razis, A.F., 2016. Polycyclic aromatic hydrocarbons (PAHs) and their bioaccessibility in meat: a tool for assessing human cancer risk. Asian Pac. J. Cancer Prev. 17 (1), 15–23.

Harris, T.O., Hufnagle, W.O., Betley, M.J., 1993. Staphylococcal enterotoxin type A internal deletion mutants: serological activity and induction of T-cell proliferation. Infect. Imm. 61 (5), 2059–2068.

Hartwig, A., Pelzer, A., Burnouf, D., Titéca, H., Delincée, H., Briviba, K., et al., 2007. Toxicological potential of 2-alkycyclobutanones—specific radiolytic products in irradiated fat-containing food—in bacteria and human cell lines. Food Chem. Toxicol. 45, 2581–2591.

Hayes, W.J., Laws, E.R., 1990. Handbook of Pesticide Toxicology: Classes of Pesticides, vol. 3. Academic Press, New York.

Heinonen, M., Ollilainen, V., Linkola, E., Varo, P., Koivistoinen, P., 1989. Carotenoids in finnish foods—vegetables, fruits, and berries. J. Agric. Food Chem. 37 (3), 655–659.

Hobbs, B.C., 1983. *Clostridium perfringens* food poisoning. In: Rechcigl Jr., M. (Ed.), Handbook of Foodborne Diseases of Biological Origin. CRC Press, Boca Raton, FL, p. 295.

Hrelia, R., 1996. The genetic and non-genetic toxicity of the fungicide. Vinclozolin Mutagenesis 11, 445–453.

Hutchings, J.B., 2003. Expectations and the Food Industry: The Impact of Color and Appearance. Kluwer Academic/Plenum Publisher, New York.

IARC (International Agency for Research on Cancer), 1995. IARC Monographs on the Evaluation of Carcinogenic Risks to Humans: Dry Cleaning, Some Chlorinated Solvents and Other Industrial Chemicals, 63. IARC, Lyon, pp. 393–407.

IFT, 1998. Radiation preservation of foods. A scientific status summary by the Institute of Food Technologists' Expert Panel on Food Safety and Nutrition. J. Food Tech. 52 (1), 55–62.

Incledon, B.J., Hall, J.C., 1997. Acetyl-coenzyme a carboxylase: quarternary structure and inhibition by graminicidal herbicides. Pestic. Biochem. Physiol. 57, 74–83.

IOM (Institute of Medicine), 1998. Dietary Reference Intakes for Thiamin, Riboflavin, Niacin, Vitamin B6, Folate, Vitamin B12, Pantothenic Acid, Biotin, and Choline. National Academy Press, Washington, DC.

IOM (Institute of Medicine), 2001. Dietary Reference Intakes for Vitamin A, Vitamin K, Arsenic, Boron, Chromium, Copper, Iodine, Iron, Manganese, Molybdenum, Nickel, Silicon, Vanadium, and Zinc. The National Academies Press, <http://www.nap.edu/openbook.php?isbn = 0309072794>.

Ito, N., Hasegawa, R., Sano, M., et al., 1991. A new colon and mammary carcinogen in cooked food, 2-amino-1-methyl-6-phenylimidazo[4,5-b]pyridine (PhIP). Carcinogenesis 12 (8), 1503–1506.

Iwata, N.G., Pham, M., Rizzo, N.O., Cheng, A.M., Maloney, E., Kim, F., 2011. Trans fatty acids induce vascular inflammation and reduce vascular nitric oxide production in endothelial cells. PLoS One 6 (12), e29600.

Jimenez-Aguilar, D.M., Ortega-Regules, A.E., Lozada-Ramirez, J.D., Perez-Perez, M.C.I., Vernon-Carter, E.J., Welti-Chanes, J., 2011. Color and chemical stability of spray-dried blueberry extract using mesquite gum as wall material. J. Food Comp. Anal. 24 (6), 889–894.

John, E.M., 2003. Ionizing radiations sources. Biol. Effects Emission Exposures 83, 1766.

Johnson, I., 2001. Antioxidants and antitumour properties. In: Pokorny, J., Yanishlieva, N., Gordon, M. (Eds.), Antioxidants in Food: Practical Applications. CRC Press, Cambridge, pp. 100–123.

Jones, J.M., 1993. Food Safety. Eagan Press, St. Paul, MN.

Josephson, E.S., Thomas, M.H., Calhoun, W.K., 1975. Effects of treatments of foods with ionizing radiation. In: Harris, R.S., Karmas, E. (Eds.), Nutritional Evaluation of Food Processing. AVI, Westport, CT, pp. 393–411.

Kahkonen, M., Hopia, A., Heinonen, M., 2001. Berry phenolics and their antioxidant activity. J. Agric. Food Chem. 49 (8), 4076–4082.

Kant, R., 2005. Sweet proteins-potential replacement for artificial low calorie sweeteners. Nutr. J. 4, 5. Available from: https://doi.org/10.1186/1475-2891-4-5.

Karthikeyan, R., Manivasagam, T., Anantharaman, P., Balasubramanian, T., Somasundaram, S.T., 2011. Chemopreventive effect of *Padina boergesenii* extracts on ferric nitrilotriacetate (Fe-NTA)-induced oxidative damage in Wistar rats. J. Appl. Phycol. 23 (2), 257–263.

Kato, T., Ohgaki, H., Hasegawa, H., et al., 1988. Carcinogenicity in rats of a mutagenic compound, 2-amino-3,8-dimethylimidazo[4,5-f]quinoxaline. Carcinogenesis 9 (1), 71–73.

Khattab, S.N., Massoud, M.I., Abd El-Razek, A.M., El-Faham, A., 2017. Physico-chemical and sensory characteristics of steviolbioside synthesized from stevioside and its application in fruit drinks and food. J. Food Sci. Technol. 54 (1), 185–195.

Kinderlerer, J.L., Hatton, P., 1990. Fungal metabolites of sorbic acid. Food Addit. Contam. 7 (5), 657–669.

Klug, C., Lipinski, G.R., 2011. Acesulfame potassium. In: Nabors, L.O. (Ed.), Alternative Sweeteners. CRC Press, Boca Raton, FL, pp. 20–25.

Kroger, M., Meister, K., Kava, R., 2006. Low-calorie sweeteners and other sugar substitutes: a review of the safety issues. Compr. Rev. Food Sci. Food Safety 5 (2), 35–47.

Kumari, A., Choudhary, S., Arora, S., Sharma, V., 2016. Stability of aspartame and neotame in pasteurized and in-bottle sterilized flavoured milk. Food Chem. 196, 533–538.

Lee, J.K., 2012. Regulatory frameworks of food additives in Korea. International Conference for Sharing Information on Food Standards in Asia. Hanyang Women's University.

LeTellier, P.R., Nawar, W.W., 1972. 2-Alkylcyclobutanones from radiolysis of lipids. Lipids 7, 75–76.

Li, Z.H., Velisek, J., Zlabek, V., Grabic, R., Machova, J., Kolarova, J., et al., 2011. Chronic toxicity of verapamil on juvenile rainbow trout (*Oncorhynchus mykiss*): effects on morphological indices, hematological parameters and antioxidant responses. J. Hazard. Mater. 185 (2–3), 870–880.

Lind, T., Sundqvist, A., Hu, L., Pejler, G., Andersson, G., Jacobson, A., et al., 2013. Vitamin A is a negative regulator of osteoblast mineralization. PLoS One 8 (12), e82388. Available from: https://doi.org/10.1371/journal.pone.0082388.

Litovitz, T.A., Felberg, L., Soloway, R.A., 1994. Annual reports of the American Association of Poison Control Centers. Toxic exposure surveillance system. Am. J. Emerg. Med. 13, 551–597.

Lopez-Garcia, E., Schulze, M.B., Meigs, J.B., Manson, J.E., Rifai, N., Stampfer, M.J., et al., 2005. Consumption of *trans* fatty acids is related to plasma biomarkers of inflammation and endothelial dysfunction. J. Nutr. 135 (3), 562–566.

Madden, J.M., MacCardell, B.A., Morris Jr., J.G., 1989. Vibrio cholerae. In: Doyle, M.P. (Ed.), Foodborne Bacterial Pathogens, 20. Marcel Dekker, New York, pp. 525–542.

Maher, L.K., Escott-Stump, S., 2004. Krause's Food, Nutrition, and Diet Therapy, eleventh ed. Elsevier, USA.

Marchioni, E., Raul, F., Burnouf, D., Miesch, M., Delincée, H., Hartwig, A., et al., 2004. Toxicological study on 2-alkycyclobutanones—results of a collaborative study. Radiat. Phys. Chem. 71, 145–148.

Markowski, J., Plocharski, W., 2006. Determination of phenolic compounds in apples and processed apple products. J. Fruits Ornamental Plant Res. 14 (2), 133–141.

McCarthy, T., Kerry, J., Kerry, J., Lynch, P., Buckley, D., 2001. Evaluation of the antioxidant potential of natural food/plant extracts as compared with synthetic antioxidants and vitamin E in raw and cooked pork patties. Meat Sci. 58 (1), 45–52.

McClain, R.M., Jochen Bausch, J., 2003. Summary of safety studies conducted with synthetic lycopene. Regul. Toxicol. Pharmacol. 37, 274–285.

McNeal, T.P., Nyman, P.J., Diachenko, G.W., Hollifield, H.C., 1993. Survey of benzene in foods by using head-space concentration techniques and capillary gas-chromatography. J. AOAC Int. 76, 1213–1219.

Mead, P.S., Slutsker, L., Dietz, V., McCaig, L.F., Bresee, J.S., Shapiro, C., et al., 1999. Food-related illness and death in the United States. Emerg. Infect. Dis. 5, 607–625.

Mordukhovich, I., Rossner Jr., P., Terry, M.B., Santella, R., Zhang, Y.J., Hibshoosh, H., et al., 2010. Associations between polycyclic aromatic hydrocarbon-related exposures and p53 mutations in breast tumors. Environ. Health Perspect. 118 (4), 511–518.

Mukherjee, I., Gopal, M., Chatterjee, S.C., 2003. Persistence and effectiveness of iprodione against Alternaria blight in mustard. Bull. Environ. Contam. Toxicol. 70, 586–591.

NAS, 1987. Regulating pesticides in food. The delaney paradox. Report of Committee on Scientific and Regulatory Issues. Unlikely Pesticide Use Patterns. National Academy of Sciences, National Academy Press, Washington, DC.

Ndiaye, B., Jamet, G., Miesch, M., Hasselmann, C., Marchioni, E., 1999. 2-Alkylcyclobutanones as markers for irradiated foodstuff, (II) The CEN (European Committee for Standardisation) method filed of application and limit of utilisation. Radiat. Phys. Chem. 55, 437–445.

Nelson, G., Hoon, M.A., Chandrashekar, J., Zhang, Y., Ryba, N.J., Zuker, C.S., 2001. Mammalian sweet taste receptors. Cell 106 (3), 381–390.

NIN, 1994. Studies on newer adulterants and contaminants. Annual Report 1993–94 of the National Institute of Nutrition. Indian Council of Medical Research, Hyderabad.

Nontasan, S., Moongngarm, A., Deeseenthum, S., 2012. Application of functional colorant prepared from black rice bran in yogurt. Procedia APCBEE 2, 62–67.

Nordlee, J.A., Taylor, S.L., Townsend, J.A., Thomas, L.A., Bush, R.K., 1996. Identification of a Brazil nut allergen in transgenic soybeans. N. Engl. J. Med. 334, 688–692.

O'Bryan, C.A., Crandall, P.G., Ricke, S.C., Olson, D.G., 2008. Impact of irradiation on the safety and quality of poultry and meat products: a review. Crit. Rev. Food Sci. Nutr. 48, 442–457.

Oerke, E.C., 2006. Crop losses to pests. J. Agric. Sci. 144, 31–43.

Ogamba, E.N., Izah, S.C., Omonibo, E., 2016. Bioaccumulation of hydrocarbon, heavy metals and minerals in *Tympanotonus fuscatus* from coastal region of Bayelsa state, Nigeria. Int. J. Hydrol. Res. 1, 1–7.

Ozsoy, N., Candoken, E., Akev, N., 2009. Implications for degenerative disorders: antioxidative activity, total phenols, flavonoids, ascorbic acid, betacarotene and beta-tocopherol in *Aloe vera*. Oxid. Med. Cell. Long. 2 (2), 99–106.

Padhye, N.V., Doyle, M.P., 1992. Rapid procedure for detecting enterohemorrhagic *Escherichia coli* O157:H7 in food. Appl. Environ. Microbiol. 57 (9), 2693–2698.

Pauli, G.H., 2006. Regulation of irradiated foods and packaging. In: Sommers, C.H., Fan, X. (Eds.), Food Irradiation Research and Technology. Blackwell Publishing, Iowa, pp. 37–41.

Peraica, M., Radic, B., Lucic, A., Pavlovic, M., 1999. Toxic effects of mycotoxins in humans. Bull. World Health Organ. 77 (9), 754–766.

PFA-Prevention of Food Adulteration, 2008. PFA Seventh Amendment Rules. The Universal Law Publishing Co. Pvt. Ltd. 102–103.

Pierre, F.H., Martin, O.C., Santarelli, R.L., Taché, S., Naud, N., Guéraud, F., et al., 2013. Calcium and α-tocopherol suppress cured-meat promotion of chemically induced colon carcinogenesis in rats and reduce associated biomarkers in human volunteers. Am. J. Clin. Nutr. 98 (5), 1255–1262.

Pingali, P.L., Rola, A.C., 1995. Public regulatory roles in developing markets: the case of Philippines. In: Pingali, P.L., Roger, P. (Eds.), Impact of Pesticides on Farmer Health and the Rice Environment. Kluwer Academic Publishers, Boston, MA, pp. 391–411.

Plumb, G., Chambers, S., Lambert, N., Wani- gatunga, S., Williamson, G., 1997. Influence of fruit and vegetable extracts on lipid peroxidation in microsomes containing specific cytochrome P450s. Food Chem. 60 (2), 161–164.

Popp, J., Peto, K., Nagy, J., 2013. Pesticide productivity and food security – a review. Agron. Sustain. Dev. 33, 243–255.

Priya, K., Gupta, V.R.M., Srikanth, K., 2011. Natural sweeteners: a complete review. J. Pharm. Res. 4 (7), 2034–2039.

Pryor, W.A. (Ed.), 1984. Free Radicals in Biology. Academic Press, London.

Raben, A., Vasilaras, T.H., Moller, A.C., Astrup, A., 2002. Sucrose compared with artificial sweeteners I: different effects on adlibitum food intake and body weight after 10 week of supplementation in overweight subjects. Am. J. Clin. Nutr. 76 (4), 721–729.

Raffi, J.J., Agnel, J.P.L., Frejaville, C.M., Saintlebe, L.R., 1981a. Radio-induced products in maize starch -glyceraldehyde, dihydroxyacetone, and 2-hydroxymalonaldehyde. J. Agric. Food Chem. 29, 548–550.

Raffi, J.J., Agnel, J.P.L., Thiery, C.J., Frejaville, C.M., Saintlebe, L.R., 1981b. Study of gamma-irradiated starches derived from different foodstuffs - a way for extrapolating wholesomeness data. J. Agric. Food Chem. 29, 1227–1232.

Rahimi, R., Nikfar, S., Larijani, B., Abdollahi, M., 2005. A review on the role of antioxidants in the management of diabetes and its complications. Biomed. Pharmacother. 59, 365–373.

Ram, S., Bajpai, P., Singh, R.L., Shanker, R., 2009. Surface water of a perennial river exhibits multi-antimicrobial resistant shiga toxin and enterotoxin producing *Escherichia coli*. Ecotoxicol. Environ. Safety 72 (2), 490–495.

Rao, V.S., 2000. Principles of Weed Science, second ed. CRC Press, New Hampshire.

Rao, S., Kiran, V., 2011. Effect of lycopene in tomato as a natural antioxidant in osteoarthritis- a pilot study. Int. J. Acad. Res. 3 (2), 1076–1079.

Ribeiro, G., Santos, D., 2013. Recompensa alimentar I: mecanismos envolvidos e implicações ara a obesidade. Revista Portuguesa de Endocrinologia. Diabetes Metab. 8 (2), 82–88.

Riera, C.E., Vogel, H., Simon, S.A., le Coutre, J., 2007. Artificial sweeteners and salts producing a metallic taste sensation activate TRPV1 receptors. Am. J. Physiol. Regul. Integrat. Comp. Physiol. 293 (2), 626–634.

Roberts, H.R., 1981. Foodsafety in perspective. In: Raben, H.R. (Ed.), Food Safety. Wiley, New York.

Rodrigues, J.F., Andrade, R.D.S., Bastos, S.C., Coelho, S.B., Pinheiro, A.C.M., 2016. Miracle fruit: an alternative sugar substitute in sour beverages. Appetite 107, 645–653.

Rodriguez-Amaya, D.B., 2016. Natural food pigments and colorants. Curr. Opin. Food Sci. 7, 20–26.

Sakaguchi, G., Ohishi, I., Kosaki, S., 1988. Botulism—structure and chemistry of botulinum. In: Hardegree, M.C., Tu, A.T. (Eds.), Handbook of Natural Toxins, Vol. 4: Bacterial Toxins. Marcel Dekker, New York.

Sanchez-Echaniz, J., Benito-Fernandez, J., Mintegui-Raso, S., 2001. Methemoglobinemia and consumption of vegetables in infants. Pediatrics 107 (5), 1024–1028.

Santos-Buelga, C., Scalbert, A., 2000. Proanthocyanidins and tannin like compounds—nature, occurrence, dietary intake and effects on nutrition and health. J. Sci. Food Agric. 80 (7), 1094–1117.

Sayed, H.M., Fouad, D., Ataya, F.S., Hassan, N.H., Fahmy, M.A., 2012. The modifying effect of selenium and vitamins A, C, and E on the genotoxicity induced by sunset yellow in male mice. Mutat. Res. 744, 145–153.

Scallan, E., Hoekstra, R.M., Angulo, F.J., Tauxe, R.V., Widdowson, M.A., Roy, S.L., et al., 2011. Foodborne illness required in the United States major pathogen. Emerg. Infect. Dis. 17 (1), 7–15.

Schoefs, B., 2002. Chlorophyll and carotenoid analysis in food products. Properties of the pigments and methods of analysis. Trends Food Sci. Technol. 13, 361–371.

Scott, P.M., 1994. *Penicillium* and *Aspergillus* toxins. In: Miller, J.D., Trenholm, H.L. (Eds.), Mycotoxins in Grains: Compounds Other Than Aflatoxin. Eagan Press, St. Paul, MN, pp. 261–285.

Shahidi, F., Ho, C., 2005. Phenolics in food and natural health products: an overview. ACS Symp. Series 909, 1–8.

Shaner, D.L., 2003. Herbicide safety relative to common targets in plants and mammals. Pest. Manag. Sci. 60, 17–24.

Sharma, S., Goyal, R.P., Chakravarty, G., Sharma, A., 2008. Toxicity of tomato red, a popular food dye blend on male albino mice. Exp. Toxicol. Pathol. 60, 51–57.

Shim, S.M., Seo, S.H., Lee, Y., Moon, G.I., Kim, M.S., Park, J.H., 2011. Consumers' knowledge and safety perceptions of food additives: evaluation on the effectiveness of transmitting information on preservatives. Food Control. 22 (7), 1054–1060.

Siekmeier, R., Steffen, C., Maerz, W., 2007. Role of oxidants and antioxidants in atherosclerosis: results of in vitro and in vivo investigations. J. Cardiovasc. Pharmacol. Ther. 12 (4), 265–282.

Singh, R.L., 1989. Metabolic disposition of (14-C) Metanill Yellow in rats. Biochem. Int. 19, 1109–1116.

Singh, P.K., Singh, R.L., 2017. Bio-removal of azo dyes: a review. Int. J. Appl. Sci. Biotechnol. 5 (2), 108–126.

Singh, R.L., Khanna, S.K., Singh, G.B., 1987. Safety evaluation studies on pure Metanil Yellow acute and sub-chronic exposure responses. Bev. Food World 74, 9–13.

Singh, R.L., Khanna, S.K., Singh, G.B., 1988. Acute and short term toxicity of popular blend of Metanil Yellow and Orange II in albino rats. Ind. J. Exp. Biol. 26, 105–111.

Singh, R.L., Khanna, S.K., Singh, G.B., 1991a. Metabolic disposition of (14-C) Metanil Yellow in guinea pigs. Vet. Hum. Toxicol. 33, 220–223.

Singh, R.L., Singh, S., Khanna, S.K., Singh, G.B., 1991b. Metabolic disposition of Metanil Yellow, Orange II and their blend by caecal microflora. Int. J. Toxicol. Occup. Environ. Health 1, 250.

Singh, R.L., Singh, P.K., Singh, R.P., 2015. Enzymatic decolorization and degradation of azo dyes – a review. Int. Biodeter. Biodegr. 104, 21–31.

Smith, J.E., Moss, M.O., 1985. Mycotoxins: Formation, Analysis and Significance. John Wiley & Sons, New York.

Sorensen, M., Jensen, B., Poulsen, H., Deng, X., Tygstrup, N., Dalho, K., et al., 2001. Effects of a Brussels sprouts extract on oxidative DNA damage and metabolizing enzymes in rat liver. Food Chem. Toxicol. 39 (6), 533–540.

Speijers, G.J.A., Van Egmond, H.P., 1993. Worldwide ochratoxin A levels in food and feeds. In: Creppy, E.E., et al., (Eds.), Human Ochratoxicosis and Its Pathologies, 231. Montrouge, Colloque INSERM/John Libbey Eurotext Ltd., pp. 85–100.

Spence, J.D., 2006. Nutrition and stroke prevention. Stroke 37 (9), 2430–2435.

Stefanova, R., Vasilev, N.V., Spassov, S.L., 2010. Irradiation of food, current legislation framework, and detection of irradiated foods. Food Anal. Methods 3, 225–252.

Sugimura, T., Wakabayashi, K., Nakagama, H., Nagao, M., 2004. Heterocyclic amines: mutagens/carcinogens produced during cooking of meat and fish. Cancer Sci. 95 (4), 290–299.

Sun, H., Cui, M.L., Ma, B., Ezura, H., 2006. Functional expression of the taste modifying protein miraculin, in transgenic lettuce. FEBS J. 580 (2), 620–626.

Swithers, S.E., Martin, A.A., Davidson, T.L., 2010. High intensity sweeteners and energy balance. Physiol. Behav. 100 (1), 55–62.

Tandel, K.R., 2011. Sugar substitutes I: health controversy over perceived benefits. J. Pharmacol. Pharmacother. 2 (4), 236–243.

Tareke, E., Rydberg, P., Karlsson, P., Eriksson, S., Toernqvist, M., 2000. Acrylamide: a cooking carcinogen? Chem. Res. Toxicol. 13, 517–522.

Tchounwou, P.B., Yedjou, C.G., Patlolla, A.K., Sutton, D.J., 2012. Heavy metals toxicity and the environment. Natl. Inst. Health 101, 133–164.

Thayer, D.W., Christopher, J.P., Campbell, L.A., Ronning, D.C., Dahlgren, R.R., Thomson, G.M., et al., 1987. Toxicology studies of irradiation-sterilized chicken. J. Food Prot. 50, 278–288.

Tomas-Barberan, F., Gil, M., Cremin, P., Waterhouse, A., Hess-Pierce, B., Kader, A., 2001. HPLC-DAD-ESIMS analysis of phenolic compounds in nectarines, peaches, and plums. J. Agric. Food Chem. 49 (10), 4748–4760.

US Environmental Protectin Agency (EPA), 1999. Reregistration Eligibility Decision (RED), Captan, EPA-738-R-99-015, Office of Pesticide Programs. US Government Printing Office, Washington, DC.

Urbach, G., 1997. The flavor of milk and dairy products: II. Cheese: contribution of volatile compounds. Int. J. Dairy Technol. 50, 79–89.

Van den Berg, H., Faulks, R., Granado, H., Hirschberg, J., Olmedilla, B., Sandmann, G., et al., 2000. The potential for the improvement of carotenoid levels in foods and the likely systemic effects. J. Sci. Food Agric. 80 (7), 880–912.

Vaya, J., Aviram, M., 1999. Nutritional antioxidants: mechanism of action, analyses of activities and medical application. Nutrition 49, 1–7.

Verger, P.J.P., Boobis, A.R., 2013. Global food supply: reevaluate pesticides for food security and safety. Science 341, 717–718.

Verma, C., Nanda, S., Singh, R.K., Singh, R.B., Mishra, S., 2011. A review on impacts of genetically modified food on human health. Open Nutr. J. 4, 3–11.

White, A.J., Bradshaw, P.T., Herring, A.H., Teitelbaum, S.L., Beyea, J., Stellman, S.D., et al., 2016. Exposure to multiple sources of polycyclic aromatic hydrocarbons and breast cancer incidence. Environ. Int. 89–90, 185–192.

WHO, 1994. Safety and Nutritional Adequacy of Irradiated Food. World Health Organization, Geneva.

WHO, 1999. High-Dose Irradiation: Wholesomeness of Food Irradiated With Doses Above 10 kGy. Report of a Joint FAO/IAEA/WHO Study Group. WHO Technical Report Series 890. World Health Organization, Geneva.

WHO, 2005. How Safe Is Your Food? From Farm to Plate, Make Food Safe. Available on: <http://www.afro.who.int/en/rdo/speeches/4599-message-of-the-who-regional-director-for-africa-dr-matshidiso-moeti-for-world-health-day-2015.html> (accessed 26.01.16.).

Wieczorek, A., 2013. History of Agricultural Biotechnology: How Crop Development has Evolved. <http://www.nature.com/scitable/knowledge/library/history-of-agricultural-biotechnology-how-crop-development-25885295>.

Williams, R.D., 1991. FDA helps prevent spread of cholera epidemic to U.S., FDA consumer, 9, 14. In: Morello, J. A., Domer, J.E. (Eds.), Abstracts of the 90th Annual Meeting of the American Society for Microbiology. Annual Meeting Program Committee, Washington, DC.

Wilson, D.M., Payne, G.A., 1994. Factors affecting *A. flavus* group infection and aflatoxin contamination of crops. In: Eaton, D.L., Groupman, J.D. (Eds.), The Toxicology of Aflatoxins: Human Health and Agricultural Significance. Academic Press, London.

Wink, M., 1988. Plant breeding importance of secondary metabolites for protection against pathogens and herbivores. Theor. Appl. Genet. 75, 225–233.

Wong, P., Carito, S., 1990. The use of monoclonal antibody for the detection and neutralization of *Clostridium botulinum* type E toxin. In: Morello, J.A., Domer, J.E. (Eds.), Abstracts of the 90th Annual Meeting of the American Society for Microbiology. Annual Meeting Program Committee, Washington, DC.

World Bank, 2006. Toxic Pollution fromAgriculture: An Emerging Story. Available online: <http://econ.worldbank.org/WBSITE/EXTERNAL/EXTDEC/0,contentMDK:21139876~pagePK:64165401~piPK:64165026~theSitePK:469372,00.html>

Woyessa, G.W., Kassa, S.B., Demissie, E.G., Srivastava, B.B.L., 2015. Determination of the level of some trace and heavy metals in some soft drinks of Ethiopia. Int. J. Curr. Res. Chem. Pharm. Sci. 2, 84–88.

Yanishlieva-Maslarova, N., 2001. Inhibition oxidation. In: Pokorny, J., Yanishlieva, N., Gordon, M. (Eds.), Antioxidants in Food Practical Applications. CRC Press, Woodhead Publishing Ltd., Cambridge, pp. 35–59.

Yao, L.H., Jiang, Y.M., Shi, J., Tomasbarberan, F.A., Datta, N., Singanusong, R., et al., 2004. Flavonoids in food and their health benefits. Plant Foods Hum. Nutr. 59, 113–122.

Yaping, Z., Suping, Q., Yu, W., Zheng, X., Hong, S., Side, Y., et al., 2002. Antioxidant activity of lycopene extracted from tomato paste towards trichloromethyl peroxyl radical CCl_3O_2. Food Chem. 77, 209–212.

Yaylayan, V.A., 2006. Precursors, formation and determination of furan in food. Journal für Verbraucherschutz und Lebensmittelsicherheit 1 (1), 5–9.

Yoshizawa, T., Swanson, S.P., Mirocha, C.J., Behrens, J.C., Swanson, S.P., 1981. Metabolic fate of T-2 toxin in a lactating cow. Food Cosmet. Toxicol. 19 (1), 31–39.

Young, J.H., 1989. Pure Food: Securing the Pure Food and Drug Act of 1906. Princeton University Press, Princeton, NJ.

Zegota, H., 1988. Suitability of Dukat strawberries for studying effects on shelf life of irradiation combined with cold storage. Z. Lebensm. Unters. Forsch. 187, 111–114.

Zhang, M., Wilhoit, L., Geiger, C., 2005. Assessing dormant season organophosphate use in California almonds. Agric. Ecosyst. Environ. 105, 41–58.

Zhu, M.J., Mendonca, A., Ismail, H.A., Du, M., Lee, E.J., Ahn, D.U., 2005. Impact of antimicrobial ingredients and irradiation on the survival of Listeria monocytogenes and the quality of ready-to-eat turkey ham. Poult. Sci. 84, 613–620.

Further Reading

Radice, S., Marabini, L., Gervasoni, M., Ferraris, M., Chiesara, E., 1998. Adaptation to oxidative stress: effects of vinclozolin and iprodione on the HepG2 cell line. Toxicology 129, 183–191.

CHAPTER

3

Toxicity of Food Additives

Neeraj Kumar[1], *Anita Singh*[2], *Dinesh Kumar Sharma*[3] *and Kamal Kishore*[4]

[1]Dr. R. M. L. Institute of Pharmacy, Kunwarpur Badagaon, Powayan, India [2]Department of Pharmacy, Kumaun University, Bhimtal, India [3]Amrapali Institute of Pharmacy and Sciences, Lamachaur, Haldwani, Uttarakhand, India [4]Department of Pharmacy, M.J.P. Rohilkhand University, Bareilly, India

OUTLINE

DOI: https://doi.org/10.1016/B978-0-12-816333-7.00003-5

INTRODUCTION

Thousands of food additives are used in foodstuffs for imparting various properties during food processing. The European Union Regulation EC 1333/2008 defined food additives as substances that are not intended for food individually but are added to impart technological features like color, flavor, preservation, and acidity regulation. Some food additives also functions as thickeners, stabilizers, and emulsifier or anticaking agents. They may be natural, synthetic, or semisynthetic in nature and sometimes xenobiotic substances that are not present in the human body (Mepham, 2011). Food safety is important for health, control of illness, and quality of livelihood and it is critical for safer food development. The main food contaminants come from anthropogenic substances like pharmaceuticals, pesticide residues, Maillard reaction products, and organic pollutants, and heavy metals, metalloids, marine biotoxins, and mycotoxins come from natural sources (Alvito et al., 2016). Legislation for food safety starts with the framing of the Pure Food and Drugs Act in 1906 in the United States, and in 1938, the Food, Drugs and Cosmetic Act was introduced to ensure the identity, quality of ingredients, and standard for packing of finished products. The U.S. Food Safety Law 1958 was introduced along with the Food Additive Amendment in 1958. It came in existence to provide rules and regulation for the U.S. Food and Drug Administration (FDA). In 1982, the FDA framed Food Safety Regulation for food additives in the U.S. FDA's "Red Book" (Pressman et al., 2017).

CONTAMINANTS IN FOOD ADDITIVES

Chemical Contaminants

Chemical contaminants in food and food additives mainly contain traces of heavy metals such as lead, cadmium, nickel, mercury, and arsenic. Some forms of nitrates, organic environmental contaminants like organochlorides (polychlorinated biphenols), and pesticides such as dichlorodiphenyltrichloroethane may also present in food and food additives. Some other preparations were also reported for their health hazardous effect on consumers (Larsen et al., 2001). The main sources of chemical contaminants are soil, personal care products, disinfectant byproducts, water, air, and material used for packaging of products. Such contaminants reach systemic circulation of humans by consumption, use of plastic containers, disinfectants, deodorants, detergents, pesticides, herbicides, weedicides, etc. (Rather et al., 2017).

Another category of chemical contaminants belongs to mycotoxins like aflatoxins, ochratoxin A, patulin, and trichothecenes produced by various fungi, which can produce disorders related to liver, kidneys, and nervous system (Larsen et al., 2001).

Some food additives such as salicylates, artificial colors, and flavors directly or by reacting with other food ingredients produce various physiological disorders that may cause hypersensitivity reactions or hyperactivity and neurophysiological disturbances especially in children (Wroblewska, 2009).

Specific Environmental Contaminants

Specific environmental contaminants are bioaccumulative and entirely different from chemical contaminants. Most of these contaminants are known as emerging contaminants because they are present for many decades in the environment but cannot be estimated earlier due to unavailability of sensitive and appropriate instruments. Some environmental contaminants are summarized in Table 3.1 with their reported locations. Per fluorinated compounds, also known as per and polyfluoroalkyl substances, were introduced in the 1950s and reported as contaminants in 2000. Most of the environmental contaminants are transported by wastewater from households and only partially removed in recycling processes before reuse. These contaminants can be transformed during treatment of wastewater by the process of photolysis, degradation by microbes, and hydrolysis. Oxypurinol is an example of such a transformation, which was detected in different streams, rivers, and groundwater. Another example is formation of nitrosodimethylamine, a carcinogenic substance formed by the reaction between azithromycin and monochloramine. Iodoacetic acid, a genotoxic substance, is formed by reaction of iopamidol and chlorine, which are used in medical imaging (Richardson and Kimura, 2017).

The serious concern was reported for pesticides in vegetables, fruits, and cereals, and antibiotics and hormones in meat may result in poisoning and weight gain. Packaged food and beverages expose humans to polycarbonate plastics and epoxy, a resin that binds to estrogen receptors and produces toxicity (Oskarssonm, 2012).

Microbial Contaminants

Microbial contamination in food and food additives results in infection of the gastrointestinal tract, diarrhea, and poor nutrition in adults and retards growth in infants and children (Oluwafemi and Ibeh, 2011). Microbial contaminants found in water, soil, animals, and plants mainly originated from different bacterial categories such as pseudomonas, coliforms, and micrococci penetrated either by skin or gastrointestinal tract (Elshafei, 2017). About 70%–80% of food-related illness in humans is due to *Staphylococcus aureus*, *Clostridium perfringens*, and *Salmonella* species infections or toxins (Williams, 2012). Contamination of food or food additives by microorganisms is due to mishandling, eating foods very late after cooking, improper cooling of food, and contaminated ingredients. The main reasons of increase in food-borne diseases are the addition of preservatives to increase the shelf life of food, emerging new pathogens, and other contaminants (Rawat, 2015; Williams, 2012). The microbial contaminants are listed in Table 3.2 with their sources and symptoms of infection.

Contaminants From Packaging Materials

Packaging is the indispensable part of any preparation, and polymeric packaging materials are presently used extensively. To improve the performance of such packaging materials, various packaging additives like antioxidants, antistatic, antiblocking agents, lubricants, and stabilizers are used (Lau and Wong, 2000). The Commission of the European Community regulates the rules and regulations regarding packaging materials

TABLE 3.1 Environmental Contaminants and Their Occurrence

S. No.	Category	Example	Reported From	Reported Place	Reference
1.	Per- and polyfluoroalkyl substances (PFAS)	Perfluorooctane sulfonate, perfluorooctanoic acid	Human blood, drinking water	The United States, Germany	Begley et al. (2005), Skutlaek et al. (2006)
2.	Pharmaceuticals	Erythromycin, nitroglycerin, 17α-ethinyl estradiol	Drinking water	The United States	USEPA (2018)
		Diclofenac	Water, Dead animals	Europe	Richardson and Kimura (2017), Oaks et al. (2004)
3.	Illicit drugs	Methamphetamine and ecstasy	Water	The United States	Jones-Lepp et al. (2004)
		Cocaine	Water	Italy	Anonymous (2013)
4.	Antibacterial	Triclosan	Streams and rivers	Europe	Kolpin et al. (2002)
5.	Hormones	17α-Estradiol, 17β-estradiol, equilenin, equilin, estriol, estrone, mestranol	Drinking water	The United States	USEPA (2012)
6.	Nanomaterials	Zinc nanoparticle sunscreens	Blood, urine	–	Gulson et al. (2010)
7.	Disinfection byproducts	Trihalomethanes, haloacetic acids, bromate, and chlorite	Swimming pools	The United States	Zwiener et al. (2007)
8.	Brominated and emerging flame retardants	Polybrominated diphenyl ethers	Human blood, milk, and tissues	–	Vikesland et al. (2013)
9.	Artificial sweeteners	Sucralose	River water	Europe	Loos et al. (2009)
10.	Benzotriazoles	Benzotriazoles	Wastewater, rivers	The United Kingdom	Janna et al. (2011)
11.	Dioxane	1,4-Dioxane	Cape Fear River Basin	North Carolina, Germany	NSF (2009), Mohr et al. (2010)
12.	Algal toxins	Microcystins, anatoxins, nodularins, saxitoxins, brevetoxins, cylindrospermopsin	Wastewater, agricultural runoff	New Zealand, Italy, France	Richardson and Ternes (2018)

within the European community. The present limit for total migration is 60 mg/kg or 10 μg/dm^2 of food supplements. In the United States, such regulations are more complex and are regulated by the FDA by the Commission and Council Directives and adopt the threshold policy (CEC, 1976).

TABLE 3.2 Microbial Contaminants and Their Occurrence

S. No.	Microorganism	Source	Disease	Reference
1.	*Bacillus cereus*	Sauces, puddings	Vomiting, diarrhea	Peshin et al. (2002)
2.	*Campylobacter jejuni*	Poultry, eggs, meat	Fever, diarrhea	Williams (2012)
3.	*Clostridium botulinum*	Home-canned meat and vegetables	Difficulty in swallowing, speaking	Rawat (2015)
4.	*Clostridium perfringens*	Partially cooked meat products	Diarrhea and severe pain	Peshin et al. (2002)
5.	*Escherichia coli*	Fecal contamination of food	Fever, diarrhea	Williams (2012)
6.	*Listeria monocytogenes*	Raw meat, poultry, and eggs	Meningitis, encephalitis	Rawat (2015)
7.	*Salmonella* spp.	Undercooked poultry, reheated food	Diarrhea, fever	Williams (2012)
8.	*Shigella* spp.	Fecal contamination of food	Bloody stools with mucus	Peshin et al. (2002)
9.	*Staphylococcus aureus*	Custard and cream-filled food, cold meats	Diarrhea, vomiting	Rawat (2015)
10.	*Vibrio parahaemolyticus*	Fish, crustaceans	Diarrhea, vomiting	Williams (2012)
11.	*Yersinia enterocolytica*	Pork and beef	Fever, abdominal pain	Peshin et al. (2002)

Plasticizers are the additives that modify the polymeric packaging properties. The low-toxicity plasticizers are butyl stearate, alkyl sebacates, acetyl-tributyl citrate, and adipates. Phthalate plasticizers use is discontinuing due to its estrogenic, carcinogenic, and human antifertility activity (Arvanitoyannis and Bosnea, 2004).

A group of antioxidants are used to prevent oxidation/degradation of packaging materials belongs to Tinuvin, Chimassorb, Irganox, and Irgafos. The aryl-substituted phosphate and trisubstituted derivatives are toxic in nature (Carocho et al., 2014). Styrene used in packaging as a monomer that undergoes metabolism involving phenyloxirane is a mutagenic substance. Another material, polyvinylchloride, is used in packaging, but its monomer, vinyl chloride, is highly toxic in nature. Epoxy resins like bisphenol A diglycidyl ether are used in coating of packets, and epoxy compounds are alkylating agents that show cytotoxic effects. In polyurethane and adhesives, isocyanates are used, and the toxicity of isocyanates is reported in many studies. Nylon, the material to pack the food during cooking, contains caprolactam, which gives a bitter taste to food after migration from packaging material to food chain. Polyethylene terephthalate (PET) plastic containers are used for packaging of edible oils and beverages and also used in microwaves.

TABLE 3.3 Packaging Additives and Their Toxicity

S. No.	Packaging Additive	Examples/Metabolite	Toxicity Reported	Reference
1.	Plasticizer	Butyl stearate, acetyl tributyl citrate, alkyl sebacates, adipates	Phthalate: carcinogenic, estrogenic	NTP (1982), USDHHS (1982)
2.	Thermal stabilizers	Poly(vinyl chloride), poly (vinylidene) chloride, and polystryrene	Cellular toxicity	Hine et al. (1958), CEC (1990)
3.	Slip additives	Polyolefins, polystyrene, and polyvinyl chloride	Residue migration calculated	Cooper and Tice (1995)
4.	Light stabilizers	Tinuvin770 and Chimasorb 944	Cardiac myocytes	Sotonyi et al. (2001)
5.	Antioxidants	Aryl substituted phosphites, triphenyl phosphate	Highly toxic	Lefaux and Technica (1968)
6.	Styrene	Phenyloxirane	Mutagenic	Bond (1989), ECETOC (1993)
7.	Polyvinyl chloride	Venyl chloride	Highly toxic	WHO (1974, 1975)
8.	Epoxy resins	Bisphenol A diglycidyl ether	Cytotoxic	Lau and Wong (2000)
9.	Polyurethane polymers and adhesives	Isocyanate	Toxic compound	Lau and Wong (2000)
10.	Polyamides	Caprolactam	Bitter taste	Stepek et al. (1987)
11.	Polyethylene terephthalate	Migrants: Formaldehyde, acetaldehyde, antimony	Endocrine disorder	Castle et al. (1989)
12.	Polyethylene bags	Heavy metals	Abdominal pain, anemia, ataxia, and memory loss	Musoke et al. (2015)

PET migrants are formaldehyde, acetaldehyde, and antimony, which were reported to produce endocrine disorders in humans (Castle et al., 1989). A list of some packaging additives and their migrants are summarized in Table 3.3 with their toxicity.

HISTORY OF FOOD ADDITIVE SAFETY REGULATIONS

About 2500 substances are added in food processing to impart or modify various properties. The Food and Drugs Act of 1906 was set to control adulteration in food with the help of the U.S. Department of Agriculture. Until 1937, the only labeling law was in existence to prevent the sale of misbranded packaged food and drugs. Later on, after the death of about 73 persons in a sulfanilamide tragedy in 1938, the Food, Drug and Cosmetic Act (FDCA) was introduced to provide standards for identification, quality, and packaging

(Pressman et al., 2017; Krewski et al., 2010). In 1958, the U.S. Food Safety Law was introduced, which says that no additive shall be deemed to be safe if it is carcinogenic or causes any other medical problem when ingested by man or animals. The Food Additive Amendment of 1958 was also introduced to ensure safe use of food additives. The Organization for Economic Cooperation and Development was established in 1961, which provided guidelines for different type of toxicity studies (Gibb, 2008). The Kefauver Harris Amendments (1962) in the FDCA say that with toxicity or drug safety, drug efficacy data is also required. The U.S. Environmental Protection Agency in 1970 was established to ensure healthy and natural environments, on which human life depends (Krewski et al., 2010). In 1982 the "Red Book" was compiled for food safety by USFDA, which was an effort for safety assessment of food additives and colors. In 1997, the FDA Modernization Act was established, a food contact notification to address high safety concerns (Rowlands and Hoadley, 2006).

TOXICITY OF FOOD ADDITIVES

Coloring Materials

Normally natural color additives rarely produce any adverse reactions but it was found that natural colors also produce many physiological dysfunctions in body. One study suggested that persons suffering from angioedema and urticaria showed various allergic reactions against carotene and canthaxanthin. A carotene-based dye, Annatto, also reported for anaphylactic shock and confirmed the presence of an Annatto-specific IgE antibody. Some other dyes like saffron, carmine, curcumin, and enocianina were also reported for specific IgE antibodies against these dyes. Asthma, urticaria, and hypersensitivity were reported in 1959 due to use of aniline dye tartrazine, an artificial coloring material. Another study also showed that such coloring substances may provoke to migraine, blurred vision, itching, rhinitis, suffocation, weakness, heat sensation, palpitation, pruritus, and urticaria (Arora et al., 2009).

Brilliant blue FCF used in some dairy products, sweets, and drinks was banned in most of European countries due to its carcinogenic effect shown during tar-induced tumor study in rats. Fast green FCF, which provides green color to green peas, vegetables, fish, desserts, dry bakery mixes, and sauces, showed chromosomal aberrations in mice and neurotransmitter release inhibition in rats after absorption through the intestines. Indigotine, used as a coloring material in tablets and capsules, coating, ice creams, confectionary, cookies, sweets, and some baked goods, was shown to be an allergin, like occupational asthma (Mepham, 2011).

Antioxidants

Natural and synthetic antioxidants are used in the food industry for prolonging the appearance and shelf life of foodstuffs. Natural antioxidants like vitamin C, vitamin E, and some spices and herbs such as oregano, basil, rosemary, pepper, nutmeg, cinnamon, and thyme are used normally while synthetic antioxidants, which are mostly phenolic in nature, such as butylated hydroxyl anisole, butylated hydroxyl toluene, and propyl gallate, are used due to their wide availability and good performance. Various studies suggested that longer use of synthetic antioxidants can produce various diseases or physiological disorders like asthma, joint pain, dermatitis, and stomach and eye problems. Sometimes obesity, urticaria, and excessive sweating were also reported. A human-based study of butylated hydroxyl anisole and butylated hydroxyl toluene reported rhinitis, headache, asthma, back pain, diaphoresis, or somnolence (Anbudhasan et al., 2014). Another study in rats, mice, pigs, and monkeys showed carcinogenicity of the liver. Some synthetic antioxidants produce toxic metabolites after thermal treatment of foodstuffs, like gallates, which decompose over 148°C. The legal aspects for use of natural antioxidants states that only a few products are generally recognized as safe (GRAS) by various governing bodies like the Joint Expert Committee for Food Additives (JECFA) and the European Community's Scientific Committee for Food (ECSCF) and must be free from carcinogenicity and used within acceptable daily intake (ADI) limits. The synthetic compounds such as butylated hydroxyl anisole, nordihydroguaiaretic acid, hydroquinone, citric acid, and ascorbyl palmitate are strictly used under the observation of Prevention of Food Adulteration Act (PFA) of 2008 guidelines (Carocho et al., 2014).

In past years, natural antioxidants were considered safe, but some studies showed that they also have limitations, like vitamin E undergoes omega and beta oxidation to produce several metabolites, and when vitamin E accumulates in lipid bilayers, it reduces membrane fluidity. Similar effects were shown as in the case of higher cholesterol levels. Tocopherol stimulates protein phosphatase 2 A in a concentration dependent manner, which results in dephosphorylation and inactivation of protein kinase C, an important enzyme of the cell proliferation process (Bast and Haenen, 2002).

Sweeteners

Natural sweeteners are carbohydrates obtained from vegetables, trees, seeds, roots, and nuts. The commonly used natural sweeteners are honey, molasses, maple syrup, coconut sugar, agave nectar, date sugar, and xylitol. Artificial sweeteners comprise carbohydrate substitutes that replace natural sweeteners in beverages and food due to their very low or no energy value and cost-effective availability with higher sweetening value than natural sweeteners.

Artificial sweeteners are widely used in baking, soft drinks, candy, canned food, powdered drink mixes, jams, pudding, dairy products, and jellies. According to the FDA, the five main artificial sweeteners are aspartame, neotame, saccharin, acesulfame potassium, and sucralose (Neacsu and Madar, 2014).

A study on saccharin showed its strong association with leukemia, lymphoma, and myeloma in humans and bladder cancer in rats. A combined study of toxicity of all five FDA-approved artificial sweeteners was conducted on colon and renal cell lines and the results indicated that colon cells are more susceptible than renal cells to artificial sweeteners while saccharin and sucralose cause more DNA damage. In two other studies, artificial sugars were found to potentiate the effects of type 2 diabetes (Qurrat-Ul-Ain and Khan, 2015).

Nasal-associated Lymphoid Tissue (NALT) Toxicological Program conducted a study on aspartame in transgenic mice and concluded that its exposure increases the risk of cancer in mice. Acesulfame-k assimilated by the body was not metabolized but breaks into acetoacetamide, which is toxic to body (Chattopadhyay et al., 2014).

Sucralose is excreted in feces and only 11%–27% is absorbed from the intestine, filtered by the kidney, and excreted from urine. According to the FDA it is safe for human use, but one study suggests that at higher doses neurotoxic alterations are induced by sucralose (Rodero et al., 2009).

Cyclamate, a synthetic sweetener, is metabolized by bacteria of the gut into cyclohexylamine, which produces toxicity. Neotame undergoes hydrolysis by esterase enzyme and produces deesterified neotame and methanol. The 3–3 dimethyl butyl group of deesterified neotame blocks peptidases and results in decreased production of phenylalanine due to inhibition of the peptide bond breakdown between aspartic acid and phenylalanine moiety (Chattopadhyay et al., 2014).

Food Preservatives

Preservatives are generally weak organic acids like acetic acid, benzoic acid, citric acid, lactic acid, sorbic acid, and propionic acid. Preservatives do not dissociate completely and acidify the cytoplasm, which alters the membrane functions and results in disruption of nutrient transport, resulting in death of the microbe. Food preservatives when used for longer duration may result in headache, gradual loss of mental concentration, and low immune response. Long-term use of these additives may increase the risk of cardiovascular and degenerative diseases and sometimes cancer. Some synthetic preservatives are reported to induce respiratory problems, allergic reactions, anaphylactic shock, and various other health complications. Boric acid is used as a food preservative at a concentration of 4 g/L, but it has been reported toxic to human health as it suppresses the release of sperm from the testis and reduces fertility by abolishing DNA synthesis in sperm cells. Vinegar has been reported to cause esophageal injury, hypokalemia, osteoporosis, and hyperreninemia in long-term exposure (Inetianbor et al., 2015). Sulfites have been reported to cause allergies, heart palpitation, headache, and cancer. Nitrates and nitrites transform to nitrous acid after digestion with food and are suspected in stomach cancer precipitation. Benzoates are also suspected for asthma, skin rashes, and allergies, and sorbates for suspected of causing urticaria and dermatitis (Sharma, 2015).

Flavoring Agents

Many substances are chemically defined for use as flavoring agents in the United States and Europe. Flavoring agents and other additives are evaluated by the Flavor Extract Manufacturers Association (FEMA) expert panel and recognized by the FDA. Moran et al. (1980) studied acute oral toxicity of 63 selected flavoring agents. In this study, it was found that 2-ethyl 4, 5-dimethyl thiazoline was very toxic due to thiazole ring in the structural moiety of thiamine pyrophosphate coenzyme. This coenzyme participates in the transfer and formation of aldehyde and ketols degradation. Furan thioesters and two other esters, ethyl and methyl hexane carboxylate, also showed serious toxicity. Sales et al. (2018) studied various flavorings, including strawberry, vanilla, chocolate, tutti-frutti, and cookies for their cytotoxic, genotoxic, and mutagenic potential. The results of this study reported the alteration in the number of polychromatic or immature cells in bone marrow, reduced erythropoiesis, and micronucleated erythrocyte production. Another study reported that potassium benzoate, sodium benzoate, and potassium nitrate were genotoxic and cytotoxic to human peripheral blood cells. Boric acid, sodium and potassium citrate, and citric acid were reported to be genotoxic and cytotoxic in *Allium cepa* root meristematic tissue. Natural flavoring complexes are obtained from pulp, peel, leaf, bud, flowers, bark, or vegetables by using various processing methods, and there is a need to evaluate their toxicity (Smith et al., 2004).

Emulsifiers

Emulsifiers are used to aid texture in processed food and to extend shelf life by the prevention from separation of mixtures. Emulsifiers are mainly used in creamy sauces, candy, ice cream, margarine, baked goods, and mayonnaise. The commonly used emulsifiers are polysorbate-80 and carboxy methyl cellulose in various preparations. During pharmacological toxicity studies, these emulsifiers showed toxicity like disruption of gut bacteria, delayed immune responses, obesity, and irritable bowel syndrome. Another study showed that emulsifiers promote bacterial translocation in which bacteria moves across the epithelial cells and ultimately Crohn's disease occurs. Emulsifiers increased permeability of gut by which intra-macrophage bacteria like *Escherichia coli* invades and results in formation of abscess, granulomata, and fistula. Recent research also suggested that emulsifiers promote low-grade inflammation, which alters microbiota of gut and provides sufficient conditions to develop inflammatory bowel disease or colorectal cancer (Aponso et al., 2017).

Acidifiers and Acidity Regulators

Acidifiers are mainly used in soft drinks, jelly, sweets, jams, candy, baked nutrients, fruit food, and marmalade. Various studies suggested that these additives show different types of toxicity (Abu Elala and Ragaa, 2015). Acetic acid, used as an acidifier, has been reported to cause allergy, mouth sours, epidermal reactions, acidosis, and renal failure with reduction in clotting efficiency (Shibata et al., 1992). Citric acid, a widely used acid regulator, is reported for dental cell toxicity, necrotic changes in hepatocytes, chromatin decrement, and micronucleated erythrocyte production increment. Another study states that citric acid potentiates chromosomal abrasions and decreased mitotic index (Carocho et al., 2014).

Foaming Agents

The physical and chemical foaming agents are used to produce foam. Chlorofluorocarbons and low boiling hydrocarbons are known as liquid foaming agents. Nitrogen and carbon dioxide are treated as gas physical foaming agents (Rhomie, 1998). Generally, two types of complications are associated with excessive foaming in biological processes. Foaming increases the liquid volume of bioreactors, which decreases biocatalyst concentrations and the performance of cells. Foam formation is also associated with protein and enzyme denaturation, which increases aging in cells. Sodium bicarbonate decomposes at 140°C to yield carbon dioxide and water, used with citric acid or sodium citrate as a foaming agent, but it may result in mild hypertension due to increased sodium concentration in the body (Vardar-Sukan, 1998).

Gelling Agents

Normally, hydrocolloids are used as gelling agents to impart quality improvement and increase shelf life. Gelling agents are mainly used in jam, jelly, marmalade, and restructured foods and also have the ability to change rheology of food systems like flow behavior and texture (Xiaolong et al., 2015). In some formulations, such as soups, sauces, toppings, gravies, and salad dressings, these agents are used to impart viscosity and mouth feel (Saha and Bhattacharya, 2010). Gelling agent's toxicity studies showed that after long-term use, they may be responsible for increases in liver weight, lymph nodes, and the spleen. The granulomatous inflammation of the liver and reticuloendothelial cell hyperplasia of mesenteric lymph node is also reported (Aguilar et al., 2007).

Humectants

Humectants are mainly used to maintain moisture in preparations. Commonly used humectants are glycerin and propyl glycol. The toxicity study carried out by Heck et al. (2002) showed chronic obstructive lung disease and cancer but was not sure about results that showed these medical conditions are due to smoking or humectants, as studies were conducted in smoke-exposed animal.

Propellants

Propellants are used in the preparation of aerosols, but due to abnormal cardiac and respiratory responses reported in aerosol users, their use was restricted. Halogenated hydrocarbons like chlorofluorocarbons were used as propellants because they were inert and in liquid state at low pressure. Findings of FC11 in the blood of 12 asthma patients and change of heart functions in some animal studies prompted toxicity screening of propellants. Further studies reported that use of propellants may cause hypotension, decreased tidal volume, bradycardia, or tachycardia (Olson, 1977). Another study on fluoroalkane, which is also used as propellant in aerosols, found it very toxic to the heart and produces atrioventricular block, T wave depression, and asphyxia-induced sinus bradycardia (Taylor and Harris, 1970). The detailed list of food additives is summarized in Table 3.4 with their uses and toxicity.

TABLE 3.4 Toxicity of Food Additives

S. No.	Functional Class	Use	Example	Toxic Effect Reported	Reference
1.	Acidifiers	Acidity, sour taste	Ammonium hydroxide, calcium sulfate, citric acid, water, sodium diacetate	Weight gain, acidity	Shibata et al. (1992)
2.	Acidity regulators	pH regulator	Sorbic acid, acetic acid, benzoic acid, propionic acid, citric acid	Chromosomal aberration, mutation, dental cell toxicity	Carocho et al. (2014)
3.	Anticaking agents	Lowers molecules adherence	Sodium ferrocyanide and ferric ferrocyanide, calcium silicate, sodium aluminosilicate	Neuronal toxicity	Dorazio and Bruckner (2015)
4.	Antifoaming agents	Foaming prevention	Silicone fluids	neurotoxic, focal lesions, pulmonary collapse, hemorrhage	Harington (1961)
5.	Antioxidants	Deterioration protection	Oregano, basil, rosemary, pepper, nutmeg, cinnamon and thyme, BHA, BHT, and propyl gallate	Asthma, joint pain, dermatitis, stomach and eye problems	Anbudhasan et al. (2014)
6.	Colorants	Coloration of food	Erythrosine, Tartrazine, Quinoline Yellow, Carmosine	Cancer, hyperactivity, asthma, migraine, headaches, DNA damage	Pandey and Upadhyay (2012)
7.	Color retentioners	Color stabilization	Ascorbic acid	Aging, cancer	Eylar et al. (1996)
8.	Emulsifiers	Uniformity of mixtures	Polysorbate-80 and carboxy methyl cellulose	Disruption of gut bacteria, obesity, and irritable bowel syndrome	Aponso et al. (2017)
9.	Flavor enhancers	Enhancement of taste and color	Monosodium glutamate, aspartame, acesulfame K, saccharine,	Cancer, DNA damage, fetal abnormalities, lung tumors	Pandey and Upadhyay (2012)
10.	Foaming agents	Uniform dispersion	Sodium laureth sulfate, ammonium lauryl sulfate, sodium bicarbonate	Inactivate enzymes, aging	Rhomie (1998), Vardar-Sukan (1998)
11.	Gelling agents	Formation of gel	Norsorex	Genotoxicity	Xiaolong et al. (2015)
12.	Glazing agents	Impart shiny surface	Stearic acid, beeswax, candelilla wax	Increased liver, mesenteric lymph spleen, reticuloendothelial-cell hyperplasia	Aguilar et al. (2007)
13.	Humectants		Glycerin, propylene glycol		

(*Continued*)

TABLE 3.4 (Continued)

S. No.	Functional Class	Use	Example	Toxic Effect Reported	Reference
		Drying prevention		Chronic interstitial inflammation, squamous metaplasia, scab formation	Heck et al. (2002)
14.	Preservatives	Prevention of microorganism growth	Sodium benzoate, sodium metabisulfite, potassium nitrate, calcium benzoate, and benzoic acid	Asthma, neurotoxicity, carcinogenic, fetal abnormalities	Pandey and Upadhyay (2012)
15.	Propellants	Help expel food from its container	Freon 11, Freon 12, dichlorotetrafluoroethane	Cardiac and respiratory toxicity	Olson (1977)
16.	Sweeteners	Nonsugar to impart sweet taste	Aspartame, neotame, saccharin, acesulfame, and sucralose	Leukemia, lymphoma, myeloma, cancer	Qurrat-Ul-Ain and Khan (2015)

METHODOLOGY FOR TOXICITY EVALUATION OF FOOD ADDITIVES

In Vitro Genetic Toxicity Tests

Genetic toxicity tests can detect chromosomal destruction and gene mutations by the mutagenic chemicals that may produce adverse health problems such as cancer, cellular mutation, and hereditary diseases. Both in vitro and in vivo methods are available to detect gene mutations. These tests were used to estimate organ toxicity, genotoxicity, damage or alteration in hereditary material of cells/individuals, and to mimic target tissue. About 5000 human diseases are driven by defective genes or due to alteration in physiological processes by defective genetic material. About 20% of fetal and infant deaths, 50% of miscarriages, and 80% of mental retardation cases are due to inherited disorders. New technologies such as genomics, in vivo monitoring, and automated analyzers are standardized for genetic toxicity studies for their accuracy and precision. The new regulations like reduction of animal tests by European Registration, Evaluation, Authorization and Restriction of Chemicals (REACH) and new approaches like toxicity testing in 21st century are milestone developments in genetic toxicity studies. The FDA and EPA issued new guidelines for better data evaluation and risk management (Elespuru et al., 2009). In Table 3.5, some meetings of WHO are enlisted in which toxicity evaluation of different food additives was discussed.

In Vitro Bacterial Reverse Mutation Test

This test, also known as the Ames test, is used to identify the mutagenic substance that induces point mutations like frameshift mutations or base pair substitutions. Two bacterial strains with identified mutations in amino acids are used in this test (Mendes et al., 2013).

TABLE 3.5 WHO Meetings for Evaluation of Food Additives and Contaminants

S. No.	Year	Meeting No	Food Additives Evaluated	Reference
1.	1982	Twenty-Sixth	Anoxomer, sorbiton, stearyl monoglyceride citrate, glucose isomerase, protease, ethyl lactate, eugenol, anthocyanins, carmines, curcumin, quinolone yellow, sunset yellow, sorbitol, cyclamates, saccharin	FAO/WHO (1982)
2.	1983	Twenty-Seventh	Butylated hydroxytoluene and hydroxyanisole, dichloromethane, 1,1,2-trichloroethylene, anethole, benzyl acetate, carvone, azorubine, ponceau 4R, calcium benzoate, acesulfame, thaumatin, lactitol, xylitol, karaya, tragacanth	FAO/WHO (1983)
3.	1988	Thirty-Third	Anethole, potassium bromated, erythrosine, maltitol, trichlorogalactosucrose, karya gum, contaminants aluminum, arsenic, cadmium, phthalate, iodine, methylmercury, and tin	FAO/WHO (1989)
4.	1991	Thirty-Seventh	Annatto, lycopene, natamycin, propyl paraben	FAO/WHO (1991)
5.	1993	Forty-First	Gallates, benzyl acetate, limonene, quinine, carotene, konjac flour, propylene glycol alginate, beta cyclodextrin, urea, contaminants cadmium, chloropropanols, lead	FAO/WHO (1993)
6.	2004	Sixty-First	A-Amylase, annatto, curcumin, diacetyltartaric, d-tagatose, laccase, mixed xylanase, b-glucanase, neotame, polyvinyl alcohol, quillaia, xylanase	FAO/WHO (2004)
7.	2006	Sixty-Fifth	Beeswax, candelilla wax, quillaia extract, calcium L-5-methyltetrahydrofolate, phospholipase A1	FAO/WHO (2006)
8.	2007	Sixty-Eighth	Sodium chlorite, asparaginase, carrageenan, cyclotetraglucose, isoamylase, magnesium sulfate, phospholipase A1, EDTA, steviol glycosides	FAO/WHO (2007)
9.	2010	Seventy-Third	Activated carbon, cassia gum, indigotine, steviol glycosides, sucrose, sucrose, titanium dioxide	WHO/FAO (2011)
10.	2011	Seventy-Fourth	Benzoe tonkinensis, gum rosin, tall oil rosin, wood rosin, polydimethylsiloxane, ponceau, pullulan, quinoline yellow, sunset yellow FCF	FAO/WHO (2011)
11.	2013	Seventy-Seventh	Advantame, glucoamylase, nisin, octenyl succinic acid, modified gum arabic	FAO/WHO (2013)
12.	2014	Seventy-Ninth	Benzoe tonkinensis, carrageenan, Citrem, gardenia yellow, paprika extract, pectin	FAO/WHO (2015)
13.	2016	Eightieth	Benzoates, lipase, magnesium stearate, maltotetraose hydrolase, polyvinyl alcohol	FAO/WHO (2016)
14.	2017	Eighty-Second	Allura red AC, carob bean, pectin, quinoline yellow, rosemary, steviol glycosides, tartrazine, xanthan gum	FAO/WHO (2017)

Salmonella typhimurium with histidine and *E. coli* with tryptophan mutation were generally used. When these bacteria were grown in media containing mutagenic substance, it caused a second mutation that reversed the existing mutations by restoring synthesizing capacity of deficient amino acid. Therefore this test was also known as the reverse mutations test. The test involved addition, deletion, or substitution of one or more base pairs of DNA. This test is based on point mutation, which causes many genetic disorders in humans, like point mutation in oncogene and tumor suppressor gene in somatic cells, and ultimately results in cancer and tumors, respectively, in experimental animals and humans. These studies are fast, low cost, and easy to execute in a laboratory (Pagnout et al., 2014).

Some limitations of this test were also present like utilization of prokaryotic cells that have different structures of chromosomes, repair process of DNA, and metabolism and requirement of metabolic activator from exogenous source. The results of these studies can only prove the genotoxicity but cannot estimate the carcinogenic or mutagenic potency of test substrate in humans, so such studies were only applied to find initial toxicity data. The chemicals that are highly toxic to bacteria like antibiotics or bacteriostatic in nature cannot be tested by this method. The compounds that were active only in mammalian cells like topoisomerase inhibitors or analogs of nucleosidases did not show any result in such studies but are genotoxic in nature. The commonly used methods for bacterial reversal mutation tests were the preincubation method, fluctuation method, incorporation method, and suspension method. The chemicals or compounds like short chain aliphatic nitrosamine, aldehyde, pyrrolizidine, nitrocompounds, azo dyes, diazo compounds, alkaloids, divalent metals, and allyl compounds were more efficiently screened for their genotoxicity (Chandrasekhar et al., 2013; OECD, 1997).

In Vitro Mammalian Cell Gene Mutation Tests Using the hprt and xprt Genes

The test is only for those substances that produce gene toxicity when exposed to hypoxanthine guanine phosphoribosyltransferase (*hprt*) or xanthine guanine phosphoribosyltransferase (*xprt*) receptor gene. The OECD prescribed test guidelines in 1984 as TG476, which was revised many times until 1997. The present guidelines were developed using thymidine kinase gene to study genetic toxicity. This test measures forward mutations in receptor genes, especially hypoxanthine-guanine phosphoribosyltransferase gene (hprt in humans and xprt in rodents) and xanthine-guanine phosphoribosyltransferase transgene (gpt). The hprt test detects frameshift mutations, base pair substitution, insertions, and small deletions; gpt test (XPRT test) finds gpt transgenic autosomal location by which large deletion and mitotic recombination occurs. The hprt test is more widely used for regulatory purposes. The in vitro test requires a metabolic activation system so it does not mimic in vivo conditions. Mycoplasma contamination of medium also makes false results. The hprt enzyme activity deficient and xprt enzyme activity deficient cells are resistant to 6-thioguanine, a purine analogue. In hprt or gpt (*XPRT*) proficient cells, 6-thioguanine inhibits cellular metabolism and stops cell division. The mutant cells survive in the presence of 6-thioguanine, but proficient cells with hprt and gpt enzyme activities do not survive. The chemical is applied to cellular suspension or monolayer cultures in the presence of metabolic activator and without metabolic activator for 3–4 hours and then subcultured. The cloning efficiency is estimated just after the test and adjusted with any loss of cells with negative control. The treated cells are incubated for 7–9 days or as required in suitable medium, then colonies are counted.

The commonly used mutant cells for hprt test are CHO, CHL, V79 lines of Chinese hamster cells, TK 6 human lymphoblastoid, and L5178Y mouse lymphoma cells. For gpt or XPRT test CHO derived AS 52 mutant cells are preferred. The cytotoxicity is expressed as relative survivals (RS) and mutant frequency (OECD, 2016a).

In Vitro Mammalian Chromosomal Aberration Test

This test identifies the substances that may delete or rearrange the chromosomal structures in established cell line cultures. Mostly, chemical mutagen induces chromatid type aberrations. A chromosome aberration causes various human genetic abnormalities like alteration of tumor suppressor genes and oncogenes. On the basis of p53 status, genetic stability, DNA repair capacity, and cells of organs, culture of cell lines or culture of primary cells are selected for this test. Limitations of the test include exogenous source of metabolic activator, intrinsic mutagenicity, change in pH, and high levels of cytotoxicity, which may lead to artificial positive results. The cell lines are treated with and without metabolic activator and after predetermined intervals again treated with colcemid or colchicine, metaphase arresting agents. Cell lines are harvested and stained, then analyzed for chromosomal aberrations microscopically. There are many strains, cell lines, and primary cell cultures used to perform this test (OECD, 2010).

In Vitro Mammalian Cell Micronucleus Test

The test identifies the mutagenic materials that induce aneuploidy or chromosome breaks or both. During anaphase of cell division, when a chromosome fragment or an intact chromosome does not move to mitotic pole, then a micronuclei is formed that results in one part deficient daughter nuclei. This test detects both clastogen, a mutagenic agent that causes structural chromosomal breaks and aneugen-caused numerical abnormalities or full chromosome deficiency. Generally, mammalian peripheral blood lymphocytes are used in this test. The OECD guidelines for in vitro mammalian cell micronucleus test (MNvit) was accepted in 2010 and revised in 2016. The test was based on micronuclei detection in the cytoplasm of interphase cells that originated from chromosome fragments that lack a centromere. The protocol used in the test was with or without cytochalasin B (cytoB), which was an actin polymerization inhibitor (OECD, 2016d).

In Vivo Genetic Toxicity Tests

Transgenic Rodent Somatic and Germ Cell Gene Mutation Assay

Transgenic rodent (TGR) mutation assay is used to identify substances that induce mutations in the genes of transgenic receptors and cause chromosomal aberrations. The mutations induced by test chemicals were detected by transgenes that contain receptor genes. The recovery of transgenes and analysis of receptor gene phenotype in bacterial host (deficient in receptor gene) gives the score of mutations. The transgene responds to mutagens by base pair substitution, frameshift mutations, insertions, and small deletions. The recommended transgenic animals are *lac*Z bacteriophase mouse, gpt delta mouse, *lac* I mouse or rat, *lac* Z plasmid mouse, and commonly used mutagens were *N*-ethyl-*N*-nitrosourea, ethyl carbamate (urethane), 2,4-diaminotoluene, and benzopyrene. The possible targets for these mutagens were the liver in rats and bone marrow, colon, liver, lungs, and male germ cells. In the experiment, rodents were treated with test chemical for a specified time (mostly 28 days) by an appropriate route of administration; then after treatment, 44 paired DNA lesions were fixed to stable mutation. The reading for manifestation, fixation, expression, and sampling times were recorded. Animals were then sacrificed and the genomic DNA was isolated from tissue and purified. After study, tissue collected for mutagenic and carcinogenic toxicity study must be stored below $-70°C$ and used within 5 years for DNA isolation. The isolated DNA was stored at $4°C$ in suitable buffer and analyzed within a year. The observed data were number of plaque or colony units, number of mutants, mutant frequency, and also number of reactions per DNA sample if multiple packaging or rescue reactions were used (OECD, 2013).

Mammalian Bone Marrow Chromosomal Aberration Test

This test identifies substances that induce chromosome or chromatid type structural chromosomal aberrations in bone marrow cells. The OECD guidelines adopted as TG 475 in 1984 and the presently revised version of 1997 are adopted to regulate such studies. The limitations of in vitro studies were overcome in this study. This test is accessing the genetic toxicity with consideration of metabolism, pharmacokinetics, and ability to repair DNA directly by body so it can be considered as further evaluation of genotoxicity of the in vitro method. It is performed with exposure of test chemical and then treatment with metaphase-arresting agents such as colcemid and colchicine and isolation of the bone marrow cells, staining, and analysis of metaphase cells for chromosomal aberration. The healthy young adult animals, commonly rats, were used. One or more positive control substances like ethyl methane sulfonate, methyl methane sulfonate, ethyl nitrosourea, mitomycin C, cyclophosphamide, and triethylenemelamine were used to increase the frequency of cells with structural chromosomal aberrations (OECD, 2016b).

Mammalian Erythrocyte Micronucleus Test

The substances, which induce micronuclei in erythroblasts, are evaluated by this method and estimated as immature erythrocytes or reticulocytes. The OECD guidelines for this test were adopted in 1983, and the revised version of 1997 was accepted for toxicity study. Due to variations in genotoxicity among rodent species, the mammalian in vivo erythrocytes micronucleus test is used so it may be considered as further study of in vitro methods. The damage to chromosomes or erythroblasts mitotic apparatus due to test chemical was estimated by evaluating micronucleus formed in erythrocytes of bone marrow or peripheral blood cells of small animals or rodents. The aim of study was to test substances that cause cytogenetic damage and micronuclei formation containing lagging chromosome fragment or whole chromosome. The immature erythrocyte without main nucleus was formed from bone marrow erythroblasts. Immature erythrocyte contains micronuclei in the cytoplasm, and higher frequency of such erythrocytes was the sign of structural or numerical chromosomal aberration induction. The micronucleated erythrocytes were stained and counted to compare with controls. One or more positive control substances were used in positive control such as ethyl methane sulfonate, ethyl nitrosourea, methyl methane sulfonate, mitomycin, cyclophosphamide, colchicines, or vinblastine (OECD, 2016c).

Rodent Dominant Lethal Test

The mutagens that induce inherited dominant lethal mutations in germ cells resulting in embryonic or fetus death are tested in this study (OECD, 2015). The OECD TG 478 was adopted in 1984 and some modifications were also done. Dominant lethal test is used to identify the substance that induces mutations in germ cells.

Dominant lethal mutations generally result in embryonic or fetal death and are used to predict genetic diseases and human hazards transmitted through germ lines. The test is very expensive and time consuming due to high labor cost and the large number of animals used in the study. The dominant lethal mutation was defined as mutation occurred in germ cells or early embryo and lethal to fertilized eggs and developing embryo (Ashby and Clapp, 1995).

Generally, male mice treated with test chemicals are mated with virgin females that were untreated with any other chemical in the past. After mating, females were euthanized and uteri were examined for live and dead embryo implants. Results were calculated by comparing live and dead implants per female in test as well as control groups and the postimplantation loss was estimated. One or more positive control substances such as triethylenemelamine, cyclophosphamide, monomeric acrylamide, or chlorambucil were used. The five daily doses were administered and mating of animals were ensured and counted for mating intervals, which may be weekly. At the 13th gestation day, the second half of pregnancy, females were euthanized and uteri were examined for dominant lethal effects (OECD, 2015).

Short-Term Toxicity Studies With Rodents

Short-term toxicity tests were carried out in rodents to predict test doses of substance for chronic and subchronic toxicity study. As per study guidelines, rodents, usually rats or mice of either sex, were used for study. The 6- to 8-week-old animals were divided into groups of 10 rodents per sex per group. The mortality in test animals was not acceptable and in control group should not be more than 10%. Not more than 10% of animals lost their tissues or organs due to autolysis in whole study. Necropsy should be done soon after animal sacrification or death. The test substance may be given in the diet or dissolved in the drinking water or encapsulation/oral intubation in at least three dose levels. One sufficiently high dose should be given to produce toxicity; second, low dose in which toxicity was not reported; and third, one must be an intermediate dose to find minimum toxic effects. For generation, measurement and data assessment must be developed, validated, and maintained by computerized system and as per Good Laboratory Practice guidelines. In clinical testing, ophthalmological examination, hematology, clinical chemistry, urine analysis, neurotoxicity, and immunotoxicity tests were performed. In microscopic examination, necrosis of tissues, organ weight, microscopy of tissues, and histopathological examinations are performed (USFDA, 2003a).

Subchronic Toxicity Studies With Rodents

Subchronic toxicity studies were conducted for more than 3 months and used for dose selection for chronic and long-term toxicity study. More than 6- to 8-week-old rats or mice of either sex were used in study. The test must have less than 10% mortality in control or test groups and not more than 10% loss of animals and tissues or organs due to autolysis. The necropsy was completed just after sacrifice of animals to minimize autolysis. The test substance or mixture must be known and have a Chemical Abstract Service (CAS) regulatory number or numbers. The observations and clinical tests are the same as short-term toxicity studies, but the parameters taken in consideration were in significant numbers. In clinical chemistry, at least three out of five specific determinants of hepatocellular evaluation are taken into consideration, such as alanine aminotransferase, aspartate aminotransferase, sorbitol dehydrogenase, glutamate dehydrogenase, and total bile acids. In hepatobiliary evaluation alkaline phosphate, bilirubin, gamma-glutamyl transpeptidase, 5′ nucleotide, and total bile acids estimation are carried out. In urine analysis, urine volume, specific gravity, pH, glucose, and protein are estimated. The neurotoxic and immunotoxicity data was evaluated up to a significant level (USFDA, 2003b).

Subchronic Toxicity Studies With Nonrodents

Subchronic toxicity studies on nonrodents guidelines available as OECD TG 409 were introduced in 1981 and a revised version was adopted in 1998. These studies are carried out after short-term toxicity test in repeated doses for 28 days and provide data of toxic effects, target organs, accumulation of test chemical, and safe and toxic doses. The commonly used nonrodent was defined as breed dog (Beagle), but other species like swine or minipig may also be used. The observation period is 90 days, and data is estimated for ophthalmological examination by ophthalmoscope, body weight and food/water consumption by animals, hematological and clinical biochemistry analysis of blood sample, and urine analysis with tissue damage. In histopathological examination, all body parts and organs need to be examined. The toxicity data for each animal should be maintained and analyzed statistically. The test report contain data for change in body weight, toxicity signs, duration of clinical observation, organ/body weight ratio, biochemistry data, and statistical report (OECD, 2018a).

One-Year Toxicity Studies With Nonrodents

One-year toxicity tests are carried out usually with dogs and considered as long-term toxicity study. This test is used to find toxicity of test substances in nonrodents and the maximum dose at which no observed adverse reaction occurs in the body. Dogs should be 4–6 months old and tests should have at least four dogs per sex per dose. The animals were sacrificed for mortality, autolysis, and necropsy as in other studies. Generally, three dose levels of test substances are used: a higher dose to induce toxicity, a low dose for efficacy, and an intermediate dose for minimal toxic dose estimation. For generation, measurement, and data processing, computerized systems are used that are developed, validated, and maintained according to GLP guidelines. The data obtained from such studies cannot be used to find carcinogenicity but can provide information about carcinogenicity. The observations are recorded twice a day with a minimum interval of 6 hours. A long-term toxicity study provides data not only for pharmacologic and toxicological responses but also for behavioral changes, neurological toxicity, and autonomic dysfunctions (USFDA, 2003c).

Chronic Toxicity or Combined Chronic Toxicity/Carcinogenicity Studies

The chronic and carcinogenicity studies guidelines were summarized in OECD TG 453 and adopted in 2018 as per the original guidelines framed in 1981. These guidelines referenced the broad range of chemical toxicity studies including pesticides and chemicals used in industries (OECD, 2018b). These studies mainly were used to find possible health hazards and carcinogenicity, so the objectives of these studies were to recognized carcinogenic property, time of appearance, chronic toxicity, target organs, dose and response relationship, and idea of mode of action of test chemical. Generally, rodents were used in such studies, but if relevant data about carcinogenicity was obtained, then nonrodents could be used to study health effect. For nonrodent chronic toxicity and carcinogenicity studies, some modifications are required as per OECD TG409, OECD guidance document no. 116 and repeated dose 90 days oral toxicity in nonrodents (OECD, 2018a). The designing and conduct for such study were summarized in OECD guidance document no. 116. The study is designed in two parallel phases. One was a chronic toxicity study in which a test substance was administered in gradual doses to many groups for 1 year or as required. The second is a carcinogenicity study in which a test substance was administered in animals for their whole life. The observations are recorded separately for chronic toxicity phase and carcinogenicity phase study (OECD, 2018b).

Carcinogenicity Studies With Rodents Including in Utero Exposure Phase

In carcinogenic studies, normally rodents of either sex were used, but later the FDA recommended to add the in utero phase, which means that during the toxicity study animal reproduction is also facilitated so the toxicity of a substance on the uterus and fetus is also studied. These guidelines provide specific guidance to design and perform in utero exposure phase to bioassay and other chronic toxicity/carcinogenicity studies of food additives. Normally, rats and mice selected for study must be acclimatized for 5 days, then females are treated with test material for 4 weeks before mating while males are treated by test material 10 days prior to mating. According to the FDA, at least 70 animals per sex per group should be selected for study and also ensure that at least 25 animals per sex per group must survive until the end of the study. In mating procedures, a female is placed with a single randomly selected male until pregnancy occurs or evidence of mating completion has been observed. Each female is examined for sperm presence in vaginal lavage or the presence of vaginal plug. One animal per sex per litter or two animals for single-sex litters are selected randomly. Observation and data processing are the same as in normal chronic/carcinogenic toxicity study with three dose levels (USFDA, 2017).

Reproduction Studies

The reproductive studies are used to estimate the effect of test substances on male and female reproductive systems; maturation of postnatal, offspring reproductive capacity; and cumulative effect on reproduction through several generations. According to an FDA report, a minimal two-generation reproductive study with one litter per generation must be completed. Rodents like rats and mice can be used for such studies, but rats are preferred due to small size, easy breeding, 3 weeks' gestation period, high fertility rate, and spontaneous ovulation. The animals of 5–9 weeks of age and all group animals have almost uniform weight and age and a minimum of three doses are used. Clinical observation for behavioral changes, toxicity signs, mortality, estrus cycle length, vaginal smears, and growth of offspring should be recorded for F0 and F1 generations. End point of reproductive toxicity is in terms of female fertility, gestation, live born, weaning index, and testicular spermatid numbers for males. The motility, morphology, and quantity of sperms are also measured. After completion of the test, gross necropsy and microscopic examinations for any structural abnormalities or changes are examined. Histopathology of reproductive organs of females are examined for growing follicles, *Corpus lutea*, while in males, epididymis for sperm granulomas, leukocytes infiltration, and aberrant cells formation are examined. The suitable ANOVA tests can be applied to analyze research data (USFDA, 2000a).

Developmental Toxicity Studies

This test may be performed as stand-alone or multigenerational reproductive study in rats, mice, rabbits, or hamsters for reproduction study and teratological effects of test chemicals. In this study, the treatment is given before organogenesis and continues until the day of parturition. The guidelines recommended a dose range finding study to find the most appropriate dose. The observations are monitored for maternal toxicity like mortality, body weight, organ weight, lesions, and feeding until 1 day before the expected day of parturition, which is 21 for rats, 29 for rabbits, 18 for mice, and 15 for hamsters. Dams and fetuses are also screened for abnormalities of skeleton and soft tissues (USFDA, 2000b).

Metabolism and Pharmacokinetic Studies

To determine characteristics of dose response of any test substance, the studies of metabolic and pharmacokinetic parameters in test animals are very important. In this study, the extant of absorption, distribution in tissue, metabolic pathways and rate, and elimination rate study data is observed. Usually, rodents like rats or mice or nonrodents like dogs are preferred for single-dose pharmacokinetic study. Sampling of red blood cells, plasma, serum, urine, and feces with some organs like kidney, liver, fat, and target organs are compiled during study. To identify the organ or tissue where the test substance is concentrated, whole body autoradiography is used (USFDA, 2007).

Human Studies

The FDA provides general guidelines to perform human clinical studies on food additives but generally advises not to conduct human clinical studies until a very high dose of proposed additive is to be consumed. The test is performed in all age groups, like children, mothers, and older persons. Physical examination and laboratory test include a blood test for platelet count, blood urea, and creatinine, and other tests like a liver function and renal function test should also be performed. In early clinical study absorption, biotransformation and excretion data for food additives and its metabolites are calculated. The enzymatic reactions and food additive interactions with food, nutrients, and medications, followed by long-term clinical studies, need to be performed for food additives' adverse effects (USFDA, 1993). In Table 3.6, classification of methods for toxicological evaluation of food additives is listed.

TABLE 3.6 Methods for Toxicological Evaluation of Food Additives

S. No	Toxicity Test	Methods	Effect of Test Chemical	Reference
1.	Genetic toxicity tests	Bacterial reverse mutation test	Point mutations	Parasuraman (2011)
		In vitro mammalian cell gene mutation tests using the hprt or xprt genes	hprt or xprt reporter gene mutation	
		In vitro mammalian cell gene mutation tests using the thymidine kinase gene	Thymidine kinase reporter mutation	
		In vitro mammalian chromosomal aberration test	Structural chromosomal aberrations	
		In vitro mammalian cell micronucleus test	Chromosomal breaks and aneuploidy	
		Transgenic rodent somatic and germ cell gene mutation assays	Gene mutations in transgenic reporter genes	
		Mammalian bone marrow chromosomal aberration test	Structural chromosomal aberrations	
		Mammalian erythrocyte micronucleus test	Micronuclei in erythroblasts	
		Rodent dominant lethal assay	Genetic damage causing in fetus	
		Mammalian spermatogonial chromosomal aberration test	Structural chromosomal aberrations in male germ cells	
		Mouse heritable translocation assay	Structural chromosome changes	
		Unscheduled DNA synthesis (UDS) test with mammalian liver cells in vivo	DNA damage and subsequent repair	
		In vivo mammalian alkaline comet assay	DNA damage	
2.	Short-term and subchronic toxicity studies with rodents	Body weight and feed intake data	Feed consumption	USFDA (2003a)
		Ophthalmological examination	Changes in the eyes	
		Hematology	Hemotoxicty	
		Clinical chemistry	Electrolyte balance, carbohydrate metabolism, and liver and kidney function	
		Urinalyses	Urine for sediment and presence of blood/blood cells	

(*Continued*)

TABLE 3.6 (Continued)

S. No	Toxicity Test	Methods	Effect of Test Chemical	Reference
		Neurotoxicity screening/ testing	Structural or functional integrity of the nervous system	
		Immunotoxicity	Immune toxicity	
		Gross necropsy	Structural changes in body	
		Organ weight	Organ toxicity	
		Tissues microscopy	Tissues toxicity	
		Microscopic evaluation	Cell injury	
		Histopathology of lymphoid organs	Immunotoxicity testing	
3.	Subchronic toxicity study with rodents	More than 3-month study in rats or mice	Hematology, hepatocellular and hepatobiliary study; urine analysis, neuro and immunotoxicity	USFDA (2003b)
4.	Subchronic toxicity study in nonrodents	As per OECD TG 409	Toxicity, target organs, accumulation of test chemical and dose	OECD (1998)
5.	One-year toxicity study in nonrodents	One year toxicity study in dogs at 3 dose levels	Behavioral changes, neurological toxicity, autonomic dysfunctions	USFDA (2003c)
6.	Combined chronic toxicity/ carcinogenicity study	As per OECD TG 453	Chemical toxicity, carcinogenicity, health hazards	OECD (2018b)
7.	Carcinogenicity study with in utero exposure phase	70 animals for reproductive toxicity of food additive	Carcinogenic and reproductive toxicity	USFDA (2017)
8.	Reproductive studies	At least two generations reproductive study	Sperm granuloma, leukocytes infiltration, aberrant cell formation	USFDA (2000a)
9.	Developmental toxicity study	Multigeneration toxicity study	Reproductive and teratological toxicity	USFDA (2000b)
10.	Metabolism and pharmacokinetic study	Dose response of test substance	ADME data in animals	USFDA (2007)
11.	Human studies	Human clinical studies	Hematological, liver, renal toxicity	USFDA (1993)

NEED OF ADVANCEMENTS IN TOXICITY STUDY

Presently, toxicity of medicines, consumables, chemicals, food additives, and agricultural chemicals is evaluated by using laboratory animals with some assumptions and extrapolations. But such studies are expensive and time consuming, and provide data mostly about adverse health effects without information on biological changes due to toxic effects (Rovida et al., 2015). The biology of body systems and rapid assay technologies like autoanalyzer and bioinformatics help the researcher to develop new methods for toxicity tests (Fig. 3.1).

The EPA asked the National Research Council to develop new ideas for toxicity studies with modern technologies and new standards. The National Research Council formed a committee on toxicity testing, and the assessment of environmental agents suggested that before adopting new protocols and testing strategies, one must have overlaps to verify the results. Uniform testing protocol, strategies, and mode of action of test chemicals must be considered so there is a need to develop chemical-specific testing methods. Every toxicity study for regulatory purposes must consider risk management as a primary need (Krewski et al., 2010).

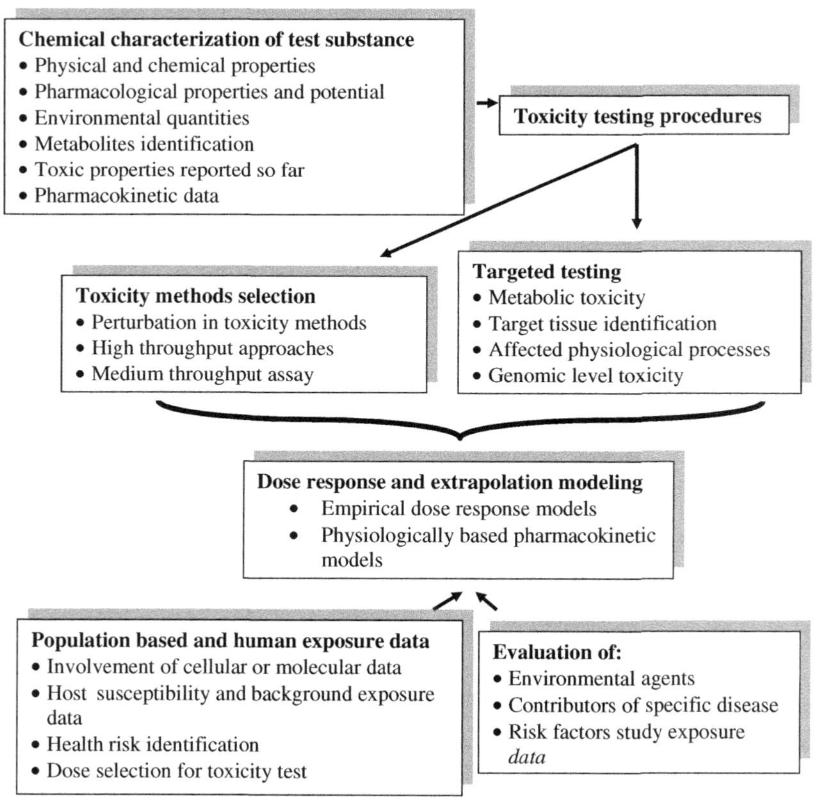

FIGURE 3.1 Proposed pathway to develop new testing strategies (Krewski et al., 2010).

NEW TECHNOLOGIES IN TOXICITY STUDIES

In present methods of toxicity studies, one common drawback is the absence of human relevant drug metabolism (Bhushan et al., 2016). New technologies are optimizing rodent study with human biological systems and provide more predictive efficacy, safety, and are less time consuming and have cost effective toxicity data. Due to major advancements in the field of molecular biology, bioinformatics and biotechnology offer new concepts of regulatory toxicology in practice, which reduces the use of animals in toxicity studies (Vliet, 2011).

Humanized mouse model, in which one or more mouse genes were replaced by human gene, known as genetically humanized mice, is used for such studies. The incorporated part of the human gene into the mouse gene is also transferred to the next generation after breeding. These mice can be used to study drug metabolism and in toxicity studies because they have some xenobiotic receptors, transporters, and cytochrome 450 as do humans. The toxicity of acetaminophen and 2-amino-1-methyl-6-phenylimidazo(4,5-b)pyridine was studied by humanized mouse model (Bhushan et al., 2016).

The in vitro microphysiological cell constructs can be used to mimic the physiological action of the liver for drugs and metabolites. The multiorgan constructs can also be seen in the near future for liver-intestine, liver-skin, and liver-neurosphere cell constructs. The iPS-derived cells are induced pluripotent stem cells derived from human hepatocytes like cells. They can be patient specific because of direct isolation of cells from the patient. Such cells for heart, kidney, and brain are in the process of development (Ware et al., 2015).

Primary monolayer cell cultures can be prepared by fleshly isolated cells of brain, skin, kidney, and liver, which shows the same morphological and biochemical characteristics as source animal or human cells. These cultures can be used for biochemical assay or imaging technology. The 3D cell culture models have enhanced functional properties (Vliet, 2011). Some new techniques/models for food additives toxicity studies are summarized in Table 3.7.

TABLE 3.7 New Technologies/Models in Toxicity Studies

S. No.	Type of Toxicity	Model (In Vitro/In Vivo)	Reference
1.	Pulmonary toxicity	MucillAir, EpiAirway	Chapman et al. (2013)
2.	Renal toxicity	Primary human proximal tubular epithelial cells	Chapman et al. (2013)
3.	Hepatotoxicity	Humanized mouse	Bhushan et al. (2016)
4.	Cardiovascular toxicity	Recombinant hES	Chapman et al. (2013)
5.	Liver metabolism	iPS-derived cells	Ware et al. (2015)
6.	Endocrine toxicity	iCells cardiomyocytes	Chapman et al. (2013)
7.	Cell specific toxicity	Primary monolayer cell culture	Vliet (2011)

1. Chemistry and specifications of additives
Identity of the substance
- *Chemical name (IUPAC)*
- *CAS number*
- *Synonyms, trade names, abbreviations*
- *Molecular and structural formulae*
- *Molecular weight or atomic weight*
- *Spectroscopic data*
- *Physical and chemical properties*
- *Solubility and effect of pH*
- *Particle size, shape and distribution*

Specifications of substance
- *Article of commerce*
- *Purity in percentage*
- *Impurities: nature, limits*

Manufacturing process of substance
Methods of analysis in food
Stabilityand reaction with food
- *Chemical/Physico chemical stability*
- *Degradation products*
- *Reactions with food constituents*

2. Existing authorizations and evaluations of additive
Body which carried out evaluations
Duration and place of study
Results of study
NOAELs/LOAELs and BMDL values
Uncertainties and health hazards data

3. Proposed Uses and Exposure assessment for additive
Proposed uses in food
Assessment of exposure
Residues or contaminants exposure

4. Toxicokinetics and toxicity of additive
Toxicokinetics
Genotoxicity
Subchronic, Chronic toxicity
Carcinogenicity
Reproductive and Developmental toxicity

FIGURE 3.2 Submission of data for new or modified authorization of additives.

SUBMISSION OF DATA FOR FOOD ADDITIVE AUTHORIZATION

The Scientific Committee for Food of European Food Safety Authority gives guidance about documents for approval of new food additives or modifications in previous authorizations. The authorization documents contain data about test substance identification, impurities, and residuals as described by the manufacturer. The previous risk management data and authorizations were processed in the past for that additive and will be applied for modifications in authorization (SCF, 2001). The age groups detail what was used in the processed food and the proposed users of food. The toxicity and health hazards data of additives are evaluated by various methods with specified standards used to perform toxicity studies. The typical document file for a new authorization or modifications in already authorized additives are arranged in sequence such as, firstly, chemistry and specifications of the proposed additives, the existing authorization, and the previously evaluated data of toxicity, if applied. The proposed users and age groups for which food preparations will be processed using test additives are from the quantity of population that will become exposed to it in the future. Finally, toxicological study data in tabular form and statistically analyzed data with details of methods are used to evaluate different types of toxicities (EFSA, 2012). The detailed checklist is summarized in Fig. 3.2.

CONCLUSION

Food additives are used in foodstuffs and play a key role to impart technological features like color, flavor, preservation, thickeners, stabilizers, taste, preservation, emulsifier, or acidity regulation. It is not possible to avoid the use of food additives in food processing. Some of the additives may be toxic or harmful to health; therefore, there is an urgent need to screen them for cytotoxicity, genotoxicity, mutagenicity, hepatotoxicity, and associated disorders. The present toxicity testing is mainly based on laboratory animals testing, which is an expensive and time-consuming process and depends on assumptions/extrapolations. To prevent the adverse effect of food additives on human health from toxicity, cost effective technologies need to be developed.

Furthermore, there should be uniform worldwide guidelines on the status of these agents.

References

Abu Elala, N.M., Ragaa, N.M., 2015. Eubiotic effect of a dietary acidifier (potassium diformate) on the health status of cultured *Oreochromisniloticus*. J. Adv. Res. 6 (4), 621–629.

Aguilar, F., Autrup, H., Barlow, S., Castle, L., et al., 2007. Beeswax (E 901) as a glazing agent and as carrier for flavours. Scientific Opinion of the Panel on Food additives, Flavourings, Processing aids and Materials in Contact with Food (AFC). EFSA J. 615, 1–28.

Alvito, P., Assuncao, R., Louro, H., Silva, M.J., Vasco, E. (Eds.), 2016. Food contaminants and human health—Challenges in chemical mixtures. National Institute of Health, Doutor Ricardo Jorge, Av. Padre Cruz, 1649-016 Lisbon, Portugal.

Anbudhasan, P., Surendraraj, A., Karkuzhali, S., Sathishkumaran, P., 2014. Natural antioxidants and its benefits. Int. J. Food Nutr. Sci. 3 (6), 225–232.

Anonymous, 2013. Directive 2013/39/EU of the European Parliament and of the Council of 12 August 2013 amending Directives 2000/60/EC and 2008/105/EC as regards priority substances in the field of water policy. Official Journal of the European Union L 226/1. Available at http://goo.gl/b0xcE1 (accessed March 2018.).

Aponso, M.M.W., De Silva, G.O., Abeysundara, A.T., 2017. Emulsifiers as food additives: an overview on the impact to obesity and gut diseases. J. Pharmacognosy Phytochem. 6 (3), 485–487.

Arora, A.K., Mahajan, S., Gupta, P., Kapoor, S.S., 2009. Is chemical safety to food hazardous? Dangers of food preservatives. J. Indian Acad. Forensic Med. 31 (4), 419–422.

Arvanitoyannis, I.S., Bosnea, L., 2004. Migration of substances from food packaging materials to foods. Crit. Rev. Food Sci. Nutr. 44 (2), 63–76.

Ashby, J., Clapp, M.J., 1995. The rodent dominant lethal assay: a proposed format for data presentation that alerts to pseudo-dominant lethal effects. Mutat. Res. 330 (1–2), 209–218.

Bast, A., Haenen, G.R., 2002. The toxicity of antioxidants and their metabolites. Environ. Toxicol. Pharmacol. 11 (3-4), 251–258.

Begley, T.H., White, K., Honigfort, P., Twaroski, M.L., Neches, R., Walker, R.A., 2005. Perfluorochemicals: potential sources of and migration from food packaging. Food Addit. Contam. 22 (10), 1023–1031.

Bhushan, A., Martucci, N.J., Usta, O.B., Yarmush, M.L., 2016. New technologies in drug metabolism and toxicity screening: organ-to-organ interaction. Expert Opin. Drug Metab. Toxicol. 12 (5), 475–477.

Bond, J.A., 1989. Review of the toxicology of styrene. Crit. Rev. Toxicol. 19 (3), 227–249.

CEC, 1976. Directive 76/ 893/EEC relating to plastics materials and articles intended to come into contact with foodstuffs. Commission of the European Communities 23, November 1976, Brussels.

CEC, 1990. Plastics in contact with food. Industrie Alimentari 29 (282), 477–485.

Carocho, M., Barreiro, M.F., Morales, P., Ferreira, Isabel, C.F.R., 2014. Adding molecules to food, pros and cons: a review on synthetic and natural food additives. Comprehen. Rev. Food Sci. Food Safety 13 (4), 377–399.

Castle, L., Mayo, A., Crews, C., Gilbert, J., 1989. Migration of poly(ethylene terephthalate) (PET) oligomers from PET plastics into foods during microwave and conventional cooking and into bottled beverages. J. Food Protect. 52 (5), 337–342.

Chandrasekhar, P., Lakshmi, T., Suresh, S., 2013. Bacterial reverse mutation test with *Nardostachys jatamansi*. Int. J. Pharm. Pharm. Sci. 5 (3), 262–266.

Chapman, K.L., Holzgrefe, H., Black, L.E., Brown, M., Chellman, G., Copeman, C., et al., 2013. Pharmaceutical toxicology: designing studies to reduce animal use, while maximizing human translation. Regul. Toxicol. Pharmacol. 66 (1), 88–103.

Chattopadhyay, S., Raychaudhuri, U., Chakraborty, R., 2014. Artificial sweeteners: a review. J. Food Sci. Technol. 51 (4), 611–621.

Cooper, I., Tice, P.A., 1995. Migration studies on fatty acid amide slip additives from plastics into food stimulants. Food Addit. Contam. 12 (2), 235–244.

Dorazio, S.J., Bruckner, C., 2015. Why is there cyanide in my table salt? Structural chemistry of the anticaking effect of yellow prussiate of soda ($Na4[Fe(CN)6]\bullet 10H_2O$). J. Chem. Educ. 92 (6), 1121–1124.

ECETOC, 1993. Technical Report No. 52, Styrene toxicology investigations on the potential for carcinogenicity. European center for ecotoxicology and toxicology of chemicals, Brussels, Belgium. Retrieved from http://www.ecetoc.org/wp-content/uploads/2014/08/ECETOC-TR-052.pdf.

EFSA, 2012. Guidance for submission for food additive evaluations. EFSA J. 10 (7), 2760.

Elespuru, R.K., Agarwal, R., Atrakchi, A.H., Bigger, C.A., Heflich, R.H., Jagannath, D.R., et al., 2009. Current and future application of genetic toxicity assays: the role and value of in vitro mammalian assays. Toxicol. Sci. 109 (2), 172–179.

Elshafei, A.M., 2017. Role of microorganisms in food contamination, processing and safety. J. Food Microbiol. 1 (1), 9–10.

Eylar, E., Baez, I., Navas, J., Mercado, C., 1996. Sustained levels of ascorbic acid are toxic and immunosuppressive for human T cells. P. R. Health Sci. J. 15 (1), 21–26.

FAO/WHO, 1982. Joint FAO/WHO Expert Committee on Food Additives, World Health Organization & Food and Agriculture Organization of the United Nations. Evaluation of Certain Food Additives and Contaminants: Twenty-Sixth Report of the Joint FAO/WHO Expert Committee on Food Additives [meeting held in Rome from 19 to 28 April 1982]. World Health Organization, Geneva. Available from: http://www.who.int/iris/handle/10665/41546.

FAO/WHO, 1983. Evaluation of Certain Food Additives and Contaminants: Twenty-Seventh Report of the Joint FAO/WHO Expert Committee on Food Additives [meeting held in Geneva from 11 to 20 April 1983]. World Health Organization, Geneva. Available from: http://www.who.int/iris/handle/10665/39165.

FAO/WHO, 1989. Evaluation of Certain Food Additives and Contaminants: Thirty-Third Report of the Joint FAO/WHO Expert Committee on Food Additives [meeting held in Geneva from 21 to 30 March 1988]. World Health Organization, Geneva. Available from: http://www.who.int/iris/handle/10665/39252.

FAO/WHO, 1991. Joint FAO/WHO Expert Committee on Food Additives, World Health Organization & Food and Agriculture Organization of the United Nations. (1991). Evaluation of Certain Food Additives and Contaminants: Thirty-Seventh Report of the Joint FAO/WHO Expert Committee on Food Additives [meeting held in Geneva from 5 to 14 June 1990]. World Health Organization, Geneva. Available from: http://www.who.int/iris/handle/10665/40288.

FAO/WHO, 1993. Joint FAO/WHO Expert Committee on Food Additives, World Health Organization, Food and Agriculture Organization of the United Nations & International Programme on Chemical Safety. (1993). Toxicological evaluation of certain food additives and contaminants / prepared by the forty-first meeting of the Joint FAO/WHO Expert Committee on Food Additives (JEFCA) [meeting held in Geneva, from 9 to 18 February 1993]. World Health Organization, Geneva. Available from: http://www.who.int/iris/handle/10665/41742.

FAO/WHO, 2004. Joint FAO/WHO Expert Committee on Food Additives (2003: Rome, Italy), World Health Organization & Food and Agriculture Organization of the United Nations. Evaluation of Certain Food Additives and Contaminants: Sixty-First Report of the Joint FAO/WHO Expert Committee on Food Additives. World Health Organization, Geneva. Available from: http://www.who.int/iris/handle/10665/42849.

FAO/WHO, 2006. Joint FAO/WHO Expert Committee on Food Additives. Meeting (65th: 2005, Geneva, Switzerland), World Health Organization & Food and Agriculture Organization of the United Nations. Evaluation of Certain Food Additives: Sixty-Fifth Report of the Joint FAO/WHO Expert Committee on Food Additives. World Health Organization, Geneva. Available from: http://www.who.int/iris/handle/10665/43408.

FAO/WHO, 2007. Joint FAO/WHO Expert Committee on Food Additives. Meeting (68th: 2007: Geneva, Switzerland), World Health Organization & Food and Agriculture Organization of the United Nations. Evaluation of Certain Food Additives and Contaminants: Sixty-Eighth Report of the Joint FAO/WHO Expert Committee on Food Additives. World Health Organization, Geneva. Available from: http://www.who.int/iris/handle/10665/43870.

FAO/WHO, 2011. World Health Organization, Food and Agriculture Organization of the United Nations & Joint FAO/WHO Expert Committee on Food Additives. Meeting (74th: 2011: Rome, Italy). Evaluation of certain food additives and contaminants: seventy-fourth [74th] report of the Joint FAO/WHO Expert Committee on Food Additives. Geneva: World Health Organization. http://www.who.int/iris/handle/10665/44788.

FAO/WHO, 2013. World Health Organization, Food and Agriculture Organization of the United Nations & Joint FAO/WHO Expert Committee on Food Additives. Meeting (77th: 2013, Rome, Italy). Evaluation of certain food additives and contaminants: seventy-seventh report of the Joint FAO/WHO Expert Committee on Food Additives. World Health Organization. http://www.who.int/iris/handle/10665/98388.

FAO/WHO, 2015. World Health Organization & Joint FAO/WHO Expert Committee on Food Additives (79th: 2014: Geneva, Switzerland). Evaluation of Certain Food Additives: Seventy-Ninth Report of the Joint FAO/WHO Expert Committee on Food Additives. World Health Organization. Available from: http://www.who.int/iris/handle/10665/150883.

FAO/WHO, 2016. World Health Organization, Food and Agriculture Organization of the United Nations & Joint FAO/WHO Expert Committee on Food Additives. Meeting (80th: 2015, Rome, Italy). Evaluation of certain food additives and contaminants: eightieth report of the Joint FAO/WHO Expert Committee on Food Additives. World Health Organization. http://www.who.int/iris/handle/10665/204410.

FAO/WHO, 2017. Safety evaluation of certain food additives: prepared by the eighty-second meeting of the Joint FAO/WHO Expert Committee on Food Additives (JECFA). World Health Organization. http://www.who.int/iris/handle/10665/258934.

Gibb, S., 2008. Toxicity testing in the 21st century: a vision and a strategy. Reproduct. Toxicol. 25 (1), 136–138.

Gulson, B., McCall, M., Korsch, M., Gomez, L., Casey, P., Oytam, Y., et al., 2010. Small amounts of zinc from zinc oxide particles in sunscreens applied outdoors are absorbed through human skin. Toxicol. Sci. 118 (1), 140–149.

Harington, J.S., 1961. A study of the chemical composition and potential hazards of an antifoam substance used in intracardiac surgery. Thorax. 16 (2), 120–127.

Heck, J.D., Gaworski, C.L., Rajendran, N., Morrissey, R.L., 2002. Toxicologic evaluation of humectants added to cigarette tobacco: 13-week smoke inhalation study of glycerin and propylene glycol in Fischer 344 rats. Inhal. Toxicol. 14 (11), 1135–1152.

Hine, C.H., Kodama, J.K., Anderson, H.H., Simonson, D.W., Wellington, J.S., 1958. The toxicology of epoxy resins. AMA Arch. Ind. Health 17 (2), 129–144.

Inetianbor, J.E., Yakubu, J.M., Ezeonu, S.C., 2015. Effects of food additives and preservatives on man: a review. Asian J. Sci. Technol. 6 (2), 1118–1135.

Janna, H., Scrimshaw, M.D., Williams, R.J., Churchley, J., Sumpter, J.P., 2011. From dishwasher to tap?Xenobiotic substances benzotriazole and tolyltriazole in the environment. Environ. Sci. Technol. 45 (9), 3858–3864.

Jones-Lepp, T.L., Alvarez, D.A., Petty, J.D., Huckins, J.N., 2004. Polar organic chemical integrative sampling and liquid chromatography-electrospray/ion-trap mass spectrometry for assessing selected prescription and illicit drugs in treated sewage effluents. Arch. Environ. Contam. Toxicol. 47 (4), 427–439.

Kolpin, D.W., Furlong, E.T., Meyer, M.T., Thurman, E.M., Zaugg, S.D., Barber, L.B., et al., 2002. Pharmaceuticals, hormones, and other organic wastewater contaminants in U.S. streams, 1999-2000: a national reconnaissance. Environ. Sci. Technol. 36 (6), 1202–1211.

Krewski, D., Acosta Jr, D., Andersen, M., Anderson, H., Bailar 3rd, J.C., Boekelheide, K., et al., 2010. Toxicity testing in the 21st century: a vision and a strategy. J. Toxicol. Environ. Health B. Crit. Rev. 13 (2-4), 51–138.

Larsen, E.H., Petersen, A., Hilbert, G., Andersen, N.L., Larsen J.C. (Eds.), 2001. Chemical contaminants: Food monitoring, 1993-1997. Part 2, The Danish Veterinary and Food Administration, Ministry of Food, Agriculture and Fisheries, Morkhoj Bygade 19, DK-2860 Soborg.

Lau, O.W., Wong, S.K., 2000. Contamination in food from packaging material. J. Chromatogr. A. 882 (1–2), 255–270.

Lefaux, R., Technica, S., 1968. Practical Toxicology of Plastics. Cleveland, CRC Press, Iliffe, London.

Loos, R., Gawlik, B.M., Boettcher, K., Locoro, G., Contini, S., Bidoglio, G., 2009. Sucralose screening in European surface waters using a solid-phase extraction-liquid chromatography-triple quadrupole mass spectrometry method. J. Chromatogr. A 1216 (7), 1126–1131.

Mendes, B.M., Bispo, A.C.A., Campos, D.C., Silva, J.B., 2013. Preclinical Test-Bacterial Reverse Mutation Test for 18F-Fluorocholine produced in CDTN. International Nuclear Atlantic Conference – INAC, Recife, PE, Brazil.

Mepham, B., 2011. Food additives: an ethical evaluation. Br. Med. Bull. 99 (1), 7–23.

Mohr, T.K.G., Stickney, J.A., DiGuiseppi, W.H. (Eds.), 2010. Environmental Investigation and Remediation: 1, 4-Dioxane and Other Solvent Stabilizers. CRC Press, Francis & Taylor group, Boca Raton, FL.

Moran, E.J., Easterday, O.D., Oser, B.L., 1980. Acute oral toxicity of selected flavor chemicals. Drug Chem. Toxicol. 3 (3), 249–258.

Musoke, L., Banadda, N., Sempala, C., Kigozi, J., 2015. The migration of chemical contaminants from polyethylene bags into food during cooking. Open Food Sci. J. 9, 14–18.

NSF, 2009. Science Nation: Testing the waters—1, 4-dioxane in North Carolina's Cape Fear River Basin. www.nsf.gov/news/special_reports/science_nation/capefearwatershed.jsp.

NTP, 1982. Technical Report Series 217. Carcinogenesis bioassay of Di(2-ethylhexyl)phthalate in rats and mice, NIH Publication No. 82-1773.

Neacsu, N.A., Madar, A., 2014. Artificial sweeteners versus natural sweeteners. Bull. Transilvania Univ. Brasov 7 (56), 59–64 (1).

OECD, 1997. Test No. 471: Bacterial Reverse Mutation Test, OECD Guidelines for the Testing of Chemicals, Section 4. OECD Publishing, Paris. Available from: http://dx.doi.org/10.1787/9789264071247-en.

OECD, 1998. Test No. 409: Repeated Dose 90-Day Oral Toxicity Study in Non-rodents, OECD Guidelines for the Testing of Chemicals, Section 4. OECD Publishing, Paris. Available from: http://dx.doi.org/10.1787/9789264070721-en.

OECD, 2010. Proposal for Updating Guideline 473: In Vitro Mammalian Chromosome Aberration Test. Organisation for Economic Co-Operation and Development Publishing. https://www.oecd.org/chemicalsafety/testing/50108781.pdf

OECD, 2013. Test No. 488: Transgenic Rodent Somatic and Germ Cell Gene Mutation Assays, OECD Guidelines for the Testing of Chemicals, Section 4. OECD Publishing, Paris. Available from: http://dx.doi.org/10.1787/9789264203907-en.

OECD, 2015. Test No. 478: Rodent Dominant Lethal Test. OECD Publishing, Paris. Available from: http://dx.doi.org/10.1787/9789264243118-en.

OECD, 2016a. Test No. 476: In Vitro Mammalian Cell Gene Mutation Tests Using the *Hprt* and *xprt* Genes, OECD Guidelines for the Testing of Chemicals, Section 4. OECD Publishing, Paris. Available from: http://dx.doi.org/10.1787/9789264264809-en.

OECD, 2016b. Test No. 475: Mammalian Bone Marrow Chromosomal Aberration Test. OECD Publishing, Paris. Available from: http://dx.doi.org/10.1787/9789264224407-en.

OECD, 2016c. Test No. 474: Mammalian Erythrocyte Micronucleus Test. OECD Publishing, Paris. Available from: http://dx.doi.org/10.1787/9789264224292-en.

OECD, 2016d. Test No. 487: In Vitro Mammalian Cell Micronucleus Test. OECD Guidelines for the Testing of Chemicals, Section 4. OECD Publishing, Paris. Available from: http://dx.doi.org/10.1787/9789264264861-en.

OECD 2018a. Test No. 408: Repeated Dose 90-Day Oral Toxicity Study in Rodents, OECD Guidelines for the Testing of Chemicals, Section 4, OECD Publishing, Paris, https://doi.org/10.1787/9789264070707-en.

OECD, 2018b. OECD (2018), Test No. 453: Combined Chronic Toxicity/Carcinogenicity Studies. OECD Guidelines for the Testing of Chemicals, Section 4. OECD Publishing, Paris. Available from: http://dx.doi.org/10.1787/9789264071223-en.

Oaks, J.L., Gilbert, M., Virani, M.Z., Watson, R.T., Meteyer, C.U., et al., 2004. Diclofenac residues as the cause of vulture population decline in Pakistan. Nature 427, 630–633.

Olson, M.E., 1977. Aerosol propellants and safety. Lab. Med. 8 (10), 13–15.

Oluwafemi, F., Ibeh, I.N., 2011. Microbial contamination of seven major weaning foods in Nigeria. J. Health Popul. Nutr. 29 (4), 415–419.

Oskarssonm, A., 2012. Environmental contaminants and food safety. Acta. Vet. Scand. 54 (Suppl. 1), S5.

Pagnout, C., Jomini, S., Bauda, P., 2014. Revised procedure of the bacterial reverse mutation test for genotoxic evaluation of nanoparticles. In: Sierra, L., Gaivao, I. (Eds.), Genotoxicity and DNA Repair. Methods in Pharmacology and Toxicology. Humana Press, New York, NY.

Pandey, R.M., Upadhyay, S.K., 2012. Food additive. ISBN: 978-953-51-0067-6, InTech, Available from: http://www.intechopen.com/books/food-additive/food-additive.

Parasuraman, S., 2011. Toxicological screening. J. Pharmacol. Pharmacother. 2 (2), 74–79.

Peshin, S.S., Lall, S.B., Gupta, S.K., 2002. Potential food contaminants and associated health risks. Acta. Pharmacol. Sin. 23 (3), 193–202.

Pressman, P., Clemens, R., Hayes, W., Reddy, C., 2017. Food additive safety: a review of toxicologic and regulatory issues. Toxicol. Res. Applicat. 1, 1–22.

Qurrat-Ul-Ain, Khan, S.A., 2015. Artificial sweeteners: safe or unsafe? J. Pak. Med. Assoc. 65 (2), 225–227.

Rather, I.A., Koh, W.Y., Paek, W.K., Lim, J., 2017. The sources of chemical contaminants in food and their health implications. Front. Pharmacol. 8, 830.

Rawat, S., 2015. Food spoilage: microorganisms and their prevention. Asian J. Plant Sci. Res. 5 (4), 47–56.

Rhomie, L., III, Heck, 1998. A review of commercially used chemical foaming agents for thermoplastic foams. J. Vinyl Additive Technol. 4 (2), 113–116.

Richardson, S.D., Kimura, S.Y., 2017. Emerging environmental contaminants: challenges facing our next generation and potential engineering solutions. Environ. Technol. Innovation 8, 40–56.

Richardson, S.D., Ternes, T.A., 2018. Water analysis: emerging contaminants and current issues. Anal. Chem. 90 (1), 398–428.

Rodero, A.B., Rodero, L.D.S., Azoubel, R., 2009. Toxicity of sucralose in humans: a review. Int. J. Morphol. 27 (1), 239–244.

Rovida, C., Asakura, S., Daneshian, M., Hofman-Huether, H., Leist, M., Meunier, L., et al., 2015. Toxicity testing in the 21st century beyond environmental chemicals. Altex. 32 (3), 171–181.

Rowlands, J.C., Hoadley, J.E., 2006. FDA perspectives on health claims for food labels. Toxicology 221 (1), 35–43.

SCF, 2001. Guidance on submissions for food additive evaluations by the scientific committee on food. 12 July 2001. http://ec.europa.eu/food/fs/sc/scf/out98_en.pdf

Saha, D., Bhattacharya, S., 2010. Hydrocolloids as thickening and gelling agents in food: a critical review. J. Food Sci. Technol. 47 (6), 587–597.

Sales, I.M.S., Silva, J.M., Moura, E.S.R., Alves, F.D.S., Silva, F.C.C., Sousa, J.M.C., et al., 2018. Toxicity of synthetic flavorings, nature identical and artificial, to hematopoietic tissue cells of rodents. Braz. J. Biol. 78 (2), 306–310.

Sharma, S., 2015. Food Preservatives and their harmful effects. Int. J. Sci. Res. Publicat. 5 (4), 5–6.

Shibata, M., Tamano, S., Shirai, T., Kawabe, M., Fukushima, S., 1992. Inorganic alkalizers and acidifiers under conditions of high urinary $Na+$ or $K+$ on cell proliferation and two-stage carcinogenesis in the rat bladder. Jpn. J. Cancer Res. 83 (8), 821–829.

Skutlarek, D., Exner, M., Farber, J., 2006. Perfluorinated surfactants in surface and drinking waters. Environ. Sci. Pollut. Res. Int. 13, 299–307.

Smith, R.L., Adams, T.B., Cohen, S.M., Doull, J., Feron, V.J., Goodman, J.I., et al., 2004. Safety evaluation of natural flavour complexes. Toxicol. Lett. 149 (1–3), 197–207.

Sotonyi, P., Keller, E., Jaray, J., Nemes, B., Benko, T., Kovacs, A., et al., 2001. A light stabilizer Tinuvin-770 induced toxic injury of adult rat cardiac myocytes. Forensic Sci. Int. 119 (3), 322–327.

Stepek, J., Duchacek, V., Curda, D., Horacek, J., Sipek, M., 1987. Polymers as Materials for Packaging. Ellis Horwood, Chichester, p. 461.

Taylor IV, G.J., Harris, W.S., 1970. Cardiac toxicity of aerosol propellants. JAMA 214 (1), 81–85.

USDHHS, 1982. Environmental health perspectives. US Department of Health and Human Service Vol. 45, Phthalate Esters, November 1982.

USEPA, 2012. U.S. Environmental Protection Agency, The unregulated contaminant monitoring rule-3 (UCMR-3). Available at http://water.epa.gov/lawsregs/rulesregs/sdwa/ucmr/ucmr3/index.cfm.

USEPA, 2018. U.S. Environmental Protection Agency, Contaminant candidate list 3 - CCL-3. Available at http://www2.epa.gov/ccl/contaminant-candidate-list-3-ccl-3.

USFDA, 1993. Red Book II: Chapter VI Human studies. Retrieved from https://www.fda.gov/downloads/Food/GuidanceRegulation/UCM078753.pdf.

USFDA, 2000a. Redbook 2000: IV. C. 9. Guidelines for reproduction studies. Retrieved from https://www.fda.gov/Food/GuidanceRegulation/GuidanceDocumentsRegulatoryInformation/IngredientsAdditivesGRASPackaging/ucm078396.htm.

USFDA, 2000b. Redbook 2000: IV.C.9.b Guidelines for developmental toxicity studies. https://www.fda.gov/Food/GuidanceRegulation/GuidanceDocumentsRegulatoryInformation/IngredientsAdditivesGRASPackaging/ucm078399.htm.

USFDA, 2003a. Redbook 2000: IV. C. 5. A chronic toxicity studies with rodents. Retrieved from https://www.fda.gov/Food/GuidanceRegulation/GuidanceDocumentsRegulatoryInformation/IngredientsAdditivesGRASPackaging/ucm078339.htm#exp.

USFDA, 2003b. Redbook 2000: IV. C. 4. A subchronic toxicity studies with rodents toxicological principles for the safety assessment of food. Retrieved from https://www.fda.gov/Food/GuidanceRegulation/GuidanceDocumentsRegulatoryInformation/IngredientsAdditivesGRASPackaging/ucm078345.htm.

USFDA, 2003c. Redbook 2000: IV.C.5.b One-year toxicity studies with non-rodents. https://www.fda.gov/Food/GuidanceRegulation/GuidanceDocumentsRegulatoryInformation/IngredientsAdditivesGRASPackaging/ucm078348.htm.

USFDA, 2007. Redbook 2000: Metabolism and pharmacokinetic studies. Retrieved from https://www.fda.gov/downloads/Food/GuidanceRegulation/UCM078741.pdf.

USFDA, 2017. Redbook 2000: IV. C. 8 *In-utero* exposure phase for addition to carcinogenicity studies or chronic toxicity studies with rodents. Retrieved from https://www.fda.gov/Food/GuidanceRegulation/GuidanceDocumentsRegulatoryInformation/IngredientsAdditivesGRASPackaging/ucm078394.htm.

Vardar-Sukan, F., 1998. Foaming: Consequences, prevention and destruction. Biotechnol. Adv. 16 (5–6), 913–948.

Vikesland, P.J., Fiss, E.M., Wigginton, K.R., McNeill, K., Arnold, W.A., 2013. Halogenation of bisphenol-A, triclosan, and phenols in chlorinated waters containing iodide. Environ. Sci. Technol. 47, 6764–6772.

Vliet, Erwin van, 2011. Current standing and future prospects for the technologies proposed to transform toxicity testing in the 21st century. ALTEX : AlternativenZuTierexperimenten 28 (1), 17–44.

WHO, 1974. IARC Report of working group on vinyl chloride, Int. Techn. Report No.74/005, IARC, Lyon.

WHO, 1975. IARC Report of working group on epidemiological studies on vinyl chloride, Int. Techn. Report No.75/001, IARC, Lyon.

WHO/FAO, 2011. World Health Organization, Food and Agriculture Organization of the United Nations & Joint FAO/WHO Expert Committee on Food Additives. Meeting (73rd: 2010, Geneva, Switzerland). Evaluation of certain food additives and contaminants: seventy-third [73rd] report of the Joint FAO/WHO Expert Committee on Food Additives. Geneva: World Health Organization. http://www.who.int/iris/handle/10665/44515.

Ware, B.R., Berger, D.R., Khetani, S.R., 2015. Prediction of drug-induced liver injury in micropatterned co-cultures containing iPSC-derived human hepatocytes. Toxicol Sci. 145 (2), 252–262.

Williams, P., 2012. Food toxicity and safety. In: Mann, J., Truswell, A. (Eds.), Essentials of Human Nutrition. Oxford University Press, Oxford, pp. 449–466.

Wroblewska, B., 2009. Influence of food additives and contaminants (nickel and chromium) on hypersensitivity and other adverse health reactions—a review. Polish J. Food Nutr. Sci. 59 (4), 287–294.

Xiaolong, W., Zhisong, C., Jingli, M.U., Fei, J., Wei, G., Xiao, L., et al., 2015. Toxicity evaluation of 10 oil gelling agents. Chinese J. Appl. Environ. Biol. 5, 798–804.

Zwiener, C., Richardson, S.D., DeMarini, D.M., Grummt, T., Glauner, T., Frimmel, F.H., 2007. Drowning in disinfection by-products? Swimming pool water quality reconsidered. Environ. Sci. Technol. 41, 363–372.

CHAPTER

4

Food Allergies

Rasna Gupta[1], Ankit Gupta[2], Rajat Pratap Singh[3], Pradeep Kumar Singh[1] and Ram Lakhan Singh[1]

[1]Department of Biochemistry, Dr. Rammanohar Lohia Avadh University, Faizabad, India

[2]Department of Biochemistry, All India Institute of Medical Sciences (AIIMS), Raibareli, India

[3]Department of Biotechnology, Dr. Rammanohar Lohia Avadh University, Faizabad, India

OUTLINE

Food Safety and Human Health
DOI: https://doi.org/10.1016/B978-0-12-816333-7.00004-7

INTRODUCTION

Eating is compulsory for each living creature to sustain the life. But for humans, eating is not only a necessity for life but also an enjoyable experience given by the variety and abundance of food available in the marketplace. Various food industries supplying a range of food products have a great impact on human population, who consume their products based on the quality and processing of food items. Therefore food allergies began with the consumption of diverse foods. Centuries ago, a renowned Roman philosopher, Lucretius, said that food of one person could be poisonous for another. In the history of food allergies, the first case of a food-allergic patient was recorded in the 20th century. Earlier, the medical community and concerned authorities ignored many of these food allergies, but now allergies are well recognized and widely acknowledged in various continents, particularly in North America, Europe, Japan, and Australia.

Limited understanding is available about the prevalence of food allergies and their impact on humans living in various parts of world. However, some of individuals are more prone to food allergies. Sometimes these allergies result in more serious and deadly consequences to the affected individuals; therefore, food allergies need more investigation by authorities.

Current estimates about food allergies suggest approximately 25% of the US population (up to 15 million people) has been affected with food allergy. However, food allergy is prevalent in 1%–3% of adults and 3%–8% of infants in Western societies (Nwaru et al., 2014), whereas, 26.5% of the Indian population is susceptible to a high level of food sensitization, higher than European populations (Mahesh et al., 2016). An abnormal response to a particular food is known as food sensitivity. These adverse responses to foods are classified into two categories: toxic and nontoxic (Bruijnzeel-Koomen et al., 1995). The nontoxic reactions comprise both immune- and nonimmune-mediated pathogenic mechanisms (Fig. 4.1). Nonimmune mediated mechanisms, also known as food intolerance, involve pharmacological, enzymatic, and some unclear reasons, including irritants and psychosomatic responses. Whereas the immune-mediated mechanisms (food allergy) involve immunological responses, those are further classified into two groups: IgE-mediated responses, also known as type I hypersensitivity, and non-IgE-mediated responses, which are further classified into type III hypersensitivity (IgG or IgM-mediated immune responses) or type IV hypersensitivity (delayed or cell-mediated immune responses). However, some hereditary marks are also found to be involved in conferring type I hypersensitivity responses through production of excessive levels of IgE antibodies, also known as atopic syndrome.

Food-Induced Allergic Reactions

Food or food additive–induced allergic reactions are either immediate (IgE-mediated) type allergic reactions, as the symptoms manifest within minutes to a few hours, or delayed type, in which symptoms appear within a few hours to days after exposure.

Immediate Food Allergy

These types of reactions frequently arise in young children below the age of 5 years and are more commonly seen in children associated with certain atopic disorders like eczema, dermatitis, allergic rhinitis (hay fever), or allergic asthma. Milk, peanuts, and eggs are the foods that commonly cause allergic reactions in young children, whereas peanuts, seafood, and tree nuts cause allergic reactions in adult individuals.

The consumption of alcohol, exercising, and the usage of nonsteroidal antiinflammatory drugs are also considered as allergic factors in older patients. When certain individuals ingest specific foods in combination with these factors, they may experience allergic reactions. About 30% of cases of anaphylaxis may arise with sensitization to specific foods after exercising. Individuals with certain clinical complications are more prone to sensitization and allergic reactions to food.

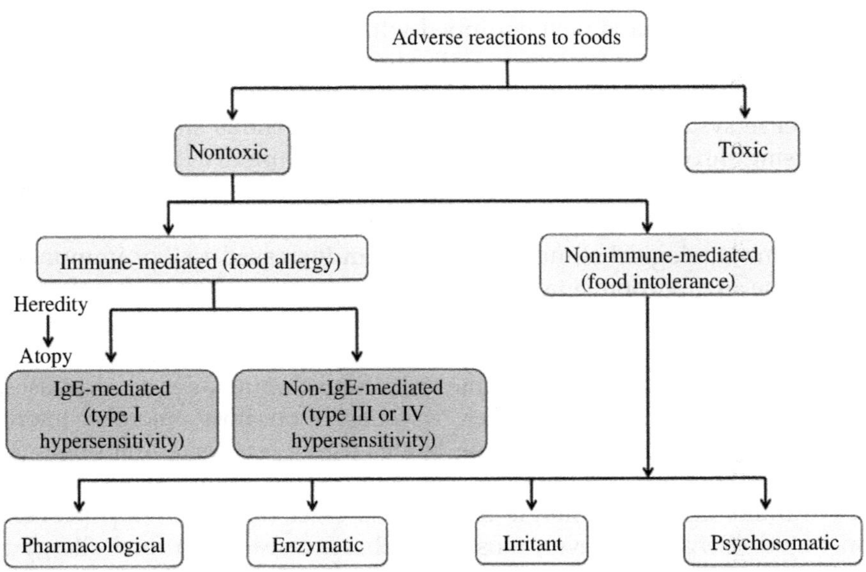

FIGURE 4.1 Classification of adverse reactions to certain foods is the basis of pathogenic mechanisms.

Delayed-Onset Food Allergy

Delayed-onset food allergies start after several hours or days, most commonly in response to dairy products, soy, or wheat. Symptoms occur due to inflammation of the skin or gastrointestinal (GI) tract and result from attraction of white cells from the blood into the tissues. The immunoglobulin G (IgG) mediated type III allergy is an example of auto immune diseases that induce the immune system inappropriately on exposure to certain foods. The digestive tract is densely packed with immune cells as compared to other parts of body (Guinane and Cotter, 2013). These immune cells recognize food particles as antigens and produce IgE antibody, histamine, and other bioactive chemicals. Cow's milk, fish, gluten, soy, yeast, corn, peanuts, and nuts such as almonds, hazelnuts, cashews, Brazil nuts, and egg whites and egg yolks are some common delayed-onset food allergens. The type III food allergies are reversible if patients have awareness to facilitate their body's healing by temporarily avoiding the injurious food, consuming raw and boiled food, and taking enzyme supplements.

Food Allergy and Associated Complexity

Infants and children are more prone to true food allergy than adults and manifest symptoms in the skin (i.e., atopic eczema or urticaria), in mucosa (i.e., angioedema), in the respiratory tract (i.e., laryngedema or bronchial obstruction), and in the digestive tract starting from the mouth [i.e., oral allergy syndrome (OAS)] to the anus (i.e., proctitis or perianal eczema) (Sicherer and Sampson, 2009a). Food allergy displays unspecific symptoms (e.g., vomiting, nausea, diarrhea, or constipation) (Zopf et al., 2009). Most of the allergic reactions to certain foods (e.g., eating apples, peaches, melons) are associated with pollen allergy due to cross-contamination. Some individuals with pollen allergy also suffer from hypersensitivity and anaphylaxis due to eating mangos (Silva et al., 2009) (Table 4.1).

Our intestinal gut mucosa is well equipped with specialized antiinflammatory immune defense machinery, including secretary antibodies (IgA) and hyporesponsiveness, which display immunological responses to several dietary antigens and gut microbiota (Turner, 2009). This defense system is stimulated by various exogenous stimuli, including a range of proteins passing through the intestinal barrier and adapted with reinforcement by IgA and immunoregulatory network. The conventional theory suggests that prevalence of food allergy has risen with extended hygienic living style. In Western societies, the infants are not exposed to immunological stimuli, leading to an increased risk of immunological disorders in response to environmental changes (Guarner et al., 2006). The altered hygienic, dietary, and medical practices have changed microbial exposure (e.g., gut microbiota in humans) resulting in allergic prevalence. An environment rich in microbes in the early life stages such as during childhood reduces the risk of developing allergic responses.

Therefore most of the recent research is focused on host–microbe interactions to strengthen the immune tolerance in human beings through reverse immunological hypersensitivity (von Hertzen et al., 2009). The complete understanding of exogenous variables involved in immune-mediated allergic response is essential to develop preventive measures against food allergy. However, most of the basic research on food allergy is concentrated on animal studies (e.g., mice), but the recent advances in its treatment are being supported by clinical trials and studies on human samples.

TABLE 4.1 Food-Induced Allergic Disorders (Sicherer and Sampson, 2009b)

Immunopathology	Disorder	Key Features	Additional Immunopathology	Typical Age	Most Common Causal Foods	Natural Course
IgE antibody dependent (acute onset)	Urticaria/ angioedema	Triggered by ingestion or direct skin contact (contact urticaria); food commonly causes acute (20%) but rarely chronic (2%) urticaria	–	Children > adults	Primarily major allergens	Depending on food
	Oral allergy syndrome (pollen–food related)	Pruritus, mild edema confined to oral cavity, uncommonly progresses beyond mouth (7%) or anaphylaxis (1%–2%); might increase after pollen season	Sensitization to pollen proteins by the respiratory route results in IgE that binds certain homologous, typically labile food proteins (in certain fruits/vegetables)	Onset after pollen allergy established (adult > young child)	Raw fruit/vegetables, Cooked forms tolerated Examples of relationships: birch (apple, peach, pear, carrot), ragweed (melons)	Might be long-lived and vary with seasons
	Rhinitis, asthma	Symptoms might accompany a food-induced allergic reaction but rarely an isolated or chronic symptom	–	Infant/child > adult, except for occupational disease (e.g., baker's asthma)	General: major allergens.	Depending on food
		Symptoms might also be triggered by inhalation of aerosolized food protein			Occupational: wheat, egg, and seafood, for example	
	Anaphylaxis	Rapidly progressive, multiple organ system reaction can include cardiovascular collapse	Massive release of mediators, such as histamine, although mast cell tryptase levels not always increased	Any	Any but more commonly peanut, tree nuts, shellfish, fish, milk, and egg	Depending on food
			Key role of platelet-activating factor			
	Food-associated, exercise-induced anaphylaxis	Food triggers anaphylaxis only if ingestion followed temporally by exercise	Exercise is presumed to alter gut absorption, allergen digestion, or both	Onset more commonly later childhood/adult	Wheat, shellfish, and celery are most described	Presumed persistent

(Continued)

TABLE 4.1 (Continued)

Immunopathology	Disorder	Key Features	Additional Immunopathology	Typical Age	Most Common Causal Foods	Natural Course
IgE antibody associated/cell mediated (delayed onset/chronic)	Atopic dermatitis	Associated with food in 35% of children with moderate-to-severe rash	Might relate to homing of food-responsive T cells to the skin	Infant > child > adult	Major allergens, particularly egg and milk	Typically resolves
	Eosinophilic gastroenteropathies	Symptoms vary on site(s)/degree of eosinophilic inflammation Esophageal: dysphagia and pain Generalized: ascites, weight loss, edema, and obstruction	Mediators that home and activate eosinophils play a role, such as eotaxin and IL-5	Any	Multiple	Likely Persistent
Cell-mediated (delayed onset/chronic)	Dietary protein enterocolitis	Primarily affects infants. Chronic exposure: emesis, diarrhea, poor growth, and lethargy. Reexposure after restriction: emesis, diarrhea, and hypotension (15%) 2 hours after ingestion	Increased TNF-α response, decreased response to TGF-β	Infancy	Cow's milk, soy, rice, and oat	Usually resolves
	Dietary protein proctitis	Mucus-laden, bloody stools in infants	Eosinophilic inflammation	Infancy	Milk (through breast-feeding)	Usually Resolves

Emerging Problems and Recent Recommendations Associated With Food Allergy

Most food allergies are likely to be associated with a small category of food (e.g., cow's milk, eggs, fishes, crustaceans, peanuts, soybeans, nuts, and wheat). More than 170 foods have been listed as causing food allergies. Most packaged foods have a clear statement reflecting the key ingredients of the dish in order to warn the allergic individuals before consumption. Since 1993, the Codex Committee recommended several food allergens to be shown on food labels. The Codex Alimentarius Commission (CAC) has declared a list of foods and associated ingredients causing allergic responses and ordered that they be displayed on food labels. This list includes several foods if they are present in a concentration of 10 mg/kg or more, such as gluten-containing cereals (e.g., wheat, barley, oats, and rye), crustaceans (e.g., crab), eggs and related products, fish and concerned products, peanuts and soybeans, tree nuts, and milk and its products containing lactose. Several other related issues with food allergy have also emerged, including precautionary labels and concerned allergic responses to the commercially available genetically modified foods.

The World Health Organization (WHO), the World Trade Organization (WTO), and governments must emphasize the prevalence of food allergies and concerned intolerance, affecting the health of significant proportion of the consuming population. Therefore government regulatory authorities must ensure that food labels display sufficient information for allergic individuals before consumption. The government must emphasize the most common allergic foods suggested by CAC and strictly ensure the labeling in those packaged food products. Highly refined peanut and soybean oils must be excluded from the labeling, since these products do not have sufficient protein content that confers allergic reactions.

ALLERGENS AND THEIR TYPES

Food allergens are basically naturally occurring proteins found in various food products that are capable of inducing abnormal immune responses. Milk and eggs are the most common food allergens for young children; however, most outgrow such allergies once they reach 5–7 years of age. In contrast, food allergies derived from seafood, peanuts, and tree nuts develop later, and they persist throughout their lives. Generally, if a person is sensitized to proteins present in the food, he/she develops allergic responses to mostly all the foods. Eight foods known to cause food allergies are peanuts, tree nuts, milk, eggs, wheat, soy, fish, and shellfish.

Peanuts

Peanuts are the most common food allergens that cause a severe fatal allergic reaction known as anaphylaxis. People prone to peanut allergy are advised to avoid peanuts and peanut-derived food products. These people should always read ingredient labels to ensure peanuts are not an ingredient and should always have easy and quick access to epinephrine auto-injectors (e.g., EpiPen, Auvi-Q, Adrenaclick). However, even a trace amount of peanuts is capable of inducing allergy, such as simply touching peanuts, and its derived products are less likely to elicit allergic responses than its contact with sensitive areas including eyes, mouth, nose, etc. Children are more prone to peanut allergy, and epidemiology study suggests that the number of children having peanut allergy significantly increased during the years between 1997 and 2008 in the United States. Similarly, other studies in the United Kingdom and Canada also suggest that peanut allergy is more prevalent in school children. The peanut allergy persists throughout life, and 20% of children have been found to outgrow this allergy eventually. Unlike tree nuts, peanuts grow underground and belong to a different plant family known as legumes. Other legumes including beans, peas, lentils, and soybeans do not induce allergic responses, mostly to a person already susceptible to peanut allergy. However, some recent studies suggest that 25%–40% of the people having peanut allergy also develop allergies to tree nuts, possibly due to contact among them during manufacturing and processing.

Tree Nuts

Tree nuts, including walnuts, almonds, cashew nuts, pistachio, hazelnuts, etc., are also a most common food allergen that causes severe and fatal allergic responses. The allergy to tree nuts may persist for lifetime, and recent reports suggest that approximately 9% of the susceptible children eventually outgrow their allergy. If a person has an allergy to one type of tree nut, they would have a higher possibility to be allergic to another type. Therefore these allergic persons are advised to avoid other nuts due to the likelihood of cross contact with tree nuts.

Milk

Cow's milk is one of the most common food allergens whose sensitivity varies from person to person. Mostly infants and children below 3 years of age (approximately 2.5%) are susceptible to milk allergy. The allergic responses to milk may range from mild (e.g., hives) to severe (e.g., anaphylaxis). Therefore sensitive individuals are advised to avoid cow's milk and derived dairy products by simply reading labels on the food products. Mostly infants develop allergy to milk during their first year of age. The children prone to milk allergy have higher levels of cow's milk antibodies in their blood, and the simple blood test detecting such antibodies confers whether the child is likely to outgrow a milk allergy or not.

There should be no confusion between milk allergy and lactose intolerance. A food allergy is an abnormal immune response to milk protein, whereas lactose intolerance does not involve any immune reaction. The person with lactose intolerance does not have the enzyme lactase that metabolizes a major sugar (e.g., lactose present in the milk); therefore a susceptible person is unable to digest lactose and experiences discomfort with symptoms such as nausea, gas, and diarrhea.

Egg

Egg allergy is also a common allergy in children. Persons with egg allergy may develop mild to severe symptoms; therefore they should avoid eggs and egg products.

The egg white contains protein that causes allergic responses. Some vaccines [e.g., MMR (measles–mumps–rubella) vaccine, influenza vaccine] contain egg protein and can be safely administered to susceptible patients.

Wheat

Wheat is one of the most important grains whose products are consumed worldwide. Wheat allergy is a most common allergy that occurs in children by the age of 3 and is outgrown by adulthood. The individuals with wheat allergy should avoid wheat products, but they may eat alternate grains such as barley, corn, oat, rye, rice, and tapioca. Wheat proteins are classified on the basis of their solubilities: albumins soluble in water; globulins soluble in salt solutions; gliadins soluble in ethanol; and glutenins soluble in urea, KOH, and detergents. These proteins may elicit immune reactions and therefore a person develops allergic responses against them. Some reports suggest that hypoallergenic wheat flour may be used to overcome wheat allergy because 50% of the patients taking this flour are hyposensitized and gain an ability to eat normal wheat products.

Soy

Babies and children (approximately 0.4%) are more prone to soy allergy, and allergic responses may range from mild to severe symptoms. The allergic responses against soy are developed early in childhood and often are outgrown by the age of 3 or mostly by the age of 10. Soybeans belong to the legume family bearing seed pods that split upon ripening. The persons susceptible to soy allergy are not necessarily allergic to other legumes, but they should avoid soybeans and derived products.

Fish

Fish allergy usually persists throughout life, and 40% of the affected individuals have experienced the first allergic reactions during their adulthood. Finned fishes (e.g., salmon, tuna, and halibut) cause severe allergic reactions (e.g., anaphylaxis). More than 50% of the individuals having allergy to one type of fish may also develop allergic responses to other types of fish. Therefore susceptible persons should completely avoid fishes and derived products.

Shellfish

Shellfish are categorized into two types: crustacean (e.g., shrimp, crab, and lobster) and mollusks (e.g., mussels, oysters, scallops, and clams). Crustacean shellfish mostly cause severe allergic responses (e.g., anaphylaxis). A person having an allergy for one group of shellfish may eat varieties belonging to other groups; however, some persons exhibit allergic responses to all types of shellfish. Approximately 60% of the susceptible individuals experience first-time allergic responses to shellfish during their adulthood.

Sesame

Sesame allergy is prevalent in hundreds of thousands of individuals living in the United States as declared by a 2010 survey. However, the exact reasons behind its prevalence are still unknown. The prevalence of sesame allergy has increased significantly in the worldwide populations over the past two decades, as evident by several reports. Sesame is present in many food products such as sesame oil, flavor or spice blends, baked items, bread crumbs, and ethnic foods. Some nonfood items such as cosmetics, medications, and nutritional supplements also contain sesame as an ingredient and may also cause allergy.

Other Allergens

Approximately 90% of food allergies belong to the foods described above, but a person may develop allergic responses virtually to any food. Among other food allergens, corn, gelatin, meat (beef, chicken, pork, etc.), seeds (sunflower, poppy, etc.), spices (garlic, mustard, coriander, etc.) may also cause food allergy. Allergic responses to fresh fruits and vegetables (apple, carrot, peach, tomato, banana, etc.) are named and diagnosed as oral allergy syndrome (OAS).

IMMUNOLOGY OF FOOD ALLERGY

During the course of evolution, human immunogenetics has evolved in response to several elements (e.g., various allergens present in dirty environments) and innate and adaptive immunity have been found regulated by several genes. This evolutionary process of the mucosal immune system has adapted two antiinflammatory responses: immune exclusion by IgA to regulate microbial colonization on the epithelium layer and thereby preventing entry of the potent harmful agents; and suppression of the local and peripheral hypersensitivity against contiguous antigens. These strategies of immune suppression are consistent with the notion why apparent and persistent food allergy is comparatively rare. While considering such immune mechanisms, here we will focus on three important processes: oral tolerance, nonresponsive tolerance, and food sensitization and anaphylactic reactivity to concerned food allergens.

Oral Tolerance

More than 100 years ago, the first case of oral tolerance was reported for egg proteins (Gupta et al., 2011). The oral tolerance is an induced tolerance of mucosa in response to antigens present in the gut (Brandtzaeg, 1996). The identical mechanisms to downregulate the immune responses to commensal microbiota also exist in humans (Duchmann et al., 1997). Reports suggest the clear differences in these mechanisms between human and laboratory mice since gut microbiota are more strictly protected from systemic immune responses in humans (Macpherson et al., 2005). Therefore oral tolerance is a robust adaptive immune function, since a considerable amount of food proteins (130–190 g) are absorbed by gut mucosa after food intake.

Several reports describe the various possible routes of antigen sampling and its presentation (Pabst and Mowat, 2012) [e.g., sampling by dendritic cells (DCs)] across epithelium, antigen presentation by M or goblet cells to DCs, or direct entry of soluble antigens through paracellular or transcellular routes. Basically, resident regulatory T cells ($Foxp3^+$Treg) play a key role in oral tolerance (Fig. 4.2). The intestinal lamina propria in mucosa is a residence of various immune effector cells. The two tolerance-associated diverse subsets of $CD11C^+$ DCs are found in lamina propria and express either CX_3CR1 or CD103. CX3CR1 has been found to play a key role in oral tolerance to food allergy through IL-10 production and expanding T regulatory cells by a study on CX3CR1 knockout mice (Hadis et al., 2011). However, $CX_3CR1^+CD103^-$ cells are found to be involved in intestinal inflammation by eliciting prime effector T cells and in the development of effective oral vaccines (Cerovic et al., 2013). Similarly, $CD11c^-CDb^+F4/80^+$ macrophages also produce IL-10 and carry an antiinflammatory gene signature (Denning et al., 2007).

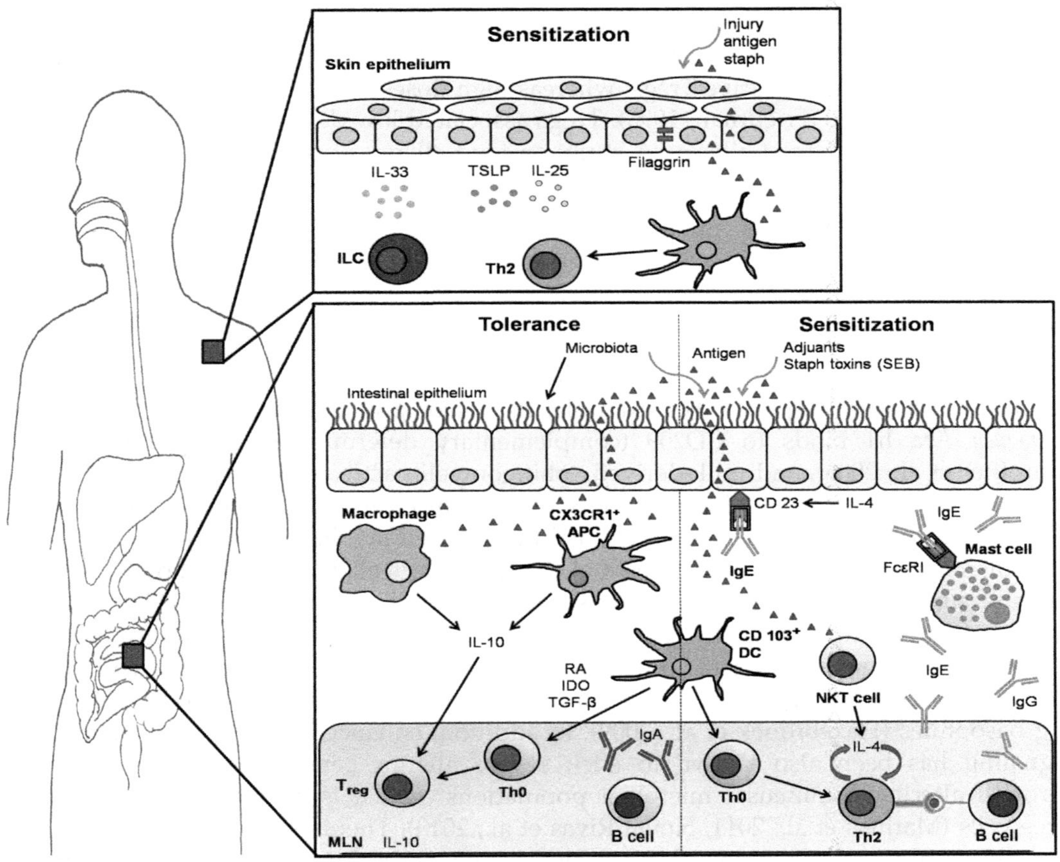

FIGURE 4.2 Immune mechanisms depicting sensitization and tolerance to various allergens in the skin and intestinal mucosal epithelium.

There are also a number of evidences supporting the function of $CX_3CR1^-CD103^+$ DCs in oral tolerance. These cells inhabit lymph nodes and activate naïve T cells. They also induce $FoxP3^+$Treg cells through retinoic acid and TGF-β-dependent mechanism (Coombes et al., 2007). Both Treg cells and IgA-secreting B cells contribute to oral tolerance by displaying key gut-homing receptors CCR9 and α4β7 (Mora et al., 2006). Indoleamine 2,3-dioxygenase (IDO) is an enzyme responsible for degradation of the essential amino acid tryptophan and created a new paradigm in immunology (Mellor and Munn, 2004). The DCs expressing IDO have been found to promote tolerance through suppression of T-cell response (Matteoli et al., 2010). A recent finding implicates that intestinal goblet cells secrete a mucin known as MUC2 that enhances oral tolerance by supporting $CD103^+$ DCs delivering immunoregulatory signals (Shan et al., 2013). The role of Tregs in food allergy has been addressed by several reports. The patients with IPEX (immune dysregulation, polyendocrinopathy, enteropathy, X-linked) syndrome have a mutation in FOXP3 locus and develop severe food allergy (Bennett et al., 2001) and several other disorders such as autoimmunity, enteropathy, and atopic dermatitis (AD), suggesting the role of Tregs in intolerance. The importance of Tregs in food allergy has been also displayed through diphtheria toxin treatment of Foxp3 mutant and DEREG (DEpletion of REGulatory T cells) mice (Kim et al., 2007). The antigen (Ag) concentration also modulates oral tolerance because low doses of Ag drive Tregs, whereas high doses lead to deletion of T cells or loss of immune response (Friedman, 1996). Tregs associated low-dose response has been found critical in displaying food allergy. A study on food allergy of peanuts showed loss of oral tolerance due to low Treg responses, whereas high-dose feeding helps in conquering allergic responses.

Food Sensitization

Some food antigens promote innate immunity, whereas some elicit a type 2 cytokine associated immune response. For example, the major glycoprotein allergen from *Arachis hypogaea* Ara h1 binds to CD209 (complementary determining 209) present on DCs (Shreffler et al., 2006), and milk-derived sphingomyelin activates invariant natural killer T cells to trigger type 2 cytokine responses (Jyonouchi et al., 2011). Food sensitization has been found to be associated with microbial flora since microbial population (e.g., bacteria and their by-products) sustain intestinal T regulatory cell (Treg cells) populations and thereby suppress food-derived allergic sensitivity (Berin and Sampson, 2013). There are some evidences supporting the role of commensal bacteria in food sensitivity (e.g., mice having lower commensal bacterial colonies) including germ-free and treated with antibiotics, display enhanced food sensitivity through increased levels of serum IgE, and circulating basophiles (Hazebrouck et al., 2009). In addition, enhanced IL-4Rα receptor mediated signaling has been also known to elicit severe allergic sensitization and anaphylaxis through altering commensal microbial populations, which is normalized by transfer of Treg cells (Mathias et al., 2011; Noval Rivas et al., 2013). Therefore the interaction between host immunity and commensal microbiota appears bidirectional since host immune response not only responds to the presence of microbial flora but also microbiota structure is also associated with concerned pathology.

In response to allergic sensitization, skin epithelium responds via secreting thymic stromal lymphopoietin (TSLP) and tissue-derived cytokines (i.e., IL-25 and IL-33), which perhaps activate innate lymphoid cells. In contrast, a recent study has shown that only IL-33 is required for allergic sensitization, not two other components (described above) on the basis a cholera toxin derived from oral peanut model (Chu et al., 2013). In addition, IL-33 is also found to be associated with increased mucosal permeability and promotes Th2 cells via DCs (Turnquist and Thomson, 2009). IL-33 is a key cytokine, constitutively expressed by epithelial cells and evident by inducible expression by DCs during allergic asthma and hookworm infection (Wills-Karp et al., 2012); however, a key molecule to produce IL-33 in response to food allergy still remains to be elucidated. Some environmental interactions [e.g., staphylococcal enterotoxin B (SEB)] are also known to induce food allergy and suppress tolerance by inducing TIM4 expression in DCs resulting in food antigen–specific Th2 differentiation and intestinal allergic responses (Yang et al., 2007). SEB are also able to elicit Th2-associated responses through deletion or impaired function of Treg cells (Ganeshan and Bryce, 2012). In contrast, some innate signals may protect against food sensitization through impairment of IgE and IgA responses (Berin and Wang, 2013). Allergic sensitization is increased during intestinal penetrance of allergens. During entry of allergens through intestinal epithelium, IL-4 promotes upregulation of CD23 (a low-affinity IgE receptor) that facilitates antigen uptake through binding with Ag-specific IgE (Yu et al., 2001). In contrast, large or low-solubility antigens pass through the epithelium thereby triggering systemic responses (Berin and Mayer, 2009). In addition, Ag also penetrates across the epithelium when tight junction integrity is compromised. One example is seen in patients with eosinophilic esophagitis (EoE) where desmosomal ICAM desmoglein-1 expression is reduced in tissues resulting in profound allergic responses (Sherrill et al., 2014). Therefore skin is a prime site for antigen penetrance, and integrity of the barrier is also critical since it results in sensitization upon oral challenge (Fig. 4.2). For example, filaggrin-deficient mice having a weak epithelial barrier are more prone to sensitization during contact of proteins with skin, and this epicutaneous sensitization thereby induces anaphylaxis. A limited number of genes are found to be involved in the regulation of skin homeostasis and mutations in desmoglein-1, filaggrin, and TSLP are reported to elicit food allergy or EoE in various cohorts (Liegel et al., 2013; Sherrill et al., 2014). However, their precise role in food allergy or eczema is still unclear since both are coincident in children (Tan et al., 2012).

Anaphylaxis

Anaphylaxis is a life-threating allergic reaction that is usually biphasic, which means it has two episodes. The first episode is an acute reaction that occurs immediately due to release of preformed mediators during exposure to an allergen, whereas the second episode is a late-phase reaction that occurs several hours later involving the influx of inflammatory cells. However, both of the reactions have been shown to display clinical heterogeneity in their responses because some patients encounter either acute or late-phase reactions, whereas some are prone to both acute and late-phase allergic responses. There are multiple mechanisms that result in anaphylactic responses (Fig. 4.3).

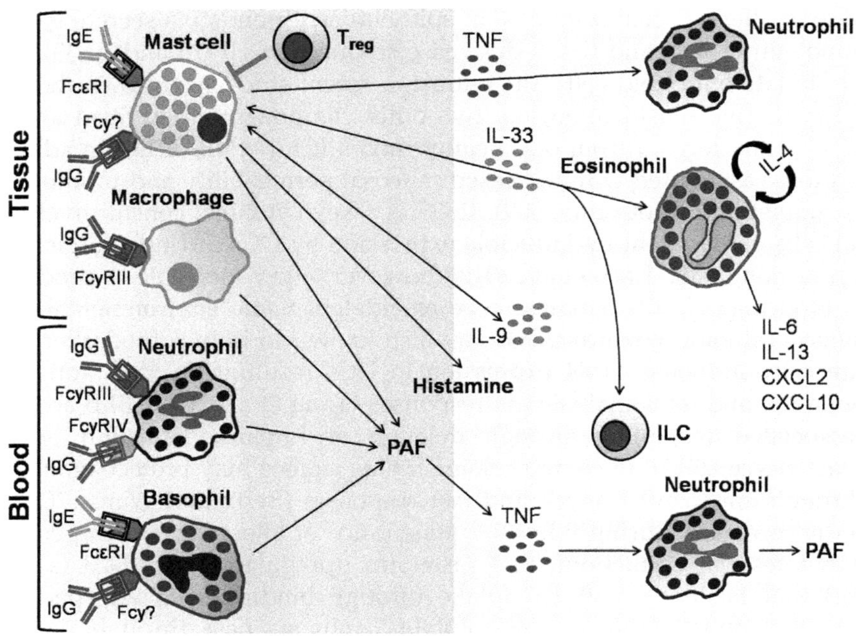

FIGURE 4.3 Multiple pathways of anaphylaxis.

Immunoglobulins in Anaphylaxis

Anaphylactic responses are found to be mediated by two antibodies (e.g., IgG and IgE), which have been shown to play key roles in anaphylaxis. IgE acts through a high-affinity FcεRI receptor that is expressed abundantly on mast cells and basophiles (Fig. 4.3). There are several studies that suggest the role of FcεRI receptor in IgE-mediated systematic anaphylaxis results in food-allergic responses (Strait et al., 2002). However, IgG has two types of receptors: high-affinity FcγRI and FcγRIV receptors and low-affinity FcγRIIB and FcγRIII receptors. In addition, the various cells types involved in anaphylaxis also express these receptors. The role of FcγRII/FcγRIII receptors has been shown in IgG-mediated anaphylaxis associated with a drop in temperature, whereas FcγRIV are involved in systemic anaphylactic responses (Strait et al., 2002; Jönsson et al., 2011). Some reports suggest that IgE and IgG knockout mice exhibit only partial protection against peanut-induced anaphylaxis, whereas complete blockade of IgG in IgE knockout mice results in total protection against anaphylaxis, implicating their role in anaphylaxis (Arias et al., 2009). In addition, studies on humanized mice also implicate the potential role of IgG via FcγRI (CD64) receptors in anaphylaxis (Mancardi et al., 2013).

Mediators of Anaphylaxis

There are several mediators known to be involved in anaphylaxis in early phases [e.g., histamine, serotonin, and platelet-activating factor (PAF)]. Histamine is a well-known mediator responsible for IgE-mediated systemic anaphylaxis that acts via activation of H1 and H2 receptors. These receptors have an important therapeutic application in treatment of patients suffered from acute allergic responses. In addition to this, serotonin and PAF are also involved in anaphylaxis. Within tissues, mast cells and macrophages synthesize PAF, whereas in blood, neutrophils and basophiles produce PAF that not only activates platelets but also activates inflammatory cells (e.g., neutrophils) resulting in increased vascular permeability (Fig. 4.3).

Apart from these mediators (listed above), there are also some less-defined mediators that exist such as performed mediators (e.g., tryptase, chymase, and heparin), lipid-derived mediators (e.g., prostaglandins, leukotriene LTC_4, LTD_4, LTE_4), and several inflammatory cytokines (e.g., tumor necrotic factor TNF-α, IL-9, IL-33). IgE-mediated mast cell activation leads to secretion of various cytokines (e.g., TNF and IL-33) (Hsu et al., 2010). TNF is a key mediator that directs late-phase inflammation through activation of neutrophils (Wershil et al., 1991) and increased levels of PAF in serum (Choi et al., 2003). IL-33 is a cytokine belonging to type 1 family and important for eliciting IgE-mediated tissue inflammatory responses through induction of mast cells and eosinophils to produce type 2 cytokines (e.g., IL-6 and IL-13) (Bouffi et al., 2013). In addition, mast cells also produce IL-9, which is also known to be involved in systemic and oral antigen-induced food-derived anaphylaxis by inducing abnormal accumulation of mast cells in mucosal surfaces of the digestive tract, also known as intestinal mastocytosis (Osterfeld et al., 2010).

Pathways for Anaphylaxis and Concerned Food Allergy

There are four distinct pathways known to elicit anaphylactic responses: Classic pathway, alternative pathway, IgG–basophil–PAF pathway, IgG–neutrophil–PAF pathway. The classic pathway involves activation of mast cells and neutrophils through cross-linking of IgE and aggregation of high-affinity IgE receptors, FcεRI and mast cell degranulation resulting in release of performed mediators including histamine, PAF, etc. triggering anaphylaxis. The alternative pathway includes IgG, IgG receptor FcγRIII, and the release of PAF as a key mediator for anaphylaxis. However, this pathway is not well documented and understood. The third pathway is reported as dispensable for IgE-mediated anaphylaxis because basophiles release PAF, leading to increased vascular permeability upon contact with IgG-allergen complexes. It has been shown that mice with depleted basophils are rescued from death in active anaphylaxis, suggesting IgG and PAF mediated role of basophils in eliciting anaphylaxis (Tsujimura et al., 2008). In addition, basophiles along with mast cells are also found to confer protection from peanut-induced anaphylaxis (Reber et al., 2013). The fourth pathway involves IgG, IgG receptors, and neutrophils in anaphylaxis also mediated by PAF (Fig. 4.3). Neutrophils have been also reported to induce anaphylaxis because mice depleted in neutrophils are protected from both active as well as passive systemic anaphylaxis (Jönsson et al., 2011).

Nonresponsive Tolerance

Food allergen–specific IgE is 10-fold more prevalent compared to food allergy advocating that an additional level of tolerance regulation exists above routine immunological priming toward IgE and Th2. The patients with hyper-IgE syndrome display diminished allergic response to food allergens if they have a Stat3 mutation supporting the role of

STAT3 signaling in mast cell degranulation (Siegel et al., 2013). Recent reports also provide the evidences for the role of Tregs in suppressing IgE-primed mast cell degranulation during exposure to an allergen through OX40–OX40L interaction (Gri et al., 2008). Treg transfer is well understood to be associated with suppressing anaphylaxis and restoring intestinal Th17 homeostasis via induction of mast cell–derived IL-6 (Ganeshan and Bryce, 2012). However, this process is mediated by TGF-β instead of being dependent on OX40 (Ganeshan and Bryce, 2012). Some reports also suggest the role of Treg in downregulation of FcεRI present on mast cells in vitro (Kashyap et al., 2008). Therefore this nonresponsive tolerance has been emerged in spite of prevalence of IgE-mediated immune priming, and this active form of tolerance is distinct form antigen desensitization (internalization of antigen-specific IgE on mast cells).

ASSOCIATED FOOD ALLERGY CONDITIONS

Gastrointestinal Food-Allergic Conditions

Various disorders resulting from the adverse immunological responses to dietary antigens arise from GI food allergies. Despite significant overlap between these conditions, several specific syndromes have been described as follows.

Immediate Gastrointestinal Hypersensitivity

This type of hypersensitivity is grouped under IgE-mediated allergic response, in which upper GI symptoms may arise within minutes whereas lower GI symptoms arise with the delay of several hours (Sicherer, 2003; Jones, 2008). The most common immunological IgE-mediated response associated with this type of hypersensitivity is acute immediate vomiting.

Eosinophilic Esophagitis

EoE involves localized eosinophilic inflammation of the esophagus (Chehade and Sampson, 2009; Rothenberg, 2009). In some patients, avoidance of specific foods is the only way to normalize histopathology. Both IgE- and non-IgE-mediated mechanisms appear to be involved. Feeding disorders, reflux symptoms, vomiting, and abdominal pain are the major symptoms of EoE-affected children, whereas dysphagia and esophageal food impactions type symptoms arise during adolescence and adult EoE.

Food Protein–Induced Allergic Proctocolitis

Allergic proctocolitis typically presents in infants who seem generally healthy but have visible blood mixed with mucus in the stool (Sicherer, 2003). Absence of immunoglobulin E and symptoms like vomiting and diarrhea are helpful in determining this disorder from other GI disorders. Due to the absence of a proper diagnostic test, the contributory role of milk or soy food allergen is concluded from an exposure-based characteristic history.

Food Protein–Induced Enterocolitis Syndrome

Food protein–induced enterocolitis syndrome is also a non-IgE-mediated disorder that is usually associated with symptoms like diarrhea and chronic emesis in young infants. On repeated exposure of food allergen, a subacute syndrome with repetitive emesis and dehydration may arise (Nowak-Wegrzyn et al., 2003; Sicherer, 2003). Milk and soy protein are the most causative agents despite some studies that also report reactions to other foods, including rice, oats, or other cereal grains. Similarly, identical cases have been reported in adults, most often related to crustacean shellfish ingestion.

Oral Allergy Syndrome

OAS, also referred to as pollen-associated FA syndrome, is a form of localized IgE-mediated allergy, usually to raw fruits or vegetables. OAS-associated symptoms like itching of the lips, tongue, roof of the mouth, and throat, with or without swelling, are commonly seen in individuals who are allergic to pollens.

Cutaneous Reactions

This type of reaction is very common in food allergy, which includes cell-mediated (contact dermatitis, dermatitis herpetiformis), IgE-mediated (urticaria, angioedema, flushing, pruritus), and mixed cell-mediated (AD) and IgE reactions. These are characterized in the following sections.

Acute Urticaria

This is a common reaction produced in response to IgE-mediated food allergy. Food allergy sometimes leads to acute urticaria and hardly ever is the cause of chronic urticaria (Burks, 2003). A few millimeters to several centimeters polymorphic, round, or irregular-shaped pruritic wheals emerge on the intake of injurious food.

Angioedema

Another IgE-mediated food allergy with symptoms of nonpruritic, nonpitting, clear edematous swelling of abdominal organs, nasal passages, and subcutaneous tissue (face, hand, genitals, and buttocks) commonly persists in the population (Burks, 2003). When nasal passages are involved, a medical emergency, laryngeal angioedema, arises, which needs immediate assessment. Anaphylaxis is well characterized by both urticaria and acute angioedema.

Atopic Dermatitis

AD, also known as atopic eczema, arises due to interaction between skin barrier dysfunction and environmental factors including microbes, irritants, and allergens (Lack, 2008). It results in itchy, red, swollen, and cracked skin. Clear liquid may originate from the influenced areas, which regularly thicken after some time. The allergic condition typically appears during childhood with increasing severity over age. Null mutations of the skin barrier protein filaggrin have been found to be linked with increased incidence of transcutaneous allergen sensitization and the progress of food allergy in response to AD (Marenholz et al., 2009; Leung, 2009; Van den Oord and Sheikh, 2009).

Allergic Contact Dermatitis

The common clinical features of allergic contact dermatitis include erythema, papules, marked pruritus, vesicles, and edema. Actually, it is a variant of eczema, which arises due to cell-mediated allergic response for haptens, which are simply food additives or naturally occurring in foods such as mangos (Warshaw et al., 2008).

Respiratory Reactions

This reaction is the result of IgE-mediated FA, which arises during systemic allergic response and is considered as one of the key indicators of severe anaphylaxis (James, 2003).

Heiner Syndrome

It is a pulmonary syndrome that mostly arises in response to hypersensitive foods such as cow's milk in infants and young children. It is a rare disease with unique chronic or recurrent lower respiratory symptoms that are linked with upper respiratory tract, GI, pulmonary infiltrates, and iron deficient anemia.

This syndrome is an outcome of non-IgE-mediated immune response and arises due to milk protein fraction induced precipitating antibodies. In some cases, iron deficiency, peripheral eosinophilia, and deposition of antibodies and C3 protein are easily detected through lung biopsies. Milk elimination from the diet would remove symptoms within days, and also pulmonary infiltrates clear within a week (del Savio and Sherertz, 1994).

FOOD ALLERGY VERSUS FOOD INTOLERANCES

In food allergy, antibodies are actively produced against the allergic food by the abdominal immune system. Some individuals produce few symptoms on consumption of a particular food while they are not producing food-associated antibodies. This nonimmune reaction is attributed to food intolerance. However, it is more common than FA and the immune system plays no role in it. The commencement of symptoms is initially slower or delayed on consumption of allergic foods, and once they appear they persist for longer times. It is difficult to find out the real cause of food intolerance as it may be the result of chronic illness or be food-borne.

The range of symptoms arises during food intolerance, including GI symptoms such as nausea, vomiting, diarrhea, bloating, irritable bowel, night sweats, fatigue, joint pains, skin rashes, and dark circles under the eyes. Food intolerance is the result of various causes mentioned as follows.

Lactose Intolerance

Lactose is a disaccharide found only in milk and other dairy products. Lactase is an enzyme found in the gut lining and responsible for the breakdown of lactose. The lack of lactase leads to lactose intolerance; in that condition, bacteria present in the gut break down the sugar and lead to formation of gas, which in turn causes symptoms of bloating, abdominal pain, and sometimes diarrhea. Since lactase levels decline as people get older, lactose intolerance becomes more common with age. Therefore lactose intolerance is not prevalent in babies and young children below the age of 5 years. Lactose intolerance also depends upon racial and ethnic background, which is truly diagnosed through laboratory tests.

Food Poisoning

Food-borne illness or food poisoning mostly results from eating contaminated, spoiled, or toxic food. Nausea, vomiting, and diarrhea are some common symptoms of food poisoning. The Centers for Disease Control (CDC) data suggest that one in six Americans suffer from some form of food poisoning each year. Food poisoning is mostly caused by the pathogenic bacteria (e.g., *Escherichia coli*, *Listeria*, and *Salmonella*) *Salmonella* is one of the main culprits causing serious food poisoning cases in the United States. Food poisoning caused by parasites is common compared to bacteria, and parasites spread through food are very dangerous. Toxoplasma is a common food poisoning parasite. Virus-mediated food poisoning is also very common in populations. Norovirus or Norwalk virus is responsible for approximately 19 million cases of food poisoning worldwide every year, but it can be rarely fatal. Sapovirus, rotavirus, and astrovirus are some other viruses that cause food poisoning, but they are less common. Hepatitis A is a serious condition spread through food.

Irritable Bowel Syndrome

A number of symptoms including abdominal pain and changes in the pattern of bowel movements without any evidence of underlying damage are displayed by irritable bowel syndrome (IBS) (NIDDKD, 2016). People with IBS have various symptoms including diarrhea, constipation, belly pains or cramps, bloating, and harder or looser stools.

Carbohydrate Malabsorption

The malabsorption of a dietary sugar in a person having GI disorders leads to fermentation of the sugar, which results in gas production and a potential laxative effect. Symptoms associated with these events may be abdominal pain, flatulence, bloating, and altered bowel habits. Other rare symptoms may include fatigue, nausea, heartburn, and urgency with bowel movements. Patients should be screened for sugar malabsorption to determine dietary restrictions that will improve symptoms and quality of life.

Scombroid

Scombroid poisoning is mainly caused by ingestion of contaminated food (mainly fish). Improper storage of the dark meat of the fish leads to bacterial growth that produces scombroid toxin, which is a combination of histamine and histamine-like chemicals.

The ingestion of this toxin is not harmful to every person, and diagnosis of toxin or poison in fish is not 100% reliable. Bacteria are inactivated or killed during cooking, but the toxin remains in the tissue and enters into the body after ingestion. Scombroid poisoning symptoms such as nausea, vomiting, diarrhea, headache, flushing, and abdominal cramps may arise within 30–60 minutes after ingestion. Some other symptoms like itching, burning sensation in the mouth, hives, and fever are also visible in this poisoning.

PRECAUTIONS FOR FOOD ALLERGY

Avoid the Offending Food for At Least 3 Months

In the absence of sophisticated blood assays against a number of reactive foods, an allergen-free diet up to 2 weeks would help to predict the causes responsible for discomforts. Listing or documentation of food items and the resulting symptoms/experiences for the next few days would help to predict reactive food. Liking and satisfactory response to certain foods may also help in determining allergy against them. Elimination of suspected food up to 2 weeks and their reintroduction in the diet may help to find the food item causing allergic responses. Thereafter, completely avoid that food up to 5 days, and if again the allergy appears then it means that a person is truly allergic to that food.

Healing of Digestive System With Enzymes and Nutrients

One of the biggest factors for food allergies is a leaky gut that lets undigested food particles to pass through the intestines and eventually enter into the bloodstream. These undigested food particles persist due to less secretion of digestive enzymes (e.g., proteolytic enzymes such as protease, bromelain, and papain), lack of proper breakdown into simple, usable, and highly assimilable substances. These enzymes also enhance the immunity by promoting the growth of gut microbes. Vitamins and minerals also assist in healing of a leaky gut. In fact, the enzymes rely upon nutrients to carry out their tasks effectively. Poor nutrition leads to various diseases (e.g., allergies to cancer that resulted from enzymatic deficiency and poor assimilation of nutrients). For example, vitamin A helps in eliminating free radicals and the regulation of prostaglandins secretion during allergic reactions. Vitamin B6 is useful in the treatment of food allergy–related problems such as migraines, depression, and attention deficit hyperactivity disorder. Zinc increases the digestive ability of the body by producing hydrochloric acid in the stomach. Magnesium is also useful in reducing symptoms such as migraines, asthma, and depression. Vitamin C has a free radical scavenging property, which is produced during food allergy. Vitamins A, C, B6, and zinc all have a positive effect on main immune gland (i.e., thymus gland). Omega-3 fatty acid, quercetin, and methylsulfonylmethane possess antiinflammatory properties, which helps in reversion of leaky gut. Probiotics and glutamine are involved in the improvement of the body's immune system and help in healing of permeable gut.

Acupressure

Acupressure is an alternative way to cure stress and emotional issues linked to various diseases (e.g., cancer, cardiovascular disease, hypertension, and allergies). Physical pressure is applied to a number of acupuncture points (up to hundred) by the hand, elbow, or with various devices. Acupressure corrects the energy disruption resulting from negative emotions and eventually leads to immediate relief and eliminates both the emotional trigger and the physical disease.

DIAGNOSIS FOR FOOD ALLERGENS

Skin Prick Tests

This test is used to measure immunoglobulin IgE against suspected allergic food. It is a cost-effective test that gives quick results. During the test, the allergist places a drop of food allergen solution on the forearm or back of the allergic person by scratching the skin with the help of needle or plastic probe. The most important characteristic of this method is that it is very simple and there is no bleeding and pain. After about 30 minutes, we can get result: the positive result is confirmed by a wheal—a raised white bump encircled by itchy red skin. Usually, a large wheal is considered a positive result, but accuracy does not always depend on the size.

Blood Tests

The presence of IgE antibodies in response to certain foods is measured by specific blood tests, also known as radioallergosorbent tests because they use radioactivity, but modern tests do not. Different laboratories sometimes utilize different "brands" of the blood test and may report results utilizing distinctive scoring systems or units.

Oral Food Challenge

Sometimes, food allergy is not diagnosed properly even after skin prick and blood tests. In such cases, an oral food challenge (OFC) is applied to the patients. During OFC, the allergist asks patients to take reactive food in measured doses starting from a very low amount. Thereafter, patients are kept under observation for any symptoms or signs in response to such food after each dose. Large amounts of doses are applied during the absence of any symptoms, but the food challenge is stopped after any sign appears. With this treatment, most reactions are mild, such as flushing or hives, and severe reactions are uncommon. At the extreme, medications, most often antihistamines, are provided to provide relief from allergic symptoms.

Food Elimination Diet

Temporary elimination of reactive foods from the diet aids in diagnosis of food allergy.

This method is helpful in diagnosing both IgE-mediated food allergy and other related disorders such as gut-associated allergies when combined with the blood or skin test. This test generally takes 2–4 weeks. During this period, individuals are advised to avoid the suspect food and doctors regularly monitor the symptoms. If one or more foods are associated with allergic reaction, then the symptoms related to them will disappear by at the end of this period.

TECHNIQUES FOR DETECTION OF ALLERGENS

Enzyme-Linked Immune Sorbent Assay

An allergen protein molecule is detected through the enzyme-linked immune sorbent assay (ELISA) method that allows binding of antibodies to the allergen followed by a colorimetric change resulting from an enzyme-linked conjugate. The drawback of the ELISA method is cross-reactivity due to interference with matrices (e.g., chocolate and different types of nuts). In addition, ELISA is also not suitable for cooked or heated food products due to denaturation of protein molecules and resulting unavailability of the allergen despite being reactive to sensitive individuals.

Lateral Flow Devices and Dipstick Tests

Lateral flow tests (LFD) detect allergen traces on the surfaces and in foods. Being faster and easier handling gives an advantage to LFD over ELISA. In addition, LFD are also able to detect allergens in both raw items and processed food products (e.g., RIDAQUICK Soya) may detect heated soy proteins from meat or cereal products with higher sensitivity and a detection limit of 0.5 $\mu g/100\ cm^2$.

Polymerase Chain Reaction

Detection of nucleic acid molecules of allergens in both raw and cooked food products makes polymerase chain reaction (PCR) a more sensitive method because DNA molecules typically are found intact even during higher temperature. In addition, PCR methods do not have any interference in contrast to ELISA-based methods due to purification of DNA before analysis that eliminates these inhibitors. Oils and other products, such as milk or egg whites, cannot be tested by PCR because they do not contain DNA; therefore ELISA must be preferred for their detection.

Mass Spectrometry

Liquid chromatography–mass spectrometry (LC–MS) method is used for quantitative measurement of allergenic proteins. In this method, first, the protein is extracted from food and then alkylated, reduced, and enzymatically digested into small peptides. An endopeptidase trypsin is frequently used in selective cleavage of protein C terminal at arginine and lysine residues by producing small accessible fragment for MS analysis. The complex mixture of peptides generated after cleavage is temporally separated by liquid chromatography. The recovered small peptides are then ionized through electrospray ionization and further used for MS analysis.

Spectroscopy

The newest analytical approach for allergen analysis that is gaining traction in global laboratories uses liquid chromatography–tandem mass spectrometry (LC–MS–MS) for the identification of allergenic peptides in food products. Rather than analyzing the full intact allergenic protein (as done with ELISA), LC–MS–MS allows for the direct analysis of digested peptide fragments of allergenic proteins, which are detected using their characteristic molecular masses.

Biosensors

An optical biosensor detects proteins from milk, eggs, hazelnuts, peanuts, shellfish, and sesame in food samples by both direct and sandwich immunoassays. Affinity-purified polyclonal antibodies raised against the proteins are immobilized on the biosensor chip, and food samples are injected, and the proteins bound to the antibodies on the surface are detected by a shift in the resonance angle.

GUIDELINES FOR MANAGEMENT OF FOOD ALLERGIES

The guidelines are intended to assist health-care professionals in making appropriate decisions about patient care in all developing countries. In the United States, the diagnosis and management-related guidelines were developed in 2008 to meet a well-awaited need for harmonization of the best clinical practices linked to food allergy across medical specialties. The recommendations are not fixed protocols that have to necessarily be followed. Health-care professionals are supposed to be take these guidelines into consideration when exercising their clinical decisions. The resulting guidelines reflect considerable effort by a wide range of participants to establish consensus and consistency in definitions, diagnostic criteria, and management practices. In addition, brief recommendations are provided on diagnosis and management of food allergy, and further treatment of acute food allergy responses. These documents are intended as a resource to guide clinical practice and to develop educational materials for patients, their families, and the public as well. They are not official regulatory documents of any government or agency. They provide guidance on addressing points of debate in patient management and also finding the gaps in our current knowledge, which will help focus the success of future research in this area. These guidelines are as follows:

1. Written action plan with provision of two doses of epinephrine (first-line treatment for anaphylaxis and delays in administration associated with increased mortality).
2. Patient/family education (teach food allergy basics).
3. Emotional and social impact of food allergy (fear of adverse events and death).

4. Each label on food should be read every time.
5. Understand labeling laws (FALCPA) and their limitations.
6. Be familiar with hidden ingredients.

FOOD LABELING

Individuals must be careful of the foods they are allergic to, and therefore labeling of food is extremely necessary to protect allergic individuals from allergy-derived serious consequences. Food labeling aids in changing consumer's behavior and various food production practices. In addition, food labeling also relates consumers with specific food products with an interest in health, the environment, culture, and social well-being.

Food labeling fulfills a dual purpose: first to protect consumers, and second to ensure fair marketing. Food labeling does not restrict a food product from being marketed.

In theory, the market will determine whether a technology will succeed since labeling provides information to buyers, and their actions give a signal to sellers about consumer preferences.

Food labeling enables consumers to reject a product, with the loss in sales causing the producer to remove the product from the market. In the case of genetically modified foods, labeling has been proposed as a way to allow consumers to demonstrate their views about the technology.

CONCLUSION

Food allergy is an important clinical problem of increasing prevalence, and its further assessment is key for proper diagnosis and treatment. Currently, diagnosis relies on a careful history and diagnostic tests (e.g., skin prick tests, serum-specific IgE testing where appropriate and, if indicated, OFC). The treatment is based on avoidance of reactive foods and appropriate prompt response to allergic reactions with epinephrine. The development of improved methods for prevention, diagnosis, and management of the disorder is a prerequisite and requires further insight into the pathophysiology of food allergies.

References

Arias, K., Baig, M., Colangelo, M., Chu, D., Walker, T., Goncharova, S., et al., 2009. Concurrent blockade of platelet-activating factor and histamine prevents life-threatening peanut-induced anaphylactic reactions. J. Allergy Clin. Immunol. 124 (2), 307–314.

Bennett, C.L., Christie, J., Ramsdell, F., Brunkow, M.E., Ferguson, P.J., Whitesell, L., et al., 2001. The immune dysregulation, polyendocrinopathy, enteropathy, X-linked syndrome (IPEX) is caused by mutations of FOXP3. Nat. Genet. 27 (1), 20–21.

Berin, M.C., Mayer, L., 2009. Immunophysiology of experimental food allergy. Mucosal Immunol. 2 (1), 24–32.

Berin, M.C., Sampson, H.A., 2013. Mucosal immunology of food allergy. Curr. Biol. 23 (9), 389–400.

Berin, M.C., Wang, W., 2013. Reduced severity of peanut-induced anaphylaxis in TLR9-deficient mice is associated with selective defects in humoral immunity. Mucosal Immunol. 6 (1), 114–121.

Bouffi, C., Rochman, M., Zust, C.B., Stucke, E.M., Kartashov, A., Fulkerson, P.C., et al., 2013. IL-33 markedly activates murine eosinophils by an NF-κB-dependent mechanism differentially dependent upon an IL-4-driven autoinflammatory loop. J. Immunol. 191 (8), 4317–4325.

Brandtzaeg, P., 1996. History of oral tolerance and mucosal immunity. Ann. N.Y. Acad. Sci. 778, 1–27.

Bruijnzeel-Koomen, C., Ortolani, C., Aas, K., Bindslev-Jensen, C., Björkstén, B., Moneret-Vautrin, D., et al., 1995. Adverse reactions to food. European Academy of Allergology and Clinical Immunology Subcommittee. Allergy. 50 (8), 623–635.

Burks, W., 2003. Skin manifestations of food allergy. Pediatrics 111 (6), 1617–1624.

Cerovic, V., Houston, S.A., Scott, C.L., Aumeunier, A., Yrlid, U., Mowat, A.M., et al., 2013. Intestinal CD103(-) dendritic cells migrate in lymph and prime effector T cells. Mucosal Immunol. 6 (1), 104–113.

Chehade, M., Sampson, H.A., 2009. The role of lymphocytes in eosinophilic gastrointestinal disorders. Immunol. Allergy Clin. North Am. 29 (1), 149–158.

Choi, I.W., Kim, Y.S., Kim, D.K., Choi, J.H., Seo, K.H., Im, S.Y., et al., 2003. Platelet-activating factor-mediated NF-kappaB dependency of a late anaphylactic reaction. J. Exp. Med. 198 (1), 145–151.

Chu, D.K., Llop-Guevara, A., Walker, T.D., Flader, K., Goncharova, S., Boudreau, J.E., et al., 2013. IL-33, but not thymic stromal lymphopoietin or IL-25, is central to mite and peanut allergic sensitization. J. Allergy Clin. Immunol. 131 (1), 187–200.

Coombes, J.L., Siddiqui, K.R., Arancibia-Cárcamo, C.V., Hall, J., Sun, C.M., Belkaid, Y., et al., 2007. A functionally specialized population of mucosal CD103 + DCs induces Foxp3 + regulatory T cells via a TGF-beta and retinoic acid-dependent mechanism. J. Exp. Med. 204 (8), 1757–1764.

del Savio, B., Sherertz, E.F., 1994. Is allergic contact dermatitis being overlooked? Arch. Fam. Med. 3 (6), 537–543.

Denning, T.L., Wang, Y.C., Patel, S.R., Williams, I.R., Pulendran, B., 2007. Lamina propria macrophages and dendritic cells differentially induce regulatory and interleukin 17-producing T cell responses. Nat. Immunol. 8 (10), 1086–1094.

Duchmann, R., Neurath, M., Märker-Hermann, E., Meyer Zum Büschenfelde, K.H., 1997. Immune responses towards intestinal bacteria—current concepts and future perspectives. Z. Gastroenterol. 35 (5), 337–346.

Friedman, A., 1996. Induction of anergy in Th1 lymphocytes by oral tolerance. Importance of antigen dosage and frequency of feeding. Ann. N.Y. Acad. Sci. 778, 103–110.

Ganeshan, K., Bryce, P.J., 2012. Regulatory T cells enhance mast cell production of IL-6 via surface-bound TGF-β. J. Immunol. 188 (2), 594–603.

Gri, G., Piconese, S., Frossi, B., Manfroi, V., Merluzzi, S., Tripodo, C., et al., 2008. $CD4^+$ $CD25^+$ regulatory T cells suppress mast cell degranulation and allergic responses through OX40-OX40L interaction. Immunity 29 (5), 771–781.

Guarner, F., Bourdet-Sicard, R., Brandtzaeg, P., Gill, H.S., McGuirk, P., van Eden, W., et al., 2006. Mechanisms of disease: the hygiene hypothesis revisited. Nat. Clin. Pract. Gastroenterol. Hepatol. 3 (5), 275–284.

Guinane, C.M., Cotter, P.D., 2013. Role of the gut microbiota in health and chronic gastrointestinal disease: understanding a hidden metabolic organ. Therap. Adv. Gastroenterol. 6 (4), 295–308.

Gupta, R.S., Springston, E.E., Warrier, M.R., Smith, B., Kumar, R., Pongracic, J., et al., 2011. The prevalence, severity, and distribution of childhood food allergy in the United States. Pediatrics 128 (1), 9–17.

Hadis, U., Wahl, B., Schulz, O., Hardtke-Wolenski, M., Schippers, A., Wagner, N., et al., 2011. Intestinal tolerance requires gut homing and expansion of Fox $P3^+$ regulatory T cells in the lamina propria. Immunity 34 (2), 237–246.

Hazebrouck, S., Przybylski-Nicaise, L., Ah-Leung, S., Adel-Patient, K., Corthier, G., Wal, J.M., et al., 2009. Allergic sensitization to bovine beta-lactoglobulin: comparison between germ-free and conventional BALB/c mice. Int. Arch. Allergy Immunol. 148 (1), 65–72.

Hsu, C.L., Neilsen, C.V., Bryce, P.J., 2010. IL-33 is produced by mast cells and regulates IgE-dependent inflammation. PLoS One 5 (8), 1–9.

James, J.M., 2003. Respiratory manifestations of food allergy. Pediatrics 111 (6), 1625–1630.

Jones, S.M., 2008. The spectrum of allergic reactions to foods. In: Metcalfe, D.D., Sampson, H.A., Simon, R.A. (Eds.), Food Allergy: Adverse Reactions to Foods and Food Additives, fourth ed. Wiley-Blackwell, Malden, MA, pp. 101–109.

Jönsson, F., Mancardi, D.A., Kita, Y., Karasuyama, H., Iannascoli, B., Rooijen, N.V., et al., 2011. Mouse and human neutrophils induce anaphylaxis. J. Clin. Invest. 121 (4), 1484–1496.

Jyonouchi, S., Abraham, V., Orange, J.S., Spergel, J.M., Gober, L., Dudek, E., et al., 2011. Invariant natural killer T cells from children with versus without food allergy exhibit differential responsiveness to milk-derived sphingomyelin. J. Allergy Clin. Immunol. 128 (1), 102–109.

Kashyap, M., Thornton, A.M., Norton, S.K., Barntein, B., Macey, M., Brenjovich, J., et al., 2008. Cutting edge: CD4 T cell-mast cell interactions alter IgE receptor expression and signaling. J. Immunol. 180 (4), 2039–2043.

Kim, J.M., Rasmussen, J.P., Rudensky, A.Y., 2007. Regulatory T cells prevent catastrophic autoimmunity throughout the lifespan of mice. Nat. Immunol. 8 (2), 191–197.

Lack, G., 2008. Epidemiologic risks for food allergy. J. Allergy Clin. Immunol. 121 (6), 1331–1336.

Leung, D.Y., 2009. Our evolving understanding of the functional role of filaggrin in atopic dermatitis. J. Allergy Clin. Immunol. 124 (3), 494–495.

Liegel, R.P., Handley, M.T., Ronchetti, A., Brown, S., Langemeyer, L., Linford, A., et al., 2013. Loss-of-function mutations in TBC1D20 cause cataracts and male infertility in blind sterile mice and Warburg micro syndrome in humans. Am. J. Hum. Genet. 93 (6), 1001–1014.

Macpherson, A.J., Geuking, M.B., McCoy, K.D., 2005. Immune responses that adapt the intestinal mucosa to commensal intestinal bacteria. Immunology 115 (2), 153–162.

Mahesh, P.A., Wong, G.W., Ogorodova, L., Potts, J., Leung, T.F., Fedorova, O., et al., 2016. Prevalence of food sensitization and probable food allergy among adults in India: the EuroPrevall INCO study. Allergy 71 (7), 1010–1019.

Mancardi, D.A., Albanesi, M., Jönsson, F., Iannascoli, B., Van Rooijen, N., Kang, X., et al., 2013. The high-affinity human IgG receptor FcγRI (CD64) promotes IgG-mediated inflammation, anaphylaxis, and antitumor immunotherapy. Blood 121 (9), 1563–1573.

Marenholz, I., Kerscher, T., Bauerfeind, A., Esparza-Gordillo, J., Nickel, R., Keil, T., et al., 2009. An interaction between filaggrin mutations and early food sensitization improves the prediction of childhood asthma. J. Allergy Clin. Immunol. 123 (4), 911–916.

Mathias, C.B., Hobson, S.A., Garcia-Lloret, M., Lawson, G., Poddighe, D., Freyschmidt, E.J., et al., 2011. IgE-mediated systemic anaphylaxis and impaired tolerance to food antigens in mice with enhanced IL-4 receptor signaling. J. Allergy Clin. Immunol. 127 (3), 795–805.

Matteoli, G., Mazzini, E., Iliev, I.D., Mileti, E., Fallarino, F., Puccetti, P., et al., 2010. Gut CD103$^+$ dendritic cells express indoleamine 2,3-dioxygenase which influences T regulatory/T effector cell balance and oral tolerance induction. Gut 59 (5), 595–604.

Mellor, A.L., Munn, D.H., 2004. IDO expression by dendritic cells: tolerance and tryptophan catabolism. Nat. Rev. Immunol. 4 (10), 762–774.

Mora, J.R., Iwata, M., Eksteen, B., Song, S.Y., Junt, T., Senman, B., et al., 2006. Generation of gut-homing IgA-secreting B cells by intestinal dendritic cells. Science (New York) 314 (5802), 1157–1160.

NIDDKD, 2016. Definition and Facts for Irritable Bowel Syndrome. 23 February 2015.

Noval Rivas, M., Burton, O.T., Wise, P., Zhang, Y.Q., Hobson, S.A., Garcia Lloret, M., et al., 2013. A microbiota signature associated with experimental food allergy promotes allergic sensitization and anaphylaxis. J. Allergy Clin. Immunol. 131 (1), 201–212.

Nowak-Wegrzyn, A., Sampson, H.A., Wood, R.A., Sicherer, S.H., 2003. Food protein-induced enterocolitis syndrome caused by solid food proteins. Pediatrics 111 (4), 829–835.

Nwaru, B.I., Hickstein, L., Panesar, S.S., Roberts, G., Muraro, A., Sheikh, A., 2014. EAACI Food Allergy and Anaphylaxis Guidelines Group. Prevalence of common food allergies in Europe: a systematic review and meta-analysis. Allergy. 69 (8), 992–1007.

Osterfeld, H., Ahrens, R., Strait, R., Finkelman, F.D., Renauld, J.C., Hogan, S.P., 2010. Differential roles for the IL-9/IL-9 receptor alpha-chain pathway in systemic and oral antigen-induced anaphylaxis. J. Allergy Clin. Immunol. 125 (2), 469–476.

Pabst, O., Mowat, A.M., 2012. Oral tolerance to food protein. Mucosal Immunol. 5 (3), 232–239.

Reber, L.L., Marichal, T., Mukai, K., Kita, Y., Tokuoka, S.M., Roers, A., et al., 2013. Selective ablation of mast cells or basophils reduces peanut-induced anaphylaxis in mice. J. Allergy Clin. Immunol. 132 (4), 881–888.

Rothenberg, M.E., 2009. Biology and treatment of eosinophilic esophagitis. Gastroenterology 137 (4), 1238–1249.

Shan, M., Gentile, M., Yeiser, J.R., Walland, A.C., Bornstein, V.U., Chen, K., et al., 2013. Mucus enhances gut homeostasis and oral tolerance by delivering immunoregulatory signals. Science (New York) 342 (6157), 447–453.

Sherrill, J.D., Kc, K., Wu, D., Djukic, Z., Caldwell, J.M., Stucke, E.M., et al., 2014. Desmoglein-1 regulates esophageal epithelial barrier function and immune responses in eosinophilic esophagitis. Mucosal Immunol. 7 (3), 718–729.

Shreffler, W.G., Castro, R.R., Kucuk, Z.Y., Charlop-Powers, Z., Grishina, G., Yoo, S., et al., 2006. The major glycoprotein allergen from *Arachis hypogaea*, Ara h 1, is a ligand of dendritic cell-specific ICAM-grabbing nonintegrin and acts as a Th2 adjuvant in vitro. J. Immunol. 177 (6), 3677–3685.

Sicherer, S.H., 2003. Clinical aspects of gastrointestinal food allergy in childhood. Pediatrics 111 (6), 1609–1616.

Sicherer, S.H., Sampson, H.A., 2009a. Food allergy: recent advances in pathophysiology and treatment. Annu. Rev. Med. 60 (10), 261–277.

Sicherer, S.H., Sampson, H.A., 2009b. Food allergy. J. Allergy Clin. Immunol. 125 (2), S116–S125.

Siegel, A.M., Stone, K.D., Cruse, G., Lawrence, M.G., Olivera, A., Jung, M.Y., et al., 2013. Diminished allergic disease in patients with STAT3 mutations reveals a role for STAT3 signaling in mast cell degranulation. J. Allergy Clin. Immunol. 132 (6), 1388–1396.

Silva, R., Lopes, C., Castro, E., de Oliveira, J.F., Bartolomé, B., 2009. Anaphylaxis to mango fruit and crossreactivity with Artemisia vulgaris pollen. J. Investig. Allergol. Clin. Immunol. 19 (5), 420–422.

Strait, R.T., Morris, S.C., Yang, M., Qu, X.W., Finkelman, F.D., 2002. Pathways of anaphylaxis in the mouse. J. Allergy Clin. Immunol. 109 (4), 658–668.

Tan, T.H., Ellis, J.A., Saffery, R., Allen, K.J., 2012. The role of genetics and environment in the rise of childhood food allergy. Clin. Exp. Allergy 42 (1), 20–29.

Tsujimura, Y., Obata, K., Mukai, K., Shindou, H., Yoshida, M., Nishikado, H., et al., 2008. Basophils play a pivotal role in immunoglobulin-G-mediated but not immunoglobulin-E-mediated systemic anaphylaxis. Immunity 28 (4), 581–589.

Turner, J.R., 2009. Intestinal mucosal barrier function in health and disease. Nat. Rev. Immunol. 9 (11), 799–809.

Turnquist, H.R., Thomson, A.W., 2009. IL-33 broadens its repertoire to affect DC. Eur. J. Immunol. 39 (12), 3292–3295.

Van den Oord, R.A., Sheikh, A., 2009. Filaggrin gene defects and risk of developing allergic sensitisation and allergic disorders: systematic review and meta-analysis. Br. Med. J. 339, b2433. Available from: https://doi.org/10.1136/bmj.b2433.

von Hertzen, L.C., Savolainen, J., Hannuksela, M., Klaukka, T., Lauerma, A., Mäkelä, M.J., et al., 2009. Scientific rationale for the Finnish Allergy Programme 2008-2018: emphasis on prevention and endorsing tolerance. Allergy. 64 (5), 678–701.

Warshaw, E.M., Belsito, D.V., DeLeo, V.A., Fowler, J.F., Maibach, H.I., Marks, J.G., et al., 2008. North American Contact Dermatitis Group patch-test results, 2003-2004 study period. Dermatitis. 19 (3), 129–136.

Wershil, B.K., Wang, Z.S., Gordon, J.R., Galli, S.J., 1991. Recruitment of neutrophils during IgE-dependent cutaneous late phase reactions in the mouse is mast cell-dependent. Partial inhibition of the reaction with antiserum against tumor necrosis factor-alpha. J. Clin. Invest. 87 (2), 446–453.

Wills-Karp, M., Rani, R., Dienger, K., Lewkowich, I., Fox, J.G., Perkins, C., et al., 2012. Trefoil factor 2 rapidly induces interleukin 33 to promote type 2 immunity during allergic asthma and hookworm infection. J. Exp. Med. 209 (3), 607–622.

Yang, P.C., Xing, Z., Berin, C.M., Soderholm, J.D., Feng, B.S., Wu, L., et al., 2007. TIM-4 expressed by mucosal dendritic cells plays a critical role in food antigen-specific Th2 differentiation and intestinal allergy. Gastroenterology 133 (5), 1522–1533.

Yu, L.C., Yang, P.C., Berin, M.C., Di Leo, V., Conrad, D.H., McKay, D.M., et al., 2001. Enhanced transepithelial antigen transport in intestine of allergic mice is mediated by IgE/CD23 and regulated by interleukin-4. Gastroenterology 121 (2), 370–381.

Zopf, Y., Hahn, E.G., Raithel, M., Baenkler, H.W., Silbermann, A., 2009. The differential diagnosis of food intolerance. Deutsches Arzteblatt international 106 (21), 359–370.

CHAPTER

5

Safety of Milk and Dairy Products

Mozammel Hoque[1] *and Sukanta Mondal*[2]

[1]ICAR-Indian Veterinary Research Institute, Bareilly, India [2]Animal Physiology Division, ICAR-National Institute of Animal Nutrition and Physiology, Adugodi, India

OUTLINE

INTRODUCTION

Milk is secreted in the mammary glands of mammals. It is a natural food supplement for the infants of mammals during their early stage of life when other types of food ingredients could not be digested by them. Milk is obtained from dairy animals during or soon after pregnancy and it is considered as an agricultural product for human consumption. Milk that is obtained immediately after parturition is called as colostrum, which is rich in antibodies and meant to be fed to the young ones to provide passive immunity against many diseases. Global cow milk production is 510.09 million metric tons in 2018. Not only is India credited to be the largest producer of milk in the world, but also it is the leading exporter of skimmed milk powder. Milk is consumed throughout the world, and it is estimated that

Food Safety and Human Health
DOI: https://doi.org/10.1016/B978-0-12-816333-7.00005-9

more than 6 billion people consume milk and milk products and more than 750 million people live in dairy farming households (Hemme and Otte, 2010). Raw or unpasteurized milk refers to a dairy product that has received no heat treatment to destroy pathogens or spoilage organisms, whereas pasteurized milk refers to milk that has passed through the process of pasteurization. Pasteurization is a universally accepted technique in which milk or milk products are subjected to heating to a certain temperature for a specific period of time with the purpose of destroying disease-causing and spoilage organisms. Various protocols for pasteurization of milk for different combinations of time and temperature are summarized as the following: high temperature short time (HTST) uses metal plates and hot water to increase milk temperatures to at least 161°F for not less than 15 seconds followed by rapid cooling; high heat short time (HHST) is similar to HTST but uses slightly different equipment and higher temperatures for a shorter time; ultra-pasteurized (UP) milk is heated to not less than 280°F for 2 seconds; and ultra-high temperature (UHT) milk is heated until sterile. Among these methods of pasteurization, milk processed by UHT is sterile and refrigeration is not required for its storage. All the organisms are not killed by the other methods of pasteurization, so these milks, whether raw or pasteurized, get spoiled; to avoid spoilage it should be in refrigerated temperature. Fat globules in milk are broken into smaller particles by a process termed homogenization, which prevents the cream layer from separating and floating to the top of the milk. Most conventional pasteurized milk is homogenized, whereas organic pasteurized milk and raw milk are often nonhomogenized. Colostrum is the "first milk" produced by the mammary gland of an animal after parturition. As the colostrum has high food value and disease-resistant ingredients, people prefer consumption of raw bovine colostrum, which sometimes poses a health hazards to the consumers when the dairy animals are affected with diseases. A vast spectrum of choices of milk and dairy products are available to consumers worldwide. Raw milk, pasteurized raw milk, homogenized or nonhomogenized milk, milk labeled with different levels of fat content, organic milk, and plant-based milk are all available to consumers. Wide varieties of milk products like butter, cheese, cream, ice cream, colostrum, yogurt, kefir, and other fermented milk products are available in the market in addition to liquid milk.

Milk and milk products are very important and quality food for human consumption throughout the world. Milk contains carbohydrates, protein, fat, and nine essential nutrients, making it a complete food. But if not handled properly, milk can also pose various health hazards.

Yogurt and other cultured dairy products provide many health benefits. Sustained research and development have led to wider uses of cultured dairy products and new processing methods for enhanced shelf-life and safety. Future research directions will likely include investigating the effects of probiotic dairy products on gut microbiota and overall health (Aryana and Olson, 2017).

MILK-BORNE DISEASES

A variety of microorganisms are present in milk and its products and serve as important sources of food-borne pathogens. They include *Bacillus cereus*, *Campylobacter jejuni*, *Escherichia coli* O157:H7, *Listeria monocytogenes*, and *Salmonella typhimurium*. The possibility of transmission of diseases to humans by consumption of raw milk from animals affected with bovine tuberculosis and brucellosis cannot be ruled out. Food-borne pathogens primarily enter from excretion from the udder of an infected animal. Direct contact with contaminated sources in the dairy farm environment may also contribute in food-borne pathogens. Pasteurizing milk destroys most of the organisms present in milk and thereby reduces the health hazards to the consumers; however, there are many occasions where milk-borne diseases pose a great health concern. Those included consumption of unpasteurized raw milk and milk products, inadequate or faulty pasteurization, and contamination of milk and milk products in the food chain following post-pasteurization. Moreover, some organism like *L. monocytogenes* can survive even after pasteurization and thrive in processing environments, which may contaminate the dairy products again. Thereby, the consumers are vulnerable to the risk from the direct exposure to food-borne pathogens present in unpasteurized dairy products as well as dairy products that become recontaminated after pasteurization. Microbial load may come into milk and milk products from contaminated feed, water, and animal excreta in the farm environment. Constantly maintained reservoirs of food-borne pathogens can reach humans by direct contact, ingestion of raw contaminated milk or cheese, or contamination during the processing of milk products. Isolation of bacterial pathogens with similar biotypes from dairy farms and from outbreaks of human disease substantiates this hypothesis (Oliver et al., 2005).

Pasteurization eliminates pathogenic organisms that may be present in raw milk, but post-pasteurization contamination is an ongoing challenge. Pasteurization is of paramount importance and also very effective from a safety point of view. Legal pasteurization, coupled with a comprehensive food safety program, can greatly reduce or eliminate the possibility of food-borne illness resulting from dairy products. There is a comprehensive set of requirements that deals with milk production, milk hauling, pasteurization, food safety, equipment sanitation, and labeling (Saue, 2009). The most recent assessment by the World Cancer Research Fund and the American Institute for Cancer Research found that most individual epidemiological studies showed increased risk of prostate cancer with increased intake of milk or dairy products. There is limited evidence suggesting that milk and dairy products are a cause of prostate cancer (Giovannucci et al., 1998; Chan et al., 2001, 2005).

In our country, preventive measures of milk-borne diseases are very difficult to implement due to wide verities of inherent problems. Milk is produced by farmers and small-scale unorganized dairy units and distributed to the consumers. Standard protocols like pasteurization, processing, packaging, transportation, and cold chain maintenance are not practiced. The only preventive measures followed by the consumers are boiling for long periods before consumption. Incorrect processing and storage of milk and dairy products can transmit a large number of diseases and even cause outbreaks of brucellosis, listeriosis, tuberculosis, etc. In the organized dairy sector, standard protocols of milk processing like pasteurization, packaging, transportation, and storage and distribution to the consumers under the cold chain are maintained and thus minimize the chance of occurrence of milk-borne diseases. However, lapses in any of the steps in milk processing, storage, and distribution may lead to incidences of milk-borne diseases. Individual measures to prevent milk-borne infections include not consuming raw milk, consuming milk from reliable sources, and pasteurizing with standard protocol; avoiding taking home-made cheeses, creams, yogurts; and maintaining the cold chain for milk-based products, etc., in order to avoid bacterial, viral, fungal, or parasitic contamination of milk. Strict sanitary preventive measures should be adopted in the dairy sector to prevent milk-borne infections. They included regular health check-up of dairy animals and personnel involved in handling of milk and milk products; adoption of hygienic measures in the dairy sector like use of face masks, hair covers, and hand sanitizers at regular intervals; and strict hygienic measures and cleanliness of containers, utensils, and other appliances used in milking, storage, packaging, refrigeration, and transportation. Besides regular testing of milk and milk products, food safety norms should be adhered to.

An interim report of the National Milk Quality Survey, 2018, released by the Food Safety and Standards Authority of India (FSSAI) in 2018 revealed that milk in India is largely safe. A few samples were found to be adulterated from among a large number of samples collected. The survey, however, found slightly less than 10% of samples had contaminants coming mainly from poor farm practices and over 90% of the samples were found to be safe in the survey (Sharma, 2018).

LACTOSE INTOLERANCE

Lactose, the carbohydrate present in milk, may be a medical concern that is referred to as lactose intolerance for many consumers. The condition is developed as the inability to metabolize lactose due to the absence or very low amount of the enzyme lactase in the intestinal tract of the consumer. It is estimated that about 75% of adults have some decrease in lactase activity during adulthood throughout the world. Abdominal pain, flatulence, bloating, abdominal distention, diarrhea, and nausea are generally observed as clinical symptoms. These clinical symptoms develop as a result of the breakdown of lactose into simple sugars (glucose and galactose) in the colon. Moreover, a lot of hydrogen gas is accumulated in the large intestine, causing flatulence and abdominal pain. It also causes loose stools and diarrhea as the lactose present in the milk osmotically draws into the colon. Affected individuals prefer alternate products such as water ices, sorbets, and juices in place of consuming milk. All of these items are produced in plants that may also produce milk products, so, much as with allergen control, separation of product types and proper sanitation are critical. Techniques have been developed to solve the problem by reduction or elimination of lactose content in the milk. Milk is passed over lactase enzyme bound to an inert carrier to make lactose-free milk. No lactose adverse effects will be there once the molecule is cleaved. By this process, milk with reduced amounts of lactose from 0% to 30% may be produced. Due to the formation of glucose by lactose cleavage, the lactose-free milk is slightly sweeter than that of regular milk. However, it does not contain more glucose and is no different than regular milk from a nutritional point of view. About 17% of the Finnish-speaking population has hypolactasia in Finland (Sahi, 1974) and has had "HYLA" (*hydrolyzed lactose*) products available for many years. Low-lactose level dairy products ranging from ice cream to cheese are prepared by enzymatic hydrolysis of lactose into glucose and galactose. A long shelf-life of the product is achieved by an ultra-pasteurization process combined with aseptic packaging. Valio in the year 2001 marketed a lactose-free milk drink in Sweden, Estonia, Belgium, and the United States. The chromatographic separation method to remove lactose from milk was patented by Valio. The lactose-free milk is not so sweet as HYLA milk but has the fresh taste of normal milk. In the United Kingdom, it is estimated that 4.7% of the population is affected by lactose intolerance in United Kingdom (Dairy Council, 2000) Lactose-free products milk, cheese, and yogurt contain only 0.03% lactose. To aid digestion in those with lactose intolerance, milk with added bacterial cultures such as *Lactobacillus acidophilus* (acidophilus milk) and bifidobacteria (a/B milk) is available in some areas (NDC, 2000). *Lactococcus lactis* cultured milk or cultured buttermilk often is used in cooking (Rombauer and Becker, 1975). Alternately, an ultra-filtration technique also produces lactose-free and lactose-reduced milk by removing smaller molecules such as lactose and water while leaving calcium and proteins behind. Milk produced via these methods has lower sugar content than regular milk (Hayley, 2015).

SOURCING OF MILK AND MILK PROTEINS

Raw milk turns sour when left standing for a while. This is the fermentation of lactose in the milk into lactic acid by lactic acid bacteria. Such milk becomes unpleasant to the consumer when fermentation goes on for a prolonged period. However, this process of fermentation is being utilized with the introduction of beneficial bacterial cultures (e.g., *Lactobacilli* sp., *Streptococcus* sp., *Leuconostoc* sp., etc.) to produce a variety of fermented milk products. The reduced pH from lactic acid accumulation denatures proteins and causes the milk to undergo a variety of different transformations in appearance and texture, ranging from an aggregate to smooth consistency. Varieties of products as a result of fermentation of milk include sour cream, cheese, buttermilk, yogurt, viili, kefir, kumis, etc. Most of the harmful pathogens present in the milk are destroyed by pasteurization of milk and thereby increase the shelf-life, but the spoilage makes it unfit for consumption ultimately, with unpleasant odor and taste, and may cause food poisoning. Lactic acid–producing bacteria ferments the lactose in milk to lactic acid under normal conditions. The milk thus becomes acidic and in turn prevents or slows the growth of other organisms. The lactic acid–producing bacteria are mostly killed on pasteurization of milk. Milk is generally refrigerated in tanks at a temperature between 1 and 4°C in order to prevent spoilage. Most milk is pasteurized by heating briefly and is then refrigerated to allow transport from factory farms to local markets. UHT treatment prevents the spoilage of milk. Milk treated by UHT can be stored unrefrigerated for several months. The moisture content is usually less than 5% in both drum- and spray-dried powdered milk. Freezing of milk can cause fat globule aggregation upon thawing, resulting in milky layers and butterfat lumps. These can be dispersed again by warming and stirring the milk. It can change the taste by destruction of milk-fat globule membranes, releasing oxidized flavors (Wiki-milk, Hui, 2006).

MILK ALLERGY

This allergy is an adverse immune reaction to one or more proteins in cow's milk and dairy products. These allergy symptoms may develop rapidly, termed anaphylaxis, a potentially life-threatening condition that requires treatment with epinephrine among other measures or with a gradual onset taking hours to days with symptoms including atopic dermatitis, inflammation of the esophagus, enteropathy, and proctocolitis (Caffarelli et al., 2010). Milk allergy is the most common food allergy recorded in early childhood and affects somewhere between 2% and 3% of infants in developed countries. In the United States, 90% of allergic responses to foods are caused by eight foods, with cow's milk being the most common. It is recommended that human babies should exclusively be breastfed preferably for 6 months before introduction of cow's milk to reduce the risk of developing a milk allergy. Management is by avoiding eating any dairy foods or foods that contain dairy ingredients (Lifschitz and Szajewska, 2015).

ANTIBIOTIC RESIDUE IN MILK

Mastitis is the infection of the mammary gland commonly affecting dairy cows and it is treated using antibiotics. As some people are seriously allergic to certain antibiotics (e.g., penicillin), dairy products need to have antibiotic content below the standard limit. When cows are given antibiotics, their milk is not usable until they are cured of illness and the antibiotic is no longer present in the body. There are times when a cow's milk is accidentally reintroduced into the market prematurely, and the antibiotic content is too high in the milk. This milk will often mix with milk from other farms and might require that an entire tanker load be dumped. To mitigate the problem, standard protocol should be followed, such as milk from the antibiotic treated animals should not be included in the milk supply pool and the withdrawal period of a particular antibiotic should be strictly observed. The milk sample must be tested for the presence of antibiotic residues, and milk of the affected animals should be included only when no traces of antibiotic would be detected in the milk.

HEAVY METALS AND OTHER POLLUTANTS IN MILK

A serious health concern to the consumers of milk today is the presence of heavy metals and other pollutants in milk and its products. Heavy metals from industrial sites reach canals and rivers through direct discharges and runoff of contaminated sites. Water storm runoff from city roads may also contain significant amounts of heavy metals. These heavy metals ultimately lead to farmland that grows animal feed. Dairy animals consume food and water contaminated by these metals and pollutants which makes their way to the milk. Milk may also be contaminated with these metals and pollutants during processing, transportation, storage, and consumption. Some heavy metals have been potentially cheap, including chemical acids, which are waste products of some industrial processes. By adopting drastic changes in the milk production process, a considerable reduction in the level of contamination of heavy metals and pollutants in milk may also be achieved. Water used for drinking and forage used for feeding animals should be regularly inspected to monitor the level of heavy metals in milk. Attempts should be taken that no traces of metal are present in any step during handling, processing, and storage of milk and dairy products, and if they are added, it should be used at permissible limits and that too only of food grade materials to prevent the contamination of metals. Dairy farm and land for cultivation of fodder should be away from polluting industries or heavy traffic areas to avoid contamination of the pollutants. More precautions in the milk and dairy production should be taken, and the safe water from deep and clean wells or portable water for dairy animals must be provided, and the heavy metal content tests and their determination have to be monitored by supervising authorities. The Public Health and Environmental Department in every province in all countries should be established to detect the contaminant contents in tap and well water (Zirati et al., 2018). The infected hand pumps and tube wells, which were being used for domestic usages in the lead/mercury/arsenic contaminated areas, must be identified and milk-processing units should be under strict observation (Iftakharul et al., 2016; Pajohi-Alamoti et al., 2017; Saeed Faraji et al., 2017). Removal of heavy metals from aqueous solution using fruit peel as a low-cost adsorbent is reported (Jena and Sahu, 2017).

Flavored milk also contains artificial color and chemicals that may have health concern. Food grade color and chemicals with permissible limits are to be incorporated with strict compliance.

ADULTERATION OF MILK

Milk adulteration is a serious concern throughout the world and more so in developing countries. Milk is a very popular nutritious and perishable diet in great demand, which prompts unscrupulous persons to adulterate it easily, and it is difficult to detect it.

Adulteration of milk by adding water, whey, vegetable oil, protein, and milk from different species with the intension of making profit does not pose health hazards, whereas adulterants by adding urea, formalin, detergents, alkaline and acidic compounds, hydrogen peroxide, and melamine do pose serious health hazards to the consumers.

Melamine toxicity: Melamine is chemically 2,4,6-triamino-1,3,5-triazine (C3H6N6), a nitrogen-rich compound used for producing plastics, adhesives, laminates, paints, permanent-press fabrics, flame retardants, textile finishes, tarnish inhibitors, paper coatings, and fertilizer mixtures. Melamine is a harmful adulteration in milk in order to falsely mimic the protein content of the milk. In 2008, in China, over 50,000 infants were hospitalized and 6 died from baby food that contained powdered milk tainted with melamine (Keuhn, 2009), raising global concern to ban Chinese dairy products and milk proteins. Following this incident, the U.S. Food and Drug Administration issued a Health Information Advisory regarding the melamine-contaminated infant formula. Subsequent to the melamine episode in these infant formulas in China, international food regulation authorities investigated other food products containing milk, including milk and whey powder, casein, milk-based candies, instant powdered coffee products, biscuits, chocolates, milk-based drinks, and cakes and found them also contaminated with melamine (WHO, 2009). It is not unusual to find melamine at trace levels in dairy products because of its presence in the environment or its presence in animal feeds that have been treated with melamine-containing substances, such as fertilizers and pesticide. Additionally, melamine is added into milk, wheat gluten, and other protein sources to raise the quantity of protein at a low cost (Liu et al., 2011). Melamine is added to the diluted milk by the dairy farmers and milk producer in order to maintain the false standard level of protein content in milk on test. Therefore, infants and children are impacted most often due to their dependence on milk for nutrition. The immaturity of their organs puts infants and children at an even greater risk of becoming ill (Hau et al., 2009).

In 2007, in North America, widespread pet illnesses and deaths were linked to the formation of melamine-cyanurate crystals in the kidneys of these animals (Puschner et al., 2007). Melamine is reported to be harmful when ingested, inhaled, or absorbed through the skin (Honkar et al., 2015). Melamine has low oral acute toxicity and was found to be a noncarcinogenic compound. However, it can lead to renal and urinary problems and even cause infant death when it reacts with cyanuric acid inside the body (Cheng et al., 2010). Melamine becomes more harmful when combined with its analogs, particularly cyanuric acid. Research revealed that when melamine and cyanuric acid enter the blood, they form large numbers of yellow crystals, which then block and cause damage to the renal cells (FDA, 2008). The harmful dose of melamine is comparable to that of common table salt with an LD50 of more than 3 g/kg of bodyweight (Honkar et al., 2015). Codex Alimentarius has set the maximum melamine limit at 1 mg/kg for powdered infant formula, 0.15 mg/kg for liquid infant formula, and 2.5 mg/kg for milk and products containing milk or milk-derived ingredients (Codex Alimentarius, 1995).

Sources of melamine contamination into milk: Melamine can contaminate milk and dairy products through various routes, which include adulteration to milk product to enhance protein content; use of cyromazine pesticide on crops (cyromazine is metabolized into melamine in the animal's body and therefore could contaminate their milk); use of nitrogen-containing fertilizers, if they contain melamine as a source of nitrogen; consumption of cyromazine or melamine-contaminated crops; and transfer of melamine from plastic milk packaging materials (Honkar et al., 2015). Melamine may also get into milk and tissue of the dairy cows grazing melamine-fertilized pasture. It is found that melamine may appear in milk from grass as quickly as 8 hours after the animal consumes it (Cruywagen et al., 2009). It is reported that the concentration of melamine in dairy products was present in the following increasing order: milk, yogurt, coffee mate, cheese, and infant formula (Poorjafari et al., 2015).

Considering the adverse effect of melamine on human health, particularly on infants, it needs strict controlling measures to stop the entry of melamine in milk and dairy products. The mitigative measures should include good manufacturing practice (GMP) and good quality control programs in controlling the level of melamine in milk. In addition to that, more effort is to be exerted to measure the amount of melamine in edible parts of crops and feed, as well as in agricultural fertilizers. These quantities need to be controlled and regulated to prevent the future use of melamine as an ingredient in fertilizers (Jalili, 2017).

ARTIFICIAL COW'S MILK

A synthetic dairy start-up called Muufri manufactures an artificial cow's milk made from a special variety of yeast that has been genetically engineered to produce milk proteins. The artificial milk is produced following the similar protocol used by the pharmaceutical laboratory to prepare insulin, wherein DNA is extracted from a milch cow for sequencing and subsequently selected sequences are inserted into yeast cells for synthesis of milk. The yeast culture is then grown in industrial-sized Petri dishes at just the right temperature and concentrations, and within a few days, the yeast will have produced enough milk for harvesting. The taste and food value of the synthesized milk are comparable to real milk. The artificial milk derived its protein from yeast and fat and flavor from vegetables. Further, to make it parallel to natural milk, it is fortified with minerals such as calcium and potassium, and sugar. Once the composition is fine-tuned, the ingredients emulsify naturally into milk. This of course means that all of the nutritional values can be tweaked and the artificial milk could potentially be even better than regular milk. Initially, Muufri milk will be more expensive to buy than regular milk, but it is expected to eventually be cheaper as the production is scaled up. As the artificial milk does not contain microbial load, its shelf-life is longer compared to natural milk. The product of technological innovation in synthesizing cow's milk comprises less than twenty components and about 87% water. If we want the world to change its diet from a product that isn't sustainable to something, that is, it has to be identical to, or better than, the original product.

The artificial milk is expected to overcome many ethical and animal activists' reservations against natural milk. The dairy cows are kept overcrowded in barns where cows are fed a constant cocktail of growth hormones and antibiotics and have their tails docked and their horns removed and their milk extracted. Moreover, according to Food and Agriculture Organization, dairy production is responsible for 3% of the world's annual greenhouse gas emissions, consumes a lot of water resources, and competes with human food grains (Gandhi and Pandya, 2014).

PLANT-BASED MILK (PLANT MILK, NONDAIRY MILK, VEGAN MILK)

Plant-based milk both as a regular drink and as a substitute for dairy milk has been consumed for centuries. To name a few, soy milk, almond milk, rice milk, and coconut milk are the most commonly used ones. The protein content varies from product to product. The plant-based milk has the advantage of being lactose and cholesterol free, and the drink is fortified with calcium and vitamins, particularly vitamin B_{12}. Besides the use of plant-based milk as a drink, various plant-based milk products like ice cream, cheese, and yogurt are equally popular. The plant-based milk and its products have a great demand as a large section of vegetarian consumers prefer them due to ethical, religious, environmental, and health reasons. In the United States, soy milk, almond milk, and rice and coconut milk are very popular as nondairy milk products. It is estimated that plant-based milk constituted 9.3% of the total U.S. milk market (Purdy, 2017). Soy and oat milk are very popular in European countries. Besides, the market is flooded with a wide variety of nondairy milk that includes flax milk, cashew milk, hemp milk, hazelnut milk, and milk from peas and lupin. The global dairy alternative market is set to be worth US$16.3 billion in 2018 (Innova Market Insights, 2017). In India, various plant-based milks are locally available and consumed; however, these vegan milks are not easily available in India. Abhay Rangan, an animal rights activist and entrepreneur from Bangalore, India, is bringing affordable plant-based milk to customers across the country through his company Veganarke, which currently manufactures a line of plant-based, shelf-stable almond and coconut milk under the brand GoodMylk. Currently, he has launched his plant-based milks throughout India via online sales and is the first Indian company to make vegan yogurt available across India (Smith, 2018).

SYNTHETIC MILK

It is an excellent imitation of natural milk. Every component of the milk is mimicked by an adulterant in synthetic milk like fat by vegetable oil, the nitrogen component by urea, and detergents are added to make it frothy. Thus, the cocktail product is composed of water, detergent or soap, sodium hydroxide, vegetable oil, salt, and urea. The process is very simple, fast, and cheap, and all the ingredients are locally available and thus tempted the unscrupulous milk vendors to indulge in this fraudulent practice to make profit. This synthetic milk is then disposed of in the market or adulterated with natural milk in varying proportions. This adulterated milk has a deleterious effect on human health.

Components of synthetic milk: Water is added as a staple component used in the preparation of synthetic milk. Mostly substandard water is used in synthetic milk. All other components are added in water medium to get equivalent consistency and composition of fat protein and sugar and appearance like natural milk. To adjust the sweetness of milk and to counter the sour taste formed due to the acidity in stored milk, cane sugar is added in the synthetic milk. This cane sugar in milk could be detected as low as 0.1% cane sugar as adulterant by the available chemical test on inspection. To adjust and or to increase the consistency and viscosity, starch is added in synthetic as well as natural milk. Iodine reagent test is able to detect as low as 0.1% of starch as adulterant in milk. Urea is a source of nitrogen, thus, it is generally added in synthetic milk to increase its nitrogen content and hence the level of the protein in milk. The presence of urea in milk could be detected up to 0.1% urea as adulterant in milk by dimethyl amino benzaldehyde (DMAB) test. Practically the test has a limitation, as the normal milk also contains low levels of urea, displaying a faint yellow color when tested with DMAB test. Thus, at lower levels of urea adulteration, it is very difficult to find whether the positive test for urea is due to naturally present urea in milk or adulterated urea in synthetic milk. Glucose is also added in synthetic milk to increase sweetness. The added glucose in milk could be detected up to 0.5% by chemical test. Neutralizers are also added in synthetic milk to counter acidity. When milk is stored for a prolonged time it becomes acidic. This happens due to breakdown of milk lactose into lactic acid by the growth of bacteria. While boiling, the fermented milk clots easily and becomes unsuitable for consumption. Development of acidity in milk can be countered by the addition of neutralizers. Detection of neutralizers in milk could be identified up to the level of 0.1% sodium carbonate by Rosalic acid test. The frothiness of milk like natural milk could be achieved by the addition of detergents. The detergents in milk could be detected as low as 0.1% of adulterant by available chemical test.

Effect of synthetic milk components on human health: Water is the chief adulterant used in milk. It increases the volume of milk and thus lowers the nutritional value per unit of volume. If the milk used in the adulteration is contaminated, it may cause many water-borne diseases like cholera, typhoid, shigella, polio, meningitis, and hepatitis A and E. Varieties of organisms like bacteria, protozoa, and viruses are involved for the causation of the diseases. Cane sugar causes sweetness in synthetic milk. Generally, the cane sugar added in small amounts to sweeten the milk does not cause major health concern. But poor quality of sugar in synthetic or adulterated milk can cause decreased wholesomeness of milk. Lactose, the sugar in natural milk, does not contribute to diabetes; on the contrary, cane sugar may lead to diabetes. Urea is an organic water soluble molecule composed of carbon, nitrogen, oxygen, and hydrogen of chemical formula CO (NH2)2 and is also known as carbamide. Urea is commonly added to increase the milk solid nonfat (SNF) content or its total nitrogen content. Normal milk may contain small amounts of urea and reflects to the nonprotein nitrogen (NPN) content in normal milk. When urea is added in milk with low fat and SNF, the milk appears thick and concentrated, giving a feeling of rich milk, and becomes toxic due to the presence of excess urea. A recent Indian Council of Medical Research report has suggested that urea adulterated items have a cancerous effect on the human system. During preparation of synthetic milk, detergents are mixed to develop frothiness like natural milk and to make use of its emulsification action. Most of the detergent contains dioxane, which is carcinogenic in nature and can cause cancer on consumption for prolonged periods. Detergents may also contain other toxic ingredients like nonylphenol ethoxylate, sodium lauryl sulfate, and phosphates. Results from various studies showed that sodium lauryl sulfate in any form causes eye and skin irritation, organ toxicity, neurotoxicity, developmental and reproductive toxicity, endocrine disruption, mutations, and cancer. It is reported that nonylphenol ethoxylate may cause various pathological events like kidney and liver damage, decreased testicular growth and sperm count, disrupted growth and metabolic development, and increased mortality. The main component present in detergents and household cleaners is phosphates, which can cause nausea, diarrhea, and skin irritations. All these hazards make detergents toxic, and the synthetic milk containing these detergents becomes unfit for consumption. Neutralizers are added to mask the developed acidity or bitter taste in synthetic milk. Among these neutralizers, the most common is sodium hydroxide, which can be very harmful if ingested. Its ingestion may result in a burning sensation, abdominal pain, shock, or collapse. Washing soda (sodium carbonate) used as a detergent is capable of causing adverse health concern when ingested, such as gastritis and vomiting. Its ingestion may also cause diarrhea, which may further result in frequent, loose bowel movements (Mudgil and Barak, 2013).

Synthetic milk menace has thrived in our country due to lack of strict regulatory bodies and a large portion of milk is produced in unorganized sectors. Local cooperative and private dairies handle 20% of the milk in the country, and 66% of milk is sold in organized markets as compared to about 90% in most developed countries. In rural India, local milkmen delivered milk at home daily by carrying bulk quantities in a metal container, usually on a bicycle.

Modus operandi for milk supply chain in major cities in India includes milk from the milk producer collected by collection agents at a milk chilling center and transported in bulk to a processing plant, and packaging is done in plastic packs or cartons and distributed to consumers through vendors and shops. Detergent and chemical used in the process may erroneously enter into the milk, causing health hazards. Breach in sanitation, hygiene, and cold chain in the process of handling and packaging of milk may also pose serious health concern. About 8% of samples in the survey were found to have detergents, which are hazardous to health (CSE, 2015). Synthetic milk is a menace to human health in our country and a great challenge to combat. The seriousness of the problem can be gauged by the fact that WHO has issued a health advisory to India. This stated that failing to immediately check the adulteration of milk and milk products would see 87% of Indian citizens suffering serious diseases like cancer by 2025. Roughly 68% of all milk and milk products in India have been found to be in violation of FSSAI standards. To mitigate the problem, FSSAI has developed a kit, the Detect Adulteration with Rapid Test, to enable citizen to detect food adulteration on their own. Strict regulation, motoring, and supervising activities are to be adapted to eradicate the malpractice.

ORGANIC MILK

Organic milk is produced with the practice of organic dairy farming. Organic dairy farming is a holistic approach where animals are reared on pasture or organic feed and fodder cultivated without the use of chemical fertilizers, pesticides, antibiotics, and hormones. Organic dairy farmers maintain a healthy livestock herd with great emphasis on preventive measures against diseases like proper vaccinations, parasitic control, and as well as treatment of disease. Withdrawal period following use of antibiotic is restricted at least for 30 days or double specified time mentioned for the particular drug. Treatment with alternate medical therapies like herbal, naturopathy, and homeopathy is permitted in organic animal husbandry practices. However, use of any kind of hormones is not permitted. Sanitation approaches with standard antiseptic chemicals like teat dips and washing of hands, utensils, and equipment are allowed. Equipment and utensils should be thoroughly washed with clean water prior to milking.

Standard operative procedures (SOP) in an organic dairy farm: The following SOPs are to be followed (Greene, 2002):

Feed and fodder fed to cows and calves should be exclusively organic in origin. Organic crops, hay, and pasture grown without the use of synthetic fertilizers and pesticides are used in organic farming. No prohibited materials should be used on the land earmarked for production of organic crops for a minimum of 3 years to get the first organic harvest. Approved feed additives and supplements such as vitamins and minerals should only be used. Feed of animals under organic farming should not contain any abattoir byproducts, urea, growth promoter, or manure. Genetically modified organisms (GMOs) and synthetic milk replacers are not permitted under organic farming. Calves should be reared on organic milk to be included in the organic chain. Intensive housing of dairy cows should not be there, and all the animals must have an outdoor exposure, preferably with grazing on natural pasture. Judicious usage of antibiotics with proper withdrawal period may be used to treat infection under strict veterinary supervision. Animal welfare protocols should be strictly followed. Certain procedures, such as tail docking, are prohibited. Other procedures, such as dehorning and hoof trimming, are done so as to minimize the stress to the animal. An organic farmer must keep sufficient records to verify the compliance with the standards. Each farm is inspected and audited every year.

Organic dairy products: Dairy products are certified in case those are prepared from milk from the dairy animals that have been reared under continuous organic farming system for at least 1 year.

Superior characteristic of organic milk and dairy products: Milk and a wide variety of dairy products like cheese, yogurt, milk powder, ghee, butter, butter oil, buttermilk, and ice cream are extensively used as food items. Milk produced in conventional dairy farm may contain pesticides, antibiotics, hormone, heavy metals, melamine, and urea residues, which may cause serious health hazards to consumers. Milk is a perfect indicator that reflects the level of pollutants and pesticides that persist in the environment and get into the milk food chain. In addition, unethical practices are common to obtain more milk beyond their natural capacity. To reduce the feed cost, quality diet is not provided, leading to production of inferior quality milk. Consumption of such milk may lead to early puberty, hypersensitivity, hormonal imbalance, and certain types of cancer in humans (Hallam, 2002; Heckman, 2006). Milk produced under organic dairy farming protocol take all the mitigative approaches to produce organic milk, which are better for health and superior quality that nonorganic milk. It is reported that conjugated linoleic acid (CLA) levels in organic milk are higher because the dairy cows have greater access to eat more amounts of grass, hay, and silage (McBride and Greene, 2009). CLA boosts the immune system in the human body and reduces the growth of tumors.

Constraints for organic dairy farming in India: The organic dairy sector is gaining popularity, and there is a potential demand for such products. Many dairy farmers are coming forward and are ready to accept the challenges to start-up with organic dairy farm. Organic dairy farmers get their product certified on compliance of organic dairy farming protocol like raising the dairy animals on organic feed and fodder, not using any chemicals, pesticides, hormones, or GMOs. A rigorous system for inspection, certification, and verification of organic practices are followed to qualify the milk and dairy product leveled as organic to protect the consumers' right. The future challenges are to keep pace with the production vis-a-vis with the demand and to lower the costs of production faced by organic farmers (Pierce and Tilth, 2009).

Challenges: Organic dairy farmers face various challenges in the process of producing organic milk. These may be summarized as paperwork for certification and compliance cost, provision of organic inputs such as feed and fodder, feed supplements, replacement of heifers, high costs of maintaining animal health, and production.

A1 and A2 milk: On the basis of type of beta-casein present in the milk, it is graded as A1 and A2 type of milk. It is the type of beta-casein and the primary difference being the 67th position of 209 amino acid chains which determine the A1 and A2 type. A1 type has histidine at 67th position, whereas the A2 type has proline at that position. It has been found that A2 milk contains more Omega-3 fatty acids, which are good for health. It is thought that the problematic component in A1 milk is the BCM7 (beta-casomorphin), which is a 7 amino acid long chain produced during the digestion of A1 beta-casein. It is observed that BCM7 is not well absorbed by the human body and causes health hazards. It is reported that BCM7 passes into the blood of babies who consumed the infant formula, causing delayed psycho-motor (brain-to-muscle) development. Further, it is reported that it is likely to poses a risk factor for type-1 diabetes, coronary heart disease, and mental disorders like autism and schizophrenia because it may enter the brain through the blood. However, most of the evidence is based on animal trials, and these diseases have a wide range of contributing factors. A significant difference was reported in digestive symptoms between milks containing A1 and A2 beta-casein a human clinical trial. It has been approved that certain unwanted proteins or peptides that do not naturally occur in the human body may cause digestive disorders, like irritable bowel syndrome or a weak gut. On the contrary, research revealed no convincing or probable evidence that the A1 beta-casein protein in cow's milk is a factor causing type-1 diabetes or heart disease. Further, the evidence for A1 beta-casein proteins in milk causing autism and schizophrenia is more speculative and unsubstantial than that for type-1 diabetes and heart disease (Truswell, 2005). Subsequently, European Food Safety Authority published an even more comprehensive 107-page scientific review that said the same thing but much more definitely (EFSA Report, 2009). The A2 milk phenomenon is not limited to only indigenous Indian cows. A lot of foreign breeds like Guernsey, Jersey from Channel area, and Asian and African cattle have a predominant A2 gene frequency. Milk from the water buffalo (*Bubalus bubalis*) is reported to be predominantly A2 type; all buffalo breeds of India produce only A2 milk. The A1 beta-casein type is the most common type found in cow's milk in Europe (excluding France), the United States, Australia, and New Zealand including breeds like Holstein. Scientists of National Bureau of Animal Genetic Research, Karnal, India, screened 22 cattle breeds and concluded that in five high milk-yielding native breeds—Red Sindhi, Sahiwal, Tharparkar, Rathi, and Gir—the status of A2 allele of the beta-casein gene was 100% and in other Indian breeds it was around 94%, compared to only 60% in exotic breeds like Jersey and HF (Reddy, 2017). Market price of A2 milk is much more expensive than normal milk, despite production cost being the same under natural dairy farm practices. In 2000, the a2 Corporation (The a2 Milk Co. Ltd) was formed to commercialize the A2 protein is in a brand of milk and related products like infant formula. A genetic method for identifying cattle for producing A2 milk was patented by the firm. This genetic testing makes A2 milk more expensive under commercial production practices. Milk is a very important source of nutrition in our country and so it is paramount important to a right choice for the best quality. If one feels bloated or uncomfortable with A1 milk, alternatively an option is to switch over to A2 milk that is reported to be easier to digest (Kruszelnicki, 2018).

SUMMARY

Milk and dairy products are widely consumed as a nutritious diet. A variety of microorganisms are present in milk and its products and serves as important sources of foodborne pathogens. Pasteurization can destroy the pathogenic organisms present in raw milk and thus help prevent milk-borne diseases and prevent spoilage of milk. However, post-pasteurization contamination of milk also poses a potential challenge. Most nutrients and beneficial bacteria in milk are lost due to processes like homogenization and pasteurization. So many consumers prefer to buy unprocessed whole fat milk from a local farmer who produced wholesome milk. Lactose intolerance and milk allergy are also health concerned for many. Presence of antibiotic and hormone residues, heavy metals, melamine, and other pollutants in milk are considered as the major health concerns to the consumers. These contaminants pose health hazards and need to be addressed to prevent their entry into the food chain or remain within permissible limit. Organic milk derived from organic dairy farming coming into market has a great demand with high cost. On the basis of type of beta-casein present in the milk, it is graded as A1 and A2 type of milk and it is thought that A2 milk is superior for health. Many consumers prefer plant-based milk like soy milk, almond milk, rice milk, and coconut milk due to ethical and health reasons. Synthetic milk menace has thrived in our country due to lack of strict regulatory bodies and a large portion of milk is produced in unorganized sector. Synthetic milk is a menace to human health in our country and a great challenge to combat it. Strict regulation, motoring, and supervising activities are to be adapted to eradicate the malpractice.

References

Aryana, K.J., Olson, D.W., 2017. A 100-year review: yogurt and other cultured dairy products. J. Dairy Sci. 100 (12), 9987–100013. Available from: https://doi.org/10.3168/jds.2017-12981.

Caffarelli, C., Baldi, F., Bendandi, B., Calzone, L., Marani, M., Pasquinelli, P., 2010. Cow's milk protein allergy in children: a practical guide. Ital. J. Pediatr. 36, 5.

Centre for Science and Environment, 2015. Adulterated milk is what Indians is drinking. CSE Report June 25. New Delhi.

Chan, J.M., Stampfer, M.J., Ma, J., Gann, P.H., Gaziano, J.M., Giovannucci, E.L., 2001. Dairy products, calcium, and prostate cancer risk in the Physicians' Health StudyAm. J. Clin. Nutr. 74 (4), 549–554. Available from: https://doi.org/10.1093/ajcn/74.4.549. PMID. Available from: 11566656.

Chan, J.M., Gann, P.H., Giovannucci, E.L., 2005. Role of diet in prostate cancer development and progressionJ. Clin. Oncol. 23 (32), 8152–8160. Available from: https://doi.org/10.1200/JCO.2005.03.1492. PMID. Available from: 16278466.

Cheng, Y., Dong, Y., Wu, J., Yang, X., Bai, H., et al., 2010. Screening melamine adulterant in milk powder with laser Raman spectrometry. J. Food Compos. Anal. 23 (2), 199–202.

Codex Alimentarius, 1995. General standard for contaminants and toxins in food and feed International Food Standards (CODEX STAN 193-1995), FAO/WHO.

Cruywagen, C.W., Stander, M.A., Adonis, M., Calitz, T., 2009. Hot topic: pathway confirmed for the transmission of melamine from feed to cow's milk. J. Dairy Sci. 92 (5), 2046–2050.

FDA, 2008. Issues interim safety and risk assessment of melamine and melamine-related compounds in food, Food and Drug Administration, USA.

Gandhi, P., Pandya, R., 2014. Lab-grown- milk- biotechnology-gmo-food–climate. https://news.nationalgeographic.com/news/2014/10/141022.

Giovannucci, E., Rimm, E.B., Wolk, A., Ascherio, A., Stampfer, M.J., Colditz, G.A., et al., 1998. Calcium and fructose intake in relation to risk of prostate cancer. Cancer Res. 58 (3), 442–447. PMID 9458087.

Greene, C., 2002. U.S. organic agriculture gaining ground. USDA-ERS Agric. Outlook April, 9–14.

Hallam, D., 2002. The organic market in OECD countries: past growth, current status and future potential. In OECD NPOP Organic Agriculture: Sustainability, Markets and Policies, Proceedings from an OECD Workshop, Washington DC, United States, September 2002, pp. 179–186.

Hau, A.K., Kwan, T.H., Li, P.K., 2009. Melamine toxicity and the kidney. J. Am. Soc. Nephrol. 20, 245–250.

Hayley, P., 2015. We tried Coca-Cola's new milk that costs twice as much as regular—here's the verdict. Business Insider February 3, 2015.

Heckman, J., 2006. A history of organic farming: transitions from Sir Albert Howard's War in the Soil to USDA National Organic Program. Renew. Agric. Food Systems 21, 143–150. Available from: https://doi.org/10.1079/RAF2005126.

Hemme, T., Otte, J. (Eds.), 2010. Status and Prospects for Smallholder Milk Production: A Global Perspective. FAO.

Honkar, A.S., Landge, S.N., Kele, V.D., 2015. Impact of adulteration of milk with melamine: a case of protein replacement. Int. J. Recent Sci. Res. 6 (2), 2883–2885.

Hui, Y.H., 2006. Handbook of food science, Technology and Engineering, vol. 2. CRC Press, Taylor & Francis, p. 58, ISBN 9780849398483.

Iftakharul, M.M., Chowdhury, M.A.Z., Easha, N.J., Rahaman, M.M., Shammi, M., Fardous, Z., et al., 2016. Investigation of heavy metal contents in cow milk samples from area of Dhaka, Bangladesh. Int. J. Food Contamination 3 (1), 16.

Innova Market Insights, 2017. Global Plant Milk Market to Top US $16 Billion in 2018: Dairy Alternative Drinks Are Booming. https://www.prnewswire.com.

Jalili, M., 2017. A review paper on melamine in milk and dairy products. Dairy Vet. Sci. J. 1 (4). Available from: https://doi.org/10.19080/JDVS.2017.01.555566.

Jena, S., Sahoo, R.K., 2017. Removal of Pb(li) from aquous solution using fruit peel as a low cost adsorbent. Int. J. Sci. Eng. Technol. 5 (1), 5–13.

Keuhn, B.M., 2009. Melamine scandals highlight hazards of increasingly globalized food chain. J. Am. Med. Assoc. 301 (5), 473–475.

Kruszelnicki, K., 2018. The difference between A1 and A2 milk. In: Great Moment in Science, 19 June 2018. https://www.abc.net.au/news/science/2018-06-19/dr-karl-a1-vs-a2-milk/9879800.

Lifschitz, C., Szajewska, H., 2015. Cow's milk allergy: evidence-based diagnosis and management for the practitioner. Eur. J. Pediatr. 174 (2), 141–150.

Liu, Y., Deng, J., An, L., Liang, J., Chen, F., et al., 2011. Spectrophotometric determination of melamine in milk by rank annihilation factor analysis based on pH gradual change-UV spectral data. Food. Chem. 126 (2), 745–750.

McBride, W.D., Greene, C., 2009. Characteristics, Costs, and Issues for Organic Dairy Farming. Unites States Department of Agriculture (USDA), p. 50, Economic Research Report No. 82.

Mudgil, D., Barak, S., 2013. Synthetic milk: a threat to Indian dairy industry. Carpathian J. Food Sci. Technol. 5 (1), 64–68.

National Dairy Council, 2000. Yogurt and Other Cultured Dairy Products.

Oliver, S.P., Jayarao, B.M., Almeida, R.A., 2005. Food borne pathogens in milk and the dairy farm environment: food safety and public health implications. Food Borne Pathogens Dis. 2 (2), 115–129.

Pajohi-Alamoti, M.R., Mahmoudi, R., Sari, A.A., Valizadeh, S., Kiani, R., 2017. Lead and cadmium contamination in raw milk and some of the dairy products of hamadan province in 2013–2014. J. Health 8 (1), 27–34.

Pierce, J., Tilth, O., 2009. Introduction to organic dairy farming. Available online at http://www.extension.org/article/18325.

Poorjafari, N., Zamani, A., Mohseni, M., Parizanganeh, A., 2015. Assessment of residue melamine in dairy products exhibited in Zanjan market, Iran by high-performance liquid chromatography method. Int. J. Environ. Sci. Technol. 12 (3), 1003–1010.

Purdy, C., 2017. A tech startup is making convincing cow-free milk by genetically engineering yeast. Quartz . Available from: https://qz.com/1161985.

Puschner, B., Poppenga, R.H., Lowenstine, L.J., Filignezi, M.S., Pesavento, P.A., 2007. Assessment of melamine and cyanuric acid toxicity in cats. J. Vet. Diagn. Invest. 19 (6), 616–624.

Reddy, O.S.K., 2017. Difference between Indian native A2 cow milk and cross bred A1 cow milk and health benefit of A2 cow milk. https://www.linkedin.com.

Rombauer, I.S., Becker, M.R., 1975. The Joy of Cooking. Bobbs Merrill Co. Inc., Indianapolis, 0-672-51831-7p. 533.

Saeed Faraji, M., Sani, A.M., Taylani, F., Golestani, A., 2017. Removal of lead by using surface absorption method. Quart. J. Environ. Sci. Technol. 19 (4), 51–59. Available from: https://doi.org/10.22034/jest.2017.10702.

Sahi, T., 1974. Lactose malabsorption in Finnish-speaking and Swedish-speaking populations in Finland. Scand. J. Gastroenterol. 9 (3), 303–308. PMID 4852638.

Saue, T.M., 2009. Food safety challenges in the dairy industry. Food Safety Magazine. October–Novenber, 2009.

Sharma, R., 2018. Milk in India is largely safe, even though quality issues persist. FSSAI Press Release. New Delhi, November 13, 2018.

Smith, K., 2018. This mother-son duo bringing affordable plant bases milk to India. Livekindly, April 14, 2018. https://www.livekindly.co.

Truswell, A.S., 2005. The A2 milk case:a critical review. Eur. J. Clin. Nutr. 59, 623–631.

WHO, 2009. Background Paper on Occurrence of Melamine in Foods and Feed Prepared for the WHO Expert Meeting on Toxicological and Health Aspects of Melamine and Cyanuric Acid. World Health Organisations, Canada, pp. 1–45.

Zarati, P., Shirkhan, F., Mostafidi, M., Zahedi, M.T., 2018. An overview of the heavy metal contamination in milk and dairy products. Acta Sci. Pharm. Sci. 2 (7), 8–21.

Further Reading

Ananthakumar, T., Suresh, A.J., Niraimathi, V., 2015. Identification and estimation of melamine residue in powdered milk by RP-HPLC. Int. J. Pharm. Res. 2 (4), 122–128.

Dairy Council, October 26, 2015. Lactose intolerance: prevalence, symptoms and diagnosis.

De Noni, I., FitzGerald, R.J., Korhonen, H.J.T., Le Roux, Y., Livesey, C.T., Thorsdottir, I., et al., 2009. Review of the potential health impact of beta-casomorphins and related peptide. EFSA Rep. 231, 1–107.

Dhanashekar, R., Akkinapalli, S., Nellutla, A., 2012. Milk-borne infections: an analysis of their potential effect on the milk industry. Germ 2 (3), 101–109.

https://en.wikipedia.org/wiki/Milk.

https://en.wikipedia.org/wiki/The_a2_Milk_Company.

Maji, S., Meena, B.S., Paul, P., Rudroju, V., 2017. Prospect of organic dairy farming in India: a review. Asian J. Dairy Food Res. 36 (1), 1–8.

Mulloy, A., 2010. Gluten-free and casein-free diets in the treatment of autism spectrum disorders: a systematic review. Res. Autism. Spectr. Discord. 4 (3), 328–339. Available from: https://doi.org/10.1016/j.rasd.2009.10.008.

Neo, P., 2018. India's dairy taint: over two-thirds of all milk and milk products violate standard. 12 September 2018. https://www.foodnavigator-asia.com.

Oruganti, M., 2011. Organic dairy farming—a new trend in dairy sector. Vet. World 4 (3), 118–130.

Ruegg, P.L., 2003. Practical food safety interventions for dairy production. J. Dairy Sci. 86 (E. Suppl.), E1–E9.

Subrahmanyeswari, B., Chander, M., 2008. Animal husbandry practices of organic farmers: an appraisal. Vet. World 1 (10), 303–305.

Wikipedia. Lactose intolerance. https://en.wikipedia.org.

World Cancer Research Fund/American Institute for Cancer Research, 2007. Food, Nutrition, Physical Activity, and the Prevention of Cancer: a Global Perspective. American Institute for Cancer Research, Washington DC978-0-9722522-2-5.

Yang, R., Huang, W., Zhang, L., Thomas, M., Pei, X., 2009. Milk adulteration with melamine in China: crisis and response. Qual. Assurance Safety Crops Foods 1 (2), 111–116.

CHAPTER

6

Hazards and Safety Issues of Meat and Meat Products

Arun K. Das[1], *P.K. Nanda*[1], *Annada Das*[2] *and S. Biswas*[2]

[1]Eastern Regional Station, ICAR-Indian Veterinary Research Institute, Kolkata, India
[2]Department of Livestock Products Technology, West Bengal University of Animal and Fishery Sciences, Kolkata, India

OUTLINE

INTRODUCTION

Meat and meat products have special position in our diet because of their rich nutrient content and desirable sensory attributes. However, meat and meat products can readily be contaminated or become unacceptable due to spoilage if not processed hygienically or handled properly at various points starting from production, storage, distribution, retail/sale, and preparation until consumption. Consumption of contaminated meat and meat products is an important source of human infections, leading to meat-borne illness

DOI: https://doi.org/10.1016/B978-0-12-816333-7.00006-0

(Biswas et al., 2008). The challenges of providing quality and safe meat to people often end up compromising safety issues, particularly in developing countries. The factors associated with meat safety are not limited to animal production but also cover complex processing, chilling/freezing, and transport right through to the retail level and then consumption (Sofos, 2008; Sofos and Geornaras, 2010; Zhou et al., 2017). The meat animals in some countries are slaughtered and sold in ill-equipped, unauthorized, and unmonitored roadside stands, where proper hygiene and processing standards are not followed. The carcass is often seen hanging in open air with flies buzzing around it, posing a high risk to public health. The food or meat safety regulatory bodies in such countries lack enough control and vigilance over the meat trade.

As per the World Health Organization (WHO), food safety is described as a process wherein the food is stored, prepared, and handled in such a way that it prevents illness caused by food. A broad view of the definition in terms of meat safety is that meat and meat products should not contain any hazardous components injurious to human health or cause any harm to the consumer when processed, transported, stored, and marketed (Sofos, 2008). Further, consumption of meat and meat products must not lead to acute or chronic poisoning, cause infection with pathogenic bacteria or viruses, or in any way harm the health of consumers or their offsprings (Zhou et al., 2017). Therefore, every step in a production chain must adhere to a complete understanding of meat safety principles coupled with appropriate and effective supervision, monitoring, and management measures in order to ensure that the overall safety of meat and meat products is not adversely affected at any stage. In fact, the right to safe food in human rights law is encompassed by both the right to health and the right to food (Leary, 1994; Mechlem, 2004). This is required to ensure consumers' health and to maintain consumers' confidence and satisfaction (Andrée et al., 2010). Realizing the importance of meat safety and in its effort to improve food safety issues, WHO rightly coined the slogan "**FROM FARM TO PLATE, MAKE FOOD SAFE**" as its theme for World Health Day (WHO, 2015).

The issue of meat safety has, therefore, emerged as a serious public health concern in recent years worldwide because of societal concerns. Health conscious consumers are now focusing more on meat adulteration and contamination stories that have created a wave of attention in social media. With increased awareness and public concern about food safety; providing healthy, hygienic, and wholesome meat and meat products without compromising safety issues remains a significant challenge of the industry in both the developing and developed nations.

SOURCES OF CONTAMINATION AT DIFFERENT STAGES OF PRODUCTION

The production of meat involves rearing of animals, transportation, slaughtering, carcass dressing, chilling, packaging, and storage (Sofos, 2005a). Meat can easily be contaminated at each stage of the production chain with different hazards (chemical, physical, and biological), which may affect consumer health, if ingested. Generally, the origin of meat safety issues can be divided into three stages of the meat production chain: pre-harvest,

harvest, and post-harvest stage. Control of meat safety hazards requires application of interventions at pre-harvest, post-harvest, processing, storage, distribution, merchandizing, preparation, food service, and consumption (Sofos and Geornaras, 2010).

Contamination at Pre-harvest Stage

The first stage at which meat safety problems are encountered is during the animal breeding/production process, where there may have been an inappropriate use of feedstuffs or feed additives, or the use of veterinary pharmaceuticals, leading to the presence or accumulation of hormone/drug residues, heavy metals, and pesticide residues in the body (Andrée et al., 2010; Sofos and Geornaras, 2010; Zhou et al., 2017). Transportation of meat animals in an unhygienic way with high stocking density favours cross contamination with pathogens. Inhumane treatment of animals during the preslaughter period again deteriorates meat quality and safety. Unless precautionary measures are taken at the pre-harvest stage, many of the microbial pathogens of concern can survive in the environment *viz.*, in water, in bedding, in housing, on equipment, on pastures, in feed, etc., where the meat animals are reared. Improved meat safety can only be achieved through pathogen reduction programs or various interventions such as diet manipulation, use of feed additives or supplements, antibiotics, bacteriophage therapy, administration of vaccines or immunization, competitive exclusion, prebiotics, or probiotics (Sofos and Geornaras, 2010). Proper animal management practices, such as pen management, feed, clean and chlorinated water, are needed at the farm level in order to reduce water and feed-borne contaminants and decrease the probability of pathogens in animals (Huffman, 2002; Sofos, 2005a,b).

Contamination at Harvest Stage

Slaughtering or harvesting processes usually include stunning, sticking, and bleeding of the animal. This is the stage of slaughtering and processing of food animals, where physical and chemical contaminations might occur. Bacteria could also be introduced into the carcass through the stunning or sticking wounds, possibly passing to the muscle tissues. It is commonly assumed that the enteric pathogens found in meat are largely derived directly from fecal matter of infected animals (Heuvelink et al., 2001). The muscle tissues of animals are generally sterile until meat surfaces are contaminated with gastrointestinal pathogens during the evisceration process. Pathogenic microorganisms (i.e., those that cause human disease or food-borne illness) are found in the digestive tract of healthy food animals. These microorganisms are excreted in the feces and can be found on the hides and fleeces of the live animal. Pathogenic microorganisms are then transferred from contaminated fleece/hide onto previously sterile meat surfaces during the slaughter and dressing process, and chances of contamination of the meat surfaces occur. Cross-contamination of carcasses of animals occurs by contact with contaminated hands of workers or equipment, or by direct contact with contaminated carcasses.

Visible cleanliness of the live animal is believed to be directly related to carcass hygiene and, hence, can be used as one of the control points for improving the safety of meat

(FSS, 2006). Limiting the microbial contamination at harvest and processing steps, implementation of meat safety principals such as decontamination or sanitization procedures, or antimicrobial interventions for inhibition of microbial growth during subsequent distribution and storage of products are needed (Koutsoumanis et al., 2006; Sofos, 2009; Stopforth and Sofos, 2006). In order to control hazardous microbiological contamination of meat, food animals coming to the slaughterhouse need to be inspected for a number of conditions that may be detected either at ante-mortem and/or at post-mortem inspection (Pennington, 1997). Proper care should be taken to avoid puncture of the viscera, and leakage from the oesophagus/gastrointestinal tract and anus should be avoided by tying or clamping the alimentary canal at either end. Animal processing operations have to (1) establish sanitation standard operating procedures; (2) operate under the hazard analysis critical control point (HACCP) system; and (3) meet microbiological performance criteria and standards for *Escherichia coli* biotype I and *Salmonella*, as a verification of HACCP (FSIS, 1996).

Contamination at Post-harvest Stage

The shelf-life and quality of meat is strongly influenced by its initial quality, package parameters, and storage conditions (Zhao et al., 1994). In fact, various factors like transportation (distance and climatic conditions including temperature), type of packaging, storage (refrigeration and retention), display, and bacterial growth affect post-slaughter meat safety in the supply chain. Therefore, the third and final stage in the meat distribution chain that includes packaging, storage, transportation, sales, and food services is equally important. Poor handling and transportation of meat at post-harvest stage may result in contamination and spoilage, causing serious illness to consumers. The supply chain that stretches from abattoir to final retail display is critical for meat quality and its shelf-life throughout the distribution chain (Richardson et al., 2009; Rosenvold and Andersen, 2003). The management of cold chain facilities is imperative not only to ensure quality meat with adequate shelf-life but also to control meat safety hazards, especially biological hazards. The control of biological hazards, such as spoilage and pathogenic bacteria, viruses, and parasites, is also particularly important at this stage, where any inappropriate practices will have an impact on meat safety (Jenson et al., 2014). Therefore, the cold chain must not be interrupted, as bacteria multiply rapidly in higher temperatures (Richardson et al., 2009).

Temperature affects not only the microbial condition of the meat but also the color stability, with retail appearance being significantly affected by storage at 2°C and 5°C, compared to storage at below-zero temperatures (Jeremiah and Gibson, 2001). Temperature conditions in retail cabinets play an important role in extending shelf-life and safety of meat and meat products (Koutsoumanis and Taoukis, 2005).

Proper packaging of the meat and meat products is important because it provides protection against environmental effects, communicates with consumers as a marketing tool, saves time, and brings convenience, thus safeguarding meat quality (Yam et al., 2005). The best way to control pathogen at this stage of food distribution (packaging, storage, transportation, sales, food services, and even consumer level) is by preventing transfer of contamination among meats and meat contact surfaces, cross-contamination, and by inactivating or inhibiting the growth of existing microbial contamination. To keep meat safety problems under control, ensuring proper sanitation and hygiene during storage, cooking ready-to-eat meat and meat products, and adopting robust monitoring process are vital approaches at this stage (Lianou and Sofos, 2007).

HAZARDS ASSOCIATED WITH MEAT AND MEAT PRODUCTS

Safety problems associated with food products of animal sources can be divided into physical, chemical, and biological hazards.

Physical Hazards

A physical hazard contaminates meat or its products at any stage of the production chain and may cause injury, but seldom causes death (Stopforth et al., 2012). Meat products could get contaminated with physical hazards from several sources such as contaminated raw materials, improper and poorly designed or maintained facilities and equipment, faulty procedure during processing, and improper employee training and practices. These are foreign materials (e.g., metal fragments in ground meat) and may be introduced to meat at any stage of production. Some common physical hazards in meat and meat products and their causes or sources are listed in Table 6.1.

Potential physical hazards can cause cuts or tears in the buccal cavity and may damage the gastrointestinal tract or inner lining of the stomach. Various substances like glass (light bulbs and glass food containers), metals (fragments from equipment such as splinters, blades, needles, utensils, and staples), plastics (packaging materials, fragments of utensils

TABLE 6.1 Some Common Physical Hazards Associated With Meat and Meat Products

Source of Physical Hazards	Type	Causes	Control Measures
Raw materials	Bullets, needles, wire, meat hooks, glass and metal fragments, bone, hard plastic, wood, etc.	Animals shot in the field, hypodermic needles used for injections, improper processing conditions in the plant, poorly designed or maintained facilities, and faulty equipment/ machinery	Ante-mortem and in-house inspections of animals/raw materials, in-line magnets/metal detectors; screens, traps, good agricultural practices at farm level, good hygiene practices at the lairage, implementation of HACCP
Packaging materials, retail store and marketing	Insects, pests, and other filth, polyentrapment	Plant postprocess entry, unhygienic storing conditions	Product inspection, good sanitation and quality control programs, precautions while removing film from liner
Employees/ workers	Hair, jewellery, buttons, nail trimmings, tobacco products, etc.	Unskilled workers/ employees	Awareness and training to meat handlers, adopting good personal hygiene

used for cleaning equipment), stones, wood (splinters from wood structures used to store or transport ingredients or food products), and natural components of food like hard or sharp parts (bone fragments in minced meat) are the most commonly found sources of physical hazards in meat. There are many ways including good manufacturing practices (GMP) through which processors can prevent physical hazards in meat and meat products during production, processing, storage, and transportation, and at meat service sites.

Chemical Hazards

Food animals are constantly exposed to thousands of natural and man-made, nonnutritive feed components. Exposure of animals to certain harmful chemicals, either through direct or indirect means, causes harm to recipient animal (toxic effects) and if not cleared they may leave their residues in meat, milk, or eggs (Smith and Kim, 2017). Likewise, meat and meat products can be contaminated with a wide variety of chemicals and/or additives during any stage of production and processing. Consumers have become more health conscious and have expressed their apprehension regarding the presence of various chemicals, toxicants, adulterants, or residues of pesticides, antibiotics, drug, antimicrobials, heavy metals, and hormones in meats of animal origin (Tilahun et al., 2016). Chemical hazards associated with meat and meat products can originate from three general sources (FISS, 1997):

1. Unintentionally added chemicals
   a. Agriculture chemicals: pesticides, herbicides, animal drugs, fertilizers, etc.
   b. Plant chemicals: cleaners, sanitizers, oils, lubricants, paints, pesticides, etc.
   c. Environmental contaminants: lead, cadmium, mercury, arsenic, polychlorinated biphenyls (PCBs).
2. Naturally occurring chemical hazards: products of plant, animal, or microbial metabolisms such as aflatoxins.
3. Intentionally added chemicals: preservatives, acids, food additives, processing aids, etc.

Smith and Kim (2017) classified chemical hazards (contaminants and residues) of meat and meat products into man-made, nonagricultural environmental contaminants (dioxins, polychlorinated phenols, and brominated flame retardants), man-made environmental residues and contaminants of agricultural origin (pesticides), natural products (mycotoxins and plant toxins), toxic metals, or veterinary drugs. Stopforth et al. (2012) reported that naturally occurring substances and manufactured chemicals are two principal sources of chemicals or contaminants that may contaminate meat and meat products. Regardless of how or on what basis a given contaminant or residue is classified, the focus should be on hazard identification and assessment so that entry of these chemicals or contaminants into the food chain through various means is prevented.

Veterinary Drugs, Growth Promoters, and Antibiotic Residues

Veterinary drugs, which include antibiotics, hormones, and antibacterial and antiparasitic agents, have long been in use in animal production as therapeutic agents to control various diseases (infectious or noninfectious) or as prophylactic agents to prevent disease outbreaks as well as to control parasitic infections (Dixon, 2001). The very purpose of these drugs is to keep an animal in good health, thereby reducing the possibility of a disease being transmitted from animals to humans (Lozano and Trujillo, 2012). These drugs or their residues thus have the potential to enter the food chain of animal origin that could increase the risk of diseases in people. Similarly, growth-promoting agents, which include β-agonists (Ractopromine, Clenbuterol, Salbutamol, Mebuterol), stilbens (Diethylstilbesterol), steroids (Trenbolon acetate, 17-β-estradiol, Melengestrol acetate), and resorcycilic acid lactones (Zeranol) are used in feed to improve the feed conversion efficiency and weight gain, reduce fat deposition, and increase lean meat yield by partitioning nutrients for muscle growth rather than fat. There are many reports that suggest β-agonists can cause a reduction in the overall eating quality of meat, particularly tenderness and juiciness (Schiavetta et al., 1990).

Likewise, antibiotics or antimicrobial agents such as aminoglycosides, ß-lactams, macrolides and lincosamides, quinolones, sulfonamides, and tetracyclines are added to make more nutrients available to the animal and not to the gut bacteria. In recent years, there has been increasing concern regarding the development of increased bacterial resistance to certain antibiotics due to their inappropriate and excessive use (Butaye et al., 2001). This not only creates problems for food animal welfare in the long run, but presence of residual amounts of these antibiotics or their metabolites in meat and meat derived foods may be harmful to human health. Few antibiotics also have the potential to exert allergic reactions in sensitized individuals (penicillin) and toxicity such as aplasia of the bone marrow (chloramphenicol), alter human gut microbial populations, and transfer antibiotic resistance genes to human pathogens (Franco et al., 1990; Mitchell et al., 1998).

The reasons for the presence of antibiotic residues in meat and meat products could be due to poor management, illegal and indiscriminate use of antimicrobials, accessibility to unauthorized person, extended usage or inappropriate dosages, violation of regulations regarding drug withdrawal times prior to slaughter, and above all, the lack of consumer awareness about the magnitude of human health hazards associated with antimicrobial residues consumption through meat and meat products (Kaneene and Miller, 1997; Zhou et al., 2017).

Different guidelines and mechanisms are promulgated by various countries to control and regulate the proper use of drugs in animal production. Similarly, a number of national and international organizations, like the Codex Alimentarius Commission, whose guidelines are set by the Codex Committee on Residues of Veterinary Drugs in Food, based on the scientific advice of the Joint WHO/FAO Expert Committee on Food Additives (Mitchell et al., 1998), are also involved in the formulation of control strategies for preventing misuse of veterinary pharmaceuticals in animal production. Rational use of veterinary drugs, control of their distribution, proper doses as prescribed by veterinarians, and proper drug-withdrawal and residue detection technologies are a few of the control mechanisms (Andrée et al., 2010).

Environmental Contaminants

The potentially harmful environmental pollutants and toxins produced by humans include inorganic elements such as arsenic, cadmium, mercury, lead, polycyclic aromatic hydrocarbons (PAHs), heterocyclic amines, dioxins, pesticides, and other persistent organic pollutants such as PCBs, and industrial chemicals (Domingo and Nadal, 2016). Safety of meat and meat products is also compromised by numerous contaminants that may come from contaminants formed during production and cooking; contaminants arising from food packaging, or natural toxins in food (Oskarsson, 2012). These environmental chemicals or contaminants from man-made sources accumulate in foods of animal origin, including meat and meat products. However, the level of risk to consumer health from environmental contaminants is dependent on the actual level of contaminants in the meat and meat products as well as the amount consumed by individuals.

Contamination with heavy metals is a serious threat to animals because of their toxicity, bioaccumulation, and biomagnifications in the food chain (Eisler, 1988). Heavy metals such as lead, iron, cadmium, mercury, chromium, copper, arsenic, and nickel originating from industrial production process are released into the terrestrial and aquatic ecosystems. These metals accumulate in different tissues including the liver, muscles, and kidneys of animals through drinking water, feed additives, and feeds. Higher concentration of heavy metallic residues in meats can cause negative health effect to consumers, such as disruption of various metabolic processes and disorders of the gastrointestinal, urinary, and nervous systems (Reyes-Herrera and Donoghue, 2012). Hence, meat and meat products should be regularly screened for heavy metals. Likewise, various drugs, antibiotics, and growth promoters used in the livestock sector need to be monitored to minimize the entry of these chemical items into the food chain and hence also to humans and animals. The regulations and resultant hazards should be monitored in a synchronized way to enable a judicial route of ridding such situations within a very minimal stipulated time frame.

Food Additives

Food additives have continuously been used since time immemorial for processing and preservation of various food products and the meat sector is no exception to this. In fact, additives are used to achieve certain technical benefits such as for preservation by pickling and/or salting, or by enhancing the functionality of muscle protein (Yu et al., 2012), or modifying texture of meat products. One of the examples of an additive is polyphosphate. In adequate quantities, it provides a wide range of benefits starting from maintaining texture and also increasing the nutritional profile by holding higher amounts of water in the products. Likewise, nitrite, one of the important chemical additives, is used to enhance colour—a major sensory attribute in meat-processing and product development. In spite of these facts, the additives of chemical origin, when used consistently over a longer period of time at a concentration other than prescribed, often pose questions of doubts in terms of their effect on human health. Sometimes they are even collated with a reduction of oxidation of certain proteins, resulting in the genesis of certain dangerous health hazards, like proliferation of cancer cells (Kushi et al., 2012). Often, some chemicals that are not on the approved list are used erroneously, inadvertently, or sometimes intentionally for a very minor gain without considering the level of ill effects of such chemicals on consumers. Sudan red, lead nitrite, and aluminum sulfates are some of the chemicals used without understanding the series of ill effects these chemicals have on humans. Sudan red, as an illegal food additive and widely used for improving meat color and appeal is typically classified as an azo dye with a carcinogen of category 3 by the International Agency of Research on Cancer (Ahmed Refat et al., 2008). Therefore, the action of such chemicals exposed to different levels and types of heat treatments while processing meat items needs further study and investigation. Further, additives as per the approved list (GRAS) and generally allowed by the regulation of food safety at their level of permissibility, should be used in the processing sector, so that their resultant hazards can safely be minimized or avoided.

Naturally Occurring Toxicants

In nature, a wide range of flora and fauna exists. Some of these items may be toxicants to living objects when ingested unintentionally. These toxicants ingested by food animals while grazing in the natural pasture are often passed through milk, meat, and eggs (Beier, 2000). Important among them are alkaloids, the natural chemicals present in weeds, grasses, plants, and herbs. Pyrolizidine alkaloid (PA) is also a naturally occurring contaminant found in some plants, herbs, leaves, and flowers (Wiedenfeld, 2011). Animals exposed to such plants, herbs, leaves, and flowers may be a source of deporting the alkaloids into their products. People may get serious illnesses through PAs directly or indirectly through the consumption of products like meat and milk (Mulder et al., 2018).

Bacterial toxins (Staphylococcal enterotoxins, Botulinum toxins) and viable pathogenic bacteria exert a wide range of toxic infestation in the animals. Botulinum toxins are potent neurotoxins and grow in anaerobic condition, typically in canned food, and such toxins are even stable to a certain extent with digestion and heat (Stopforth et al., 2012). That's why the poisoning effect of these types of toxins are very much evident in cooked food stored at higher temperatures. Likewise mycotoxins are toxic secondary metabolites produced by certain species of fungi and molds (Chu, 2002). Mold growth and toxin production in food and feed grains, nuts, oilseeds, forage, etc., largely depend on environmental and storage conditions. Their presence in feed material intended for animal consumption can lead to the development of toxic symptoms in the infected animals. Residues of important mycotoxins like aflatoxins (B1) and ochratoxin A are also reported in various meat and meat products. Presence of such toxins in meat and meat products at a certain level causes a number of food-borne illnesses in humans, if ingested.

Biological Hazards

Biological hazards still remain a global concern despite technological advances, extensive scientific progress, education, and public awareness of safe food production (Mead et al., 1999). This is mainly due to unhygienic and unwholesome practices followed during production resulting in contamination of meat and meat products with a wide range of biological hazards such as bacteria, parasites, viruses, and prions. A compilation of such biological hazards is presented in Table 6.2.

The muscle tissue of healthy animals is sterile and free from microorganisms immediately after slaughter. But during the long chain of slaughtering, evisceration, washing, and deboning, the under skin of carcasses is often exposed to ambient environmental hazards or comes in contact with microorganisms present in hair, gastrointestinal and respiratory tracts, etc. (McEvoy et al., 2000; Sofos, 2014). Apart from feces, the hide, water, air, intestinal contents, and lymph nodes, insufficiently cleaned and sanitized processing equipment

TABLE 6.2 Sources of Biological Hazards in Meat and Meat Products and Their Control Approaches

Biological Hazards	Type of Meat	Type of Illness	Source	Control Approach
BACTERIA				
Bacillus cereus	Beef, pork, lamb, poultry	Toxicoinfection (diarrheal) in meat products	Soil, animals	Inactivate spores (heat); temperature control ($>60°C$ or $<10°C$) to prevent spore germination and growth; control pH to prevent growth; cook foods when needed for consumption
Campylobacter spp. (thermophilic)	Beef, pork, lamb, poultry	Invasive infection	Animals	Inactivation of cells (pasteurization, cooking); hygienic slaughter and processing procedures; irradiation of meat and poultry; treatment of water; prevention of cross-contamination of contact surfaces; personal hygiene in food preparation; keeping pets away from food-handling areas; avoidance of eating raw or partially cooked meat
Clostridium botulinum	Beef, pork, lamb, poultry	Intoxication (toxicoinfection, infant botulism)	Soil, water, animals	Inactivate spores (canning/sterilization); control cell growth (refrigeration); thorough cooking of home-canned food (boiling or stirring for 15 min); acid-preservation ($pH < 4.6$); discard swollen cans
Escherichia coli (STEC/EHEC and other pathogenic groups)	Beef, pork, lamb	Toxicoinfection	Animals	Prevent fecal contamination of food and water; inactivation of cells (pasteurization, cooking); practice of good personal hygiene; irradiation of meat; separation of raw and cooked foods; avoidance of eating raw or partially cooked meat and meat products; refrigeration
Listeria monocytogenes	Ready-to-eat beef, pork, lamb, poultry	Invasive infection	Processing environment, soil, water, animals	Inactivate cells (pasteurization, cooking); control growth (freezing, no long-term refrigerated storage); avoid cross-contamination; thoroughly reheat before consumption
Salmonella enterica	Poultry, beef, pork, lamb	Invasive infection	Animals	Inactivate cells (pasteurization, cooking); control growth (refrigeration); irradiate meat and poultry; reheating of food; prevention of cross-contamination; cleaning and sanitation of food preparation surfaces; exclusion of pets and other animals from food-handling areas; vulnerable consumers should avoid raw and undercooked meat and poultry and other animal origin foods

(Continued)

TABLE 6.2 (Continued)

Biological Hazards	Type of Meat	Type of Illness	Source	Control Approach
Staphylococcus aureus	Ready-to-eat beef, pork, lamb, poultry	Intoxication (heat-resistant toxin)	Humans, processing environment, animals	Exclusion of infected food handlers; inactivate cells (pasteurization, cooking); control food handlers (skin lesions, boils, cuts, etc.); good personal hygiene of workers; no time–temperature abuse of cooked/ready-to-eat foods (refrigeration)
Yersinia enterocolitica	Poultry, pork	Invasive infection	Animals, soil, water	Inactivate cells (pasteurization, cooking); control growth (freezing); prevention of cross-contamination
PARASITES				
Cryptosporidium parvum	Beef, poultry	Invasive infection	Meat, water, human (fecal–oral transmission)	Hand washing; pasteurization/cooking; irradiation; filtration and disinfection of water; sanitary disposal of excreta, sewage, and wastewater
Giardia duodenalis	Beef, poultry	Invasive infection	Meat, water, human (fecal–oral transmission)	Hand washing; pasteurization/cooking; irradiation; filtration and disinfection of water; sanitary disposal of excreta, sewage water; good hand hygiene
Sarcocystis spp.	Beef, pork	Invasive infection	Meat	Salting; pasteurization; cooking; irradiation
Taenia spp. (*cysticercosis, taeniasis*)	Beef, pork	Invasive infection	Meat	Prevention of fecal contamination; safe sewage disposal; heating (pasteurization or cooking); freezing; irradiation; early diagnosis and treatment
Toxoplasma gondii	Beef, pork, poultry (raised outdoors)	Invasive infection	Meat	Heating (pasteurization, cooking); irradiation; good personal hygiene after contact with cats and before food preparation; safe disposal of cat feces
Trichinella spp.	Pork	Invasive infection	Meat	Freezing; pasteurization/cooking; irradiation
VIRUSES				
Hepatitis E	Beef	Invasive infection	Animals, water, human (fecal–oral transmission)	Hand washing; sanitation; cooking
PRIONS (ENCEPHALOPATHIES)				
Bovine spongiform encephalopathy (BSE)	Beef	Invasive infection	Animals	Special controls during animal growth (e.g., control of safety of animal feed) and slaughter (e.g., control of animal health and removal of specified risk material)

Source: *Adapted from Adams, M.R., Motarjemi, Y., 1999. Basic Food Safety for Health Workers. World Health Organization, Geneva; Sofos, J.N., 2014. Meat and Meat Products. Food Safety Management. Elsevier, pp. 119–162.*

such as slicers, dicers, utensils, and packaging machineries used during animal slaughtering and meat processing can still be contaminated by microorganisms and act as sources of biological hazards (Sofos, 2014). When meat comes in contact with insufficiently cleaned and sanitized equipment and utensils during slaughtering and processing, chances of disease outbreaks in humans after consumption of such meat cannot be ruled out. Cross-contamination of the carcass or meat may also occur with additional sources of biological hazards from animal feeds, rodents, birds, insects, vehicles, and crates/containers used for animal transportation. The biological hazards may also be found in food ingredients such as salt and other additives used in meat processing and formulation of various products. Table 6.3 illustrates the equipment, machineries, and utensils widely used in the butcheries and the source of prevailing microbial contamination in them.

Meat, being a very potent congenial media for microbial growth, attracts microbes from primary sources like hide/skin, feces, soil, and water. It can also get contamination from the environment during and after transportation, in the lairage, and also at the time of processing.

The equipment, appliances used for animal rearing and slaughter, and overall human exposure increase the risk of cross-contamination to the animals, leading to microbial infestation. During slaughter processing, particularly skinning, cutting, and deboning, the carcass surface could adhere to dirt and dust, and these could be a source of various types of microorganisms. Initial contaminations of microbes to a level as low as 10^2 and as high as 10^7 colony forming units (CFU) per cm^2 is considered as permissible depending on processing operations and carcass site (Koutsoumanis et al., 2006). Contamination and

TABLE 6.3 Common Microorganisms Prevailing in Equipment and Utensils Used During Processing of Meat and Meat Products

Equipment and Utensils	Uses	Prevailing Microorganisms	References
Knives	Deboning, cutting, slicing and dicing	*Escherichia coli* and *Listeria monocytogenes*	Rivera-Betancourt et al. (2004)
Band saws	Shearing tough muscles, carcasses, and cutting of frozen meat	*Salmonella* sp., *E. coli* and *L. monocytogenes*	Warriner et al. (2002)
Bowl cutters	Chopping meat into small pieces and mincing	*Staphylococcus aureus*	Downes and Ito (2001)
Chopping boards	Slicing meat and meat products	*Salmonella* spp., *S. aureus*, *Pseudomonas aeruginosa*, and *Clostridium* spp.	Ak et al. (1994)
Meat slicers	Cutting ready-to-eat meat into desirable slices.	*L. monocytogenes*	Blackburn and McClure (2009)
Meat grinders	Mincing meat	*L. monocytogenes*	Shilenge et al. (2017)
Cold room	Storing chilled meat to prevent the growth of microorganisms	*L. monocytogenes*	Shilenge et al. (2017)
Freezer room	Freezing the meat (−18°C)	*L. monocytogenes*	Shilenge et al. (2017)

cross-contamination may also depend on the individual animal, its geographical origin, and during the particular time of a year.

While considering the aspects of animal in general and meat in particular, the abattoir and its premises are very important in possible contamination of the carcasses and meat. This not only enhances the total quantum of microorganisms over the surface of the meat and carcasses, but it also acts as the source of microbial infection for its consumers. To keep the microbial load of raw meat under control, different food safety measures/guidelines should be followed strictly, identifying and monitoring the critical control points (CCPs) of each operations of animal products and processing in accordance with HACCP (Bhandari et al., 2013).

Further, application of modern techniques along with observance of time/temperature integrators, combining different hurdles in the preservation system, and cleaning and disinfections techniques can be effectively used to raise the nutritional quality and safety of meat and meat products (Brown, 2000; Buncic et al., 2014).

Spoilage and Pathogenic Bacteria

Approximately 90% of food-borne illness in human is caused by bacteria alone. Meat is a very good medium of microbial growth, as it contains basic nutrients, which favor the growth of microorganisms. Besides, handling the raw meat in unhygienic manner creates an environment conducive enough for the growth of microorganisms. Spoilage microorganisms belong to genera of both Gram negative and Gram positive. A compilation of such spoilage and pathogenic microorganisms is given in Table 6.4. Spoilage of meat occurs when the total number of such microbes reaches 10^7 CFU per gram, thereby deteriorating its quality, sensory attributes, and shelf-life (Brown, 2000). Microorganisms, such as *Enterobacteriaceae* spp., *Acinetobacter* spp., *Aeromonas* spp., *Alcaligenes* spp., *Moraxella* spp., *Flavobacterium* spp., *Staphylococcus* spp., *Micrococcus* spp., and *Pseudomonas* spp., to name a few, are responsible for spoilage as well as loss of acceptable eating quality of meat (Dave and Ghaly, 2011; Ercolini et al., 2009; Sofos, 2014). In processed meat products, *Micrococci, Streptococci, Lactobacilli,* and *Bacillus* spp. are the important microbes. The spoilage processes by the microorganism consist of lipid oxidation and protein degradation along with the loss of important molecules. The spoilage in meat is evident with observation of increased pH values beyond the ultimate pH of the meat apart from development of off flavor, gas, and slime production due to the breakdown of fat, complex protein, polypeptides, carbohydrates, and other nutrients. However, the changes in meat quality due to spoilage is variable with the types of microorganisms and their way of action, such as altering pH, enzymatic activity, and sugar and lipid content, and these are associated with temperature, packaging, and time-length of storage (Dave and Ghaly, 2011). Further, the type of meat product, like cooked, cured, heat processed, fermented, or dried, reacts differently with spoilage microorganisms in various ways due to their inherent level of water activity and pH of the individual type of products. In case of meat stored at low temperature, microorganisms involved in spoilage rely on oxidative metabolism (Sofos, 2014). Health hazards often develop out of such stored meat due to the presence of biogenic amines where apart from microbial infestation, other factors like moisture, time, and temperature of the stored products play a vital role.

TABLE 6.4 Spoilage and Pathogenic Microorganisms Commonly Associated With Meat and Meat Products

Spoilage microorganisms	Both Gram-negative and Gram-positive bacteria, such as *Pseudomonas* spp., *Enterobacter* spp., *Acinetobacter* spp., *Aeromonas* spp., *Alcaligenes* spp., *Moraxella* spp., *Flavobacterium* spp., *Staphylococcus* spp., *Micrococcus* spp., *Serratia* spp., lactic acid bacteria, *Clostridium* spp., *Bacillus* spp., coliforms, yeasts, and molds		
	Bacteria	*Type of meat*	*References*
	Pseudomonas, Acinetobacter, Alcaligenes and *Moraxella* spp.	Unpreserved meat products stored at chilled temperatures (4°C ± 1°C)	Brown (2000), Doulgeraki and Nychas (2013)
	Enterobacter spp.	Refrigerated meat product	Dave and Ghaly (2011)
	Enterobacter and *Pseudomonas* sp.	Modified atmosphere packed meat (especially pork)	Dave and Ghaly (2011)
	The lactic acid bacteria, *Enterococci, Micrococci* and yeasts	Raw, salted, and cured products such as corned beef, uncooked hams, and bacon	Brown (2000), Dave and Ghaly (2011)
Pathogenic microorganisms	*Bacillus cereus, Campylobacter* spp., *Clostridium perfringens, C. botulinum, C. difficile, Escherichia coli, Listeria monocytogenes, Salmonella* spp., *Staphylococcus aureus, Aeromonas, Brucella, Enterobacter, Shigella, Yersinia enterocolitica*, etc.		
	Bacteria	*Type of meat*	*References*
	Campylobacter jejuni and *Campylobacter coli*	Raw poultry products	Mataragas et al. (2008), Sofos (2014)
	C. botulinum	Beef, pork, and poultry products	Akhtar et al. (2009), Sofos (2014)
	Escherichia spp.	Undercooked poultry-meat products, nonintact meat products such as beef	Sofos and Geornaras (2010); Sofos (2014)
	L. monocytogenes	Ready-to-eat pork or poultry-meat products, or reheated meat products, partially cooked pork products, and poultry foods	Mataragas et al. (2008), Sofos (2014)
	Salmonella spp.	Processed pork or poultry-meat products (ready-to-eat or to be reheated) and partially cooked pork products	Sofos and Geornaras (2010), Sofos (2014)
	S. aureus	Processed meat products, especially ready-to-eat pork products	Mataragas et al. (2008), Sofos (2014)
	Y. enterocolitica	Cooked, reheated pork products or improper cooking products, contaminated water	Mataragas et al. (2008), Sofos (2014)

Microbial contamination of meat is considered as one of the major sources of causing food-borne diseases. Even mild heat treatment cannot guarantee the complete inactivation of some deadly microbes posing for developing diseases (Sofos and Geornaras, 2010). Besides, the microorganisms associated with meat and meat products release certain toxins during germination and multiplication of spores. An elaborated illustration of the type of meat and sources of microorganisms with possible control measures is presented in Table 6.2.

In general, contamination of meat is associated with inappropriate farming practices (Shilenge et al., 2017) as well as unhygienic processing, improper storage, and packaging and transportation of products. Thus, it is crucial to improve the "farm-to-fork" practices in order to prevent, reduce, and/or at least control a number of food-borne diseases related to meat production (Jacob et al., 2010). To achieve this, control of the pathogens responsible for contamination, starting from farm to fork, should be given utmost priority. This is possible by identifying critical control points (CCPs) of each operation toward the production of wholesome and hygienic meat through implementation of HACCP. The protocol for each operation should be followed according to standard operating procedures for each facet of the elements of hygienic meat production, starting from selection of the animal at the farm level by the raisers and ending at the table of the consumers.

Viruses

Although it is difficult to ascertain the stage at which meat is contaminated with a virus, it is believed that poor evisceration (loosened, tied-off rectum, cut bladder, punctured viscera) and adoption of unhygienic practices by infected meat handlers during processing are some of the causes of meat contamination. Risk level is high when meat contaminated with norovirus, hepatovirus A, and orthohepevirus A is transmitted to humans and causes illness like gastroenteritis, hepatitis A, and hepatitis E (Smulders et al., 2013). Although enteric viruses, like rotavirus, adenovirus, sapovirus, aichivirus, parvovirus, and poliovirus, develop low to moderate illnesses in humans, their direct relationship with meat and meat products is yet to be ascertained. Avian influenza, one of the virulent pathogenic viruses, resulting in almost 100% death in birds, can also cause human infection with certain levels of mortality. According to a joint statement by the United Nations Food and Agriculture Organization (FAO) and WHO, there is no direct epidemiological evidence of avian influenza and its transmission to humans through meat and meat products that have been properly cooked (Rassool, 2006; WHO, 2005). But a recent study indicates that the tissue type (feather, muscle, and liver tissues) and temperature (4°C and 20°C) can greatly influence the survival of highly pathogenic avian influenza (HPAI) H5N1 virus in the tissues of infected chickens (Yamamoto et al., 2017). A transmissible bovine spongiform encephalopathy (BSE) or prion disease is a major animal health problem with neuro-generative changes in animals as well as humans (Sofos, 2014). This outbreak has caused a huge loss in the animal industry as well as evoked serious threats on consumption of meat, especially beef (Sofos, 2014). However, this could be averted by following certain specific preventive control measures like eliminating body parts of diseased or suspected animals. These body parts, like the brain, eyes, and spinal cord are often termed as risk materials, and all animals slaughtered beyond 30 months of age should be looked at with disdain (Chesebro, 2003). Therefore, developing advance meat recovery systems against such diseases is one way of controlling such an outbreak.

As the transmission of virus is associated with poor sanitation, inadequate cooking, and cross contamination, control can be managed through awareness of the workers relating to food services with some skills in sanitation and hygiene practices. Introducing a ban on the use of risky materials in the feed industry should be followed along with other provisions to control such diseases effectively.

Parasites

The infestation of parasites in meat has existed since time immemorial, and humans can acquire parasitic infections through the consumption of raw or undercooked meat or its products. The parasites that are important in regard to pork are *Taenia solium, Trichinella spiralis, Sarcocystis suihominis*, and *Toxoplasma gondii*, causing taeniasis, trichinosis, sarcocystosis, and toxoplamosis, respectively (Zhou et al., 2017). With regard to beef, *Taenia saginata* and *Staphylococcus hominis* are important parasites to be considered. *Cryptosporidium* sp. and *T. gondii* can be transmitted from poultry meat mainly when reared in free-range system. The involvement of some parasites like *Echinococcus granulosus, E. multilocularis, Angiostrongylus cantonesis*, and *Cysticercus* sp. in the recent years has brought parasite and meat into focus. However, the infestation of parasites to humans through meat is generally remote, as meat is cooked (heating to 70°C core temperature) thoroughly, resulting in infectiveness of such parasites, even if embedded within the meat. Furthermore, freezing, salting, chemical treatment, and ionizing radiations readily destroy or inactivate the parasitic infestations in meat and meat products (Sofos, 2014). Alternatively, adherence to guidelines during management of farms, maintaining a high level of sanitation, and inspection during ante-mortem and post-mortem inspection and during processing of meat are some of the effective methods to reduce the risks of human exposure.

MANAGEMENT AND CONTROL OF HAZARDS ASSOCIATED WITH MEAT AND MEAT PRODUCTS

The physical, chemical, or biological hazards associated with meat animals may cause the meat products to be unsafe for human consumption. Therefore, the source and types of hazards associated with meat and meat products at different stages of production (farm, processing, storage, and packaging) need to be identified and characterized in order to evaluate their potentiality on consumers' health. It is necessary to implement various plans and programs in order to control such hazards in meat and meat products.

Elimination of Physical Hazards

Eliminating physical hazards can be achieved through GMP by installing the elimination system like the use of metal detectors, magnets, etc. Some common physical hazards like stones, bones, plastic materials, or metals in carcasses or meat can be checked by using X-rays within the live animal body, and visual examination of the carcass, body parts, and viscera. Identification can also be done by radar system through transmission of low-power microwaves. However, effective maintenance of different facets of an animal-rearing system is the basis to eliminate physical hazards in the meat-processing industry.

Management of Chemical Hazards

Use of fertilizers, pesticides, weedicides, herbicides, and insecticides in the field of agriculture for optimization in agricultural production invariably pollutes the soil, plants, and water bodies and has become a potent source of chemical hazards in animal products starting from milk, meat, and eggs (Smith and Kim, 2017; Tilahun et al., 2016). The situation becomes more critical when effluents from industries in the form of heavy metals and toxic materials contaminate the fields and water bodies. In spite of many regulations by international organizations, the drug residues in meat and meat products are a potential area of chemical contamination causing multidimensional health issues in the human population. Standardization of maximum residual level, surveillance programs, and effective promulgation of regulations are some of the immediate interventions to get rid of the situation. However, for addressing the chemical hazards in a most effective way, each and every step of the production process should be rigorously monitored and analyzed through identification of CCPs. Further, enforcement of food safety regulations should be done on the basis of screening meat samples and standardizing maximum residue level (MRLs) to protect human health from the potent risk of chemical contaminants.

Controlling Biological Hazards

Controlling biological hazards is a very challenging job, as meat is a very good media and is vulnerable to the growth and multiplication of microbes, if hygienic processing, handling, packaging, transportation, and preservation methods are not adopted in totality. Traditionally, methods to control temperature, moisture or application of chemicals, ionizing radiation are often employed for controlling microbes in meat and meat products (Sofos, 2008). Subsequently, when it was felt that individual intervention in controlling microbes might not be always sufficient to get optimum results, the concept of hurdle technology came up. Hurdle technology is a concept where a right combination of a number of techniques/factors (water activity, temperature, pH, competitive flora, redox potential, preservatives etc.) at a time is used for controlling microbes with a satisfactory level of success (Sofos, 2005a).

However, in order to get satisfactory achievement in controlling microbes, sufficient attention should be given during ante-mortem and post-mortem examination of animals and the approaches should be employed (Stopforth and Sofos, 2006) in the following ways:

1. Animals should be provided with potable water and clean and wholesome feed free from physical hazards, microbes, and animal residues. It is advisable to incorporate high roughage diet and supplements in order to minimize carriage and shedding of certain pathogens.
2. Prebiotics, probiotics, and competitive exclusion have the potential for control of food-borne pathogen in animals, as feeding of beneficial bacteria competes with the hazardous pathogen in guts and in live animals.
3. Cross-contamination among the animals through excreta, blood and viscera, and equipment used during processing must be checked.

4. Controlling sources and quantum of microbes during rearing of livestock and at the point of slaughter site within the abattoir must be done.
5. Decontamination of animal carcasses should be practiced by using potable water with pressure jet, or incorporating some antimicrobial agents like solutions of organic acids, chlorine, etc.
6. Washing and decontamination of animal carcasses (intervention after flaying and evisceration) can be achieved through carcass cleaning with different pressure, thermal, and chemical decontaminants (Table 6.5).
7. Pest management is required to prevent cross-contamination and dissemination of certain diseases within animals.
8. Use of feed additives, vaccines, and antibiotics is helpful to reduce microbes in the animal body, especially in rumen and in feces. However, the question of antibiotic resistance pathogen is a matter of concern in this regard.
9. Although antimicrobial applied in fresh meat is not considered as a routine practice, antimicrobial compounds (both chemical and natural) of organic and phyto and phenolic origin compounds can be applied to fresh whole muscle. Some recent strategies include enhancement of pH using ammonia gas, antimicrobial protein from porcine leucocytes, antimicrobial agents from organic acids, calcium alginate gel, treatment with phosphates compound, chlorine, acetic acid, etc.
10. Thermal interventions like radiant heating, flush steam heating, hot water or steam pasteurization, and immersion in hot water are the most commonly used methods in controlling contamination in ready-to-eat (RTE) meat products.
11. Nonthermal physical interventions in the form of high-pressure process, pulse-electric field pasteurization, sonication, UV light, oscillating magnetic field, and various irradiations processes in the form of cold sterilization are gaining popularity and can be used commercially.
12. Antimicrobials such as pure LAB culture as biopreservative, addition of crude antimicrobial metabolites, essential oil, etc., can be used to control microbes in RTE meat products.
13. Microbial cross-contamination along with protection from other hazardous materials can be minimized effectively through vacuum, modified atmospheric packaging of meat and meat products.

MEAT SAFETY AND HUMAN HEALTH

The importance of meat in human nourishment cannot be ignored owing to the fact that meat is a very vital food component with proteins, vitamins, minerals, micronutrients, and fat. In fact, the proteins from animal sources are believed to have higher biological value than proteins from other sources, including plants. In spite of these, meat is a potential carrier of a number of bacterial and parasitic diseases and naturally occurring toxins such as mycotoxins, if it gets contaminated through pollution of the air, water, and soil. Persistent exposure to organic pollutants like dioxins and metals through industrial contaminants often make the animal and its meat a prominent source of human illness. Further, important zoonotic diseases caused by bacteria, viruses, and parasites in humans

TABLE 6.5 Technological Interventions for Microbial Safety of Meat and Meat Products

Interventions	Applications	Treatment Time	Approx. Microbial Reduction	Advantages	Disadvantages/Limitations
DURING SLAUGHTER AND DRESSING					
Hot water/steam pasteurization	Carcasses, primals	10–15 s at 75°C–85°C	1–3 logs	Can be used in combination with chemicals for greater effect	Possible discoloration of lean meat
Rinse and Chill	Carcasses	10–15 s	0.2–2 logs	Improvement in meat quality	Capital outlay
Steam Vacuum	Carcasses	Seconds	1–3 logs	Directed at visible contamination	Labor costs, some bleaching of meat surface
Acidified sodium chlorite	Carcasses, effective for vacuum packed primals	Up to 60 s	Up to 4 logs	Not affected by organic load. Possible continual effect on stored product	If using strong acids as the activator, storage and operator safety required
Electrolyzed water	Carcasses, poultry, meat surfaces	Spray or dip	1.5–3 log on inert surfaces 2–2.5 log on poultry	Salt is the only chemical used	Initial capital needed—but may be substantially cheaper than other methods
Cetylpyridiumchloride (CPC)	Carcasses, hide, trimmings	15–30 s at 1% CPC	1.5–2 logs on hides 2.1 logs on beef tissue	Effect on hide maintained up to 4 h (1 study); does not impact flavor, texture, appearance, or the odor of foods	Residual levels, if used on meat at 1% CPC
DURING POST SLAUGHTER STAGE					
Irradiation (gamma rays)	Primals, ground beef	Several mins	2–6 logs	Able to treat packaged food	Expensive to install; consumer acceptance issues
Irradiation (electron beam)	Primals, ground beef	Seconds	Up to 4 logs	Able to treat packaged food	Expensive to install–central treatment facility only; consumer acceptance issues
Organic acids (acetic, lactic acid)	Carcasses, primals, livers, lips, cheek meat, tongues	10–30 s	1–3 logs	Applied by spray or immersion; can be used with other interventions	Possible discoloration of lean meat, organoleptic problems; concerns about acid-resistant pathogens, corrosion of equipment

(Continued)

TABLE 6.5 (Continued)

Interventions	Applications	Treatment Time	Approx. Microbial Reduction	Advantages	Disadvantages/Limitations
Acidic calcium sulfate	Ground beef or other meat, ready-to-eat products			Makes pathogens (Listeria) more sensitive to heat, for example, during temp abuse/cooking.	Additive yet to be approved
Natural antimicrobials	Primals, ground beef, processed meats	Residual effect	1–2 logs	Spray application, then VP chilled storage, used as a surface coating (in case of alginate)	Effective only on Gram positive microbes
Pulsed electric fields	Ground beef, steaks	<1 s	1 log		Limited applications in meat sector as works best for liquids
Pulsed light	Primals	<1–10 s	1–3 logs	Can be used on packaged product	Probably not suitable for opaque foods; yet to be commercially viable
Ultrasound	Primals	0.25–3 min	0.5–2 logs	Possible to treat vacuum packed food	High power equipment required. Commercial development incomplete
Ultraviolet light	Meat marinades and brine	10 s–10 min	Up to 2 logs	Can be used on packaged product	Limited to surface sterilization or liquids
High pressure	Primals, ground beef, processed meats	0.5–5 min	Up to 4 logs	Increase shelf-life by reducing initial microbial count, treat in-pack	Expensive; systems not yet large enough; possiblilites of meat colour/textural changes

Source: *Adapted from Midgley, J., Small, A., 2006. Review of new and emerging technologies for red meat safety. Meat & livestock; McCann, M., Sheridan, J., McDowell, D., Blair, I., 2006. Effects of steam pasteurization on Salmonella Typhimurium DT104 and* Escherichia coli *O157: H7 surface inoculated onto beef, pork and chicken. J. Food Eng., 76(1), 32–40; Chen, J., Ren, Y., Seow, J., Liu, T., Bang, W., Yuk, H., 2012. Intervention technologies for ensuring microbiological safety of meat: current and future trends. Compreh. Rev. Food Sci. Food Safety, 11(2), 119–132.*

are attracting global attention and are often labeled as meat-borne zoonotic diseases. In this era of increased international trade and tariff and competitive markets, the issue of meat safety has become a global one and therefore requires stringent standards and procedures to be adopted worldwide in order to maintain safety and quality standards. As the safety and quality issues have become more relevant than the quantity of meat and meat products to society, it has become a subject of extensive scientific debate. Hence, research is needed to study the ill effects of the residues of different chemicals, toxins, and pathogenic microorganisms often encountered during rearing and processing of meat animals.

CONCLUSION

The challenges meat food safety facing today are manifold due to the complexity within itself (i.e., the diverse form of species, processing conditions, variety of products, and consumers with different tastes and preferences). In fact, the whole system, starting with animal production, product processing, and distribution, coupled with ever-increasing international trade, changing consumers' needs and preferences, increased awareness of consumer activities groups and scrutiny by news media to get meat that is safe, maintaining nutritional quality, and sensory attributes make the whole job challenging for the meat scientists and processors. The safety aspect of meat and meat products has become even more daunting due to invasion of classical, new, emerging, and reemerging meat-borne organisms along with the question of developing resistance to antibiotics. Besides, chemical residues and ill or side effects arising out of food additives are becoming a potent area of concern. To overcome the challenges and to fulfill the basic objectives of meat safety, adoption of good farm management practices coupled with ante- and post-mortem examination of animals, assessment of risk factors, and implementation of HACCP in production and processing levels are the needs of the hour.

References

Adams, M.R., Motarjemi, Y., 1999. Basic Food Safety for Health Workers. World Health Organization, Geneva.

Ahmed Refat, N.A., Ibrahim, Z.S., Moustafa, G.G., Sakamoto, K.Q., Ishizuka, M., Fujita, S., 2008. The induction of cytochrome P450 1A1 by Sudan dyes. J. Biochem. Mol. Toxicol. 22 (2), 77–84.

Ak, N.O., Cliver, D.O., Kaspar, C.W., 1994. Cutting boards of plastic and wood contaminated experimentally with bacteria. J. Food Prot. 57 (1), 16–22.

Akhtar, S., Paredes-Sabja, D., Torres, J.A., Sarker, M.R., 2009. Strategy to inactivate *Clostridium perfringens* spores in meat products. Food Microbiol. 26 (3), 272–277.

Andrée, S., Jira, W., Schwind, K.H., Wagner, H., Schwägele, F., 2010. Chemical safety of meat and meat products. Meat Sci. 86 (1), 38–48.

Beier, R.C., 2000. Toxicology of naturally occurring chemicals in food. Foodborne Disease Handbook, vol. 3: Plant Toxicants, p. 37.

Bhandari, N., Nepali, D., Paudyal, S., 2013. Assessment of bacterial load in broiler chicken meat from the retail meat shops in Chitwan, Nepal. Int. J. Infect. Microbiol. 2 (3), 99–104.

Biswas, A., Kondaiah, N., Bheilegaonkar, K., Anjaneyulu, A., Mendiratta, S., Jana, C., et al., 2008. Microbial profiles of frozen trimmings and silver sides prepared at Indian buffalo meat packing plants. Meat Sci. 80 (2), 418–422.

Blackburn, C.W., McClure, P.J., 2009. Foodborne Pathogens: Hazards, Risk Analysis and Control. CRC Press, ILC Bocao Raton, FL, Washington, DC, Chapter 5.

Brown, M.H., 2000. Processed meat products. In: Lund, B., Baird-Parker, A.C., Gould, G.W. (Eds.), The Microbiological Safety and Quality of Food. Aspen Publishers, Gaithersburg.

Buncic, S., Nychas, G.J., Lee, M.R., Koutsoumanis, K., Hébraud, M., Desvaux, M., et al., 2014. Microbial pathogen control in the beef chain: recent research advances. Meat Sci. 97 (3), 288–297.

Butaye, P., Devriese, L.A., Haesebrouck, F., 2001. Differences in antibiotic resistance patterns of *Enterococcus faecalis* and *Enterococcus faecium* strains isolated from farm and pet animals. Antimicrobial Agents Chemother. 45 (5), 1374–1378.

Chen, J., Ren, Y., Seow, J., Liu, T., Bang, W., Yuk, H., 2012. Intervention technologies for ensuring microbiological safety of meat: current and future trends. Compreh. Rev. Food Sci. Food Safety 11 (2), 119–132.

Chesebro, B., 2003. Introduction to the transmissible spongiform encephalopathies or prion diseases. Br. Med. Bull. 66 (1), 1–20.

Chu, F.S., 2002. Mycotoxins. In: Cliver, D.O., Riemann, H.P. (Eds.), Foodborne Diseases, second ed Academic Press, San Diego, CA, pp. 271–303.

Dave, D., Ghaly, A.E., 2011. Meat spoilage mechanisms and preservation techniques: a critical review. Am. J. Agric. Biol. Sci. 6 (4), 486–510.

Dixon, S.N., 2001. Veterinary drug residues. In: Watson, D.H. (Ed.), Food Chemical Safety, vol. 1. Woodhead, Cambridge, UK, pp. 109–147.

Domingo, J.L., Nadal, M., 2016. Carcinogenicity of consumption of red and processed meat: what about environmental contaminants? Environ. Res. 145, 109–115.

Doulgeraki, A.I., Nychas, G.J.E., 2013. Monitoring the succession of the biota grown on a selective medium for pseudomonads during storage of minced beef with molecular-based methods. Food Microbiol. 34 (1), 62–69.

Downes, F., Ito, K., 2001. Compendium of Methods for the Microbiological Examinations of Foods, fourth ed American Public Health Association Press, Washington, DC.

Eisler, R., 1988. Zinc hazards to fish, Wildlife and Invertebrates: a synoptic review. US Fish Wildlife Service. Biol. Reprod. 85, 6–8.

Ercolini, D., Russo, F., Nasi, A., Ferranti, P., Villani, F., 2009. Mesophilic and psychrotrophic bacteria from meat and their spoilage potential in vitro and in beef. Appl. Environ. Microbiol. 75 (7), 1990–2001.

FISS, 1997. Meat and Poultry Products Hazards Control Guide. United States Department of Agriculture (USDA).

FSIS, 1996. Pathogen Reduction: Hazard Analysis and Critical Control Point (HACCP) Systems; Final rule, 9CFR Part 304 (Food Safety and Inspection Service), 61. Federal Register, pp. 38806–38989.

FSS, 2006. Red Meat Safety & Clean Livestock-Food Standard Scotland. http://www.foodstandards.gov.scot/downloads/Red_meat_safety_and_clean_livestock.pdf.

Franco, D.A., Webb, J., Taylor, C.E., 1990. Antibiotic and sulfonamide residues in meat: implications for human health. J. Food Prot. 53 (2), 178–185.

Heuvelink, A.E., Roessink, G.L., Bosboom, K., de Boer, E., 2001. Zero-tolerance for faecal contamination of carcasses as a tool in the control of O157 VTEC infections. Int. J. Food Microbiol. 66 (1–2), 13–20.

Huffman, R., 2002. Current and future technologies for the decontamination of carcasses and fresh meat. Meat Sci. 62 (3), 285–294.

Jacob, C., Mathiasen, L., Powell, D., 2010. Designing effective messages for microbial food safety hazards. Food Control 21 (1), 1–6.

Jenson, I., Vanderlinde, P., Langbridge, J., Sumner, J., 2014. Safety of meat and meat products in the twenty-first century. Pract. Food Safety: Contemporary Issues and Future Directions 553–575.

Jeremiah, L., Gibson, L., 2001. The influence of packaging and storage time on the retail properties and case-life of retail-ready beef. Food Res. Int. 34 (7), 621–631.

Kaneene, J., Miller, R., 1997. Problems associated with drug residues in beef from feeds and therapy. Revue Scientifique et Technique de l'OIE (France) 16, 694–708.

Koutsoumanis, K., Taoukis, P., 2005. Meat safety, refrigerated storage and transport: modeling and management. In: Sofos, J.N. (Ed.), Improving the Safety of Fresh Meat. Woodhead Publishing, Ltd, Cambridge, UK, pp. 503–561.

Koutsoumanis, K.P., Geornaras, I., Sofos, J.N., 2006. Microbiology of land muscle food. In: Hui, H.Y. (Ed.), Handbook of Food Science, Technology and Engineering, vol. 1. CRC Press Taylor & Francis Group, Boca Raton, FL, 52.51-52.43.

Kushi, L.H., Doyle, C., McCullough, M., Rock, C.L., Demark-Wahnefried, W., Bandera, E.V., et al., 2012. American cancer society guidelines on nutrition and physical activity for cancer prevention: reducing the risk of cancer with healthy food choices and physical activity. CA: Cancer J. Clin. 62 (1), 30–67.

Leary, V.A., 1994. The right to health in international human rights law. Health Hum. Rights 1, 24–56.

Lianou, A., Sofos, J.N., 2007. A review of the incidence and transmission of *Listeria monocytogenes* in ready-to-eat products in retail and food service environments. J. Food Protect. 70 (9), 2172–2198.

Lozano, M.C., Trujillo, M., 2012. Chemical residues in animal food products: an issue of public health. Public Health–Methodology Environmental Systems Issues pp. 163–188.

Mataragas, M., Skandamis, P., Drosinos, E., 2008. Risk profiles of pork and poultry meat and risk ratings of various pathogen/product combinations. Int. J. Food Microbiol. 126 (1–2), 1–12.

McCann, M., Sheridan, J., McDowell, D., Blair, I., 2006. Effects of steam pasteurisation on *Salmonella typhimurium* DT104 and *Escherichia coli* O157: H7 surface inoculated onto beef, pork and chicken. J. Food Eng. 76 (1), 32–40.

McEvoy, J., Doherty, A., Finnerty, M., Sheridan, J., McGuire, L., Blair, I., et al., 2000. The relationship between hide cleanliness and bacterial numbers on beef carcasses at a commercial abattoir. Lett. Appl. Microbiol. 30 (5), 390–395.

Mead, P.S., Slutsker, L., Dietz, V., McCaig, L.F., Bresee, J.S., Shapiro, C., et al., 1999. Food-related illness and death in the United States. Emerging Infect. Dis. 5 (5), 607.

Mechlem, K., 2004. Food security and the right to food in the discourse of the United Nations. Eur. Law J. 10 (5), 631–648.

Midgley, J., Small, A., 2006. Review of new and emerging technologies for red meat safety. Meat & Livestock Australia, 1–73.

Mitchell, J., Griffiths, M., McEwen, S., McNab, W., Yee, A., 1998. Antimicrobial drug residues in milk and meat: causes, concerns, prevalence, regulations, tests, and test performance. J. Food Prot. 61 (6), 742–756.

Mulder, P.P., López, P., Castelari, M., Bodi, D., Ronczka, S., Preiss-Weigert, A., et al., 2018. Occurrence of pyrrolizidine alkaloids in animal-and plant-derived food: results of a survey across Europe. Food Addit. Contaminants: Part A 35 (1), 118–133.

Oskarsson, A., 2012. Environmental contaminants and food safety. Acta Veterinaria Scandinavica vol. 54, S5. BioMed Central.

Pennington, T., 1997. Report on the circumstances leading to the 1996 outbreak of infection with *Escherichia coli* O157 in Central Scotland, the implications for food safety and the lessons to be learned. The Stationery Office, London, UK.

Rassool, G., 2006. No Bird Flu Risk for Consumers From Properly Cooked Poultry and Eggs. Blackwell Publishing, Oxon.

Reyes-Herrera, I., Donoghue, D., 2012. Chemical contamination of poultry meat and eggs. Chemical Contaminants and Residues in Food. Elsevier, pp. 469–497.

Richardson, M., Matthews, K., Lloyd, C., Brian, K., 2009. Meat quality and shelf life. Better Returns Programme EBLEX Agriculture and Horticulture Development Board. http://www.eblex.org.uk.

Rivera-Betancourt, M., Shackelford, S.D., Arthur, T.M., Westmoreland, K.E., Bellinger, G., Rossman, M., et al., 2004. Prevalence of *Escherichia coli* O157: H7, *Listeria monocytogenes*, and Salmonella in two geographically distant commercial beef processing plants in the United States. J. Food Protect. 67 (2), 295–302.

Rosenvold, K., Andersen, H.J., 2003. Factors of significance for pork quality—a review. Meat Sci. 64 (3), 219–237.

Schiavetta, A., Miller, M., Lunt, D., Davis, S., Smith, S., 1990. Adipose tissue cellularity and muscle growth in young steers fed the beta-adrenergic agonist clenbuterol for 50 days and after 78 days of withdrawal. J. Anim. Sci. 68 (11), 3614–3623.

Shilenge, L., Shale, K., Matodzi, T., Machete, F., Tshelane, C., 2017. A review of microbial hazards associated with meat processing in butcheries. Afr. J. Sci. Technol. Innovat. Dev. 9 (1), 1–6.

Smith, D.J., Kim, M., 2017. Chemical contamination of red meat. In: Schrenk, D., Alexander, C. (Eds.), Chemical Contaminants and Residues in Food, second ed. Woodhead Publishing, Elsevier, Oxford, UK, pp. 451–489.

Smulders, F.J., Nørrung, B., Budka, H., 2013. Food Borne Viruses and Prions and Their Significance for Public Health, vol. 6. Wageningen Academic Publishers.

Sofos, J., 2005a. Improving the Safety of Fresh Meat. CRC/Woodhead Publishing Limited, Cambridge, UK.

Sofos, J., 2009. ASAS Centennial Paper: Developments and future outlook for postslaughter food safety 1. J. Anim. Sci. 87 (7), 2448–2457.

Sofos, J.N., 2005b. Pathogens in animal products: major biological hazards. Encyclopedia of Animal Science. Taylor & Francis, pp. 698–700.

Sofos, J.N., 2008. Challenges to meat safety in the 21st century. Meat Sci. 78 (1), 3–13.

Sofos, J.N., 2014. Meat and meat products. Food Safety Management. Elsevier, pp. 119–162.

Sofos, J.N., Geornaras, I., 2010. Overview of current meat hygiene and safety risks and summary of recent studies on biofilms, and control of *Escherichia coli* O157:H7 in nonintact, and *Listeria monocytogenes* in ready-to-eat, meat products. Meat Sci. 86 (1), 2–14.

Stopforth, J., Sofos, J., 2006. Recent advances in pre- and postslaughter intervention strategies for control of meat contamination. In: Juneja, V.K., Cherry, J.P., Tunick, M.H. (Eds.), Advances in Microbial Food Safety, ACS Symposium, 931. American Chemical Society, Oxford University Press, Washington, DC, pp. 66–86.

Stopforth, J.D., Sofos, J.N., Taylor, S.J., Baumert, J.L., 2012. Food safety issues in animal source foods related to animal health and welfare. In: Pond, W.G., Bazer, F.W., Rollin, B.E. (Eds.), Animal Welfare in Animal Agriculture: Husbandry, Stewardship, and Sustainability in Animal Production. CRC Press/Taylor & Francis Group, Boca Raton, FL, pp. 205–231.

Tilahun, A., Jambare, L., Teshale, A., Getachew, A., 2016. Review on chemical and drug residue in meat. World J. Agric. Sci. 12 (3), 196–204.

WHO, 2005. No bird flu risk for consumers from properly cooked poultry and eggs. http://www.who.int/mediacentre/news/releases/2005/pr66/en/.

WHO, 2015. WHO Estimates of the Global Burden of Foodborne Diseases.

Warriner, K., Aldsworth, T., Kaur, S., Dodd, C., 2002. Cross-contamination of carcasses and equipment during pork processing. J. Appl. Microbiol. 93 (1), 169–177.

Wiedenfeld, H., 2011. Plants containing pyrrolizidine alkaloids: toxicity and problems. Food Addit Contam Part A 28 (3), 282–292.

Yam, K.L., Takhistov, P.T., Miltz, J., 2005. Intelligent packaging: concepts and applications. J. Food Sci. 70 (1), R1–R10.

Yamamoto, Y., Nakamura, K., Mase, M., 2017. Survival of H5N1 highly pathogenic avian influenza virus in tissues derived from experimentally infected chickens. Appl. Environ. Microbiol. 83 (16), e00604–e00617.

Yu, Z.F., Zhao, L.Q., Zheng, J.J., Wang, Z.A., 2012. Safety analysis and control measures of meat products. Food Sci. Technol. 37 (1), 135–137.

Zhao, Y., Wells, J.H., McMillin, K.W., 1994. Applications of dynamic modified atmosphere packaging systems for fresh red meats. Rev. J. Muscle Foods 5 (3), 299–328.

Zhou, G., Ye, K., Tume, R.K., 2017. Meat Safety in China. *Food Safety in China: Science, Technology, Management and Regulation*, pp. 453–475.

CHAPTER

7

Safety of Fish and Seafood

Mrinal Samanta and Pushpa Choudhary

ICAR-Central Institute of Freshwater Aquaculture, Bhabaneswar, India

OUTLINE

INTRODUCTION

In human diets, fish and seafood are the important components supporting the valuable source of protein, important fatty acids, vitamins, and minerals. Among these, the major benefits of fish and seafood are associated with the presence of high levels of long chain polyunsaturated fatty acids (long chain-PUFAs or Omega-3), including eicosapentaenoic acid (EPA) and docosahexaenoic acid (DHA). PUFA, EPA, and DHA play a very important role in health and immunity (FAO, 2016). Over the years, the intake of fish and seafood in Europe has been increasing, and in 2011, the per capita consumption has reached ~24.5 kg (EUMOFA, 2014). The global trend of fish and seafood products

Food Safety and Human Health
DOI: https://doi.org/10.1016/B978-0-12-816333-7.00007-2

consumption has also significantly increased from an average of 9.9 kg in the 1960s to 20.1 kg in 2014 (FAO, 2016). With the ever-increasing demand of fish and seafood products, there is also a growing concern about food safety and human health.

Fish and seafood products are a highly perishable commodity and are prone to contamination by pathogens causing various types of food-borne diseases (Darlington and Stone, 2001). The presence of human pathogenic microorganisms in fish and fishery products is influenced by various factors, such as aquaculture practices, environmental conditions, harvest, processing, and storage and distribution of products. Change in aquaculture practices from semi-intensive to intensive has resulted in the pollution and contamination of the aquatic environments. Application of excess feed and fertilizer has increased the organic biomass of the water bodies, resulting in enhanced microbial load in the water and sediments. Presence of various fish and human pathogens in fish is the result of their direct contact with contaminated aquatic environment and ingestion of bacteria from sediments or contaminated feed. In addition, various outlets of human and animal contaminants including sewage converging into the ponds, lakes, reservoirs, rivers, and the sea are also the potential threat of human pathogenic bacteria in fish and fishery products (Bottone et al., 2005). There are several evidences of fish and seafood-borne human diseases that occur when these food products are contaminated with various types of pathogens (Huss et al., 2003). Bacterial species under the genera *Vibrio, Salmonella, Listeria, Clostridium, Staphylococcus,* and *Yersinia* and various types of parasites present in the aquatic environments are sometimes associated with the fish and fishery products causing potential threat to human health (Novoslavskij et al., 2016). In addition to pathogenic contamination, some of the fishery products may contain biotoxins such as ciguatoxin or muscle paralyzing toxins, and some fish species of the families *Tetraodontidae, Molidae, Diodontidae,* and *Canthigasteridae* are highly toxic and are completely unsafe for human consumption (Civera, 2003). Shellfish, oysters or clams, and mussels can accumulate toxic metabolites of the phytoplanktons of the genera *Alexandrium, Gymnodinium, Dinophysis,* and *Pseudo-nitzschia*. Consumption of these contaminated seafood can cause a wide variety of illnesses including paralytic shellfish poisoning (PSP), diarrheic shellfish poisoning, amnesic shellfish poisoning, and also neurotoxic shellfish poisoning (Farabegoli et al., 2018).

Thus, there is a growing concern about the safety of fish and fishery products for human consumption. To avoid fish and seafood-borne illnesses, screening of pathogens and or toxins needs to be carried out in the harvested fish and sea animals, and the highest level of monitoring and screening should be undertaken in ready-to-eat (RTE) and vacuum-packed (VP) smoked fish and fishery products. The consumers should also be cautioned while consuming seafood like bivalve mollusks, as it may contain saxitoxin (STX) or PSP toxins, and food prepared from the *Scombroid* family of fish and dark flesh fish to avoid scombroid poisoning or histamine fish poisoning.

BACTERIAL CONTAMINATION OF FISH AND SEAFOOD PRODUCTS

A number of illnesses in human arise from the consumption of fish and seafood products that have either been contaminated at the source as a raw material or become contaminated during their processing. These illnesses are caused primarily by a variety of bacteria themselves or by the ingestion of their toxins formed in the foodstuff prior to consumption. Among many species of bacteria involved in fish and seafood poisoning, the most common bacteria belong to the genus of *Vibrio, Salmonella, Yersinia, Listeria, Escherichia, Staphylo, Shigella,* and *Clostridium.*

In aquatic environments, various *Vibrio* spp. are present. Many of them cause diseases in fish, and some are related to the diseases in human with the consumption of contaminated fish and fishery products. *Vibrio anguillarum, V. salmonicida, V. ordalii, V. vulnificus, V. viscous,* and *V. wodanis* are primarily important in causing serious diseases in fish (Gauthier, 2015; Callol et al., 2015). On the other hand, *V. parahemolyticus* and *V. vulnificus* are more often associated with human vibriosis in the consumption of raw, undercooked, or contaminated fish and marine products (Iwamoto et al., 2010; Gauthier, 2015; Callol et al., 2015). These bacteria are associated with ~20%−30% of food poisoning cases in Japan and seafood-borne diseases in many Asian countries (Alam et al., 2002) and in the United States (Kaysner and DePaola, 2001; Newton et al., 2012). *V. parahaemolyticus* has two virulence factors, a thermostable direct hemolysin (TDH) that makes pore-forming protein and contributes in the invasion of the bacterium and a TDH-related hemolysin (TRH) that plays a similar function as TDH and helps in pathogenesis. In addition to TDH and TRH, it also synthesizes adhesions and type III secretion systems (T3SS1 and T3SS2) for its survival in the environment. *V. vulnificus* have three biotypes (biotypes 1, 2, and 3). Among these, biotype 1 is more prevalent in water and humans and biotype 2 in fish and human (Gauthier, 2015). In the fish intestine, *V. vulnificus* and *V. parahaemolyticus* concentration is much higher as compared to water and sediments and may cause food-borne diseases in human (Givens et al., 2014).

The presence of *Listeria monocytogenes* has been detected in aquatic environments, in fish and fishery products, cold and hot-smoked salmon, marinated fish, fermented fish, frozen seafood, and fish salads (González-Rodríguez et al., 2002; Papadopoulos et al., 2010; Tocmo et al., 2014). These bacteria can grow in a wide range of temperatures and may cause listeriosis in human by the consumption of foods contaminated with this pathogen. Contamination of fish and fishery products with this organism generally occurs during the processing phase (Tompkin, 2002). Smoked VP fish products are stored under refrigeration to increase their shelf life. However, *L. monocytogenes* can grow and proliferate at 4°C as it is a psychrotrophic organism and its load may reach to an infective level during long storage at this temperature. The major safety issues with RTE and VP smoked fish products remain with the consumption of these products without prior heating or additional cooking (Kin et al., 2012). In spite of various efforts of curbing the contamination of *L. monocytogenes* in fish products, achieving nearly a zero count in RTE smoked fish is technologically difficult. Recently, various modern technologies such as irradiation and high-pressure processing have helped to improve the safety of smoked fish commodities by minimizing the contamination of *L. monocytogenes*. Chemical sanitizers, biopreservatives such as lactic acid bacteria that produce bacteriocins, and essential oils with antibacterial properties have also been used for improving safety of fish products. Novel packaging technologies such as modified atmosphere packaging (MAP) and bioactive packaging has also been introduced. Although MAP seems to be better than VP for smoked fish products, VP remains the best because of its effectiveness in reducing oxidative reactions and also for the cost efficiency. Each technology has its own effectiveness and limitations, and among all technologies, hurdle technology may be the best in controlling *L. monocytogenes* contamination in fish products.

In addition to meat, *Salmonella* sp. has been identified from fish and fishery products and also from a variety of seafood including clams, oysters, mussels, crabs, lobsters, squid, cuttlefish, and octopus (Kumar et al., 2009). *Salmonella* sp. is naturally present in birds, animals, and humans and is not an inhabitant of the aquatic environments. Therefore, the presence of these organisms in fish and seafood products is mostly due to the contamination of aquatic environments with human and animal sewage (Amagliani et al., 2012), birds and animals' offal, and hygienic failures during production and handling (Li et al., 2009; Budiati et al., 2013). Like meat, *salmonella* in fish and seafood products can also cause human salmonellosis (Brands et al., 2005; Iwamoto et al., 2010). The survival and growth of salmonella in food products depend upon temperature, pH, availability of nutrients, and the moisture content. Fish and seafood products provide these vital factors for the survival of *Salmonella*, and the bacteria survives even after prolonged storage under frozen conditions. *Salmonella* spp. can grow in the temperature range between 5.2°C and 46.2°C, but their optimum growth occurs between 35°C and 43°C (ICMSF, 1996). *Salmonella* can survive for 10–15 days in septic tanks (Parker and Mee, 1982), 54 days in water, and 119 days in the sediment samples (Chao et al., 1987; Moore et al., 2003). At a favorable temperature, significant increase in the salmonella population in fish and seafood products occurs within 24 hours (Kumar et al., 2015). The food-borne salmonellosis is primarily caused by the nontyphoidal *Salmonella* group. Presence of virulence factors and their pathogenicity level determine the ability of the strain in causing diseases. Several genes are responsible for the survival, growth, and toxicity/pathogenicity of the strain in causing diseases in humans. The *rpo* gene helps to protect environmental stress, *rpoE* and *rpoH* genes prevent thermal stress (Bang et al., 2005), *invA* gene is responsible for invasion of gut epithelial tissue, and the enterotoxin (*stn*) gene is responsible for their pathogenicity in human and animals (Asten and Dijk, 2005).

Contamination of aquatic bodies such as lakes, rivers, ponds, and wells with *Yersinia* spp. sometimes occurs due to the contact with excreta of animals and humans and from the firm and slaughterhouse waste (Bottone et al. 2005). *Yersinia* sp. can survive and grow in a wide range of temperatures, ranging from 4°C to 42°C (Bottone et al., 2005). Although the majority of the *Yersinia* spp. in the aquatic environments are nonpathogenic strains, sometimes *Yersinia enterocolitica* and *Yersinia pseudotuberculosis*, which are known human pathogens, are also detected from aquatic bodies. *Y. enterocolitica* can survive at 4°C for up to 64 weeks in stream water and almost 5 years in sterile water (Karapinar and Gonul, 1991; Liao and Shollenberger, 2003). The presence of *Y. enterocolitica* has been detected in several fish species: rainbow trout (*Oncorhynchus mykiss*), Atlantic salmon (*Salmo salar*), sardine (*Sardina pilchardus*), whiting fish (*Merlangius merlangus*), hilsa (*Tenulosailisha*), butter catfish (*Ompokbimaculatus*), magur (*Clarias batrachus*), catla (*Catla catla*), and rohu (*Labeo rohita*) (Davies et al., 2001; Kakatkar et al., 2010). Among various body parts of fish, the highest number of samples positive to *Y. enterocolitica* has been recorded from gill (50%), followed by body surface (25%) and intestine (8.3%) (Akhila et al., 2013).

Infections with *Shigella* spp. may occur by contaminated fish and seafood products, and the symptoms of illness are manifested by loose stools, abdominal pain, fever, and bloody diarrhea. Unhygienic conditions, contaminated water, and the preparation of food by an infected food handler are the source of shiegellosis. Fish and seafood products get contaminated with *Shigella* spp. when these products are harvested from sewage-contaminated water bodies. Consumption of contaminated raw oysters has caused an outbreak of shigellosis (Reeve et al., 1989). *Shigella* spp. is heat sensitive and is readily killed by cooking. The strategies to control and prevent shigellosis associated with fish and seafood products include monitoring of water samples, prohibition of sewage contamination in the water bodies, harvesting of fish and seafood from clean waters, and following hygienic measures while handling their cooking and serving.

Food-borne botulism is a rare but fatal disease caused by the botulinum neurotoxin, also called as "miracle poison" produced by *Clostridium botulinum* an anaerobic, Gram-positive, spore-forming rod-shaped bacteria commonly found in plants, soil, water, and the intestinal tracts of animals. In the improperly processed food, this toxin or the bacteria or its spores cause food-borne intoxication and botulism. Spores of *C. botulinum* exist widely in the environment and are heat resistant. In the anaerobic condition, spores germinate to grow and then excrete toxins. There are eight antigenically different exotoxins: A, B, C_1, C_2, D, E, F, and G. Among these toxins, type A is the most potent followed by types B and F. The human botulism is mostly caused by A, B, and E type toxins and rarely by the F type, whereas types C, D, and E cause illness in other mammals, birds, and fish. Generally, all botulinum neurotoxins share high homology in their amino acids sequence and are synthesized as an inactive single polypeptide chains of 150 kDa, which is comprised of a heavy (H) chain ($\sim$100 kDa) and a light (L) chain ($\sim$50 kDa) and are linked by a disulfide bond. Prevention of food-borne botulism needs good hygienic practice in food preparation, particularly during heating/sterilization. The spores are much resistant to vegetative form and remain viable in most of the products even after boiling for several hours, but a very high temperature treatment such as commercial canning inactivates the spores. Therefore, much precaution needs to be undertaken to avoid food-borne botulism.

FISH AND SEAFOOD PRODUCTS BORNE PARASITIC DISEASES

Numerous parasitic diseases of humans are food-borne and some are also zoonotic. Fish-borne parasitic diseases are mostly helminthic diseases caused by trematodes, nematodes, and cestodes. Around 18 million people in the world are infected with fish-borne trematodes, and almost half a billion people are at risk (WHO, 2004; Chai et al., 2005). Fish-borne parasitic zoonotic diseases are intestinal trematodiasis, opisthorchiasis, and anisakiasis or diphyllobothriasis. Consumption of raw or improperly cooked or unprocessed fish and fishery products are mainly responsible for the spread of food-borne parasitic diseases. Liver flukes such as *Clonorchis sinensis*, *Opisthorchis viverrini*, *Opisthorchis felineus*, and *Metorchis conjunctus* cause serious diseases like cholangitis, pancreatitis, choledocholithiasis, and cholangiocarcinoma. Liver fluke–borne diseases are mostly prevalent in southern China, Hong Kong, Korea, and Japan and in some parts of Thailand, as there is a customary practice in these countries of eating raw fish or shrimps along with rice.

NATURALLY OCCURRING TOXINS IN FISH AND SEAFOOD

Scombroid poisoning, also known as "histamine fish poisoning," is one of the most common causes of food-borne illness that happens due to the ingestion of canned, smoked, or even fresh fish such as tuna and mackerel belonging to the suborder *Scombroidea* (O'Connor and Forbes, 2000). Other nonscombroid fish species like mahi-mahi, pilchards, herring, sardines, anchovies, bluefish, Western Australian salmon, sock-eye salmon, swordfish, marlin, Cape yellow tail, and amberjack dolphin fish also cause this disease (Attaran and Probst, 2002). The flesh of these fish species contains very high levels of histidine, which is converted to histamine by decarboxylation. The process of decarboxylation is mediated by the enzyme histidine decarboxylase (HDC), produced primarily by the Gram-negative bacteria such as *Pseudomonas aeruginosa*, *Escherichia coli*, *Klebsiella* spp., and *Morganella morganii* present in fish intestines and bodies or when contamination with these bacteria occur with fish and fishery products during processing and improper storage (Jantschitsch et al., 2011; Hungerford, 2010). Although cold storage can decrease histamine synthesis, there are several psychrophilic bacteria such as *Morganella psychrotolerans* and *Photobacterium phosphoreum* that can grow at 0°C–5°C and can form histamine even in refrigerated conditions (Hungerford, 2010). The histamine is heat stable in nature. Therefore, cooking and smoking will not be able to remove this from fish and fishery products. Fishes harboring high histamine levels look normal in appearance, and there is no change in organoleptic quality such as taste and smell. The outbreak of scombroid poisoning generally occurs during summer, and the symptoms are rash, urticaria, headache, dizziness, sweating, palpitations, burning of the mouth and throat, nausea, vomiting, diarrhea, respiratory distress, and vasodilatory shock. Usually the symptoms start within 10 minutes to 1 hour of food intake and may last for 2–5 hours. The diagnosis can be made by analyzing histamine levels in plasma or histamine metabolites in urine samples of the suspected patients. To prevent occurrence of this disease, fish and fishery products should be examined for the presence of high levels of histamine, presence of histidine decarboxylase (HDC) enzyme-producing bacteria, and rapid screening of samples by PCR amplification of HDC gene of bacteria.

Molluscan bivalvs such as mussels, oysters, and clams sometimes get contaminated with paralytic shellfish toxins (PSTs) produced by marine dinoflagellates of the genus *Alexandrium*, *Pyrodinium*, and *Gymnodinium*. Consumption of these contaminated mussels may cause PSP, a type of food-borne illness with the symptoms of nausea, vomiting, tingling of the mouth, to paralysis and in some cases fatality leading to death (Shin et al., 2018). PSTs are classified into carbamoyl, N-sulfocarbamoyl, and decarbamoyl groups depending on the chemical structures of their side chain. Among these, carbamoyl group of toxins such as STX and gonyautoxin are most toxic. The approved limit of PSTs in shellfish is 80 μg STX eq/100 g meat.

Tetradotoxin (TTX) is one of the most potent toxins found in pufferfish, toadfish, globefish, octopus, and some species of amphibian and shellfish. TTX is mainly present in the liver and sex organs of fish. The flesh containing TTX of these fish species or animals is poisonous to human. Extreme care should be taken in identifying these fish species and their organs for the preparation of fish and seafood (Farabegoli et al., 2018).

EMERGING FISH AND SEAFOOD-BORNE DISEASES

Several diverse bacterial species have been identified and isolated from fish and seafood products causing food-borne illnesses in humans without having any definitive etiological connection to fish or shellfish or any other aquatic animals. These are emerging diseases and are of great concern. Some organisms belonging to *Aeromonas* spp., *Pleisomonas* spp., and *Pseudomonas* spp. are natural flora of fish and shellfish and are considered as nonpathogenic. But these organisms can be a potential health risk in susceptible groups and immune-compromised individuals when they are transmitted through fish and seafood. Some bacteria that are usually related to food spoilage and are considered harmless can also be the potential threat to human health. Recently, emerging pathogen *Stenotrophomonas maltophilia* has received attention as an etiological agent of bacteraemia, urinary infection, skin and soft tissue infection, endocarditis, ocular infection, and cystic fibrosis (Ben-Gigirey et al., 2000, 2002). Bacterial counts in seawater increase with temperature, resulting in increased consumer risk from seafood-associated bacterial pathogens during summer months. International trade of fish and seafood products facilitates the spread of pathogens into a novel geographic area and enhances the risk of food-borne illness in human. In addition to new pathogens, some of the antimicrobial resistant bacteria also get transferred to a new region through trade in aquacultural commodities, making the population susceptible and vulnerable to the emerging diseases.

SPOILAGE OF FISH AND SEAFOOD

Consumption of spoiled food is one of the major causes of food-borne illnesses. Transport of food from the source of raw material to the processing unit and thereafter to the retailer and ultimately to the consumer is a chain process. It needs proper methods of preservation to maintain the nutritional value, flavor, taste, and texture and also to increase the shelf-life of food. Several factors including chemical, enzymatic, or microbial activities are involved in the spoilage of food, including fish and fishery products. Due to nonavailability of the onsite storage facility ~4–5 million tons of shrimp and trawled fish are lost every year due to enzymatic and microbial spoilage (Unklesbay, 1992). Just after capture and death of the fish, crabs, mussels, etc., the autolytic enzymatic spoilage starts. At the early stage, it only reduces textural quality without spoilage, off-odors, and off-flavors (Hansen et al., 1995). But in prolonged exposure, there are chemical and biological changes in dead fish due to enzymatic breakdown of major fish molecules and leading to bad odor and spoilage (FAO, 2005). Proteolytic degradation and also lipid oxidation in pelagic fish species such as mackerel and herring with high oil/fat content in their flesh are the major cause of spoilage (Fraser and Sumar, 1998).

Microbial spoilage is another major concern of fish and seafood products. Around 30% of the landed fish biomass is lost through microbial spoilage (Amos, 2007). Microbial composition of aquatic bodies deeply influences the microflora in fish gut and body. Common fish microflora includes bacterial species of the genus *Vibrio, Alcaligenes, Micrococcus, Pseudomonas,* and *Serratia* (Gram and Huss, 2000). Growth and metabolism of microbes in fish and seafood produce alcohols, aldehydes and ketones, putrescine, histamine and cadaverine, organic acids, and sulfides, resulting in unpleasant odor and unacceptable off-flavors (Gram and Dalgaard, 2002; Emborg et al., 2005; Dalgaard et al., 2006).

DETECTION OF PATHOGENS IN FISH AND SEAFOOD

Fish and seafood harboring pathogens could be a source of human illness (Butt et al., 2004; Huss, 1993). In addition to the infected fish, processing practices and handling of RTE fish products, such as smoked or marinated seafood, are also a potential source of public health hazard. An effective program of controlling pathogens in seafood needs very reliable, accurate, and sensitive methods of detection. Conventional diagnoses of food-borne illnesses are based on history, clinical signs, pathological lesions, and detection of the pathogen.

In shellfish, contamination of *Vibrio parahaemolyticus* has been detected following PCR-amplification of TDH and TRH haemolysin genes (Bej et al., 1999). In the living bivalve mollusks, contamination of *V. Parahaemolyticus* and *V. cholera* can also be detected by PCR, and it has shown the advantage over the standard ISO/TS 21872-1 culture method (Rosec et al., 2012). Application of multiplex PCR, targeting *tox*R (Bauer and Rorvik, 2007), *toxR* and *vvhA* (Neogi et al., 2010), *atpA* (Izumiya et al., 2011) and *groEL* (Hossain et al., 2013) genes have also been used for the detection of pathogenic *Vibrio* species in fish and shellfish products. For the detection of *E. coli* O157:H7, *Salmonella* spp., *V. Parahaemolyticus,* and *V. cholerae* in fish and seafood products, application of multiplex PCR assay has also been reported (Fakruddin et al., 2013).

Salmonella contamination is commonly detected following the UNIEN ISO 6579:2004 method, which generally takes several days (Anonymous, 2002). However, fish and fishery products are perishable in nature and require rapid detection of contaminants. Comparatively rapid methods for the detection of salmonella include membrane filtration, automated electrical techniques, and immunological assays (Martinez and Liebama, 2005), and the fastest method with specificity and sensitivity is based on nucleic acid–based detection with PCR and real-time PCR (Amagliani et al., 2010; De-Paola et al., 2010; Kumar et al., 2008a,b; Kumar et al., 2010; Minami et al., 2010; Shabarinath et al., 2007). However, in nucleic acid–based detection methods, positive reaction may also be obtained from DNA contamination of the nonliving microorganisms, which may not be the potential risk in food products. To overcome it, an additional step of growing microorganisms in selective media may be used, which may ensure the detection of PCR bands from the viable cells. In epidemiological surveillance, application of molecular typing such as PCR-ribotyping and ERIC-PCR is useful in tracing strains and to prevent the spread of infection (Kumar et al., 2009).

In seafood products and oysters, very low ($\sim$10 cells) contamination of *Salmonella* sp. can be detected by first enrichment of its growth followed by PCR-mediated amplification of the *invA* gene (Bej et al., 1994; Vantarakis et al., 2000). In the naturally contaminated fish, squid, cuttlefish, octopus, mussels, crabs, clams, and oysters, application of PCR targeting *invA* gene of *Salmonella* has shown higher sensitivity than ELISA and methods of U.S. Food and Drug Administration *Bacteriological Analytical Manual* (FDA-BAM)

(Kumar et al., 2008a). The *hns* gene, which encodes a DNA binding-protein, has been found to be useful for the detection of *Salmonella* spp. in shrimps, finfish, and clams through PCR (Sanath et al., 2003).

Isolation of *L. monocytogenes* following the method of U.S. Food and Drug Administration (USFDA), Association of Official Analytical Chemists, ISO 11290 standards, U.S. Department of Agriculture (USDA) Food Safety and Inspection Service (FSIS), and the French Standards (OIE, 2014) are required for international regulatory purposes. However, conventional methods are time consuming and laborious and may exclude the detection of the organism at low levels (Aono et al., 1997). Quick detection of pathogens in the raw material by nucleic acid–based assay will facilitate adequate control measures and will prevent postharvest processing and contamination of food products. The BAX system PCR assay has shown its equal potential comparable to the traditional culture-based method for the detection of *L. monocytogenes* in smoked fish (Norton et al., 2000), raw fish (Hoffman and Wiedmann, 2001), and blue crabs (Pagadala et al., 2011).

PRESERVATION AND PACKAGING

To enhance shelf-life and maintenance of nutritious components and to prevent spoilage, different types of preservation methods such as chilling, freezing, smoking, drying, brining, fermentation, and canning have been developed. Among these, the most common methods used in the seafood industry are low-temperature storage and chemical techniques that control water activity, oxidative, enzymatic, and microbial spoilage of food (Akinola et al., 2006; Berkel et al., 2004). Packaging technologies, such as modified atmosphere and active and intelligent packaging, are useful for fish preservation (Fellows, 2000; Da-Wen, 2005).

MAP has been developed as a supplement to ice and mechanical refrigeration (Mendes et al., 2008) to extend the shelf-life of fish and seafood products. MAP in combination with refrigeration increases 30%–100% shelf-life of raw fish and 100%–200% of cooked shellfish (Kuswandi et al., 2011; Gornik et al., 2013). In MAP, depending upon the type of fish, such as oily or lean, replacement of air is carried out with different gas mixtures with a fixed proportion. For lean fish, nitrogen: oxygen: carbon dioxide are used in a ratio of 30%:30%:40%, and in oily fish, nitrogen percent is increased up to 60% with proportionate reduction of oxygen in the mixture (Ravishankar, 2016). Reduction of oxygen supplementation also helps in decreasing the oxidative rancidity in fatty fish. With the length of storage, chemical, enzymatic, and microbial activity of the products get changed and also the composition of the gas mixture. The primary change in the composition of gas is enhancement of CO_2 content, which in turn reduces the growth of microbial content, especially *Pseudomonads* in fish, resulting in enhanced shelf-life of the products (Yesudhason et al., 2014). However, enhancement of oxygen supplementation is useful in suppressing the growth of strictly anaerobic bacteria like *C. botulinum*. Another method of packaging is active packaging, which involves incorporation of certain additives into the packaging containers with the aim of extending product shelf-life (Rooney, 1995) while maintaining the quality, sensory properties, and safety of foodstuffs (Ozdemir and Floros, 2004). Vacuum packaging is another method of extending shelf-life of foodstuffs, which involves removal of air from the package and application of hermetic seal. Removal of air reduces the growth of microorganisms and oxidative deteriorative reactions, and the foodstuffs maintain their texture, color, appearance, and taste for three to five times longer than conventional storage (Ravishankar, 2016). Among various types of packaging materials, high-barrier film is a better choice. MAP with high-barrier film effectively controls CO_2 and prevents microbial growth in packaged fishery products (Farber, 1991).

Coating of foodstuffs reduces contamination with microorganisms and thereby increases safety and shelf life. Any type of thin material that can wrap the food, extending its shelf-life, and that can be safely consumed together with the food is considered an edible coating or film (Bourtoom, 2008). Certain edible coating material can also act as a natural layer at the product surfaces and can prevent moisture losses, gas aromas, and solute movements out of the food but can selectively allow the exchange of gases involved in food product respiration such as oxygen, carbon dioxide, and ethylene (Embuscado and Huber, 2009). These materials are highly preferred as an alternative to plastics, because they are environmental friendly, nontoxic, biodegradable, and sometimes made up from by-products of the food industry (Rodriguez-Turienzo et al., 2013). Various types of films and coatings made with polysaccharides, proteins, and lipids are in use. Gums and chitosan are commonly used polysaccharide-based coatings (Sanchez-Ortega et al., 2014); casein, whey, collagen, wheat gluten, gelatin, keratin, and egg albumen are protein-based coatings; and acetylated monoglycerides are lipid-based coating materials (Dehghani et al., 2018).

ANTIMICROBIAL COMPOUNDS USED IN FISH AND SEAFOOD PRODUCTS

To increase the shelf-life, antimicrobial compounds are added to food during processing if they are safe and qualify by the guidelines of the specific country's legislation. Commonly used antimicrobial compounds are nitrites, sulfides, and organic acids (Archer, 2002; Ray, 2004; Chipley, 2005).

Nitrites are added to fish products as sodium nitrite or potassium nitrite along with NaCl, ascorbate, and erythorbate. In addition to its antimicrobial property of controlling *Staphylococcus aureus*, *Y. enterocolitica*, and *C. botulinum*, it also helps in maintaining color and odor and prevents lipid oxidation (Sindelar and Houser, 2009). As per the USFDA, the regulatory limit of nitrite as a food additive is 156 ppm (Ryser and Marth, 1999). In Canada, it is banned as a fish additive, but in meat, the limit is 200 ppm (DJC, 2009).

Sodium sulfite acts against yeasts, molds, and aerobic Gram-negative *Bacilli* and is used as an antimicrobial agent in fish and fishery products, sausage, pickles, fruit juices, wines, and beverages. Microbial activity of this compound rests on sulfurous acid, which reacts with thiol groups of proteins, enzymes, and cofactors inside the cells (Ray, 2004) and had inhibitory effects on coli-aerogenes and other Gram-negative microorganisms at room temperature (Omojowo et al., 2009). However, the application of sulfiting agents to meat and fish products is limited because of its destructive effect on thiamine (Walker, 1985).

Lactic acid and other organic acids inhibit the growth of undesirable microorganisms including *L. monocytogenes* in cold-smoked salmon and various other food products

(Matamoros et al., 2009). Cold-smoked salmon treated with divergicin M35-produced by *Carnobacterium divergens* M35 is effective in inhibiting the growth of *L. monocytogenes* (Tahiri et al., 2009) and reuterin or b-hydroxypropionaldehyde (b-HPA) of *Lactobacillus reuteri* has antimicrobial activity against a wide range of food-borne pathogens, including *L. monocytogenes* in fish and fishery products (Montiel et al., 2014).

Antimicrobial activities of sulfites and nitrites are enhanced by ascorbic acid and sodium ascorbate and D-isoascorbate (erythorbate) (Baird-Parker and Baillie, 1974) due to sequestration of iron and antioxidant properties (Tompkin et al., 2007). Recommended dose of ascorbic acid and erythorbate for antimicrobial function is 0.2% (Undeland et al., 2005). Benzoic acid and sodium benzoate are used as a preservative to inhibit the growth of yeasts and fungi (Chipley, 2005). Benzoic acid at 0.8% level kills *Pseudomonas* like bacteria in shrimp. In Canada, its usable limit is 1000 ppm for marinated or cold-processed packaged fish and meat.

Foods preserved with natural additives are gaining importance because of the consumer awareness, concern, and preference over the synthetic chemical additives. Some of the natural compounds used as food preservatives because of their antimicrobial effects are plant-derived essential oils such as cloves, rosemary, basil, cinnamon, thyme, and oregano; enzymes obtained from animal sources like lactoferrin and lysozyme; microbial products like bacteriocins; organic acids like citric acid, propionic, and sorbic acid; and naturally occurring polymers such as chitosan. Among the natural products, plant-derived essential oils render a wide spectrum of antimicrobial activity against food-borne pathogens and spoilage bacteria (Gutierrez et al., 2008, 2009) and are generally recognized as safe and are widely used in the food industry. In plant-derived oil, generally hydrophilic functional groups play a critical role in antimicrobial activity (Dorman and Deans, 2000). The presence of phenolic groups in some oils (clove, rosemary, thyme, oregano, etc.) plays a more effective role against Gram-positive than Gram-negative bacteria (Skandamis and Nychas, 2001).

The essential oil of thyme and oregano at 0.05% slows the process of fish spoilage. Fish and seafood products treated with these oils have extended shelf-life with good quality and taste and are suitable for human consumption even after 33 days of storage (Harpaz et al., 2003). Fried mullet fish fillets treated with edible coating solution mixed with marjoram (2.5, 5%) and thyme (2.5, 5%) (Yasin and Abou-Taleb, 2007), sea bass fillets treated with essential oils in combination with 60% CO_2–30% N_2–10% O_2 (Kostaki et al., 2009), rainbow trout fillets treated with the oregano oil (0.2%) and packed under vacuum (Frangos et al., 2010), and fish burgers treated with 0.4% and 0.8% rosemary extract and packaged under vacuum (Ucak et al., 2011) have significantly longer shelf-life than control package. Recent reports also show that essential oils and plant-derived products are effective against a wide range of microorganisms like *S. typhimurium*, *S. enteritidis*, *S. typhi*, *Y. enterocolitica*, *S. aureus*, *L. monocytogenes*, *E. coli*, and *Campylobacter* sp. (Esmail et al., 2014; Calo et al., 2015).

CONTROL OF PATHOGENS IN FISH AND SEAFOOD

To control pathogens in fish and fishery products, several strategies may be applied following the hazard analysis critical control point (HACCP) system. Each industry/processing unit generally follows a standard operating procedure. However, the following points are very critical in controlling fish and seafood-borne illnesses.

- Fish and mollusks should be harvested from a clean and certified or good management practices (GMP) accorded water bodies.
- At the time of harvest, naturally occurring pathogens in fish and molluscan shellfish may be present in relatively low numbers, but with the advancement of time the number may increase to hazardous levels. As compared to finfish, spoilage of shellfish occurs very quickly. Therefore, quick refrigeration/freezing of the harvest is required.
- Special attention should be given for most raw molluscan shellfish products, as very often they are consumed raw. The product must be screened to certify them free from at least PSP, *V. vulnificus* and *V. parahaemolyticus* pathogens.
- Quick processing of raw materials into product following pasteurization or cooking and reduction of moisture content in the product by drying are needed.
- Application of salt and other preservatives, proper packaging, followed by refrigeration or freezing are needed.

THE ROLE OF HAZARD ANALYSIS CRITICAL CONTROL POINT IN SAFETY OF FISH AND SEAFOOD

The safety of fish and seafood products is governed by several factors. Among them, three major and important factors are environment of their culture or harvest, the process of their production and handling, and the preservation of products during storage and supply chain. Food-borne pathogens survive in raw or undercooked food products, and also they may grow during storage. Therefore, consumption of raw or undercooked fish and seafood products such as bivalve molluscs may be potentially risky. The time and temperature between harvest and refrigeration also play a very important role in determining food-borne poisoning (Iwamoto et al., 2010).

With the aim to minimize food-borne illness, every fish and seafood harvester and processing unit is required to follow HACCP guidelines. The FDA plays an important role in preparing guidelines for the safety of fish and fishery products. The major components of the safety management systems in the food supply chain are GMP, good hygienic practices, sanitation standard operating procedures, and HACCP (Arvanitoyannis and Varzakas, 2009; Aruoma, 2006). Several countries including Australia, Japan, New Zealand, Canada, the United States, and countries under the European Union have endorsed the implementation of HACCP for food safety. In addition to HACCP, there are the National Shellfish Sanitation Program guidelines in the United States that regulate the harvesting of shellfish, its processing, and interstate shipping and also the safety of molluscan shellfish (Iwamoto et al., 2010).

Strategies to control and prevent fish and seafood-associated illnesses are monitoring of the aquatic bodies and harvest, implementation of process controls, and education to the consumer. Among many factors, the most common factors of seafood-borne poisoning are improper cooking, use of raw ingredients in the preparation, cross-contamination, and inadequate storage. Control of food-borne pathogens irrespective of the source from marine or aquacultural product includes immediate chilling at the harvest site and for cooked items rapid chilling followed by plate freezing and frozen storage (Arvanitoyannis and Varzakas, 2009). There is always an additional risk of cross-contamination of raw and cooked product during processing at the plant (Norhana et al., 2010a). Cooked shrimp and crabmeat, hot-smoked fish, pasteurized or cooked and breaded fish fillets are produced through heat treatment. All of these products also pass through further processes before packing, storing, and distribution. Some products that are made as RTE are highly sensitive and must be very safe to the consumers. These products come under the stringent application of the HACCP system. In the RTE seafood products, the occurrence of *Salmonella* sp. has been detected even in the dried/salted fish (Heinitz et al., 2000). *Salmonella* spp. are resistant to many stressful conditions (Arkoudelos et al., 2003; Ristori et al., 2007), and their long-term survivability is known in many food products including salted sardines at 0.69 a_w after 60 days (Arkoudelos et al., 2003). *Salmonella* spp. can grow at 4°C and can tolerate an acidic environment. Many shrimp and shrimp products such as shrimp salad and marinated or brined shrimp and RTE shrimp get contaminated with *Salmonella* and can cause food-borne diseases outbreaks (NACMCF, 2008; Norhana et al., 2010b). Therefore, elimination of *Salmonella* from foodstuffs is very important and critical for food safety and public health. In order to protect the public and economic health of a nation, food safety in the global market place is very important. Governments and private enterprises both must share the responsibility in reducing food-borne illnesses.

CONCLUSION

Internationally, fish and seafood products consumption is constantly increasing, and consumers' preferences are moving toward minimally processed and RTE products. However, these food products may be contaminated by various groups of food-borne pathogens. Among them, bacteria belonging to the genus *Salmonella*, *Vibrio*, and *Listeria* are a major concern in causing food-borne illnesses. Therefore, strict surveillance systems adopted by governments and competent regulatory authorities are fundamental to prevent contamination of foodstuffs with pathogens and for protection of public health. Coordinated efforts of the public health, veterinary, and fishery experts and food safety officers of the public and private enterprises in monitoring water quality, harvesting of fish and seafood, processing, preservation, and marketing are essential to ensure the safety of fish and seafood for human consumption.

References

Akhila, S., Shanmuga, P.S., Senthil, M.T., Tha, T., 2013. Molecular diversity analysis of *Yersinia enterocolitica* isolated from marine marketed fish. Int. J. Curr. Microbiol. Appl. Sci. 2, 204–214.

Akinola, O.A., Akinyemi, A.A., Bolaji, B.O., 2006. Evaluation of traditional and solar drying systems towards enhancing fish storage and preservation in Nigeria Abeokuta local government as a case study. J. Fisheries Int. 1, 44–49.

Alam, M.J., Tomochika, K.I., Miyoshi, S.I., Shinoda, S., 2002. Environmental investigation of potentially pathogenic *Vibrio parahaemolyticus* in the Seto-Inland Sea, Japan. FEMS Microbiol. Lett. 208, 83–87.

Amagliani, G., Omiccioli, E., Brandi, G., Bruce, I.J., Magnani, M., 2010. A multiplex magnetic capture hybridisation and multiplex real-time PCR protocol for pathogen detection in seafood. Food Microbiol. 27, 580–585.

Amagliani, G., Brandi, G., Schiavano, G.F., 2012. Incidence and role of *Salmonella* in seafood safety. Food Res. Int. 45, 780–788.

Amos, B., 2007. Analysis of Quality Deterioration at Critical Steps/Points in Fish Handling in Uganda and Iceland and Suggestions for Improvement. United Nations University, Uganda.

Anonymous, 2002. Microbiology of food and animal feeding stuffs- Horizontal method for the detection of Salmonella spp. ISO 6579, 4th ed. pp 1-27.

Aono, E., Sugita, H., Kawasaki, J., Sakakibara, H., Takahashi, T., Endo, K., et al., 1997. Evaluation of the polymerase chain reaction method for identification of *Vibrio vulnificus* isolated from marine environments. J. Food. Prot. 60, 81–83.

Archer, D.L., 2002. Evidence that ingested nitrate and nitrite are beneficial to health. J. Food. Prot. 65, 872–875.

Arkoudelos, J.S., Samaras, F.J., Tassou, C.C., 2003. Survival of *Staphylococcus aureus* and *Salmonella, Enteritidis* on salted sardines (*Sardinapil chardus*) during ripening. J. Food. Prot. 66, 1479–1481.

Aruoma, O.I., 2006. The impact of food regulation on the food supply chain. Toxicology 127, 119–221.

Arvanitoyannis, I.S., Varzakas, T.H., 2009. Seafood. In: Arvanitoyannis, I.S. (Ed.), HACCP and ISO 22000: Application to Foods of Animal Origin. Blackwell Publishing Ltd, Oxford, p. 377.

Asten, A.J., Dijk, J.E., 2005. Distribution of "classic" virulence factors *Salmonella* serovars. FEMS Immunol. Med Microbiol. 44, 251–259.

Attaran, R.R., Probst, F., 2002. Histamine fish poisoning: a common but frequently misdiagnosed condition. J. Emerg. Med. 19, 474–475.

Baird-Parker, A.C., Baillie, M.A.H., 1974. The inhibition of *Clostridium botulinum* by nitrite and sodium chloride. In: Proceedings of the International Symposium on Nitrite in Meat Products, Wageningen, Pudoc, Zeist, The Netherlands, p. 77.

Bang, I.S., Frye, J.G., McClelland, M., Velayudhan, J., Fang, F.C., 2005. Alternative sigma factor interactions in *Salmonella*: σ^E and σ^H promote antioxidant defences by enhancing σ^S levels. Mol. Microbiol. 56, 811–823.

Bauer, A., Rorvik, L.M., 2007. A novel multiplex PCR for the identification of *Vibrio parahaemolyticus, Vibrio cholerae* and *Vibrio vulnificus*. Lett. Appl. Microbiol. 45 (4), 371–375.

Bej, A.K., Mahbubani, M.H., Boyce, M.J., Atlas, R.M., 1994. Detection of *Salmonella* spp. in oysters by PCR. Appl. Environ. Microbiol. 60 (1), 368–373.

Bej, A.K., Patterson, D.P., Brasher, C.W., Vickery, M.C., Jones, D.D., Kaysner, C.A., 1999. Detection of total and hemolysin-producing *Vibrio parahaemolyticus* in shellfish using multiplex PCR amplification of *tl, tdh* and *trh*. J. Microbiol. Methods 36 (3), 215–225.

Ben-Gigirey, B., Vieites Baptista de Sousa, J.M., Villa, G.T., Barros-Velazquez, J., 2000. Characterization of biogenic amine-producing *Stenotrophomonas maltophilia* strains isolated from white muscle of fresh and frozen albacore tuna. Int. J. Food Microbiol. 57 (1–2), 19–31.

Ben-Gigirey, B., Vieites, J.M., Kim, S.H., Tomás, H.A., Villa, G., Barros-Velazquez, J., 2002. Specific detection of *Stenotrophomonas maltophilia* strains in albacore tuna (*Thunnus alalunga*) by reverse dot-blot hybridisation. Food. Control. 13 (4-5), 293–299.

Berkel, B.M., Boogaard, B.V., Heijnen, C., 2004. Preservation of Fish and Meat. Agromisa Foundation, Wageningen, The Netherlands, pp. 78–80, ISBN: 90-72746-01-9.

Bottone, E.J., Bercovier, H., Mollaret, H.H., 2005. Genus XLI. *Yersinia*. In: Garrity, G.M., Brenner, D.J., Krieg, N.R., Staley, J.T. (Eds.), Bergey's Manual of Systematic Bacteriology. The Proteobacteria. Part B. The Gamma Proteobacteria, second ed. Springer Press, New York, pp. 838–848.

Bourtoom, T., 2008. Edible films and coatings: characteristics and properties. Int. Food Res. J. 15 (3), 237–248.

Brands, D.A., Inman, A.E., Gerba, C.P., Mare, C.J., Billington, S.J., Saif, L.A., et al., 2005. Prevalence of *Salmonella* spp. in oysters in the United States. Appl. Environ. Microbiol. 71, 893–897.

Budiati, T., Rusul, G., Wan-Abdullah, W.N., Arip, Y.M., Ahmad, R., Thong, K.L., 2013. Prevalence, antibiotic resistance, and plasmid profiling of salmonella in catfish (*Clarias gariepinus*) and tilapia (*Tilapia mossambica*) obtained from wet markets and ponds in Malaysia. Aquaculture 372, 127–132.

Butt, A.A., Aldridge, K.E., Sanders, C.V., 2004. Infections related to the ingestion of seafood. Part I: viral and bacterial infections. Lancet Infect. Dis. 4, 201–212.

Callol, A., Pajuelo, D., Ebbesson, L., Teles, M., MacKenzie, S., Amaro, C., 2015. Early steps in the European eel (*Anguilla anguilla)-Vibrio vulnificus* interaction in the gills: role of the RtxA13 toxin. Fish Shell Fish Immunol. 43, 502–509.

Calo, J.R., Crandall, P.G., O'Bryan, C.A., Ricke, S.C., 2015. Essential oils as antimicrobials in food systems—a review. Food. Control. 54, 111–119.

Chai, J.Y., Darwin Murrell, K., Lymbery, A.J., 2005. Fish-borne parasitic zoonoses: status and issues. Int. J. Parasitol. (11–12), 1233–1254.

Chao, W., Ding, R., Chen, R., 1987. Survival of pathogenic bacteria in environmental microcosms. Chin. J. Microbiol. Immunol. 20, 339–348.

Chipley, J.R., 2005. Sodium Benzoate and Benzoic acid. Antimicrobials in Food, third ed., pp. 11–38.

Civera, T., 2003. Species identification and safety of fish products. Vet. Res. Commun. 1, 481–489.

Dalgaard, P., Madsen, H.L., Samieian, N., Emborg, J., 2006. Biogenic amine formation and microbial spoilage in chilled garfish (*Belone belone*) effect of modified atmosphere packaging and previous frozen storage. J. Appl. Microbiol. 101, 80–95.

Darlington, L., Stone, T., 2001. Antioxidants and fatty acids in the amelioration of rheumatoid artritis and related disorders. Br. J. Nutr. 85 (3), 251–269.

Davies, A.R., Capell, C., Jehanno, D., Nychas, G.J.E., Kirby, R.M., 2001. Incidence of food borne pathogens on European fish. Food Control. 12, 67–71.

Da-Wen, S., 2005. Emerging Technologies for Food Processing. Elsevier Academic Press, UK.

Dehghani, S., Hosseini, S.V., Regenstein, J.M., 2018. Edible films and coatings in seafood preservation: a review. Food. Chem. 240, 505–513.

De-Paola, A., Jones, J.L., Woods, J., Burkhardt, W., Calci, K.R., Krantz, J.A., et al., 2010. Bacterial and viral pathogens in live oysters: 2007 United States market survey. Appl. Environ. Microbiol. 76, 2754–2768.

DJC, 2009. Food and Drug Act. Department of Justice, Canada.

Dorman, H.J., Deans, S.J., 2000. Antimicrobial agents from plants: antibacterial activity of plant volatile oils. J. Appl. Microbiol. 88, 308–316.

Emborg, J., Laursen, B.G., Dalgaard, P., 2005. Significant histamine formation in tuna (*Thunnus albacares*) at 2°C: effect of vacuum-and modified atmosphere-packaging on psychrotolerant bacteria. Int. J. Food Microbiol. 101, 263–279.

Embuscado, M.E., Huber, K.C., 2009. Edible Films and Coatings for Food Applications. Springer Inc, Dordrecht.

Esmail, A., Rezaei, M., Hosseini, H., 2014. Antibacterial activity of plant essential oils and extracts: the role of thyme essential oil, nisin, and their combination to control *Listeria monocytogenes* inoculated in minced fish meat. Food. Control. 35 (1), 177–183.

EUMOFA, 2014. European market observatory for fisheries and aquaculture products. The EU Fish Market. 2014 ed., pp. 1–61.

Fakruddin, M.D., Sultana, M., Ahmed, M.M., Chowdhury, A., Choudhury, N., 2013. Multiplex PCR (polymerase chain reaction) assay for detection of *E. coli* O157:H7, *Salmonella* sp., *Vibrio cholera* and *Vibrio parahaemolyticus* in spiked shrimps (*Penaeus monodon*). Pak. J. Biol. Sci. 16 (6), 267–274.

FAO, 2005. Post-Harvest Changes in Fish. FAO Fisheries and Aquaculture Department, Food and Agriculture Organization, Rome, Italy. Available from: http://www.fao.org/fishery/topic/12320/en.

FAO, 2016. The State of World Fisheries and Aquaculture, vol. 2016. Food and Agriculture Oraganization of the United Nations, ISBN: 92-5-109185-2.

Farabegoli, F., Blanco, L., Rodriguez, L.P., Vieites, J.M., Cabado, A.G., 2018. Phycotoxins in marine shellfish: origin, occurrence and effects on humans. Mar. Drugs. 16, 188.

Farber, J.M., 1991. Microbial aspects of modified-atmosphere packaging technology—a review. J. Food. Prot. 54 (1), 58–70.

Fellows, P., 2000. Food Processing Technology: Principles and Practice, second ed. CRC, New York.

Frangos, L., Pyrgotou, N., Giatrakou, V., Ntzimani, A., Savvaidis, I.N., 2010. Combined effects of salting, oregano oil and vacuum-packaging on the shelf-life of refrigerated trout fillets. Food Microbiol. 27, 115–121.

Fraser, O., Sumar, S., 1998. Compositional changes and spoilage in fish. Nutr. Food Sci. 5, 275–279.

Gauthier, D.T., 2015. Bacterial zoonoses of fishes: a review and appraisal of evidence for linkages between fish and human infections. Vet. J. 203, 27–35.

Givens, C.E., Bowers, J.C., DePaola, A., Hollibaugh, J.T., Jones, J.L., 2014. Occurrence and distribution of *Vibrio vulnificus* and *Vibrio parahaemolyticus*—potential role for fish, oyster, sediment and water. Lett. Appl. Microbiol. 58, 503–510.

González-Rodríguez, M.N., Sanz, J.J., Santos, J.A., Otero, A., García-López, M.L., 2002. Foodborne pathogenic bacteria in prepackaged fresh retail portions of farmed rainbow trout and salmon stored at 3°C. Int. J. Food Microbiol. 76, 135–141.

Gornik, S.G., Albalat, A., Theethakaew, C., Neil, D.M., 2013. Shelf life extension of whole Norway lobster Nephrops norvegicus using modified atmosphere packaging. Int. J. Food Microbiol. 167 (3), 369–377.

Gram, L., Dalgaard, P., 2002. Fish spoilage bacteria problems and solutions. Curr. Opin. Biotechnol. 13, 262–266.

Gram, L., Huss, H.H., 2000. Fresh and processed fish and shellfish. In: Lund, B.M., Baird- Parker, A.C., Gould, G. W. (Eds.), The Microbiological Safety and Quality of Foods. Chapman and Hall, London, pp. 472–506. , ISBN: 10: 0834213230.

Gutierrez, J., Barry-Ryan, C., Bourke, P., 2008. The antimicrobial efficacy of plant essential oil combinations and interactions with food ingredients. Int. J. Food Microbiol. 124, 91–97.

Gutierrez, J., Barry-Ryan, C., Bourke, P., 2009. Antimicrobial activity of plant essential oils using food model media: efficacy, synergistic potential and interaction with food components. Food Microbiol. 26, 142–150.

Hansen, L.T., Gill, T., Huss, H.H., 1995. Effects of salt and storage-temperature on chemical, microbiological and sensory changes in cold-smoked salmon. Food Res. Int. 28, 123–130.

Harpaz, S., Glatman, L., Drabkin, V., Gelman, A., 2003. Effects of herbal essential oils used to extend the shelf life of freshwater-reared Asian sea bass fish (*Lates calcarifer*). J. Food. Prot. 66 (3), 410–417.

Heinitz, M.L., Ruble, R.D., Wagner, D.E., Tatini, S.R., 2000. Incidence of *Salmonella* in fish and seafood. J. Food. Prot. 63, 579–592.

Hoffman, A.D., Wiedmann, M., 2001. Comparative evaluation of culture- and BAX polymerase chain reaction-based detection methods for *Listeria* spp. and *Listeria monocytogenes* in environmental and raw fish samples. J. Food. Prot. 64 (10), 1521–1526.

Hossain, M.T., Kim, E.Y., Kim, Y.R., Kim, D.G., Kong, I.S., 2013. Development of a *groEL* gene-based species-specific multiplex polymerase chain reaction assay for simultaneous detection of *Vibrio cholera*, *Vibrio parahaemolyticus* and *Vibrio vulnificus*. J. Appl. Microbiol. 114 (2), 448–456.

Hungerford, J.M., 2010. Scombroid poisoning: a review. Toxicon 56 (2), 231–243.

Huss, H.H., 1993. Assurance of Seafood Quality. FAO Fisheries Technical Paper No. 334. FAO Publications Division, Rome, pp. 1–169.

Huss, H.H., Ababouch, L., Gram, L., 2003. Assessment and Management of Seafood Safety and Quality. FAO Fisheries Technical Paper No. 444. FAO, Rome, p. 230.

ICMSF, 1996. *Salmonellae*. Chapter 14. Microorganisms in Food 5: Microbiological Specifications of Food Pathogens. Blackie Academic and Professional, London, pp. 217–264.

Iwamoto, M., Ayers, T., Mahon, B.E., Swerdlow, D.L., 2010. Epidemiology of seafood-associated infections in the United States. Clin. Microbiol. Rev. 23, 399–411.

Izumiya, H., Matsumoto, K., Yahiro, S., Lee, J., Morita, M., Yamamoto, S., et al., 2011. Multiplex PCR assay for identification of three major pathogenic *Vibrio* spp., *Vibrio cholera*, *Vibrio parahaemolyticus*, and *Vibrio vulnificus*. Mol. Cell Prob. 25 (4), 174–176.

Jantschitsch, C., Kinaciyan, T., Manafi, M., et al., 2011. Severe scombroid fish poisoning: an underrecognized dermatologic emergency. J. Am. Acad. Dermato 65, 246–247.

Kakatkar, A.S., Gautamm, R.K., Nagar, V., Karani, M., Bandekar, J.R., 2010. Incidence of food borne pathogens in freshwater fish from domestic markets of Mumbai. Fish Technol. 47, 195–200.

Karapinar, M., Gonul, S.A., 1991. Survival of *Yersinia enterocolitica* and *Escherichia coli* in spring water. Int. J. Food Microbiol. 13, 315–319.

Kaysner, C.A., DePaola, A., 2001. Vibrio. In: Downes, F.P., Ito, K. (Eds.), Compendium of Methods for the Microbiological Examination of Foods, fourth ed. American Public Health Association, Washington, DC, pp. 405–420.

Kin, S., Wes Schilling, M., Kim, T., Smith, B.S., Silva, J.L., Campano, S.G., et al., 2012. Effects of potassium lactate and acetate on *Listeria monocytogenes* inhibition, physicochemical and sensory properties of smoked catfish fillets. J. Aquat. Food Product. Technol. 21, 338–350.

Kostaki, M., Giatrakou, V., Savvaidis, I.N., Kontominas, M.G., 2009. Combined effect of MAP and thyme essential oil on the microbiological, chemical and sensory attributes of organically aquacultured seabass (*Dicentrarchus labrax*) fillets. Food Microbiol. 26, 475–482.

Kumar, R., Surendran, P.K., Thampuran, N., 2008a. Evaluation of culture, ELISA and PCR assays for the detection of Salmonella in seafood. Lett. Appl. Microbiol. 46, 221–226.

Kumar, R., Surendran, P.K., Thampuran, N., 2008b. An eight-hour PCR-based technique for detection of *Salmonella serovars* in seafood. World J. Microbiol. Biotechnol. 24, 627–631.

Kumar, R., Surendran, P.K., Thampuran, N., 2009. Distribution and genotypic characterization of *Salmonella serovars* isolated from tropical seafood of Cochin. Indian J. Appl. Microbiol. 106, 515–524.

Kumar, R., Surendran, P.K., Thampuran, N., 2010. Rapid quantification of Salmonella in seafood using Real-Time PCR assay. J Microbiol Biotechnol 20, 569–573.

Kumar, R., Datta, T.K., Lalitha, K.V., 2015. *Salmonella* grows vigorously on seafood and expresses its virulence and stress genes at different temperature exposure. BMC. Microbiol. 15, 254.

Kuswandi, B., Wicaksono, Y., Abdullah, A., Lee, Y.H., Musa, A., 2011. Smart packaging: sensors for monitoring of food quality and safety. Sens. Instrumen. Food Qual. Saf. 5, 137–146.

Li, T.H., Chiu, C.H., Chen, W.C., Chen, C.M., Hsu, Y.M., Chiou, S.S., et al., 2009. Consumption of groundwater as an independent risk factor of *Salmonella Choleraesuis* infection: a case control study in Taiwan. J. Environ. Health 72, 28–31.

Liao, C.H., Shollenberger, L.M., 2003. Survivability and long-term preservation of bacteria in water and in phosphate-buffered saline. Lett. Appl. Microbiol. 37 (1), 45–50.

Martinez-Urtaza, J., Liebana, E., 2005. Use of pulsed-field gel electrophoresis to characterize the genetic diversity and clonal persistence of Salmonella senftenberg in mussel processing facilities. Int J Food Microbiol 105, 153–163.

Matamoros, S., Pilet, M.F., Gigout, F., Prevost, H., Leroi, F., 2009. Selection and evaluation of seafood-borne psychrotrophic lactic acid bacteria as inhibitors of pathogenic and spoilage bacteria. Food Microbiol. 26, 638–644.

Mendes, R., Pestana, C., Gonçalves, A., 2008. The effects of soluble gas stabilisation on the quality of packed sardine fillets (*Sardina pilchardus*) stored in air, VP and MAP. Int. J. Food Sci. Technol. 43, 2000–2009.

Minami, A., Chaicumpa, W., Chongsa-Nguan, M., Samosornsuk, S., Monden, S., Takeshi, K., et al., 2010. Prevalence of foodborne pathogens in open markets and supermarkets in Thailand. Food. Control. 21, 221–226.

Montiel, R., Martin-Cabrejas, I., Langa, S., El Aouad, N., Arques, J.L., Reyes, F., et al., 2014. Antimicrobial activity of reuterin produced by *Lactobacillus reuteri* on *Listeria monocytogenes* in cold-smoked salmon. Food Microbiol. 44, 1–5.

Moore, B.C., Martinez, E., Gay, J.M., Rice, D.H., 2003. Survival of *Salmonella enterica* in freshwater and sediments, and transmission by the aquatic midge *Chironomus Tentans* (chironomidae: diptera). Appl. Environ. Microbiol. 69, 4556–4560.

NACMCF, 2008. Response to the questions posed by the Food and Drug Administration and the National Marine Fisheries Service regarding determination of cooking parameters for safe seafood for consumers. J. Food. Prot. 71 (6), 1287–1308.

Neogi, S.,B., Chowdhury, N., Asakura, M., Hinenoya, A., Haldar, S., Saidi, S.,M., et al., 2010. A highly sensitive and specific multiplex PCR assay for simultaneous detection of *Vibrio cholera*, *Vibrio parahaemolyticus* and *Vibrio vulnificus*. Lett Appl Microbiol 51 (3), 293–300.

Newton, A., Kendall, M., Vugia, D.J., Henao, O.L., Mohan, B.E., 2012. Increasing rates of vibriosis in the United States, 1996–2010: review of surveillance data from 2 systems. Clin. Infect. Dis. 54 (0 5), S391–S395.

Norhana, M.N.W., Poole, S.E., Deeth, H.C., Dykes, G.A., 2010a. Prevalence, persistence and control of *Salmonella* and *Listeria* in shrimp and shrimp products: a review. Food. Control. 21, 343–361.

Norhana, M.N.W., Poole, S.E., Deeth, H.C., Dykes, G.A., 2010b. The effects of temperature, chlorine and acids on the survival of *Listeria* and *Salmonella* strains associated with uncooked shrimp carapace and cooked shrimp flesh. Food Microbiol. 27, 250–256.

Norton, D.M., McCamey, M., Boor, K.J., Wiedmann, M., 2000. Application of the BAX for screening/genus *Listeria* polymerase chain reaction system for monitoring *Listeria* species in cold-smoked fish and in the smoked fish processing environment. J. Food Protect. 63 (3), 343–346.

Novoslavskij, A., Terentjeva, M., Eizenberga, I., Valciņa, O., Bartkevics, V., Berziņs, A., 2016. Major foodborne pathogens in fish and fish products: a review. Ann. Microbiol. 66, 1–15.

O'Connor, M.M., Forbes, G.M., 2000. Scombroid poisoning: not fish allergy. Aust. NZ J. Med. 30, 520.

OIE, 2014. *Listeria monocytogenes*. Chapter 2. 9.7. Manual of Diagnostic Tests and Vaccines for Terrestrial Animals, p. 118.

Omojowo, F.S., Idris, G.L., Ihuahi, J.A., 2009. Comparative assessment of potassium sorbate and sodium metabisulphite on the safety and shelf life of smoked catfish. Nat. Sci. 7, 10–17.

Ozdemir, M., Floros, J.D., 2004. Active food packaging technologies. Crit. Rev. Food Sci. Nutr. 44, 185–193.

Pagadala, S., Parveen, S., Schwarz, J.G., Rippen, T., Luchansky, J.B., 2011. Comparison of automated BAX PCR and standard culture methods for detection of *Listeria monocytogenes* in blue crabmeat (*Callinectus sapidus*) and blue crab processing plants. J. Food. Prot. 74 (11), 1930–1933.

Papadopoulos, T., Abrahim, A., Sergelidis, D., Kirkoudis, I., Bitchava, K., 2010. Prevalence of *Listeria* spp. in freshwater fish (*Oncorhynchus mykiss* and *Carassius gibelio*) and the environment of fish markets in Northern Greece. J. Hellenic Vet. Med. Soc. 61, 15–22.

Parker, W.F., Mee, B.J., 1982. Survival of *Salmonella adelaide* and fecal coliforms in coarse sands of the swan coastal plain, Western Australia. Appl. Environ. Microbiol. 43, 981–986.

Ravishankar, C.N., 2016. Recent advances in processing and packaging of fishery products: a review. Aquatic Procedia 7, 201–213.

Ray, B., 2004. Fundamental of Food Microbiology, third ed. CRC Press, Boca Raton, FL, pp. 439–534.

Reeve, G., Martin, D.L., Pappas, J., Thompson, R.E., Greene, K.D., 1989. An outbreak of shigellosis associated with the consumption of raw oysters. N. Engl. J. Med. 321, 224–227.

Ristori, C.A., dos Santos Pereira, M.A., Gelli, D.S., 2007. Behaviour of Salmonella Rubislaw on ground black pepper (*Piper nigrum* L.). Food. Control. 18, 268–272.

Rodriguez-Turienzo, L., Cobos, A., Diaz, O., 2013. Effects of microbial transglutaminase added edible coatings based on heated or ultrasound-treated whey proteins in physical and chemical parameters of frozen Atlantic salmon (*Salmo salar*). Innov. Food Sci. Emerg. Technol. 119, 433–438.

Rooney, M.L., 1995. Active packaging in polymer films. In: Rooney, M.L. (Ed.), Active Food Packaging. Springer, Boston, MA, pp. 74–110.

Rosec, J.P., Causse, V., Cruz, B., Rauzier, J., Carnat, L., 2012. The international standard ISO/TS 21872-1 to study the occurence of total and pathogenic *Vibrio parahaemolyticus* and *Vibrio cholerae* in seafood: ITS improvement by use of achromogenic medium and PCR. Int. J. Food Microbiol. 157 (2), 189–194.

Ryser, E.T., Marth, E.H., 1999. *Listeria*, Listeriosis and Food Safety. Marcel Dekker, New York, p. 141, ISBN: 10: 0824702352.

Sanath, K.H., Sunil, R., Venugopal, M.N., Karunasagar, I., 2003. Detection of *Salmonella* spp. in tropical seafood by polymerase chain reaction. Int. J. Food Microbiol. 88 (1), 91–95.

Sanchez-Ortega, I., García-Almendárez, B.E., Santos-López, E.M., Amaro-Reyes, A., Barboza-Corona, J.E., Regalado, C., 2014. Antimicrobial edible films and coatings for meat and meat products preservation. SciWorld J 2014, 18.

Shabarinath, S., Kumar, H.S., Khushiramani, R., Karunasagar, I., Karunasagar, I., 2007. Detection and characterization of *Salmonella* associated with tropical seafood. Int. J. Food Microbiol. 114, 227–233.

Shin, C., Jo, H., Kim, S., Kang, G., 2018. Exposure assessment to paralytic shellfish toxins through the shellfish consumption in Korea. Food Res. Int. 108, 274–279.

Sindelar, J.J., Houser, T.A., 2009. Alternative curing systems. In: Tarte, R. (Ed.), Ingredients in Meat Products: Properties, Functionality and Applications. Springer Science and Business Media, New York, pp. 379–405.

Skandamis, P.N., Nychas, G.J.E., 2001. Effect of oregano essential oil on microbiological and physico-chemical attributes of minced meat stored in air and modified atmospheres. J. Appl. Microbiol. 91, 1011–1022.

Tahiri, I., Desbiens, M., Kheadr, E., Lacroix, C., Fliss, I., 2009. Comparison of different application strategies of divergicin M35 for inactivation of *Listeria monocytogenes* in cold-smoked wild salmon. Food Microbiol. 26 (8), 783–793.

Tocmo, R., Krizman, K., Khoo, W.J., Phua, L.K., Kim, M., Yuk, H.G., 2014. *Listeria monocytogenes* in vacuum-packed smoked fish products: occurrence, routes of contamination, and potential intervention measures. Com. Rev. Food Sci. Food Saf. 13, 172–189.

Tompkin, R.B., 2002. Control of *Listeria monocytogenes* in the food-processing environment. J. Food. Prot. 65, 709–725.

Tompkin, R.B., Christiansen, L.N., Shaparis, A.B., 2007. The effect of iron on botulinal inhibition in perishable canned cured meat. Int. J. Food Sci. Technol. 13, 521–527.

Ucak, I., Ozogul, Y., Durmus, M., 2011. The effects of rosemary extract combination with vacuum packing on the quality changes of Atlantic mackerel fishburgers. Int. J. Food Sci. Technol. 46, 1157–1163.

Undeland, I., Hall, G., Wendin, K., Gangby, I., Rutgersson, A., 2005. Preventing lipid oxidation during recovery of functional proteins from herring (*Clupea harengus*) fillets by an acid solubilisation process. J. Agric. Food Chem. 53, 5624–5634.

Unklesbay, N., 1992. World Food and You. Food Product Press, New York, 1560220104p. 251.

Vantarakis, A., Komninou, G., Venieri, D., Papapetropoulou, M., 2000. Development of a multiplex PCR detection of *Salmonella* spp. and *Shigella* spp. in mussels. Lett. Appl. Microbiol. 31 (2), 105–109.

Walker, R., 1985. Sulphiting agents in foods: some risk/benefit considerations. Food Addit. Contam. 2, 5–24.

WHO, 2004. Report of the Joint WHO/FAO Workshop on Foodborne Trematode Infections in Asia, Hanoi, Vietnam, 26–28 Nov 2002. WHO Regional Office for the Western Pacific, Manila, Philippines.

Yasin, N.M.N., Abou-Taleb, M., 2007. Antioxidant and antimicrobial effects of marjoram and thyme in coated refrigerated semi fried mullet fish fillets. World J. Dairy Food Sci. 2, 1–9.

Yesudhason, P., Lalitha, K.V., Srinivasa Gopal, T.K., Ravishankar, C.N., 2014. Retention of shelf life and microbial quality of seer fish stored in modified atmosphere packaging and sodium acetate pretreatment. Food Packaging Shelf Life 1, 123–130.

Further Reading

Alam, M.J., Miyoshi, S., Shinoda, S., 2003. Studies on pathogenic *Vibrio parahaemolyticus* during a warm weather in the Seto-Inland Sea, Japan. J. Environ. Microbiol. 5, 706–710.

CHAPTER

8

Microbial Environment of Food

Rajeeva Gaur, Anurag Singh and Ashutosh Tripathi

Department of Microbiology (Centre of Excellence), Dr. Rammanohar Lohia Avadh University, Faizabad, India

OUTLINE

INTRODUCTION

Food is an important source of nutrients, especially carbon, nitrogen, sulfur, phosphorus, and several micronutrients for all living systems: plants, microorganisms, humans, and animals. Food is a very broad category, but it is primarily classified into vegetarian and nonvegetarian foods. Vegetarian foods are the outcome of primary productivity (i.e., plants of lower and higher groups as well as algae and other photosynthetic bacteria). The chordates and nonchordates are the major category of nonvegetarian foods, including microbial foods. They all are the sources of food for specific living systems in the food web. Therefore every living system has an important component having specific food and then further transfers/transforms their nutrients to other groups of microorganisms in a specific food ecosystem. Such interactions are competitive versus noncompetitive toward

Food Safety and Human Health

DOI: https://doi.org/10.1016/B978-0-12-816333-7.00008-4

the nutrition they have for one group or for another group. Thus the transformations of nutrients among all the living systems do exist in particular tropic levels. In food system, microorganisms take nutrients and gasses like O_2, CO_2, sulfur, and nitrogen and survive for long-term existence in a particular food system. That is why the thermodynamic principles with the enthalpy and entropy exist with every group of microorganism. The nutritional status of the food and their constitutions for a particular living system, which varies according to the nutritional classes from autotrophic to heterotrophic levels with several other nutritional categories like chemolithotrophic, chemoorganotrophic microorganisms, lead to changes in the chemistry of the food.

In the nutritional cycle, microorganisms are the important components that transform complex organic matter into simple compounds to maintain the nutritional level of all the nutritional categories of microorganisms. Moreover, the important classification of food is based on perishability (i.e., higher, moderate, to less perishable foods). Food has wide categories, including fruits and vegetables, cereals, meat, milk, eggs, and several fermented foods. Such food is taken as raw or cooked to a processed level only for their long-term preservation. In all the levels, food environment is highly changeable in their nutritional levels as well as physicochemical levels, as food varies with the level of carbohydrate, protein, lipid, and nucleic acid contents along with several inorganic and organic components having antimicrobial constituents. Tannins, vitamins, as well as alkaloids are also important in the microbial scenario. These components directly/indirectly affect the microorganisms of the food as well as the consumers. Therefore some components of the food require processing or elimination or transformation of such components prior to consumption. A wide range of food processing, preservation, intoxication, detoxification, and quality standards have been worked out, but still much research is required for the food combinations and simplification of the nutrients like carbohydrate, protein, and lipid levels along with the nucleic acid detoxification, as higher nucleic acid content is not metabolized by human systems. Therefore with the help of suitable microorganisms, the compound can be processed for the gluten of cereals and some alkaloids of coffee, tea, and coca, as well as several other plant products. Every food has a different environment, especially for oxidation and reduction potentials, which change several other nutrients' status as well as the growth of microorganisms. It is therefore essential for the evaluation of the factors responsible for inhibition of microorganisms, along with simplification of protein and carbohydrate status; for example, change of pH either by food nutrients or by the existence of microorganism may change the degradation/depletion of protein to ammonia or other forms.

On the basis of these facts, this chapter discusses some of the research findings of the principal author, who has been working on food agriculture and industrial microbiology for the last 30 years, with some of the factors about the role of food fermenting microorganisms. Since the existence of microorganisms, plants of lower and higher origin to humans and animals of chordates to nonchordates (i.e., all living system of the universe) require food; therefore food is an important commodity for almost all living system. The mode of nutrition and their uptake in different forms vary in the living system depending on their nutritional categories. Food is a very broad category, from vegetarian to nonvegetarian; therefore foods provide survival to all living systems of the universe.

This chapter is mainly concentrated on human's food and its environment as the state of the science at a certain point of time requires several parameters to maintain the food chemistry and physical status (i.e., texture) along with the organoleptic specifications for particular foods. The wholesome and safe preservation of food and its supply require several skills and much knowledge of food microbiology. The food environment has wide variations at different stages from the field up to the harvesting or cooking/processing to where it is finally supplied to the public without spoilage. At every level of environment, the food must be evaluated critically.

Food spoilage, intoxication, and disease-causing microorganisms are the main components of almost all foods of human consumption. The microbes like bacteria, fungi, and yeast are sustained in food from very low to higher temperature, pH, water activity, redox potential, etc. They have a very broad nutritional category along with varying concentrations of nutrients, from low to high sugar and salt concentration and acid to alkaloid pH. Therefore food environment and the status of microorganisms in food must be studied carefully in order to achieve quality food production for safe consumption. Food can also be classified on the basis of its state: raw, semicooked, cooked, and fermented. They have different physical, chemical, and microbial status; therefore for every category of food the environment will be different.

As the history of food preservation, spoilage, and other aspects of food fermentation is concerned, Louis Pasteur, in France, gave several concepts of these aspects and proved that fermentation is not a chemical process but is achieved through microbial activity, and he also proved microbial spoilage through his famous experiment with a swan neck flask; thus he become a food microbiologist. He proved that air does contain microorganisms and the spoilage of broth could be possible by microbial activity. After the discovery of microorganisms by A.V. Leeuwenhoek, referred to as the father of microbiology, it was Pasteur who provided proof of consequences and the role of microbiology in food spoilage, food fermentation, and preservation. The first use of what we know as pasteurization, the heating of wine to destroy undesirable organisms, was introduced commercially in 1867–68. Further, in Germany, Robert Koch, a contemporary of Pasteur, established the existence of the microbial and disease relationship, and thereafter microbiologists proved several concepts of the effect of temperature on the survival and existence of microorganisms in food, as temperature is one of the important factors which affects the microbial growth. The main scientists were John Tyndall, N. F. Apart, F. Cohn, Joseph Lister, and several others who worked on different aspects of food microbiology, including the anaerobic life of microorganisms. The cause of botulism by *Clostridium botulinum* was discovered by E. Van Ermengem in 1896. This bacterium is a gram-positive, spore-forming rod, a strict anaerobic and even grows at low and higher temperatures; it causes serious foodborne intoxication. Several pioneering developments on food preservation were also established in the 1800s. Bens H. Benzamin, a British scientist, used an ice salt brine mixture for better deep freezing for better preservation of fish and meat. There are several environmental factors that affect the growth of microorganisms, but critically, food environment is ideally clarified as extrinsic and intrinsic factors that are widely studied in several parameters under these categories. It has been established that microbes are highly variable at all the physicochemical levels. Therefore, several standards for food and water born pathogens in modern standardization like analytical methods, process of pasteurization and indicator microorganisms may be assessed for quality assurance of various types of foods. Moreover, the history of food microbiology is rich and interesting with multidimensional research fields mainly in food spoilage, intoxication, disease-causing microorganisms, food fermentation, preservation, and quality assurance with the methods of detecting specific bacteria, antimicrobial constituents of food, as well as an integrated approach of microorganism for food health and hygiene. In all aspects, food environments play a very crucial role in food preservation, which is the ultimate goal of a microbiologist: to develop quality food for human consumption. The food environment is finally discussed in two major areas: extrinsic and intrinsic factors affecting food microorganisms.

FOOD ENVIRONMENT

Every food has its own ecosystem, which is diversified with several factors. The microbial interaction can also be discussed accordingly. The international commission of microbiological specifications for food has also been accepted and has shown importance for research work in this area. The principle author has also worked in the area and shares some information and findings that have a significant role in discussing the interaction of two different groups of microorganisms in the reduction of intoxication as well as microbial clumping of pathogenic microorganisms by lactic acid bacteria. The microbial existence and growth in an ecosystem are composed of the environment, organisms, and their specific growth ingredients at the level of inhibition, promoter, or having neutral effects and are therefore positive, negative, or no effect on the growth of the microorganisms in a particular food niche. The food environment consists of intrinsic factors that are mainly inside food, which generally represent the pH, water activity, nutrient of the food, antimicrobial components of the food, and specific microorganism levels and their interaction potentials.

EXTRINSIC FACTOR AFFECTING MICROBIAL GROWTH

Extrinsic factor is an external factor of food that depicts the temperature, humidity, gaseous composition, and microbial load. Both intrinsic and extrinsic factors can be manipulated during food preservation at the level of no growth by manipulating the conditions as well as use of specific microorganisms to save these food from intoxication, eliminating food-spoiling and disease-causing microorganisms. Temperature and gas composition of the atmosphere or surrounding the food is the least concerned to affect the growth of the microorganism. Temperature of the atmosphere and surrounding the food are mainly concerned with the geographical area, from tropical, subtropical, to cooler countries and its seasons. Therefore organisms with specific temperature categorization and their distribution in different foods may be considered. Moreover, every microorganism has its own temperature optima for its growth and metabolite production. The categorization of microorganisms on the basis of temperature range has been classified under psychrophiles, psychrotrophs, mesophiles, thermotolerant to thermophiles. It has been observed that most of the human disease-causing microorganisms tolerate only 55°C–60°C for a few minutes, except *Bacillus*, *Clostridium*, and a very few others, while mesophilic and thermotolerant are abundant in number. The psychrotrophs, which can resist a wide range of temperatures, is one of the important microbial components that damage a variety of food at low temperature; therefore food is preserved by the integrated approach of preservation. It is well known that with increased or decreased levels of temperature from the optimum value, the growth rate of microorganism decreases (Juneja et al., 1999; Kalniowski et al., 1999). Some of the bacteria like *Bacillus*, *Clostridium*, and *Streptococcus* can also grow at a high and very low temperatures (5°C–65°C), while *Staphylococcus aureus* has the ability to cause disease from cooler to tropical and subtropical countries. Such microorganisms require additional barriers for inhibition for safe preservation under refrigeration. Several metabolic capabilities are required for their growth in the cold. The homeoviscous adaptation enables such microorganism to maintain membrane fluidity at low temperature. At low temperature, microorganisms synthesize a high amount of mono and di-unsaturated fatty acids (Cossins and Sinensky, 1984). Therefore such metabolic capabilities of microorganisms make them resistant to cold and even at higher temperature. The double bonds in fatty acids prevent tight packing of the fatty acids into more crystalline array. The accumulation of compatible solute at low temperatures is analogous to their accumulation under conditions of low water activity. The membrane physical state can influence and/or control expression of genes, particularly those that respond to temperature (Vigh et al., 1998). The production of heat shock proteins (HSPs) contributes to an organism's ability to grow at low temperatures. Such proteins function as RNA chaperons, minimizing the folding of m-RNA, leading to the translation process. *Streptococcus thermophilus*, *Bacillus*, and *Clostridium* are thermophiles that can even grow at low temperature, and they may resist through such mechanisms. Northern blot analysis has proved that a ninefold induction of HSP m-RNA and showed its regulation at the transcriptional level (Wouters et al., 1999). Similarly, it has been now proved that bacteria have the ability to survive and multiply at very extreme cold and high temperatures and pressure. The temperature also regulates the expression of virulence genes in several pathogens. The morphological changes have also been reported in several microorganisms leading to adaptation. The cells grown at 4°C, 25°C, and 37°C have shown the synthesis of internalin, a protein required for the penetration of the host cells. The cells grown at 37°C produce hemolytic activity in some pathogens, while at 4°C such activity was suppressed at human body temperature. It is prevalent that such temperature variation may produce several types of specific proteins, fats, or specific metabolites that help the microorganisms from the changing environment for long-term existence, which is regulated by the genes of the microorganism. The time-temperature dependency in gene products may be regulated accordingly, which leads such a process to resist at −60°C to 250°C as ever reported. There are such metalloproteins that may work at very high and low temperatures, which is the magic of the biological system. Such proteins and fats are used for the preparation of such clothes that may protect us at such at a very high temperature. Such mechanisms may also work at other extreme environments like acidic, alkaline, and saline conditions. It is evident that extrinsic factors, mainly temperature, humidity of the atmosphere, and gaseous composition, affect microbial growth directly/indirectly.

The microbial ecology of food can be explained as the interactions among food nutrients; specific chemicals having antimicrobial nature like alkaloids, and several others like lysozymes, enzyme, iron chelating agents, lactoferrin or lactose peroxidase system, citrate contents, etc.; along with its specific microbial population at certain temperatures, pH, water activity, and nutritional levels. Interactions of microorganisms in food with several factors at a time in different environments may be discussed accordingly for different types of foods along with the antimicrobial compounds. The multivariable components of the environment of specific food should be noticed during long-term preservation of foods. The complex relationship between multiple environmental parameters in foods requires a specific modeling system for a specific group of microorganisms, environment, and food. Therefore an ecosystem of food has been discussed in detail. Food is a highly heterogeneous system where temperature and water activity of the food and environment vary depending on its variability; therefore nutrient status and antimicrobial constituents of food combine to affect the growth of microorganisms. Food also has several micro-environments, mainly the presence of oxygen, and therefore the growth of aerobic, facultative anaerobic to obligate anaerobic microorganisms specially, *Pseudomonas, Protean, Escherichia coli,* as well as *Clostridium,* etc. A variety of microorganisms of aerobic, facultative anaerobic, and very few obligate anaerobes cause food spoilage, intoxication, and also favor the disease-causing microorganisms. For example, some fungi (*Rhizopus, Mucor, Penicillium, Aspergillus, Paecilomyces, Fusarium,* etc.) grow fast in some foods, within 24–48 hours, as they are aerobic in nature and tend to grow at lower to higher pH.

Most of the foods favor the growth of several groups of microorganisms in raw form, but they can be preserved from spoiling and intoxicating microorganism by limiting the oxygen, mainly in packed foods. In some foods, oxygen is driven out during cooking and diffuses back very slowly resulting in the food product remaining anaerobic and it does not favor the growth of fungi. Similarly, salt and sugar also reduce the water activity as well as oxygen, which restrict the growth of bacteria and fungi efficiently, but sometimes aeration is also necessary to eliminate anaerobes' vegetative cells. Similarly, spores are also restricted by several integrated approaches of preservation by inhibiting the different stages of spore germination. This approach has a series of events in a particular food ecosystem especially with intrinsic factors only. The best example is canned food or any preserved foods.

INTRINSIC FOOD FACTORS

The intrinsic factors, which are inherent to food itself, include naturally occurring nutrients and other chemical substance of foods or microbial-produced compounds that may either stimulate or inhibit microbial growth, the chemical added for preservation, the oxidation reduction potential, water activity, and pH. These factors affect the growth of microorganisms by their specific state of the food nutrients like proteins, carbohydrates, nucleic acid, lipids, status at different pH and temperature, etc. The influence of pH on protein and nucleic acid especially on gene expression has also been established. Gene encoding amino acid decarboxylases, lactate dehydrogenase, outer membrane proteins, and virulence factors are influenced by pH. The genera responsible for proton transport, amino acid degradation, and adaptation to acidic or basic conditions and even virulence can be regulated by the external pH (Olson, 1993).

It is well documented that microbial cells sense the change of pH through several mechanism, mainly through proton gradients in the membranes. This leads to protonation/

deprotonation of amino acids, which change the protein's secondary or tertiary structures, altering the function of proteins that govern the signals to change or alter the uptake or reject the translocation of specific solute from the outside environment. Similarly, p*Ka* value for acid also plays an important role; it is a value of pH at which acid is dissociated. Every aliphatic acid has different p*Ka* value at which they dissociate. The organic acids across the cytoplasmic membrane move only in the protonated form. An increased intracellular concentration would indicate increased environmental acidity. The ability of bacteria to use different biochemical pathways, which generate different amounts of ATP and metabolites, influences their ability to grow in adverse conditions and also may support the others to grow/inhibit, which all depends on the micro and macro environment. Such conditions may initiate the proton pump to remove excess proton from the cytoplasm to maintain cytosolic pH, which is changed by the uptake of lactic acid or any other organic acids by the microbial cells; therefore specific microorganisms have specialized cell physiology to maintain or sustain the cell metabolism for their existence. Microorganisms are much diversified, and they may sometimes have special gene/plasmid governed functions to survive in any condition, like facultative anaerobes, microaerophiles, psychrotrophs, and osmotolerant, along with the specific functions at a wide range of pH. As protein function in a cell is highly pH sensitive, specific proteins and lipids work together for providing resistance to microorganisms at higher temperature. The role of ATP to maintain homeostasis as facultative anaerobes generate more ATP by aerobic respiration than by anaerobic fermentation (e.g., *S. aureus*) can grow at a lower pH and water activity under aerobic than the anaerobic condition, which solely depends on the generation and utilization of energy by the microorganism. In all the evaluations, methodology and instrumentation along with specific protocols have created new dimensions in the interpretation of facts and their consequences to establish new norms of food preservation and quality standards for safe consumption.

Food is one of the important components of the ecosystem for all living organisms, whether it is plant, microorganisms, humans, and animals. Most of the living system depends on primary producers, especially on plants, algae, and some of the photosynthetic bacteria. These groups are being consumed for energy and carbon sources for most of the living system in the food web; therefore the food classification on the basis of vegetarian and nonvegetarian foods are the main component. Further, there are a number of food materials, depending on nutritional status. The vegetarian foods generally comprises milk and milk products, cereals & grains and their products, fruits and vegetables etc. and nonvegetarian foods is fruits, vegetables, cereals, microbial foods like mushrooms and Single Cell Protein (SCP), milk, etc., and nonvegetarian, the flesh of several animals, insects, and almost all the chordates and nonchordates members are being considered as source of nutrients for living systems. Further, the state of food from raw, semicooked/processed, to complete processed foods is also very important in the light of spoilage and preservation. Food environment is one of the important areas for research and development as food spoilage, intoxication, and food-borne disease and illness must all be observed in the light of food environments. Food environment has generally the categories as extrinsic and intrinsic factors mainly for the growth of microorganisms. Quality control, food intoxication, food preservation, and food fermentation have always been discussed either for research or for industrial practice at commercial levels.

In this chapter, every component of a food environment for microbial growth and inhibition has been discussed in detail.

One of the microbial factors affecting food environment in which production of certain metabolites and agents initiating sporulation in bacteria play an important role in food sanitation and public health significance. Microbes of various types may be of spoiling origin, intoxicating and disease causing in different proportion may exist, and almost all require basic nutrients mainly carbohydrate of various types, proteins and lipids along with mineral contents. Almost all foods have such nutrients in different concentration and proportion, either in complex or simpler form along with different types of minerals. The status of microorganisms, either in stress or nonstress, vegetative to sporulating or any growth stage, but they multiply according to their optimal physicochemical levels especially for those who are spore formers or extremophiles in nature. It is prevalent that high salt and sugars along with several other preservation mode have been frequently used for food preservation, therefore significance of spore formers are important. Members of gram-positive *Bacillus* and *Clostridium* spp. and some of the other closely related genera form spores. The process of sporulation and germination require different phases, and at every phases different types of preservative approach work for the suppression of sporulation. Further, there are several other factors that affect sporulation. The starvation also initiates this process, and there are number of events and factors that work at a time in spite of the availability of nutrients. Therefore significance of spores in food is as much concerned with preservation as quality assessment and ensuring the quality of canned food or preserved foods, especially meat and milk products.

The molecular biology of sporulation and spore, resistance, and germination stages and the use of chemicals for the suppression of spores at different stages have been worked out by several researchers (Beaman and Gerhardt, 1986; Behravan et al., 2000). Scientific investigation of sporulation by specific bacterials has been widely worked out along with the role of heat shock and the role of specific proteins and genes responsible for the basic and fundamental knowledge of sporulation in *Bacillus* and *Clostridium*, which provide an interesting model of cellular differentiation. The advances in understanding the mechanisms of spore heat resistance have contributed to a greater knowledge of dormancy at the level of extrinsic and intrinsic factors, including microbial factors affecting the conditions of food for initiating/suppressing sporulation. This needs more research for complete remedial measures for preserved food and to control decreased rate of food poisoning from different foods. The remarkable resistance properties of spores and their significance on human disease, intoxication, and spoilage, especially tetanus, anthrax, and botulism, have led to the development of food microbiology, especially in medicine and industry.

Role of Spore Formers in Food Ecosystem

Several spore-forming bacterial genera have been reported (e.g., *Alicyclobacillus, Bacillus, Clostridium, Desulfotomaculum,* and *Sporolactobacillus*) that create problems in various foods. The basic knowledge of sporulation in the food, pharmaceutical, and alcohol industries is very important, as these microbes can survive in almost all environments. In sporulation, an unequal cell division takes place, which forms small spores. The vegetative cells are compartmentalized and cytoplasm condensates with several coats, resulting in the acentric or bicentric spores having complete genome, termed endospore, as it forms within the mother cell throughout the sporulation and showing definite pattern of gene expression. Some genes are expressed in mother cells. The pattern of the gene expression is controlled by the ordered synthesis and activation of a few specific sigma factors. There are remarkable physiological biochemicals, and morphological changes do occur. A number of DNA binding proteins take place in expression and activation (Galperin et al., 2012).

The spore physiology is very complicated, and how it works is still questionable, but two novel layers have been found, a peptidoglycan layer (the spore cortex) and a number of lay mass of spore coats that contains the proteins unique to the spore. The spore also contains high amounts of (10% of dry weight) pyridine-2-6 dicarboxilic acid (dipicolinic acid, DPA), which is found only in spores, along with divalent cations, mainly Ca^{++}, during spore formation, and a large amount of short fragment of acid soluble proteins, some of which coat the spore chromosome and protect the DNA from damage. The spores are metabolically dormant and extremely resistant to harsh environments or treatments including radiation, heat, and chemicals along with having long survival in the absence of exogenous nutrients. The sense of nutrients and their uptake along with physical factors initiate the spore for germination but mainly heat shock. Therefore a very small favorable exposure, germinate spores, where dipicolinic acid is lost, including the cortex to final germination into vegetative cells and both endogenous and exogenous compounds are formed to synthesizes macromolecules. This process is governed by a number of factors, one of which is the intrinsic factors.

The status of spore formers in food is one of the most important aspects of food preservation and food quality assessment along with various types of methods and instrumentation. The bacterial spores, mainly of *Bacillus* and *Clostridium*, are most prominent, as in the air, the spores of these bacteria always show their presence even more than 60% due to their long-term existence in the air. The level of stress/injured microbial levels along with their recovery on a suitable medium/media has also created research interest (Sonenshein, 2000). Research has been worked out by various scientists along with its limitations. The bacteria have greater inhibition in food with higher activity. The injured cells in most of the cases do not show their appearance of colonies on solid medium, but those cells that are live may regenerate after the exposure of some favorable conditions depending on the nature of microorganisms. The repair mechanism of the microbial cell depends on the levels and types of injury (e.g., injury at the levels of cell wall, cell membrane, inactivation of cellular protein, amino acid biosynthesis or the genetic level or likewise affecting the bimolecular phenomenon). Such conditions work in the collective factors of extrinsic and intrinsic levels.

Microbial types, mainly of bacteria, fungi, and actinomycetes, play an important role for survival of these agents, which are generally found in every food environment, as air does contain bacterial spores abundantly. Therefore its status in different foods and ecology of microbial shifting in food along with their distribution, throughout from the original source up to consumption, must be studied in detail. The roles of spores formed in food spoilage, intoxication, and disease are closely interlinked and simultaneously work within an environment with the existence of different food microbial ecosystems. The exact countereffects cannot be individually evaluated and so far may create complication to the final conclusion. The molecular mechanism behind sporulation, spore resistance, and dormancy, spore germination, and outgrowth along with the effects of certain chemicals and physical agents like temperature, pH, and radiations, at specific stages of spore germination may provide better counterpoint to more applied aspects of food preservation (Wuytack et al., 2000). The *Bacillus* sp. is widely distributed in nature from psychrophilic, psychrotrophic, and mesophilic to thermophilic levels along with alkalophilic, acidophilic, to neutrophilic range of temperature. It has been reported in a wide spectrum of various enzymes, antibiotics of antifungal to antibacterial, hormones, alkaloids, organic acid production, vitamins, and pesticidal properties. This genus is widely distributed in aquatic, terrestrial, and air systems; probably its wide existence is supported due to spore-forming ability. This is most resistance to even several chemicals and participating even in food from spoiling, intoxicating, and disease properties, and therefore this microorganism is neither an important pathogen nor an important agent of food spoilage. Moreover, its natural transformability and several other characteristics have initiated molecular biologists to complete the sequence of its genome. Such studies have explored the fundamental mechanisms regulating gene expression during sporulation and germination levels along with its resistance and dormancy. Determination of the genome sequences of at least five other gram-positive spore formers (e.g., *Bacillus anthracis, Bacillus halodurans, Bacillus stearothermophilus, Clostridium acetobutylicum, Clostridium difficile*) has been required to its genome, which indicated a tremendous degree of conservation of genes similar to *Bacillus*, which show the evolutionary pattern of such microorganisms. The sporulating bacteria have dipicolinic acid with Ca^{++} ions.

The r-RNA gene sequencing has shown that similar evolutionary patterns are quite closely related but are clearly derived from a common ancestor, most likely a spore former. A member of other genera including *Staphylococcus* is also derived from the same common ancestor, yet cannot sporulate; therefore sporulation specific genes whose sequences are highly conserved among spore formers appear to have disappeared from the later organisms. Some of these nonsporulating species (e.g., *Planococcus citreus*) are more closely related to present day spore formers than other spore formers.

Sporulation may be through nutrients (carbon and nitrogen) limitation. This is achieved by exhausting one or more nutrients during cell growth or shrinking of cells from rich to poor medium or the addition of inhibitors of nucleotide biosynthesis. It means catabolic repression is one of the factors regulating sporulation, but the exact mechanisms and the role of carbon, nitrogen, cyclic AMP, and GMP still require more research to be worked out, and the possible role for guanine nucleotides has been proved. Further, the signaling effect of sporulation by number of factors in which nutritional and several physicochemical effects are being captured by the cells is obscure. However, induction of enzymes synthesis of the TCA cycle is required (Ireton et al., 1995). During sporulation, some small molecules are secreted from the cells, which has been reported (Burkholder and Grossman, 2000).

Several growth-based phenomena have also been reported, in which the synthesis of amylases and proteases; synthesis of antibiotics such as bacitracin, surfactin, and gramicidin; and in some cases the protein toxins are active against insects, animals, and human. Developments of mobility along with genetic competence in some species of *Bacillus* have also been reported. Although, this phenomenon is not directly involved in the process of sporulation, these are protecting factors of extrinsic and intrinsic levels, which could be evolved by this genus, *Bacillus, Clostridium,* and other spore formers for safe sustainability from the environment. It clearly indicates that spore formers have been exposed in almost every harsh environment during its existence. Such microbes are also proving its potential significance in the existence of life, and that is why they survive in soil, water, air, and even food, which are excellent favorable ecosystems for this microorganism.

These are multiple genes that regulate the competence factor by positive and negative effectors; therefore multiple control systems are involved in which some of the small peptides are released, perhaps by detecting signals and finally initiating the release of specific secondary metabolites in the form of toxins and antibiotics, etc. There are drastic changes in the metabolism of the sporulating cell. The storage nutrient, minerals of the cells like poly-β hydroxybutrate, volutin granules, β-alkanolate, etc., are catabolized through the TCA cycle to meet out the energy for several sequential reactions to handle the sporulation process. Several other enzymes that are present in cells before sporulation may help in the process of sporulation; these factors have still to be observed.

Several morphological, biochemical, and physiological changes do occur during sporulation of cells. There are six to seven stages that have been identified in which spore-forming bacteria undergo such a highly complex process. The spore-forming bacteria have a number of genes that have multifunction activity depending of the regulatory protein combinations produced during every extrinsic and intrinsic parameter. All the seven stages are not discussed in this text, but it is evident that at every stage of sporulation the inhibitors/suppressive agent varies, indicating that the products are changing so fast that the molecules of signaling as well as the synthesized product vary vastly. This process is highly complex, and the exact nature of that molecule and signaling agent as well as its mechanisms are not very clear. If we could determine that gene and the regulatory proteins, several innovative microbial system could be expressed, which can sustain the life from several adverse conditions just like spores, which are highly resistant to temperature and several chemicals, etc., without losing their life. In spite of the work on the regulation of gene expression during sporulation, still several new findings are expected to solve the secret of life. Through the study of spore-forming microorganisms, mutation may occur frequently in the spore-forming bacteria as the same number of genes. Work at a time and small changes in the different types of mutation process with specific mutagens may change the activity of microbial cells. It is evident that *Bacillus* and *Clostridium* have been reported to have produced almost all enzymes like cellulases, amylases, proteases, lipases, and several exo and endo natures of almost all enzymes reported ever; likewise, they have the ability to produce various antibiotics, alkaloids, organic acids, amino acids, vitamins, and even interferons and others. A separate review can be made regarding the role that *Bacillus* and *Clostridium* contribute in the various ecosystems, especially for sporulating, degrading, intoxicating, and disease-causing ability by different spp. with their distribution and their variable functions.

Probiotic Microbial Community as Intrinsic Factor

Another important microbial factor that exists under intrinsic parameters is the presence of probiotic bacteria along with the prebiotic components of food. The existence of probiotic concept was first introduced by Russian Nobel Laureate Eli Metechnikoff. He proposed that a normal, healthy gastrointestinal microflora in humans and animals provided resistance against "putrefactive" intestinal bacterial pathogens. Probiotic bacteria have long been considered to influence general health; they have commensal association with the gastrointestinal tract and normal microflora, mainly of lactic acid bacteria such as *Lactobacillus* and *Bifidobacterium* sp. along with *Streptococcus* and *Pediococcus*. These fermentative bacteria constituting this flora produce lactic acid associated with fermented milk products and do not produce putrefactive compounds and toxins; therefore microbial shifting of wild type and pathogens facilitated by probiotic lactic acid bacteria extends several health benefits including better digestion and releasing anticancerous compounds, antibacterials, anticholesterols, and antihistamine, etc., providing healthy long life to humans and animals. Fermented dairy products like yogurt, kefir, and sour cream have been consumed and valued by many countries, including China, Japan, Indonesia, Europe, and several other of the cooler climate countries, dating from the ancient period of civilization. The growth of lactic acid bacteria in food affects other pathogenic microorganisms. These microbes considered as probiotic come under the beneficial microorganisms. There are several benefits attributed to this science, which are huge control of research and future prospects in food microbiology. Some of the aspects that are biological have been discussed in this text. It is well established that certain bacteria produce specific metabolites that inhibit or kill the pathogenic microorganisms by their metabolites as well as certain enzymes that inactivate their metabolic activities along with sustainability in food (Table 8.1).

There are several other factors of probiotic foods like pH, temperature, oxidation reduction potentials, etc., which may also be discussed specially for the probiotic microorganisms, existing in specific food. Food variation and their specific conditions may affect the overall biological quality of food. Several important aspects in this regard have been worked out by the principle author of the chapter regarding how the change of pH affects the protein configuration and release of amino acids in soya bean and ground nut proteins, and how such proteins are affecting the growth of specific microorganisms as well as inhibition of growth, etc. Therefore food R&D, especially the role of different groups of microorganisms in various foods, still require exhaustive research for better understanding of various concepts of food nutrition, before and after exposure with protein, lipids, and carbohydrate status with several combinations in raw, semiprocessed to processed, along with quality control specification should be developed for each type of food. Microbial systems are highly variable, and nature is a rich reservoir of microorganisms. The development of newer strains is always expected, as metagenomic studies have confirmed genome transformation from same group to different groups, always occurring in the environment; therefore newer strains either through mutation or natural recombination are producing new genus species and strains having high variability in their physiology, metabolic pathways, and resistance to several biotic and abiotic factors. Such conditions have opened a new challenge to set up new norms accordingly, depending on the newer findings and concepts. Therefore it is a never-ending process to identify the newer norms for many foods which should be worked out continuously for development of better concepts for quality control as well as methods for the assessment of those newer microorganisms. Further, the physiological significance and the role of specific proteins and lipids are important macromolecules that affect all the processes. Therefore specific proteins with rare amino acid sequences may affect even the genomic status in the process of adaptation, mutation, or genetic patterns with some character, and thereafter more changes may occur

TABLE 8.1 Bacteria and Yeast Associated With Food Fermentation Having Probiotic Values

Unprocessed Material	Fermented Product	Microorganism
Olives, tomatoes, cabbage, cucumbers	Fermented olives, pickles, sauerkraut	Lactic acid bacteria
Dough and pastes prepared from cereals	Sourdough, kisra, yeast dough	Lactic acid bacteria, yeasts
Malt, koji, prepared from cereals	Beer, spirits, sake	Lactic acid bacteria, yeasts, molds
Spirits, beer, and wines	Vinegar	Acetic acid bacteria
Grapes and other fruits and vegetable	Wine Fruit juice concentrated Fruit and vegetables dressed	Yeasts, lactic acid bacteria Yeast *Escherichia coli, Pseudomonas*
Carob, soya	Natto, dawadawa, soy sauce, tempeh	Lactic acid bacteria, *Bacillus* spp., yeasts, molds
Milk	Sour milk and cream, kumis, yogurt, kefir Raw milk refrigerated Raw milk normal temperature	Lactic acid bacteria, yeasts Acetic acid bacteria *Lactobacillus* spp. *Streptococcus* and *Lactobacillus*
	Sour cream, butter	Lactic acid bacteria *Pseudomonas, Bacillus*
	Cheese Hard cheese	Lactic acid bacteria, propionic acid bacteria, yeasts, molds *Clostridium, Acetobactor*
Meat	Fermented sausages	Lactic acid bacteria, micrococci, *Streptomyces*, yeasts, molds, staphylococci
	Ham	Lactic acid bacteria, Staphylococci, yeasts, molds
Fish	fermented fish, fish sauce	Staphylococci, lactic acid bacteria, *Vibrio costicola*
	Fermented products during process	*Leuconostoc, Streptocacetrobactor, Zygosaccharomyces*
	Canned foods processed	*Clostridium*

if such microorganisms, mainly bacteria, may give different responses in foods, and their assessment and significance for the special effects must be assessed with proper tools and techniques.

Bacteria have existed for about 3 billion years, since 2000 AD, and much work has been done in the area of real-time biosensors for detection of pathogens, genetic control of food-borne pathogens, foods that are bactericidal to pathogens. Further, the food engineering and food chemistry can provide new developments in human disease control and human nutrition for better health. Food has tremendous treasure of all nutrition, and if this nutrition could be formularized properly for every person according to their body weight and specific types of need, it would give a tremendous medical aid for the control of human and animal diseases. The main component of food spoilage is microorganisms, which must be discussed properly for maintaining the food quality and long-term preservation; therefore HACCP concepts are introduced to practice better quality control of various foods.

Food-borne pathogens and food-spoiling microbial assessment are important areas where we can gain more momentum to know about food-borne disease, intoxication, and spoilage. All the vegetarian and nonvegetarian foods (i.e., meats of almost all animals, vegetables and fruits) are being consumed by various parts of the world; therefore the quality standards for all the foods are strictly framed for specific meat, meat products, pork and beef, and others, more restricted with several standards like worms as well as several disease-causing bacteria and viruses. Poisoning by spoiled grains were also recognized for several fungi, and the role of specific fungi—*Aspergillus, Penicillum, Fusarium,* and *Paecilomyces*—was also recognized as producing several types of mycotoxins causing cancer and other serious diseases to humans and animals including plants and other living systems. Several processed foods like milk, meat sausages and canned meat, and other food products require different technologies for preservation, and microbial specifications as per the environmental conditions are needed. Integrated preservation approaches (i.e., physical, chemical, and biological means) have shown better and safer preservation of food for longer period (i.e., canning processes) for creating specific environments for complete suppression/killing of various groups of microorganisms. These processes generally require appertization, Tyndellization, radurization, radappartization and radicidation along with specific food preservative chemicals like sodium benzoate, lactic and acetic acids, parabeans, and nitrate/nitrite in 0.1%–2.0% level depending on their specific dose. Further, food antimicrobial properties may also play an important role in such processes. Salts and sugars, oils, and several species have also been used as preservatives in various foods and their products. Louis Pasture was the first scientist in 1854–64 to use heat for effective removal of microorganisms from food; therefore pasteurization, sterilization, and fermentation came to existence in food preservation, for which he was known as a pioneer food microbiologist. John Tyndall used discontinuous heating, the process of Tyndallization, for effective removal of microorganisms, a process of effective sterilization.

Status of Indicator Microorganisms as Intrinsic Level

Indicator microorganisms and their criteria for specific food safety assessment and processing are one of the important areas in which the indicators of food-borne pathogens and toxin detection have been investigated. Several parameters for indicator microorganisms can be used in the assessment of quality control of foods. These criteria might be used to address existing product quality or to predict shelf-life of foods. Estimation of a product for indicator microorganisms can provide simple, reliable, and rapid information about process levels mainly for the postprocessing contamination from the environment and the level of hygiene under which the food was processed and stored. Therefore in a shorter period, the indicator microorganisms can be identified on the basis of their specific media as well as protocols for detection and identification. There are certain criteria for an indicator microorganism:

1. They should be present and detectable in all foods whose quality is to be assessed.
2. They can survive well in a wide range of pH, temperature, moisture levels, and water activity (a_w).
3. Their growth and numbers should have a direct negative correlation with product quality.
4. They should be easily detectable and enumerated and be clearly distinguishable from other organisms.
5. Their growth should not be affected adversely by the presence of other food microflora.
6. They can exist for longer periods in variable conditions of water activity (a_w), temperature, pH, and O_2 concentration.
7. They must have variety of metabolic processes to metabolize various carbon, nitrogen, and mineral sources in lower to higher concentrations.

Further, there are several indicator microorganisms that have been identified and characterized for various foods according to the international quality food specification organizations like the USDA, FAO, WHO, etc. These specifications have generally been made on the basis of food nutritive status, conditions, and environment. It is well documented that seafoods and normal meat and meat products along with cereals, vegetables, and fruits have their own specific indicator microorganisms for assessment of quality. Specific bacteria are the main agent that has been considered as an indicator microorganism in various foods. Loss of quality in other products may be limited not to one organism but to a variety of microorganisms owing to the unrestricted environment of the food. Therefore such products must be examined for other groups of microorganisms, most likely to cause spoilage in that particular food (Hathaway, 1999).

The assessment of such contaminants is facilitated by various approaches: direct microbial count and microbial product assessment, indicator of food-borne pathogens and their toxins, along with the indicator microorganism. In all cases, bacteria, yeast, and fungi are the main agents, while fungi and yeast can easily be eliminated in processed/canned foods due to aerobic nature. Moreover, bacteria are the main agent that can exist in a very wide range of physicochemical conditions; therefore certain procedures, media, specific instrumentation, and protocols have been used for the detection of microorganisms and their products for assessment of food quality.

Aerobic plate count (APC) or standard plate count (SPC) is the most acceptable approach for the determination of total count of microorganism in a food product, and therefore the modification in the environment of incubation, use of different media: enrichment, selective, differential, specialized, etc. The APC can be detected specially for thermotolrent, thermophylic, psychrophile, mesophile, proteolytic, saccharolytic, and lipolytic group. Therefore spoiling, intoxicating, and disease-causing microorganisms can be assessed on their selective media along with the specific protocols developed for the detection of those microorganisms. The APC of refrigerated foods such as milk, meat, poultry, fish, and seafoods may be used to indicate the condition of equipment and utensils used, as well as the time/temperature, a_w profile of storage and distribution of the food, but every method has its own limitations that restrict to adopt another for specific microorganisms and foods. However, there are certain protocols as per the quality assessment agencies for specific food and organisms.

Direct microscopic count (DMC) is one of the important approaches, which can give a quick estimate of microorganisms and the possible genera on tentative basis that help a microbiologist to select specific medium for their enumeration. The basic problem is the detection of dead and alive status of microorganisms in food; therefore certain limitations initiate the adoption of another methodology.

These bacteria produce a variety of microbial metabolites that can also be indirectly correlated for the assessment of food quality and hygiene. In certain cases, the specific microbial metabolite also indicates the presence of a group of microorganisms, which is important in the assessment of quality food production and preservation. This is also a most significant correlation between presence of product and food quality. Therefore the products that are examined through, HPLC, GLC, or FTIR have quick examination in a few minutes for the assessment of the presence of specific microorganisms through the assessment of their metabolite; sometimes the decomposed products may also give an indication of the presence of such microorganisms. There are certain limits of the status of the by-products as well as main metabolites, as some metabolites are very stable and can give a clear picture of the presence of specific microorganisms. Some of the specific microbial metabolites that are identified as applicable for specific food have been internationally accepted for the assessment of food quality. These are lactic acid, histamines, ethanol, butanediol, citrate, total volatile fatty acids, diacetyl, cadaverine, and putrescine. Acetoin is applicable to vacuum-packed foods, frozen juice, canned food, fish, seafoods, etc. Such specifications are being used by food industries and are a well-accepted approach for assessing quality foods.

Further, the indicator of food-borne pathogens and toxins is one of the microbiological criteria as they apply to products safety. The assessment of pathogenic microorganisms and their toxins is very important for food quality as they cause serious problems in human health rather than food spoiling. Milk, meat, and several cooked cereals and their products have a variety of pathogenic microorganisms; therefore its assessment and safe removal are important aspects. The main intoxicating bacteria are *S. aureus* and *C. botulinum*, and other species along with several bacteria of disease-causing origin like *Bacillus cereus*, *Clostridium* spp., *Salmonella*, *Streptococcus*, *Yersinia*, *Vibrio*, and several others have much significance for the assessment and control for better public health and hygiene. Among these microorganisms, some are very potent in a low level and some can activate even at a high concentration or their metabolite is very important than microorganisms, like intoxicating one where only metabolites interact with the host, not microorganisms intake. Therefore assessment of methods and criteria of a sampling plan, use of certain instrumentation, and sampling methods for specific foods, have been adopted according to various international/national agencies (Jouve, 1999).

The metabolic products and populations of bacteria in total or specific bacterial appearance in a particular food are another important segment; therefore a correlation has been established between the presence of a metabolic product and product quality loss. Sometime an organoleptic test also confirms the presence of the product as well as quality assessment of food without any instrumentation. For this, various classes from higher, medium, moderate, and advanced definite media and methods have been used.

More advanced instrumentation like HPLC, GLC, and others have also been used for such work through which various toxins, alkaloids, and other organic products have also been examined. The plan stringency in relation to degree of health hazard and conditions of use has also been adopted for the assessment of the type of hazards as well as conditions in which food is expected to be handled and continued after sampling in the usual course of the events. The international commission on microbiological specification of foods (ICMSF), the U.S. National Academy of Sciences, and the U.S. National Advisory Committee on Microbiological Criteria For Foods have provided a framework for the World Trade Organization (WTO) to implement the health hazardous analysis critical control point (HACCP) to create a science-based preventive system for food control (Hathaway, 1999). Several countries have mandated HACCP requirements and established specific HACCP requirements for public sectors of their domestic food industries (Lupins et al., 1998).

The World Health Organization (WHO) and Food and Agriculture Organization (FAO) have jointly working in this direction. Meat, poultry, seafoods, eggs, and dairy products contribute major parts to the human diet and are more susceptible to spoilage. Factors associated with spoilage may include color defects or changes in texture, development of off flavor, off odor, slime formation, or any other characteristics making food unfit for consumption. The food enzymes themself play a role in the change of texture and chemical constituents of the product. Every food has its own specific microflora; therefore ecology of spoilage microflora along with the existence by intoxicating and disease causing may be discussed only by their physicochemical factors and types of nutrients available along with the antimicrobial constituents. The microbial contaminants of the environment surrounding the food as well as microbial contents already present in food affect the quality of the foods; therefore it is suggested blanching of any food so that the enzymes responsible for the change of texture may be checked along with the certain microorganisms of the surface, especially the pathogenic microorganisms. For example, a higher number of bacteria are present on the hide and hair of red meat animals, as well as in the gastrointestinal tract. Microorganism on the hide include bacteria such as *Staphylococcus, Micrococcus, Pseudomonas,* and *Bacillus,* along with some yeast and mold, which are normally associated with skin microflora; therefore washing of animal skin is suggested.

Similarly, every food has its own origin of microflora depending on its habitat and environment, for example, meat animals like poultry. The internal tissue of healthy poultry or any animal are essentially free from bacteria. The skin, feathers, and feet of the birds harbor microorganisms. The psychrotrophic bacteria consisting primarily of *Acinetobacter* and *Moraxella* spp. are primarily associated with the feathers. Therefore washing with 60°C–70°C with continuous flow of water is suggested for every animal. Likewise, origin of microflora in fishes, which varies depending on weather, mainly temperature, and O_2 available in water. Water temperature has a significant influence on the initial member and types of microorganism, mainly bacteria on the surface of fish. Higher numbers of bacteria are generally present on fish from warm subtropical or tropical waters compared with fish from colder waters. The microflora of gills and surface is greatly influenced by the dressing and method of culturing as well as types of fish. Therefore the animals/fishes/poultry are the major source of several bacteria of spoiling, intoxicating, as well as carrier of various disease-causing bacteria. The attachment of bacteria on the edible muscle tissue of healthy animals must be sterile prior to processing. Therefore different approaches of processing, physical, chemical, and fermentation, are the most common, with its integration along with the canning process have be adopted for safe preservation of such foods.

Bacterial attachment to muscle surface generally involves two stages. The first is loose reversible sorption that may be related to adsorption or other physicochemical factors. One of the important factors that influence attachment at this point is the population of bacteria in the water film. The second stage consists of an irreversible attachment to surface involving the production of an extracellular polysaccharide layer known as glycocalyx. The mobility of bacteria may also influence the attachment as well as dominating the surface microflora. The spoiling and pathogenic bacteria have competition, while spoiling is more pronounced, for example, *Pseudomonas, Proteus, Acinetobacter,* and *Bacillus,* which dominate over the pathogenic one. The muscle tissues provide good growth medium. The initial microflora of muscle foods is highly variable in and on the muscle of meat. The food handler's utensils, the environment process parameters, and the surroundings having aerosols may affect the microbial members as well as quality. Therefore it is not easy to predict qualitative and quantitative measures of microorganisms in specific flesh of various animals and types of flesh. A large member of marketed perishable meat poultry, seafoods, and their products are preserved at refrigerated temperature; microbial growth occurs during storage. The composition of microflora from initial to final stage may get shifted several times or some time without shifting, depending on the types of competitions as well as nature of microorganisms to either sustain various water activity, temperature, pH fluctuations, or ability of specific enzyme production at various levels during the course of growth period and conditions. For example, several common genera that dominate the meat at initial levels are *Pseudomonas, Lactobacillus, Moraxella, Acinetobacter, Brochothrix,* and *Thermophacta,* although these bacteria often constitute only a small portion of the initial microflora. The types of bacteria that ultimately predominate during storage are reflective of these genera as well as the characteristics of muscle tissues. Moreover, it depends upon the level of spoiling microorganisms as well as other categories. The author of this chapter has worked out that the spoiling origin by proteolytic and lipolytic microorganisms eliminate the pathogenic and intoxicating surface microorganisms. The competition of O_2 on the surface and the temperature and pH variation protect the surface from deleterious microorganisms; for example, growth of *Lactobacilli* generally protects the source from pathogenic microorganisms. *Pseudomonas* spp. is able to compete successfully on aerobically stored refrigerated meats. It is due to the aerobic nature and ability to grow at low temperature and generation time having about 25 minutes. *Moraxella* and *Acinetobacter* spp. are less capable of competing under refrigeration temperature at the lower pH in this range, while moderate high temperature than that of refrigeration initiate the growth of *Bacillus* temperature to thermophilic levels.

Lactic acid bacteria mainly have some common genera like *Streptococcus, Pediococcus, Leuconostoc, Vagococcus,* and *Lactobacillus* that play an important role in the preservation of meat by certain levels during the storage from low to moderate temperature (Venugopal et. al, 1999). The lactic acid producing bacteria have several beneficial aspects other than preservation, that is, increasing the nutritional quality of food, improvement in food flavor and aroma, and several pharmaceutical values increasing the immunogenic properties along with anticancerous and antibacterial products. Therefore lactic acid bacteria always help us in these respects as they are abundantly present in the environment. Their main habitats are plants, young leaves, flowers, and fruit surface, and they produce slime just like yeast. They utilize sugars and produce lactic acid, acetic acid, as well as ethanol by heterolactics, while homolactic produce only lactic acid. These agents help food sources from several pathogenic and intoxicating microorganisms. It is evident from the facts of several researchers that probiotic and prebiotic effects have been contributed by these groups of bacteria. Probiotic approach is being used in humans and animal health. These bacteria eliminate/check the growth of pathogenic microorganisms. It also reduces the risk of allergens along with the maintenance of pH condition to check several harmful bacteria.

The nutrients of the meat generally vary in carbohydrate, protein, lipids, and mineral contents. The higher water activity ($a_w = 0.99$) has corresponding water contents 74%–80%. The protein content varies from 15% to 22% on a wet weight basis, a lipid varies from 3.0% to 37%, and carbohydrate ranges from 0.1% to 1.5%. Hydrolysis leads to the accumulation of lactic acid, which decreases the pH 5.0 to 5.5; therefore the growth of lactic acid bacteria gets favorable conditions to overcome the surface leading to the protection from intoxicating and disease-causing microorganisms. The release of proteolytic enzymes such as cathepsins from lysosome results in a small amount of protein breakdown and the released amino acid contents may favor the growth of lactic acid bacteria and other groups of microorganisms depending on the occurrence of dominance at initial levels. Therefore microbial shifting of microorganisms in any food depends on the microbial competence for production of required enzymes depending on the substrate available and resistant to the change of physicochemical levels. It is therefore difficult to predict the microorganism of the food. It is evident from several reports that a few groups of microorganisms like *Bacillus, Pseudomonas, Streptococcus, Lactobacillus, Acinetobacter,* etc., are abundantly present, and they have the ability to utilize a variety of carbon and nitrogen source and have the ability to produce a variety of enzymes and metabolites that make them omnipresent and omnipotent. Protein-rich foods like all types of meat, eggs, and milk and their products have high protein. Therefore degradation results in the production of amino acids and degradation of amino acids by the spoilage microorganisms resulting in the production of ammonia, hydrogen sulfide, inole, skatole, amines, and other compounds resulting in undesirable odors, flavors, and colors. Some display the highest degree of spoilage of almost all protein-rich foods at refrigerated to mesophilic temperature. The food sanitation mainly eliminates these bacteria but some are even grown at a very high temperature from mesophilic range. The products of the meat of different animals like sausages, patties, tikka, and other fermented products are safe if properly processed using high temperature, deep frying with suitable thickness/bulk, along with the integrated preservation approach. The water activity (a_w) plays crucial role in the preservation, but it depends upon the curing process and method of preservation like canning.

Milk and dairy products are also highly perishable and have high and appropriate nutrients having sugar leading the fast growth of several bacteria; therefore chilling and pasteurization is the fast remedy of whole fresh milk. Modern dairy processing utilizes several other processes other than pasteurization, that is, heat, sterilization, fermentation, dehydration, refrigeration, and freezing as preventive measures. Milk has protein, fat, and carbohydrates in good proportion. Therefore for separation of the nutrients, several other processes like churning, centrifugal filtration, and coagulation are being applied to produce various products. A variety of microorganisms are involved in food fermentation, spoilage, and intoxication of milk and milk products noticed by off odors and flavors, change in texture, color, etc. Milk has been a good source of nutrients since prehistoric time. The major inhibitors in raw milk are lactoferrin and lactose peroxidase system. There are other natural inhibitors like lysozyme, specific immunoglobulins, and folate and Vitamin B-12 binding systems that contribute lesser inhibitor levels. Lactoferrin, a glycoprotein, act as an antimicrobial agent by binding iron. Human milk contains 2 mg/mL lactoferrin, but it is of lesser importance in cow milk, which contains only 20–200 μg/mL. The psychrotrophic aerobes that commonly spoil refrigerated milk are inhibited by lactoferrin, but the presence of citrate in cow's milk limits its effectiveness, as the citrate competes with lactoferrin for binding the iron.

The most effective natural microbial inhibitor in cow's milk is the lactose peroxidase system. Lactose peroxidase catalyzes the oxidation of thiocyanate and simultaneous reduction of hydrogen peroxide, resulting in the accumulation of hypothiocyanite (Wolfer and Sumner, 1993). Hypothiocyanite oxidizes sulfhydryl groups of proteins resulting in enzyme maclevation and structural damage to the microbial cytoplasmic membrane. Therefore the following reaction takes place in this event to form hypothiocyanate (OSCN):

1. $2SCN + H_2O_2 + 2H \xrightarrow{\text{lactose peroxidase}} (SCN) + H_2O$

   $(SCN)_2 + H_2O \longrightarrow HOSCN + SCN + H$

2. $SCN + H_2O_2 \xrightarrow{\text{lactose peroxidase}} OSCN + H_2O$

Therefore two reactions are important for the production of hypothiocyanite (OSCN) inhibitor in milk.

Dairy products provide different growth environment than the liquid milk. The dairy products are concentrated and only solid contents of milk with different processing parameters favor various types of fermentative spoiling and intoxicating bacteria, but integrated preservation approach limits the growth of pathogenic bacteria. The pH and water activity greatly vary and help in the preservation aspects of dairy products. The high and refrigerated temperature with low a_w and canning or hermetic sealing are the main preservative approaches used so far. The advanced canning and condensed form is most prevalent. In butter, the use of 6%–8% salt inhibits most of gram-negative food-spoiling bacteria. The salt is evenly present in the droplets of water present in the emulsion of fat-water mixture. Therefore evenly mixing of salt is essential to check the growth of several spoiling bacteria. The uneven salt favors the several psychrotrophic bacteria during refrigeration (Griffiths et al., 1990). The unsalted butter is usually prepared from the acidified cream and relies on low pH and refrigeration for preservation. The nutrient status of some dairy products varies with the pH, water content, and protein and carbohydrate levels. Some of the dairy composition in g/100 g is mentioned. Butter has water 16, fat 81, protein 3.6, carbohydrate 0.06, pH 6.3; cheddar cheese has water 37, fat 32–33, protein 24–25, carbohydrate 1.0–2.0, water activity 0.90–0.95, pH 5.2; nonfat dried milk has water 3.2, fat 0.8–1.0, protein 36–37, carbohydrate 52–53, water activity (a_w) 0.1–0.2; yogurt has water

89–90, fat 1.5–2.0, protein 3–4, carbohydrate 4.5–5.0, pH 4.3. Likewise, several dairy products vary in their nutrient status and provide a good source of nutrients and can be preserved through canning and refrigeration along with the little use of chemical preservatives, although the fermented milk products have several preservation compounds produced by lactic acid bacteria itself during fermentation.

Another major commodity of food is fruits and vegetables along with cereals, where spoilage microorganisms come from various sources like field, storage, selling, or during handling the pathogenic, and spoiling origin is much higher than the intoxicating one in fruits and vegetables. During storage, the intoxicating microorganisms do exist and may cause infection or intoxication. There are a number of bacteria, fungi, and yeast that play a role in such processes.

The plant pathogens are the microorganism, which cause disease or decay of plant. The specificity of plant pathogens are of several types, one the true pathogen or opportunistic pathogen. The true pathogen damages the plant from seedling to any stage of the growth of root, stems, and leaves leading to complete plant death, seriously affecting the system. Such pathogen produces a variety of disease by producing a variety of enzymes and toxins. The role of enzymes, mainly cellulases, proteases, and pectinases, are main, which facilitate pathogens to work properly in or on the host. Sometimes, pathogens have such an enzyme system and sometimes saprophytes help to establish pathogens on the host. Both the conditions prevail, but most of the time pathogens have such machineries to work properly to damage the plant at initial or later stage of the plant life cycle.

The spoilage of fruits and vegetables is also frequent. The term spoilage has different meanings. In a broader sense, it refers to any change that occurs in food that makes it unacceptable for human consumption. Therefore safety and quality assessment are directly correlated with the spoilage. The spoilage can be assessed by the change in color, flavor, texture, or aroma by chemical/enzymatic change by the fruits, vegetable or plant enzyme itself, or through microbial system on fruits, vegetables, and grains. Moreover, spoilage may also lead to intoxication by their specific microorganisms along with the disease-causing one, mainly *Salmonella*, *E. coli*, *Streptococcus*, *Listeria*, *Campylobacter*, and several others like *Staphylococcus* and pathogenic *Bacillus* may frequently develop during normal storage conditions leading to spoilage intoxicating as well as disease indication.

In general, three major spoilage categories exist in plants products:

1. The active spoilage caused by plant pathogenic microorganisms, initiating infection.
2. Passive or wound-induced spoiling, where opportunistic microorganisms get entry to internal tissue. Insects play a role in the damage of outer protective tissues of fruits and vegetables. Fungi play a major role in the spoilage of outer layers mainly of pectin, as water activity (a_w) on the surface favors the growth of fungi. Several types of fungal spoilage of fruits and vegetable have been reported: *Alternaria*, *Colletotrichum*, *Aspergillus*, *Fusarium*, *Rhizopus*, *Cladosporium*, *Alternaria*, *Botrytis*, *Verticillium*, etc., while bacteria causing soft rot in vegetable and fruits are *Erwinia*, *Pseudomonas*, *Xanthomonas*, etc., that spoil fruit and vegetables.
3. Enzymatic spoilage, where the plant itself produces certain enzymes responsible for autolysis of their own cells. Therefore the blanching (washing through warm water) of fruits and vegetables is suggested to maintain the product quality.

The main mechanism of spoilage in different stages in plant products are at the field level/harvesting stage; the second is wound induced, in which opportunistic microorganisms gain access to internal tissues; the spoilage at processing level; and last, during storage of the whole or at their production. The microbes may survive even at the level of advanced preservation state, such as the canning process. There are a variety of microorganisms, mainly bacteria, fungi, yeasts, and actinomycetes, which may spoil food at any stage, but the food nutrients, pH, temperature, water activity (a_w), O_2 availability, stage, and state of the food favors different genera and species for spoilage of fruits and vegetables. The fungal spoiling is more prominent in fruits and vegetables. Some of the common fungus like *Alternaria* spp., *Colletotrichum musae, Aspergillus niger, Monilinia fructicola, Fusarium, Verticillium, Ceratocystis paradoxa, Botrytis cenera, Geotrichum,* and *Penecillium,* cause various types of rots via blue, black crown, brown, and black green rots in vegetables and fruits. Further, bacteria that generally cause soft rot are *Erwinia, Pseudomonas,* and *Bacillus endosporium* along with fungal syndrome effects.

The spoilage initially may be the damage of epidermal or outermost coverage of fruits and vegetables by the insects and birds where minerals, carbohydrate, proteins, or lipid exist for fast development of microorganisms in different stages with specific microbial shifting from lactic acid bacteria to other groups. The lactic acid bacteria are safe and even help from the pathogenic groups. It is well established that lactic acid bacteria like *Pediococcus, Vagococcus, Streptococcus, Leuconostoc,* and *Lactobacillus* are the main genera and are commonly present on green leaves, flowers, and fruit surface. These are some sites that help food preservation rather than spoilage origin. Such microorganisms may spoil the fruits and vegetables, but during fermentation of fruits and vegetables they protect foods from intoxication as well as provide probiotic values to food.

Degradative Enzymes as Intrinsic Parameter

The role of degradative enzymes in the spoilage is important. Several classes of microbial enzyme such as pectinases, cellulases, proteases, phosphatidases and dehydrogenases, and lipases are responsible for the spoilage of fruit, vegetables, and other foods. The pectinases damage the pectin of the outer protective surface of fruits, vegetables, and plant cells. The pectinases and cellulases are the most effective enzymes involved in spoilage. Pectinases play a role in the depolymerization of pectin chain. Pectinases has three components depending on the sites of actions on pectin polymer. The important one is pectin methyl esterase. It hydrolyzes the ester group from the pectin chain. These enzymes affect the solubility of pectin but do not affect the chain length. This enzyme compound is produced by several filamentous fungi like *Monilinia, Penicillium, Aspergillus,* etc., while bacteria like *Erwinia, Pseudomonas,* etc., and other components of pectinases are polygalturonase and pectin lyase. The polygalacturonase and pectin lyase are chain-splitting enzymes that reduce the overall length of the pectin chain, Their mechanism of action differs. The polygalacturonase cleave the pectin chain by hydrolyzing the linkage between two galacturonan molecules, where as pectin lyase depolymerizes by β-elimination of the linkage. Both the enzymes are endopectinases that act on the middle portion of the pectin chain, but some plant uses has inhibitors of these enzymes.

Cellulases are the second major enzymes that can lead to spoilage. It degrades cellulose of any origin. Cellulases of different nature are produced by several microorganisms, even fungi, bacteria, yeast, and actenomyceties. They have exo-endo and β-glucosidase activities, mainly reducing end to middle of the chains along with cellobiose, respectively. Therefore plant products are spoiled by these enzyme systems. The spoilage by such enzyme is very fast due to release of glucose, favors other groups for even intoxication, spoilage and disease-causing microorganisms, likewise various enzymes producing microorganisms spoil almost all foods favoring the other intoxicating microorganisms along with disease-causing microbial agents in the specific microbial community.

Chemical Agents as Intrinsic Factor

Chemical preservative and natural antimicrobial compounds also play an important role in food quality preservation, safe consumption of food and beverages, along with the status of raw foods. Many food preservation technologies have been used since ancient periods in which food fermentation, the use of smokes and salts, sugars along with the antimicrobial compounds of leaves are used for preservation of various milk, meat, vegetables, and fruits. Drying and desiccation were also one of the methods under fire, as well as sunlight and smoke of some specific woods. The natural fermentation by using natural inoculums using the same utensils/wares may be of wood or soil clay materials for continuous microbial culture production and maintenance. These cultures are safe for consumption with probiotic values. There are a number of microorganisms in the natural ecosystem that show their dominance mainly by producing antimicrobial systems; therefore the antimicrobial compounds of several food products are one of the important factors appreciating human health and nutrition. Food antimicrobial agents are chemical compounds that are produced naturally during the course of various metabolic activities and biosynthesis in food, vegetable as well as nonvegetarian, having high affinity to interact with several pathogenic microorganisms, thereby restricting/killing microbial growth, thereby restricting deterioration. The major activities of antimicrobial constituents of foods are to restrict disease incidence. There are several agents like antimicrobial, antibrowning, citric acid, and antioxidants (butylated hydroxyanisole) used for food preservation. The traditional preservatives are propionate, benzoate, nitrates, sorbate, lactic acids, acetic acids, as well as other aliphatic acids. Most of antimicrobial constituents of foods are bactericidal, bacteriostatic, fungicidal to static in nature. The storage quality of mainly water activity and pH are the important factors that decide the state of the antimicrobial compounds. The antimicrobial compounds of foods alone cannot be effective for complete preservation; therefore they require additional use of some of the chemical preservative as mentioned earlier in 0.5%–1.5% level only. The integrated use of physical, chemical, as well as biological use like canning, food fermentation, and chemical use make the food fit for long-term preservation and safe consumption without affecting the nutritional status, thereby necessary for human health and hygiene.

The antimicrobial constituents may target the cell wall, cell membrane, metabolic enzyme, protein synthase, and genetic system—the combined actions of several compounds may act in this respect, therefore in most of the cases exact mechanism is observed.

The concentration dependent agents may be one of the limiting factors that initiate the integrated use of preservative mode to most of the foods. Therefore the inhibitors overall mechanisms at theoretical level may be discussed at certain levels. Further, the antimicrobial constituents have been classified based on the fruits and vegetables, meats of different animals, and milk as well as cereal and basics like garlic, onion, spices, and meat and milk antimicrobial constituents. These areas are important in the context of human health and hygiene; therefore being discussed in detail about their presence, mechanisms of action and types of microorganisms affected from in particular situations. The immunological consequences of various foods like milk and fruits are also well understood. The antimicrobial of foods are lysozymes, lactoferrin, citrate, acetic acid, benzoic acids, sorbic acid, and several others like lactose peroxidase system in bovine milk and several others having antimicrobial as well as antiviral activities. The ayurvedic medicines are based such as agents that directly or indirectly interact with pathogens or improving immune response to suppress the disease. The organic acids and esters, the organic acids are limited to pH, therefore pH and p*Ka* values must be taken in account, where p*Ka* value is dissociation constant, as every organic acid has its own p*Ka* value at which it is dissociated. It is evident from the mechanism that dissociation of acid does not work as antimicrobial, therefore the organic acid as preservative should be used in such pH of food where its dissociation is limited. The organic acids do not interact with the cell wall, rather they affect the cell membrane, cell wall, and membrane, allowing the without dissociated acid into the cell wall and membrane while the dissociated acid is not allowed to enter into the cell. The organic acids can penetrate the cell membrane lipid bilayer more easily. Organic acids are dissociated at pH 7.0, and the cytosolic pH 7.0 allows the dissociation, that is, $RCOOH \rightarrow COO^- + H^+$ resulting in the increasing hydrogen ion concentrations. Bacteria have to maintain neutral pH to maintain the conformational changes to the cell structural proteins, enzymes, nucleic acids, and phospholipids. Therefore the hydrogen ions are pumped out of the cell with the expense of higher ATP consumption, resulting in stress or death of the cells (Lambert and Stratford, 1999).

Organic acids also interfere with membrane permeability. The short chain organic acids interfere with energy metabolism by altering the structure of cytoplasmic membrane through interaction with membrane protein (Sheu and Freese, 1972). Such agents also work with several ways for the inhibitors of microbial growth and metabolism. Similarly, there are other acids like acetic acid and acetates that work with different p*Ka* values like 4.75, known as vinegar, and its sodium, potassium, and calcium salts, sodium and potassium diacetate, and dihydro acetic acid are the most used antimicrobial agents. Acetic acid is a more affective agent against yeast and bacteria than the filamentous fungi, while *Acetobacter* spp., lactic acid bacteria, and butyric acid bacteria are tolerant to these acids, while these bacteria are also considered as probiotic and also favor the additional preservative approach to most of the foods. Bacteria inhibited by acetic acid are *Bacillus, Clostridium, Pseudomonas, Salmonella, Staphylococcus,* and *Camphylobacter*. Several filamentous fungi are more resistant to acetic acid than bacteria, but *Aspergillus, Penicillium, Rhizopus, Mucor,* and *Saccharomyces* are more sensitive to acetic acid. It is used from 0.1% to 1.0% level in different foods. The use of 2% of acetic acid is more effective to eliminate *E. coli* in meat, vegetables, and fruit from the surface. The sodium acetate at 1.0% effectively reduces the filamentous fungi most occurring in an ecosystem contaminating food from air are *Penicillium, Rhizopus, Mucor,* etc. It is also used in baking systems. Yeast are also resistant to acetate sodium diacetate (p*K*a = 4.75), effective against *Pseudomonas, Salmonella, Lactobacillus, Enterococcus, Staphylococcus,* etc.; therefore milk products are processed, mainly cheese, in the presence of such agents. The benzoic acid and benzoates (p*K*a = 4.19) are most effective antimicrobial agents for food having pH 2.0–4.5, molds require 20–2000 μg/mL, but some like *Talaromyces, Zygosaccharomyces* are resistant to benzoic acids. The lactic acid/lactates (p*K*a = 3.71) produced during natural fermentation of food by lactic acid bacteria having capability to inhibit or kill several human pathogenic bacteria like *Salmonella, E. coli, Staphylococcus,* and *Camphylobacter,* etc. The concentration 0.2%–2.5% is effective to reduce several bacteria, likewise propionic acid, sorbic acid, and salts are found effective against several bacteria and fungi.

Lysozyme (1,4,-*N*-acetyl muramidase) is an enzyme in avian eggs, mammalian milk, tears, and other secretions up to the level of 3%. These enzymes are stead stable up to 80%–90% for 2–3 minutes at low pH and high salt concentration (4%–6%). The enzyme hydrolyze the bond, β-1-4 glycosidic bond between acetyl glucosamine and *N*-acetyl muramic acid of peptidoglycans; therefore most effective for gram-positive bacteria like *Bacillus, Clostridium, Staphylococcus, Streptococcus,* etc., and even effective in acidic and alkaline to saline conditions. The gram-negative susceptibility can be increased by using a chelating compound, EDTA, which binds Ca^{++} or Mg^{++} are essential for maintaining the integrity of the cell lipopolysaccharides of gram-negative bacteria that are also effectively controlled by the enzyme concentration 5000–9000 μg/mL, is also effective for fungi like *Penicillium, Paecilomyces* and *Aspergillus*. These agents are food grade preservatives in many foods, especially for cheese processing.

Certain antimicrobial agents like nutrients of their salts like sodium nitrites ($NaNO_2$) and potassium nitrite (KNO_2) are used in 0.05%–0.1% levels only, as its high concentration is carcinogenic, effectively eliminate the clostridia. It also favors the texture, color, and even the taste. The nitrates ($NaNO_3$) and KNO_3 have also been used in cured meats for controlling *Clostridium,* while several other bacteria like *E. coli, Achromobacterium, Flavobacterium, Pseudomonas, Salmonella,* and *Lactobacillus* are resistant to nitrite. Several green leaves and meats contain nitrites generally limited to 156 ppm (mg/kg). Therefore we can understand the antimicrobial constituents of various organic and inorganic compounds of food either naturally present or deliberately added to food for better human and health by eliminating several disease-causing, intoxicating, and spoiling microorganisms. Therefore food environment, mainly O_2 availability, temperature, pH, water activity (a_w), and redox potential of foods favor the growth of microorganisms and their inhibitory components interaction and kind of mechanisms for suppression and killing of the microorganisms. There is a greater number of other compounds like parabeans which have been effective up to pH 3.0–8.5 without dissociation, depending on the alkylation process.

Parabeans are phenolic derivatives that mainly work as phenolic antioxidants present in coffee, cocoa, tea, etc.; therefore phenolic antioxidants, phosphates like sodium acid pyrophosphates, tetra sodium pyrophosphate, sodium tri polyphosphate, and trisodium phosphates. These agents help in the buffering agents and are important in food processing like stabilization, acidification, and alkalization, precipitation of meats, and formation of complexes with organic polyelectrolytes like proteins, pectins, and starch. Gram-positive bacteria are more susceptible to phosphates than are gram-negative bacteria. Further, like NaCl, sugars, sulfur dioxide (SO_2), and its salts have been used for food preservation, sodium sulfite (Na_2SO_3) and potassium sulfite (K_2SO_3), potassium bisulfate ($KHSO_3$), sodium bisulfate ($NaHSO_3$), potassium, and sodium meta bisulfate are the main agents.

Natural Antimicrobial Compounds of Foods

Microbial agents as favorable factors for food hygiene and public health are another important component, which include the role of beneficial microbial groups that may grow at a wide range of pH, temperature, and carbohydrates having short generation time, that is, 20–25 minutes. This group is of lactic acid bacteria. They may be either homolactic to heterolactic producing lactic acid as well as ethanol. All they produce are preservative components along with flavor and aroma producing, and simultaneously more than ten different beneficial components as metabolites like anticancerous, compos, antihistamines, anticholesterol, antiviral, and anticarcinogenic compounds help human health and hygiene.

These bacteria dominate natural microflora, mainly that reside on young green leaves, flower parts, on stem and fruits surface, and colonize very fast with competing other groups of microorganisms due to early production of antagonizing metabolites, that is, lactic, acetic, and ethanol. Few species are capable of producing antibacterial antibiotics like bacteriocins and other components helping survive in a microbial niche and shift several human and plant pathogenic microorganisms.

Such activities have made this group for food fermentation and preservation of several milk, meat, and vegetables and fruit products. Bacteriocins have the ability to inhibit protein synthesis in bacteria, along with acidification of food at the early 18–20 hours during fermentation. The area of probiotic food and the environment especially that how a lactobacillus or other lactic acid bacterial strain harbor the ecosystem components like ecology of such strains availability from different parts of vegetables and fruits or especially their survival in a specific type of milk is still obscure. The research in the area of probiotic strain selectivity and their existence in the surrounding crops, milk utensils, along with existence with bacteriophages, and other conditions along with *Streptococcus, Pediococcus, Leuconostoc,* and *Vegococcous*, etc., are to be work out. If probiotic strains of such bacteria could be established in the various foods, then it will be helpful to human health and hygiene. They are also helpful to animal probiotic food in the form of green fodder or even with wheat and rice straw treated with probiotic bacteria with increased population of cellulolytic bacteria in the rumens. Still there are several factors, physicochemical and nutritional, which may be studied with the existence of better probiotic strains with human pathogens along with their sustainable health-associated effects. Further, the ecological aspects of probiotic strains in ecosystems as well as in the human stomach, and health associated as immunological levels may also be discussed.

The existence of such lactic acid bacteria along with intoxicating bacteria and fungi may also be studied for the level of toxin production in various foods. Such strains should be worked out for food quality up-gradation and better simulation of spore formers, which are also creating problems with food preservation. The role of such bacterial strains in controlling germination at various stages have been proved, but the probiotic effects in such area with food nutrition and inhibition of pathogens intoxicating various foods along with disease causing have not been studied combined with probiotic strains, may be an area of researchers who are pursuing their work in food and medical microbiology. Furthermore, the ecology of lactobacilli along with *S. thermophilus* in several dairy products like yogurt and other products should be worked out. The role of *Leuconostocs* and other groups of bacteria and fungi in cheese production, curing of meat, and their role in coffee and cocoa, fermentation, tobacco fermentation silage fermentation, or any fruits, vegetables, and cereal fermentation are also very important. The nisin, which is produced by several lactic acid bacteria, have antibotulinum as well as inhibitory to coagulase positive *Staphylococcus* spp. Several other pathogenic bacteria like *Listeria* and *Streptococcus* may also be suppressed. Further, there are a number of antimicrobial compounds reported from lactic acid bacteria able to inhibit or kill pathogenic bacteria. It has been proved that there are certain genes that are responsible and regulated with proper genetic modulation.

It is well documented that most of the human pathogenic microorganisms are getting resistant to several antibiotics by using three to four types of mechanisms either by modifying antibiotics or restricting its entry into the cells or modifying enzymes systems or any other mode, but it is difficult to synthesize a new drug to overcome this process. Moreover, it is natural microbial ecosystems, where several beneficial bacteria of such class may develop. Therefore the study on the screening of such lactic acid bacterial strains that can produce such compounds may be used for curing disease. There are a number of antimicrobial proteins, lactoproteins, and globulins that may inactivate the toxic metabolite of bacteria and fungi. Such strains of lactic acid bacteria should be isolated from the natural environment for better human health. The process through which the specificity of the mechanism of drug resistance like denaturation, modifications of the toxic metabolites, alteration in the receptor membrane compositional changes may be worked out with the screening of several lactic acid bacteria, which are separately put in a different class by Bergey's Mannual of Determinative Bacteriology. To work in these areas, the methodology, analytical tools like GLC, HPLC, FTIR, IR spectroscopy may be used for obtaining valuable findings.

The principal author of this chapter has started to work in this area after getting several good funding opportunities while working in the area of soil microbial ecology, where several new concepts have been developed for probiotic rhizospheric microorganisms. This book is in the process of publication. The metabolic processes like homofermentative, heterofermentative, and mixed-acid fermentation along with the production of several metabolites have not been discussed so far, but the production of aroma like diacetyl and aldehyde production, proteolytic and lipolytic systems, and their physiological role in different metabolic and biosynthetic activities along with phage resistance should be discussed for more elaboration for better understanding of the factors affecting those properties.

MICROBIAL PHYSIOLOGY AND GROWTH KINETICS FACTORS

Food materials are a good source of nutrients to complete even all four phases of microbial growth in batch culture. The growth kinetics of a batch system have lag, log, stationary, and death phages in logarithmic scale or $\log_{10}$ cell numbers versus time. In lag phase, cells adjust to their new environment by inducing or repressing enzyme synthesis and activity, initiating chromosome and plasmid replication, while in spores, differentiating into vegetative cells.

The length of lag phase depends on the temperature, inoculums size, and the physiological history of the organism. The actively growing cells are inoculated into an identical fresh medium at the same temperature resulting no lag phase, and these factors may change the lag phase. Food accelerates the exponential growth phase under spoiling conditions; therefore several preservative approaches, like physical, chemical, and biological approaches are employed either alone or in combination of several to save the food from spoilage, intoxication, and disease-causing microorganisms. Moreover, it is safer to manipulate factors to produce conditions in which the cells cannot grow regardless of time. The reduction of pH (below 4.0) inhibits most of the pathogenic bacteria including botulinum growth. In microbial growth, doubling time is important; it is also known as generation time, which is ideally at their optimal physicochemical levels. It is evident that a number of factors affect the growth of microorganisms in food, and most of them have been discussed in this chapter in order to understand the basic concepts of various factors affecting microorganisms in the variety of foods, the microbial diversity and ability of microorganisms to adopt various resistance factors make the process of preservation difficult, therefore the evaluation of microbial quality and quantity is always important for effective remedial measures. The role of stationary phase where stressed/injured microorganisms may survive and do not show their appearance on their normal medium, therefore required more specialized medium for their regeneration and finally appearing on nutrient agar medium. The status of such cells also have the ability to recover fast in the presence of specific micronutrients and surfactants, possibly increase the membrane permeability and repair of the damaged proteins from cell wall to membrane levels or modifications of the membrane translocation factors (Murthy and Gaur, 1987).

The level of injury and rate of repair mechanism in the presence of $MgCl_2$ and Tween 80 in the specific concentration has improved the recovery of microbial cells on the specific selective medium, that is, VRB (violated red bile salt agar). This finding has proved that certain supplementary metallic ions like Ca^{++}, Mg^{++}, and others in the presence of a surfactant can increase the cell permeability or modify the integral protein nature to some extent that more essential metals uptake may reach to the cells for higher recovery of the cells. Such science also works as an intrinsic factor for either cell longevity or multiplication and sensation. The role of such microorganisms may be considered in the assessment of the quality control aspects as well as preservation of foods, etc. Further, the factors that stress the microorganism may vary and injury at the level of external, internal organ as well as physiological levels may also be a component of discussion and research at the level of final assessment of microbial cells from the products to be preserved or assessed for quality control, because such cells do not show their appearance on normal selective medium. This is one of the most interesting areas to research work to explore the facts of injury levels along with the recovery of stressed cells.

CONCLUSION

Microbial growth in foods is very complex and diversified, which is governed by biochemical, environmental, and genetic factors along with their nutritional class. The major groups have been categorized in food-spoiling, intoxicating, and disease-causing bacteria with their specificity at various temperature, pH, air, O_2 requirement along with the antimicrobial compounds present in the food. There are several other biological factors and conditions that have been discussed in the extrinsic and intrinsic factors affecting the growth of microorganisms in a specific food. Every food, milk, meat, eggs, seafoods, and their products are more prone to microbial attack of several groups of microorganisms. Therefore the role of spore formers is also critical in order to achieve preservation approaches of foods. Moreover, the microbial shift from useful to harmful microorganisms just depends on the nature and state of foods, microorganisms, and factors of intrinsic and extrinsic levels. Further, the role of food fermentation and associated lactic acid bacteria, a probiotic compound, also plays an important role in the existence of pathogenic and spoiling bacteria. The intoxicating bacteria, food-spoiling, and disease-causing bacteria have successfully been controlled through food fermentation. Development of molecular biology and food microbial ecology assessment methods have made this science preventive against these to achieve better preservation, but more research is required in this area in coming decades. The scientific investigations of spore formers are enormous and have contributed to the development of microbiology for enhancement of food safety and quality. Therefore use of various chemicals for the suppression at various stages of spore germination as well as growth of microorganisms have also been used for better preservation of various foods, especially canned foods and other fermented foods. This book chapter has been constituted to have a deeper knowledge of all the range of microorganisms and various factors affecting them in various foods. Although the microbiology of food has a very wide variety of microorganisms and their action in different foods varies depending on the food nutrients and other physicochemical factors that are important in food industry by specifying the newer standards of microbial indicators and their control measures for international and national trade, guidelines, and policies.

References

Beaman, T.C., Gerhardt, P., 1986. Heat resistance of bacterial spores correlated with protoplast dehydration, mineralization, and thermal adaptation. Appl. Environ. Microbiol. 52 (6), 1242–1246.

Behravan, J., Chirakkal, H., Masson, A., Moir, A., 2000. Mutations in the gene p locus of *Bacillus subtilis* and *Bacillus cereus* affect access of germination to their targets in spores. J. Bacteriol. 182, 1987–1994.

Burkholder, W.F., Grossman, A.D., 2000. Regulation of the initiation of endospore formation in *Bacillus subtilis*. In: Brun, Y.V., Shimkets, L.J. (Eds.), Prokaryotic Development. American Society for Microbiology, Washington, DC, pp. 151–166.

Cossins, A.R., Sinensky, M., 1984. Adaptation of membranes to temperature, pressure and exogenous lipids. In: Shinitzky, M. (Ed.), Physiology of Membrane Fluidity. CRC Press, Inc., Boca Raton, FL, pp. 1–20.

Galperin, M.Y., Mekhedov, S.L., Puigbo, P., Smirnov, S., Wolf, Y.I., Rigden, D.J., 2012. Genomic determinants of sporulation in *Bacilli* and *Clostridia*: towards the minimal set of sporulation-specific genes. Environ. Microbiol. 14 (11), 2870–2890.

Griffiths, M.W., Phillips, J.D., 1990. Incidence source and some properties of psychrotrophic *bacillus* spp. Found in raw and pasteurized milk. J. Soc. Dairy Technol. 43, 62–70.

Hathaway, S., 1999. Management of food safety in international trade. Food Control. 10, 247–253.

Ireton, K., Jin, S., Grossman, A.D., Sonenshein, A.L., 1995. Kreb cycle function is required for activation of the spoOA transcription factor in *Bacillus subtilis*. Proc. Natl. Acad. Sci. U.S.A. 92, 2845–2849.

Jouve, J.L., 1999. Establishment of food safety objectives. Food Control. 10, 3003–3305.

Juneja, V.K., Marks, H.M., 1999. Proteolytic *Clostridium botulinum* growth at 12–48°C simulating the cooling of cooked meat development of predictive model. Food Microbiol. 16, 583–592.

Kalniowski, R.M., Tompkin, R.B., 1999. Psychotrophic *clostridium* causing spoilage in cooked meat and poultry products. J. Food Prot. 62, 766–772.

Lambert, R.J., Stratford, M., 1999. Weak-acid preservatives: modeling microbial inhibition and response. J. Appl. Microbiol. 86, 157–164.

Lupins, J.R., Kenny, M.F., 1998. Tolerance limits and methodology: effect on international trade. J. Food Prot. 61, 1571–1578.

Murthy, T.R.K., Gaur, R., 1987. Effect of incorporation of Tween 80 and magnesium chloride on the recovery of Coliforms in VRB medium from fresh, refrigerated and frozen minced Buffalo meats. Int. J. Food Microbiol. 4, 341–346.

Olson, E.R., 1993. Influence of pH on bacterial gene expression. Mol. Microbiol. 8, 5–14.

Sheu, C.W., Freese, E., 1972. Effects of fatty acids on growth and envelop protein of *Bacillus subtilis.* J. Bacteriol. 111, 516–524.

Sonenshein, A.L., 2000. Endospore forming bacteria—an overview. In: Shimkets, L.J., Brun, Y.V. (Eds.), Prokaryotic Development. American Socierty for Microbiology, Washington, DC, pp. 133–150.

Venugopal, R.J., Dickson, S., 1999. Growth rates of mesophilic bacteria, aerobic psychrotrophic bacteria and lactic acid bacteria in low dose irradiated pork. J. Food Protein 62, 1297–1302.

Vigh, V., Maresca, B., Harwood, J.L., 1998. Does the membranes physical state control the expression of heat shock and other genes? Trends Biochem. Sci. 23, 369–372.

Wolfer, L.M., Sumner, S.S., 1993. Antimicrobial activity of the lactose peroxisade system a review. J. Food Prot. 56, 887–892.

Wouters, J.A., Rombouts, F.M., Devos, W.M., Kuipers, O.P., Abee, T., 1999. Cold shock proteins and low temperature response of *Streptococcus thermophilus* CNRZ 302. Appl. Environ. Microbiol. 65, 4436–4442.

Wuytack, E.Y., Soons, J., Poschet, F., Michiel, C.W., 2000. Comparative study of pressure and nutrient induced germination of *Bacillus subtilis* spores. Appl. Environ. Microbiol. 66, 257–261.

CHAPTER

9

Safety of Water Used in Food Production

Vinod R. Bhagwat

Department of Biochemistry, S.B.H. Government Medical College, Dhule, India

OUTLINE

INTRODUCTION

Water is an indispensable resource for the food industry, which has many applications. In general, water is often taken lightly in most food preparation and processing operations. Water has a wide variety of uses in food production, for cleaning, sanitation, and

Food Safety and Human Health
DOI: https://doi.org/10.1016/B978-0-12-816333-7.00009-6

manufacturing purposes. In addition to being an ingredient in many foods, it may be used for various other operations, such as for growing, unloading, fluming, washing, brining, ice manufacture, and in sanitation and in hygiene programs. Water quality has great detrimental impact on products and operations in food production systems. The fundamental importance of water quality in food production is often underestimated. This underestimation generally becomes the underlying cause for several problems, such as mismanagement of water, equipment operation, and maintenance issues; loss of revenue; food safety; and product quality.

Food safety and food security are interrelated concepts that have a profound impact on the quality of human life. There are many external forces that affect both of these areas. Food safety is a very broad term. It covers many aspects of handling, preparation, and storage of food. Water is an inherent component in food production. In-depth understanding is required to know the impact of these factors on food safety. Every care in detail is essential to prevent illness and injury from the viewpoint of food production. Food production also covers chemical, microbiological, and microphysical aspects of water and food safety. Therefore, for adequate control of food chemical quality, control of allergens that can be life threatening to people who are highly sensitive to such allergens gets top priority in ensuring food safety. Any pathogenic bacteria, viruses, and microorganisms may produce toxins, which act as possible contaminants of water used in food production and that therefore affect food safety. At times, microphysical particles such as glass and metal, when present in water, can be hazardous and cause serious harm to consumers.

RESOURCES OF WATER

It is estimated that the world has about 1386 million km^3 of water. Out of this water, 35 million km^3 (2.5%) is freshwater. The significant amounts of freshwater (24.4 million km^3) are present in ice caps, glaciers, and deep in the ground, which is 10.5 million km^3 that is not accessible for use. Water which can be used comes essentially from rainfall over land, generated through repeated evaporation and periodic precipitation, the so-called hydrological cycle. Water is continuously recycled as a result of evaporation driven by solar energy. Therefore, the hydrological cycle consumes more energy each day than that used by humankind over its entire history.

Globally, the average annual rainfall over land is around 119,000 km^3. Out of this amount 74,000 km^3 evaporates back into the atmosphere. The remaining 45,000 km^3 flows into lakes, reservoirs, and streams or infiltrates into the ground to replenish the aquifers. Not all of the 45,000 km^3 is accessible for use, as part of the water flows into remote rivers and flows down during seasonal floods. An estimated 9000–14,000 km^3 is economically available for human use. This is equivalent to a teaspoon in a full bathtub, when compared to the total amount of water on earth (FAO, 2002).

When we talk about freshwater, in general, it refers to the water present in rivers, lakes, underground water, and glaciers. All this water is collectively termed as "*blue water*." Only part of the total rainfall accounts for this freshwater supply. The majority of rainfall that comes down on the Earth's surface either evaporates directly as 'nonbeneficial evaporation'

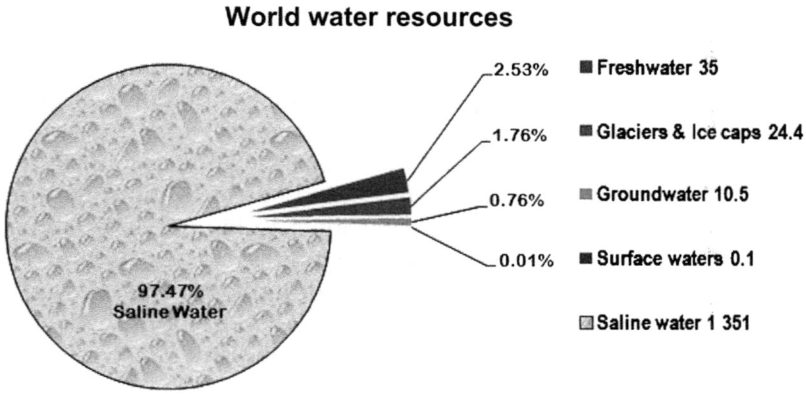

FIGURE 9.1 Distribution of water in world (figures in the legend are volume of water in millions km^3, while in the chart it is in percentage of total water).

or, after being used by plants, as 'productive transpiration'. This second type of rain water is called *green water*. The proportion of green water, out of the total available freshwater supply, varies between 55% and 80%, depending on the local wood density as well as on the region of the world. Storing of more green water in soil and plants, as well as storage as blue water, is the biggest opportunity and challenge for future water management.

Out of the total freshwater, over 60% of rain water is green water, which evaporates above savannah grazing land, forests, and agricultural land. About 40% of rain water (43,000 km^3) present in rivers, lakes, groundwater, and glaciers constitutes blue water, which then ultimately flows back into the oceans. Blue water withdrawals by humans are about 9% (or 3900 km^3) of total blue water resources. Of these, 70% is being used for irrigation (2700 km^3) and the remaining 1200 km^3 is used for industry and households, while only a very small part of this water cycle serves as drinking water (Aquastat, 2014).

The water used in food production is potable water (i.e., drinking water). It may come from a variety of possible sources including surface water such as streams, rivers, lakes, groundwater (e.g., underground natural springs, wells), rainwater, and seawater (after desalination). The quality of water is essentially dependant on the source of water. Adequate treatment of the water is necessary to ensure that it meets a drinking water standard which is safe to be used in food production (i.e., safe for human consumption). The food industry is supplied drinking water primarily in two ways. Firstly, it is by public distribution by local government authorities, and secondly through private supply by the food business itself. In European countries, the majority of drinking water supplied to the food industry comes from public distribution system by government authorities.

WATER IN FOOD PRODUCTION

In a very broad sense, there are four major uses of water in food production: (1) primary production, (2) cleaning and sanitation, (3) processing operations, and (4) as food ingredient.

Water in Primary Production

Food and agriculture sectors are the largest consumers of water. The largest use of water is in primary food production. In agriculture, water is predominantly used for crop irrigation purposes. In livestock farming, large volumes of water are used for livestock watering along with maintenance of general hygiene of the animals and equipment.

The demand of water for agriculture is one hundred times more than it is used for personal needs. Global distribution of water is shown in Fig. 9.1. The largest percentage (69%) of the global freshwater withdrawals is committed to agriculture. The industrial sector uses 19%, while a small fraction of only 12% of the water withdrawals are destined for households and municipal use. Currently, around 3763 km^3 of freshwater is withdrawn each year for human use (Aquastat, 2014). Of this, grossly half is actually consumed as a result of evaporation, incorporation into crops, and transpiration from crops. The other half recharges groundwater and surface flows or is lost in unproductive evaporation. The majority, up to 90% of the water withdrawn for domestic use, is returned to rivers and aquifers as wastewater. Industries consume only about 5% of the freshwater they withdraw.

The global nutrition has significantly improved since the 1960s, providing more food per capita at progressively lower prices. This was possible due to developments in high-yielding seeds, irrigation, and plant nutrition. The main source of food for human population in the world is agriculture, and this term also includes livestock husbandry and forestry. Food production from the livestock sector has diversity such as meat (beef, pork, poultry, and others), dairy products, and fishes, aquatic or sea food. As the population keeps increasing, more food and livestock feed has to be produced in the future, and obviously more water is required for this purpose. Thus, agriculture has to claim larger quantities of water to produce the food required to feed the world.

The pattern of global agricultural water withdrawal is uneven (Fig. 9.2). It is least in European (25%) and American (48%) countries while maximum in Asian and African countries (81%) around 2010. There is reverse trend in water withdrawal for industrial use. It is higher in Europe (54%) and America (37%), while it is least in Asia (10%) and Africa (4%) (around 2010). Asian countries use the highest total volume of water for agriculture in the world (2556 km^3/year), and the least is in the Oceania countries (25 km^3/year), which include Australia, New Zealand, and other Pacific Islands (Aquastat, 2014).

The concept of "water footprint" provides an appropriate framework for analysis to find the link between the consumption of animal products and the use of the global water resources. The water footprint is defined as "the total volume of freshwater that is used to produce the goods and services consumed by an individual or community." It is baffling but true that the water footprint of any animal product is larger than the water footprint of a wisely chosen crop product with equivalent nutritional value. The most logical reason lies in the fact that animals are secondary consumers in the food chain; therefore, in the calculation of water footprint of animals always adds up inherent increments of primary consumers. Twenty-nine percent of the total water footprint of the agricultural products in the world is related to the generation of animal products. One third of the

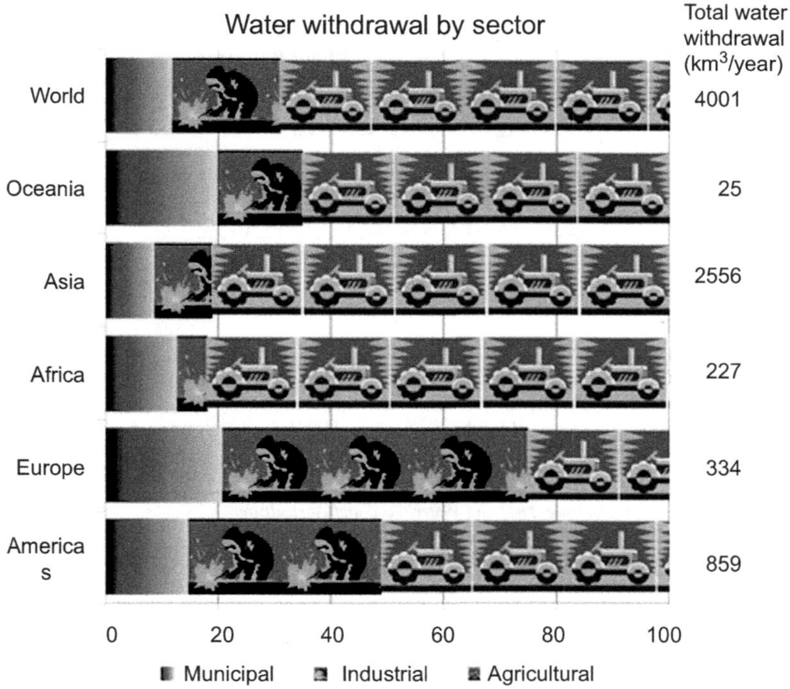

FIGURE 9.2 Comparison of sectorial water withdrawal in the world (percentages of the total water withdrawal per year. The values are the volume of water in km^3/year, around 2010).

global water footprint of animal production is related to beef cattle (Mekonnen and Hoekstra, 2010).

It is very much insidious that global meat production has almost doubled in the period from 1980 to 2004, and this trend is likely to continue in future, given the projected doubling of meat production in the period from 2000 to 2050 (Steinfeld et al., 2006). The on-going shift from traditional extensive and mixed farming to industrial farming systems is likely to continue to meet this rising demand for animal products. This intensification of animal production systems will result in increasing blue and gray water footprints per unit of animal product due to the larger dependence on concentrate feed in industrial systems. The pressure on the global freshwater resources will therefore increase both because of the increasing meat consumption and the increasing blue and gray water footprint per unit of meat consumed (Mekonnen and Hoekstra, 2010).

Plants require water in adequate quantities and at the right time for vegetative growth and development. Crops have very specific water requirements, and these vary depending on local climate conditions. The production of meat requires between 6 and 20 times more water than for cereals. An overview of the water consumption in food and agriculture is given in Table 9.1, which shows examples of water required per unit of major food products, including livestock that consume the most water per unit. Cereals, oil crops, and pulses, roots, and tubers consume far less water (FAO, 2003).

Virtual water concept emerged in the 1990s to draw increasing focus of people concerned with water management, and in particular with water related to food production.

TABLE 9.1 Water Consumption in Food and Agriculture

	Products	Unit	Water Required in m^3 Per Unit
1	Cereals	kg	1.5
2	Pulses	kg	1.0
3	Citrus, fruits	kg	1.0
4	Palm oil	kg	1.0
5	Fresh poultry	kg	6.0
6	Fresh lamb	kg	10.0
7	Fresh beef	kg	15.0
8	Sheep and goat	Head	500.0
9	Cattle	Head	4000.0

Source: *Adapted from FAO, 2003. Agriculture, Food and Water. Food and Agriculture Organization of the United Nations, Rome, Italy, <http://www.fao.org/docrep/006/Y4683E/y4683e07.htm#P0_0> (accessed 8.11.18.).*

Virtual water is defined as the water embedded in a product, that is, the water consumed during its process of production. The value is generally expressed in terms of volume (m^3), which results from multiplying the quantity of product (kg) by the unit value. On an average, it takes 2–4 L/day to satisfy the biological needs of drinking water of a human being. This may be around 1000 times as much water as required to produce the food for normal person. Therefore, the concept of virtual water is so important when discussing food production and consumption. In other words, when a country imports 1 million tons of wheat, it is also enlarging its water resource by 1 billion m^3 of water.

Specific values for the water equivalent of a selection of food products are variable, and it is highest for animal food. For instance: Cattle 4000 m^3/unit; Sheep and Goat 500 m^3/unit; as compared to plant food like palm oil 2 m^3/unit; fruits, pulses, and vegetables 1 m^3/unit of consumption. At a global level, the importance of virtual water is likely to dramatically increase as projections show that food trade will increase rapidly. It may be doubling for cereals and tripling for meat between 1993 and 2020. Hence, the transfer of virtual water embedded in the food which is traded is becoming an important component of water management on a global as well as regional level, especially in the water scarce regions (Renault, 2002).

Water in Sanitization

Water is a universal solvent. Cleaning is a very broad term. When used in the context of food handling, it implies the complete removal of food soil and associated nonfood visible components using water and detergent chemicals by appropriate methods or processes, under recommended conditions. The most important first step is flushing with water to remove visible soil. Cleaning also refers to washing of the equipment, instruments, containers, plants, associated machinery, and even the personnel who handle raw or processed food. In all practical purposes, the quality of water used for cleaning and washing should be similar to potable water.

Sanitization refers to the reduction of microorganisms to the level that is considered safe from a public health viewpoint. It is essential to differentiate and define certain terminologies in the context of food safety. *Sterilization* is the statistical destruction and removal of all living organisms, whereas *disinfection* refers to inanimate objects and the destruction of all vegetative cells (not spores). The official definition of sanitizing for food product contact surfaces as given by the Association of Official Analytical Chemists is a process that reduces the contamination level by 99.999% (5 logs) in 30 seconds (Schmidt, 1997).

Water is used in sanitization. It can be used alone or it can be used along with some chemical agent for effective sanitization. The *chemical sanitization* involves the use of chemical sanitizer at a specified concentration and contact time. In chemical sanitization, water is used as medium or vehicle for more effective and efficient cleaning and sanitization. It is imperative to understand basic water chemistry and microbiology before selecting a cleaning compound. The chemistry of waters may differ significantly from different sources. Water chemistry can also affect sanitizer performance. Water used for cleaning must be of good microbiological quality. The recommended microbiological guidelines for water destined for final cleaned water must have total plate count $<500\ mL^{-1}$, coliform $<1\ mL^{-1}$, and psychotrophs $<10\ mL^{-1}$ (ILSI, 2008).

Water is involved in approximately 95–99% of cleaning and sanitizing operations of food productions. Water functions as a vehicle to carry the detergent or the sanitizer to the surface as well as removes soil or contaminations from the food surface. The effectiveness of a detergent or a sanitizer is drastically altered by impurities in water. Water hardness is the most important chemical property that has a direct effect on cleaning and sanitizing efficiency. Hardness of water is measured in parts per million (ppm) or grains per gallon (gpg). Water having 0–60 ppm (0–3.5 gpg) hardness is soft water, 60–120 ppm (3.5–7 gpg) is moderately hard, and 120–180 ppm (7–10.5 gpg) is hard water, whereas >10 ppm (> 10.5) is considered as very hard water.

The chemistry of the water, the hardness in particular, greatly affects the performance of cleaning chemicals. Water hardness increases detergent consumption. It can cause formation of film, scale, or precipitates on equipment surfaces. Ignorance in proper understanding of water chemistry can cost money and economic losses to the processors. It may be in the form of increased use of cleaning agents or increased cleaning time.

Water pH ranges generally from 5 to 8.5. This range is of no serious consequence to most detergents and sanitizers. However, additional buffering agents may be required for highly alkaline or highly acidic water. Chlorine is more effective at lower pH. At the lower pH, more hypochlorous ions are formed, and this increases the antimicrobial activity. When the pH of water is 8.5, the efficacy of chlorination is significantly reduced. If the water used is very hard, it is necessary to treat the water adequately before use. Generally, water softening becomes essential for both processing and cleaning applications.

Water can also contain a significant number of microorganisms. Water used for cleaning and sanitizing must be potable and pathogen-free. Prior to use in cleaning regimes, standard treatments and sanitization of water may be required. Some water impurities that effect cleaning functions are presented in Table 9.2.

TABLE 9.2 Some Common Water Impurities and Their Associated Problems

Impurities	Problem Caused
COMMON IMPURITIES	
1 Oxygen	Corrosion
2 Carbon dioxide	Corrosion
3 Bicarbonates (sodium, calcium or magnesium)	Scale formation
4 Chlorides or sulfates (sodium, calcium, or magnesium)	Scale and corrosion
5 Silica	Scale formation
6 Suspended solids	Corrosion and deposition
7 Unusually high pH (above 8.5)	Mediate corrosion and deposition; alter detergent efficiency
8 Unusually low pH (below 5)	Mediate corrosion and deposition; alter detergent efficiency
LESS COMMON IMPURITIES	
9 Iron	Filming and staining
10 Copper	Filming and staining
11 Manganese	Corrosion

Thermal Sanitization

When sanitization involves the use of hot water or steam for a specified temperature and contact time, it is called thermal sanitization. It is generally adequate for most purposes. As with any heat treatment, the effectiveness of thermal sanitization is dependent upon a number of factors, such as initial contamination load, humidity, pH, temperature, and time.

STEAM

The use of steam as a sanitizing process has limited application. It has several disadvantages. It is generally expensive as compared to alternatives. Temperature is difficult to regulate and also it is practically inconvenient to monitor the contact temperature and time. Further, the by-products of steam condensation can complicate cleaning operations.

HOT WATER

In general, hot water sanitization process is commonly used. It involves immersion (of small parts, knives, etc.), spray (such as dishwashers), or circulating systems. The time required for sanitization is determined by the temperature of the water. The standards of hot water use in dishwashing and utensil sanitizing applications are specified by regulatory authorities in various countries. It is recommended that hot water immersion should be at 77°C (170°F) for at least 30 seconds for manual operations. For single tank, single temperature machines, a final rinse at temperature of 74°C (165°F) is required, and for other machines it is 82°C (180°F) (Schmidt, 1997). Many state regulations require a utensil surface temperature of 71°C (160°F) as measured by an irreversibly registering temperature indicator in ware-washing machines. Recommendations and requirements for hot-water sanitizing in food processing may vary. The Grade-A Pasteurized Milk Ordinance specifies a minimum of 77°C (170°F) for 5 minutes. Other recommendations for processing operations are 85°C (185°F) for 15 minutes or 80°C (176°F) for 20 minutes (Schmidt, 1997).

There are several advantages of hot-water sanitization. These include relatively inexpensive method, readily available, easy to apply, effective over a broad range of microorganisms, penetrates into cracks and crevices, and is relatively noncorrosive. The disadvantage of hot-water sanitization could be a slow process that requires gradual warm-up and cool-down time. In addition, it can have high energy costs and has certain safety concerns for employees. The process also has the disadvantages of film formations and subsequent shortening of the life of certain equipment or machinery parts of plants.

Water Conditioners

When mineral content of water increases, particularly calcium and magnesium, the hardness of water also increases, rendering the water unfit for cleaning. In such situations, minerals are deposited in the form of film or scales on the hard surfaces which are in contact with such water for a long time. To prevent the buildup of various such mineral deposits, water conditioners are used. These conditioners are chemicals that usually act as sequestering agents or chelating agents. These sequestering agents work by forming soluble complexes with calcium and magnesium. Examples of some common water softeners are sodium tri-polyphosphate, tetra-potassium pyrophosphate, organo-phosphates, and polyelectrolytes. Some chelating agents include sodium gluconate and ethylene diamine tetra-acetic acid (EDTA).

Water in Processing Operations

In food processing, there are broad range of possibilities with regard to water management, including increased promotion with increased efficiency of water reuse. This water reuse efficiency can be enhanced by tailoring the water quality requirements for a specific process. The major demand for water arises during diverse food processing operations such as transport of products, dissolving ingredients, treatment of products (e.g., alteration, separation), maintenance of appropriate water content in the final product, cooling processes, steam generation, and abnormal incidents (e.g., fire protection).

In general, there are two main types of water reuse that can be identified. This is relating to whether water comes into direct contact with food product(s) or not. Typical reuse applications, where water usually has no such contact, include its use in cooling and for the generation of steam. The other category is where water that does have contact with food could be at the raw product stage (e.g., washing or transport), at intermediate stages (e.g., cleaning of equipment), or in the final product itself, which is residual water (ILSI, 2008).

The reuse of water by recycling has become an increasingly vital component of food processing operations. It is a very essential way to conserve water, reduce costs, and provide security of water supplies. Under current legislation in several countries, recycled water can be used in food processing operations or as an ingredient of food, provided that it should be of the same standards as drinking water. Even though clean seawater is nonpotable, it is commonly used in processing operations such as washing whole fishery products and shellfish.

In some circumstances, nonpotable water is used by the food industry for operations such as for fire control, steam generation, and others. In such applications, the water must be clearly identified as nonpotable water and it should not be connected or mixed with the drinking water supply or used directly in food production.

Water as an Ingredient or Component of Food

Water is also used as a component of food or an ingredient of the component of food. It is a very important safety concern when one considers water as part of the food. It acts as a medium through which food can be preserved, stored and consumed by humans. Fruit juices, jams, jellies, pickles, soups, and many more such consumed forms of food products have water as an ingredient.

It is fundamental that when water is used as a component of food or as an ingredient of the food component, it must be free from undesirable color, odor, taste, and impurities that are harmful to consumers and result in low-quality products. Ordinary tap water that meets the criteria of the safe drinking water standard may not necessarily achieve these qualifications. Undesirable odor and taste can be removed, usually by using an activated carbon filter. Activated carbon particles have a massive surface area that adsorbs substances like yeast, chlorine, and those that affect odor and taste, and nonpolar materials such as mineral oil and polyaromatic hydrocarbons. The activated carbon filter helps to remove unwanted material, which might interfere in subsequent treatment steps, like ion exchange and reverse osmosis.

WATER IN FOOD STORAGE AND PRESERVATION

The role of water in food storage and preservation is an important consideration from the point of food safety and stability. In general, it is the *water activity* of food, abbreviated as a_w, and not the water content, which determines the lower limit of water for microbial growth (Sandulachi, 2012). In several food industry operations, monitoring a_w is a critical control point. The importance of a_w in food systems cannot be exaggerated in food safety. It is well known that, throughout history, several methods of food preservation like drying and the addition of sugar or salt were prevalent, and they are commonly used even today. These methods have one common principle, which is to keep low water activity. These methods prevent growth of food-spoiling microorganisms and maintain food quality.

Water activity is defined as "the ratio of the partial vapor pressure of water in equilibrium with a food to the partial saturation vapor pressure of water vapor in air at the same temperature" (Fontana, 2000). This is equivalent to the relative humidity of the air that is in equilibrium with the food. The value of a_w ranges from 0 to 1. The a_w of a food denotes the energy state of water in the food, and therefore it delineates the potential of this water to act as a solvent and participate in chemical/biochemical reactions and growth of microorganisms. It is an important property that is used to predict the stability and safety of food with respect to microbial growth, rates of deteriorative reactions, and chemical/physical properties. Water activity values are used by food designers to use water activity to formulate shelf-stable food. Mold growth is inhibited if a product is kept below a certain water activity. This results in a longer shelf-life (Sandulachi and Tataro, 2012).

In food products made with different ingredients, the water activity values can also help to limit moisture migration within a food product. For instance, if raisins of a higher a_w are packaged with bran flakes of a lower a_w, the water from the raisins migrates to the bran flakes over time, making the raisins hard and the bran flakes soggy. Food formulators generally use the water activity principle to predict how much moisture migration affects their product.

Colligative effects of dissolved species such as salt or sugar interact with water through dipole–dipole, ionic, and hydrogen bonds. Surface interactions in which water interacts directly with chemical groups on undissolved ingredients like starches and proteins are, through dipole–dipole forces, ionic bonds (H_3O or OH^-), Van der Waals forces, hydrophobic bonds, and hydrogen bonds. The permutations and combinations of these interactions, in addition to the factors in a food product, lower the energy of the water and consequently decrease the relative humidity as compared to pure water (Fontana, 2001).

The lowest a_w at which the vast majority of food spoiling bacteria will grow is about 0.9. The growth of *Staphylococcus aureus* is inhibited at an a_w of 0.91 under anaerobic conditions; however, similar inhibition occurs at a lower a_w level 0.86 under aerobic condition. The a_w for mold and yeast growth inhibition is about 0.61, whereas the lower limit for growth of mycotoxigenic molds is at 0.78 (Fig. 9.3). A list of the water activity limits for growth of microorganisms significant to public health and examples of foods in those ranges are given in Table 9.3 (Fontana, 2001).

The browning reactions in food generally increase with a_w from 0.25, it reaches the peak around a_w 0.6, and then falls down sharply until a_w 0.75. In this range, only covalent interactions of water with food component are predominant (Fig. 9.3). The lipid peroxidation reactions start increasing from a_w 0.4 and continue until the maximum value of water activity. At the higher values beyond a_w 0.8, the solute and capillary action become predominant, which promote growth of microorganisms, yeast, and molds.

The effect of temperature on the a_w of a food is product specific. In some products, a_w increase with increasing temperature, while in others a_w decrease with increasing temperature, and high-moisture foods have negligible change with temperature. Therefore, it is

TABLE 9.3 Common Food Spoiling Organisms and Their Water Activity (a_w) Limits for Growth

Microbial Group	Species	a_w	Food Products That are Affected
Normal bacteria	*Salmonella* spp. *Staphylococcus aureus, Clostridium botulinum*	0.91	Fresh milk, meat
Normal yeast	*Torulopsis* spp.	0.88	Fruit juice concentrate
Normal molds	*Aspergillus flavus*	0.80	Jams, jellies
Mycotoxigenic mold	*Aspergillus, Fusarium* and *Penicillium* spp.	0.78	Cereals, vegetable products, meat, grapes juice and wine
Halophilic bacteria	*Wallemia sebi*	0.75	Honey
Xerophilic molds	*Aspergillus echinulatus*	0.65	Flour
Osmophilic yeast	*Saccharomyces bisporus*	0.60	Dried fruits

rather difficult to predict even the direction of the change of a_w with temperature, as it depends on how temperature affects the factors that control a_w in the food.

There is no ideal device that can directly measure the a_w of a food. It is measured by an indirect method. Practically, a food sample is placed in a small air-tight chamber, and the water in the air is measured after it equilibrates with the sample. Elaborate methods for a_w determinations are described elsewhere (Fontana, 2000). Reliable laboratory instrumentation is required to guarantee the safety of food products and enforce government regulations. Recently, newer instrument technologies have been developed that have vastly improved speed, accuracy, and reliability of a_w measurements. Broadly, two types of instruments are commercially available for a_w measurement. One category of instrument uses chilled mirror dew-point technology, while the other category measures relative humidity with sensors that change electrical resistance or capacitance. Each type of instrument has certain pros and cons. The methods also vary in accuracy, repeatability, speed of measurement, stability in calibration, linearity, and convenience of use (Fontana, 2000).

The concept of water activity is very useful in predicting food safety and stability with respect to microbial growth, chemical/biochemical reaction rates, and physical properties. By measuring and controlling the a_w of foodstuffs, it is possible to predict which microorganisms will be potential sources of spoilage and infection. It also becomes easy to explain logically how to maintain the chemical stability of foods by lowering a_w, minimize nonenzymatic browning reactions, arrest spontaneous autocatalytic lipid oxidation reactions, and prolong the activity of enzymes and vitamins in food. In addition, it becomes easy to optimize the physical properties of foods, such as texture and shelf-life. A global stability map (Fig. 9.3) of foods shows these factors as a function of water activity (Fontana, 2000).

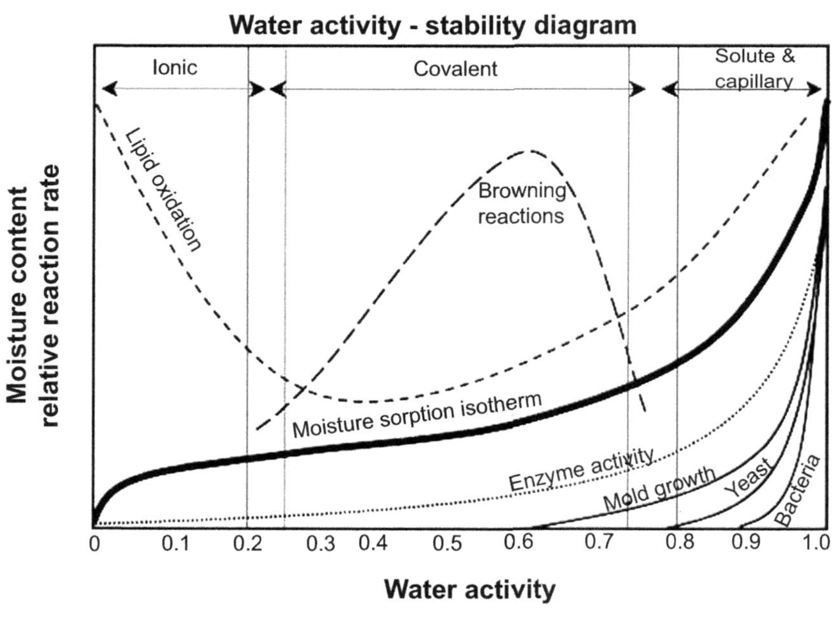

FIGURE 9.3 Global water activity-stability diagram or moisture-sorption isotherm (water activity is along the horizontal axis, while moisture content or reaction rate is plotted along the vertical axis).

The growing recognition of the a_w principle is evident from the fact that it is incorporated into U.S. Food and Drug Administration (FDA) and U.S. Department of Agriculture regulations, and most recently in NSF International Draft Standard 75 (Fontana, 2001). The major purpose of these regulations is to provide explicit requirements, critical control points, and practices to be followed by food industries and to implement and enforce these specific requirements, critical control points, and practices by industries so that products are pure, wholesome, and produced under sanitary conditions and are safe for human consumption. Recently, the instrument technologies have greatly improved speed, accuracy, and reliability of water activity measurements and are definitely an indispensable tool for food quality and safety.

WATER QUALITY

Safe and readily available water, whether it is used for drinking, domestic use, food production, or recreational purposes, is essential for public health. Improved water supply and sanitation, and better management of water resources, can boost economic growth of a country and also can contribute greatly to reduce poverty. Since the 1990s, water pollution has increased in almost all rivers in Africa, Asia, and Latin America (UNEP, 2016). The deterioration of water quality is expected to escalate over the following decades, and this will jeopardize human health, the environment, and sustainable development (Veolia/IFPRI, 2015). An estimated 80% of all industrial and municipal wastewater is released in the environment without any pretreatment. This results in an increasing deterioration of overall water quality, which has detrimental impacts on human health and ecosystems. Worldwide agricultural intensification has already resulted in increased use of chemicals to approximately 2 million tons per year (De et al., 2014). The gross impacts of this trend are largely unquantified and there are serious data gaps (WWDR, 2018).

In Europe, around 15% of groundwater monitoring stations recorded that the level of nitrates, as established by the WHO, were exceeded in drinking water. Further, the monitoring stations recorded that approximately 30% of rivers and 40% of lakes were eutrophic or hypertrophic during the period from 2008 to 2011 (WHO, 2015). The low- and lower-middle-income countries have the greatest increases in exposure to pollutants, primarily due to higher population growth and poor economic conditions in these countries, especially those in Africa (UNEP, 2016), and the lack of wastewater management systems (WWDR, 2017).

The quality of water used in food production and processing decides the food quality and its security. It is essential that to have healthy and hygienic food, the water used must be of very high quality. It is very essential that analysis of water with stringent quality control criteria must be followed at every point of food processing, packaging, and storage. In Europe, it is mandatory for the food industry to have an adequate supply of drinking water for use in food production to ensure that foods are not contaminated. Drinking water means that it is fit for human consumption, not only for drinking, but also suitable for cooking as well as for food preparation. In principle, this must be devoid of microorganisms and other contaminants that may jeopardize public health (WHO, 2017a).

Water safety can never be taken lightly. Unsafe water, which results due to direct contamination or improper or inadequate water treatment processes, generally results in a contaminated food product. Even though all types of foods are at risk, the highest among them are ready-to-eat products.

Impurities of water that are identified and measured fall into three basic categories: qualitative, general quantitative, and specific (Osmonics, 1997). Qualitative identification includes physical parameters such as turbidity, taste, color, and odor. These generally describe obvious conditions of water. Most of the qualitative parameters do not describe the concentration of the contaminants and therefore do not identify the source. However, it should be noted that taste, color, and odor evaluations should be very accurate qualitative measurements that can be instantly completed. The human nose, for instance, is sensitive to detect odors in concentrations down to the parts-per-billion level. A comprehensive quantitative water analysis, obviously, has higher precision compared to qualitative analysis (Osmonics, 1997).

WATER-BORNE FOOD CONTAMINANTS

In production of safe and ultimately secure food, safety of water assumes fundamental importance. The safety of water can be discussed by considering the possible hazards to human health. The presence of infectious agents, toxic chemicals, and radiological hazards can compromise water quality. These hazards may arise due to poor quality of water used directly or indirectly in the food production. Therefore, great attention is necessary to learn various health hazards of water such as biological, chemical, or physical pollutants that jeopardize human health.

Physical Contaminants

Physical contaminants are derived from incoming water that is not usually controlled by filtration, and it can be monitored by turbidity measurements. Microphysical particles like glass and metal microparticles are hazardous and cause serious injury to consumers. Chemical hazards include organic compounds, inorganic elements (e.g., heavy metals), and complex chemicals (e.g., pesticides), which are mentioned in detail in the EU drinking water directive and the WHO guidelines (WHO, 2017a). Chemical contaminants occur due to a variety of ways, such as environmental contamination or from a chemical spill, or due to incorrect use of pesticides or because of cross-contamination of the water supply with sewage and/or industrial waste.

Biological Contaminants

Biological contaminants not only include the organisms of concern but also the consequences of their presence (e.g., toxin formation). Water-borne microorganisms potentially causing illness include bacteria, viruses, protozoa, and helminths

TABLE 9.4 Typical Biological Contaminants of Water

Sr. No.	Biological Contaminant	Types (Examples)	Diseases Produced or Health Hazards
1	Bacterial pathogens	*Salmonella, Shigella, Campylobacter* and various pathogenic strains of *Escherichia coli*	Cholera, salmonellosis, shigellosis, leptospirosis
2	Viral pathogens	Norwalk virus, hepatitis virus, and other human enteric viruses	Hepatitis A, dengue fever, SARS CoV, legionellosis
3	Protozoan parasites	*Entamoeba histolytica, Giardia lambia, Cryptosporidium parvum, Cyclospora, Plasmodium*	Diarrhea, malaria
4	Helminthes	*Schistosomes, Tinia solium, Liver fluke. Clonorchis sinensis*	Schistosomiasis, clonorchiasis
5	Fungal species	Molds, yeast, and others	Toxins
6	Algae	Spirogyra and others	Toxins

List is illustrative only and not exclusive/exhaustive.

(Table 9.4). The resistance/susceptibility of these organisms to commonly used water treatments and the way of transmission need to be considered to ensure water quality. Most of the pathogens do not grow in water. These pathogens are introduced into the water by animal and/or human sewage. However, some are environmental pathogens that can normally grow in water. Legionella is one example, which is transmitted by inhalation/aerosols, leading to an infection of the respiratory tract (Washington, 1996). The risks related to Legionella have to be considered with respect to personnel safety (showers/washrooms), and as well to the wider environment of the plant, such as cooling towers that are used from where water can spread into the wider surrounding (Castilla et al., 2008). Water temperatures below 20°C and above 60°C would prevent multiplication in the system due to the growth characteristics of these bacteria. The European guidelines also provide information with respect to effective treatments/disinfection of water systems (EU-OSHA, 2005).

Food industries must have a system in place to ensure that they are continuously using safe/potable water in food production and processing. The Canadian Food Inspection Agency (CFIA) periodically reminds industries to follow water safety requirements for food production, processing, and handling, and maintain an action plan in the event of a major accident/water safety alert (CFIA, 2017). It is very essential that food industries have to be very vigilant and establish a communication system with the appropriate municipal, provincial, or territorial water authorities for the timely exchange of information in the event of accidents.

The food industry is ultimately responsible for safe food. Water contamination is a serious threat to human life. The food industry should be fully aware of the various possible contaminants and their potential risk to human health and must take appropriate safety precautions. Some common biological contaminants of water encountered in food industries are shown in Table 9.4.

HEALTH HAZARDS IN WATER AND FOOD CONTAMINATION

A majority of water-borne diseases are spread by pathogenic microorganisms that are transmitted through contaminated fresh water. These diseases can be transmitted while bathing, washing, or drinking water or by eating food exposed to infected water. Infection commonly results during drinking contaminated water or use of such water in the preparation of food, or the consumption of food contaminated with these pathogens. Various forms of water-borne diarrheal diseases are the most prominent examples and affect mainly children in developing countries. According to the World Health Organization, such diseases account for an estimated 4.1% of the total Disability Adjusted Life Years (DALYs) global burden of disease and cause about 1.8 million human deaths annually (Hatami, 2013). The World Health Organization estimates that 58% of that burden, or 842,000 deaths per year, are attributable to a lack of safe drinking water supply, sanitation, and hygiene (WHO, 2017b).

In terms of popularity and based on the incidence, the top most water-borne disease is the travelers' diarrhea, caused by *Escherichia coli* along with a variety of viral and parasitic enteric pathogens. Each year, between 20% and 50% of international travelers, and an estimated 10 million persons develop diarrhea. Giardiasis and cryptosporidiosis are the second largest cause of water-borne disease, due to transmission of microscopic parasites or cysts of Giardia and Cryptosporidium. Giardia is found worldwide and is within every region of Canada and the United States. The third top most water-borne disease is dysentery caused either by *Entamoeba histolytica* amebic dysentery and/or by salmonella and shigella bacillary dysentery. Dysentery kills around 700,000 people worldwide every year. *Salmonella bongori* and *Samonella enteric* are gram negative rod bacteria with >2500 serotypes or serovars known to cause the water-borne salmonellosis. Each year, almost 1 in 10 people fall sick and that results in loss of 33 million of healthy life years. It is one of the four key global causes of diarrheal diseases. Followed by this is typhoid fever due to *Salmonella typhi* bacterium at fifth place. An estimated 11–20 million people get sick from typhoid and between 128,000 and 161,000 people die per year. Next popular is cholera due to *Vibrio cholera* bacterium with 1.3–4.0 million cases per year, and 21,000–143,000 deaths worldwide are due to cholera. Viral hepatitis caused by the A virus is largely an endemic disease with transmission through water. Epidemics of A virus can be explosive and cause substantial economic loss. Similarly, the viral hepatitis caused by the E virus (HEV) can result in epidemics if common community water sources get polluted. It is estimated that there are 20 million HEV infections worldwide each year, leading to an estimated 3.3 million symptomatic cases of hepatitis-E. Worldwide, 1 in 10 people fall ill due to infection by Campylobacter, a spiral-shaped, S-shaped, or curved, rod-shaped bacteria. *C. jejuni* (subspecies *jejuni*) and *C. coli. C. lari*, and *C. upsaliensis* (less freq). A total of 33 million of healthy life years are lost and 550 million people fall ill yearly (including 220 million children under the age of 5 years).

The entire picture of water-associated diseases is complex due to various reasons. As a consequence of the emergence of newer water-borne infection diseases and the reemergence of older known diseases, the picture of water-related human health issues has become more and more comprehensive over a period of decades. Scanty data are available for some water-, hygiene-, or sanitation-related diseases like cholera, salmonellosis, or shigellosis. The analysis is yet to be performed for other water-related disease like schistosomiasis, malaria, or more modern infections such as SARS CoV or legionellosis. The burden of a variety of disease groups may only partly be attributed to water determinants. It becomes difficult to pinpoint the relative importance of water components of the local ecosystem, even where water has an essential role in the ecology of diseases.

In developing countries, four fifths of all illnesses are water-borne diseases. In this list, diarrhea is the leading cause of death among children. For approximately 1.1 billion people around the world, who still lack access to improved drinking water sources, the global picture of health and water has a strong local dimension. There is strong evidence that sanitation and water- and hygiene-related diseases account for around 2,223,000 deaths each year, as well as an annual loss of 82,196,000 disability-adjusted life years (DALY) (WHO, 2013).

Around 246.7 million people over the world are assumed to be infected by schistomiasis. Of this, 20 million people experience the full-blown infection, while another 120 million people experience only milder symptoms. An estimated 80% of transmissions occur in Africa and south of the Sahara. In Bangladesh, around 35 million people are daily exposed to arsenic. As the water they drink has elevated levels of arsenic, this will endanger their health and eventually shorten their life expectancy.

In the United States, 9.4 million people fall sick, and 1350 deaths occur each year, due to 31 major pathogens that contaminate food (Scallan et al., 2011). As relatively higher numbers of illnesses that occur relating to food are due to microorganisms, microbiological quality thus becomes the most important aspect of food safety. Therefore, food safety primarily focuses on the control of food contamination by pathogens.

In the United States, Norovirus, *Salmonella*, *Clostridium*, *Listeria*, *E. coli* O157:H7, and *Campylobacter* are the most common and leading pathogenic microorganisms in food-borne illness (Scallan et al., 2011). Among these, *Listeria* and *E. coli* O157:H7 are primarily the causative pathogen in a majority of food-borne deaths. In 1999, the CDC launched Food-Net, a useful data monitoring system, in an effort to monitor food-borne illnesses in the United States (Mead et al., 1999). In 2007, WHO initiated a system to estimate the worldwide burden of food-borne diseases (Kuchenmuller et al., 2009). These two data monitoring systems have greatly facilitated efforts to identify and subsequently decrease incidences of food-borne illness.

During the production process, foods can become contaminated at any point along the production line. Thus, programs such as Hazard Analysis and Critical Control Points (HACCP) have been designed to control food contamination (Codex, 2003). HACCP is a management system that addresses food safety through the analysis and control of biological, chemical, and physical dangers and threats from raw material production, procurement, handling, manufacturing, distribution, and even up to the consumption of the finished product. It is executed in the industry to decrease food safety risks. As the number of illnesses due to microorganisms is relatively greater than foreign objects or allergens, HACCP typically focuses on pathogen reduction and prevention (Gorham and Zurek, 2006). Some material is considered unavoidable, and the tolerance levels of such material have been set in HACCP. However, due to the high risk and potential severity of disease, certain hazards are not tolerated at all, including contamination of products by *E. coli* O157:H7 or *Listeria monocytogenes*, in the stringent criteria of HACCP.

WATER TREATMENT

The choice of the most suitable water treatment technique depends on the water source and the intended application of the water. It also depends on the design aspects of storage and distribution in a food manufacturing facility. Variable solids and chemical and microbial contents occur in water due to various factors like rainfall events, seasonal weather patterns, distribution system issues, and other factors.

The hygienic design of the water distribution and storage systems are the effective microbiological controls. The general principles, as adopted for food manufacturing equipment, should be followed in food production such that dead ends and stagnant areas are avoided, and prevention of biofilms and scaling is ensured. It is also ensured that the entire system can be cleaned and disinfected as and when needed on a regular basis. All material used for fittings, pipes, and tanks should be compatible with the conditions in the system, which includes the resistance of the material used against cleaning and disinfection agents.

Great care should be taken to prevent any backflow by providing a break at the entry point of the facility. Storage tanks should be enclosed to prevent any contamination by pests or extraneous matter. All vents should preferably be equipped with an air filter to achieve air filter class F7. At minimum, an insect screen should be installed. The tank design should favor full drainability when emptying. The entire system should be designed such that the maximum stay time of the water does not exceed 24 hours.

The water storage and distribution system design should allow for easy cleaning and disinfection. However, in reality, this is quite often an issue and cleaning is hardly possible. So typically, the water is treated at the entry point into food manufacturing units and the quality is maintained and monitored within the factory. The microbiological quality of water is monitored at various points in the plant, including last points of pipes, and appropriate treatments are applied. In a majority of cases, a combination of techniques will be necessary to meet the demands.

Ideally, water treatment processes should remove pathogens and impurities that may otherwise be harmful to human health or esthetically unpleasant. Water treatment processes may vary depending on the source water. Typically, an adsorbent material is added to the water to bind dirt and form heavy particles, which settle down after sometime to the bottom of a water storage tank for easy separation. The water is then filtered to remove even smaller particles. Finally, a small amount of disinfectant (e.g., chlorine), at a level safe for human consumption, may be added to kill any remaining viable microorganisms (WHO, 2017b).

It should be noted that the treatment of water used by the food industry is the obligatory responsibility of the concerned food business using a private water supply. Typically, private water supplies require treatment and ongoing monitoring of water quality verification following treatment. This can be carried out by laboratory testing to ensure they are fit for human consumption and can be used in food production (Kirby et al., 2003). A summary of the most common water treatment techniques in relation to the hazards that need to be controlled is given in Table 9.5.

Hardness is generally due to calcium and magnesium ions in water that may deposit in pipes, valves, and process equipment surfaces. Some food products may not dissolve well in hard water. In addition, hardness may affect flavor, aroma, and palatability of foods.

TABLE 9.5 Most Common Treatment Techniques in Relation to the Hazards That Need to be Controlled (✓Indicate Most Preferred Technique)

	Hazards/Dangers/Threats/In Water								
Treatment Method	**Solids**	**Odors/ Flavors**	**Algae**	**Protozoa**	**Bacteria**	**pH**	**Organic Chemical Residues**	**Salts, (Hardness)**	**Viruses**
Settling	✓								
Deaeration		✓					✓		
Activated carbon	✓	✓	✓[a]				✓	✓	
Neutralization						✓			
Membrane filtration (RO)	✓		✓	✓	✓		✓	✓	✓
Chlorination/ozonization				✓[b]	✓				✓[c]
Ion exchange							✓	✓	
UV radiation				✓[b]	✓				✓
Electrochemically activated water			✓	✓	✓				✓

[a]*If contamination is detected;* RO = *Reverse osmosis.*
[b]*Cryptosporadium.*
[c]*Depending on virus species.*

A water softener should be used to decrease hardness. It is a specific type of ion exchanger that is used to remove hardness. If bacteria are suspected, then a disinfection step such as ozone, chlorine, or ultraviolet treatment becomes necessary in the water treatment protocol. Filtration is recommended in all cases for all potable water use.

Filtration

Depending upon the contamination risk, the incoming water might be required to pass through a filter that retains solids (e.g., for well-water). The filtration methods vary from porous filters for larger particles to membrane filtration to retain smaller suspended particles and even microorganisms (reverse osmosis). The risk of microbiological growth needs to be considered when filters do not constantly receive water. Therefore, a continuous recirculation of the water is recommended, if no water is consumed.

Chlorination

The most common chemical oxidizing method used for disinfection of water systems is chlorination. It is used as hypochlorite solution (liquid bleach) and chloride dioxide. The major advantage of this method is that chlorine is relatively inexpensive and that automated systems and generator do not require a large capital investment. For an effective disinfection, a residual concentration of minimum 0.5 mg/L free chlorine for at least 30 minutes ($pH < 8$) should be maintained.

Hypochlorite reacts with the nitrogenous component of organic substances. This means it will be used up during this reaction. Automated systems for continuous water chlorination are available. The main disadvantages of hypochlorite are that it is highly corrosive in its undiluted form, may form unwanted by-products (chloramines, chlorophenol), and will be used up easily by organic matters.

Chloride-dioxide is a very unstable gas and must be generated at the point of use. It has some advantages over hypochlorite such as being less corrosive. The effectiveness to tackle biofilms is higher compared to hypochlorite and thus this is the preferred method for chlorination.

Electrochemically Activated Water

Treatment with electrochemically activated water has recently become increasingly popular. An electrochemical cell generates a highly oxidized fluid (anode) and a reduced fluid (cathode) using just water and salt. The cell may be separated by a diaphragm. The oxidized water has higher oxidation-reduction-potential (ORP) of up to +1300 mV, whereas, chlorine-based solutions have up to +800 mV. The solution is meta-stable for up to a couple of weeks and like chlorine-dioxide generated onsite. The disinfectant contains mainly hypochlorous acid, smaller amounts of chlorine-dioxide, and ozone. Small quantities of these will be dosed into the water system. Several manufacturers claim that this is a chemical-free method for water treatment. However, it is not true, even though concentration of chemicals is extremely low and the working principle of this method is based on the ORP rather than on the chemicals.

Ozone Treatment

Like chlorine-dioxide, it is not stable and must be generated onsite. It has a very high ORP above +2 volts and thus a high capability for disinfection with a broad spectrum of activity, including viruses. Due to its good water solubility, it is very suitable for the disinfection of water systems maintaining a concentration of 0.4 mg/L for 5 minutes (in the presence of spores 2 mg/L). The major disadvantages of this method are firstly that ozone will be used up easily by organic matter and secondly that the energy costs to produce ozone are fairly high. It should be noted that ozone should not be used in evaporative cooling systems due to its high volatility.

Ultraviolet Radiation (UV)

Microorganisms are inactivated by shortwave UV light (UV-C, 100–280 nanometer wavelengths). The nucleic acid, DNA of the microorganisms' cells is damaged during the treatment. The UV light source is usually enclosed in a transparent protective sleeve and installed in a way that water can pass through a flow chamber.

Suspended particles may protect microorganisms (shadow effect), so the effectiveness of the treatment very much depends on turbidity, adsorption, and concentration of particles and/or organic material. UV treatment could ensure up to a 4 log reduction; however, it will never lead to obtain sterile water. As the treatment is only done at the UV light source, but not throughout the system, removal of biofilms will not be possible. This treatment method is chemical free and it has a broad spectrum of activity. These are the advantages of the method.

In general, water quality controls should start at the source and also include the review of incoming and/or used municipal water supplies. The history of drinking water supplies should also be taken into account, such as known outbreaks related to the water supply or boiling water notices, when applying treatment options to ensure water quality. More detailed information on the effectiveness of treatments commonly used in the food industry can be found elsewhere (Koopmans and Duizer, 2004; Dawson, 2005; ILSI, 2008).

SOIL AND WATER QUALITY

Soils have significant influence on human health due to the amazing ability to remove contaminants from water. This power of water purification of soil is due to the removal of contaminants by physical capture as the water moves through micropore spaces in soil particles. Further removal is by chemical adsorption to solid surfaces, and by means of the biodegradation carried out by microorganisms living in the soil (Helmke and Losco, 2013). Soils that have desirable properties such as well-developed structure and sufficient organic matter act as efficient filters and purify the surface water as it percolates to lower strata. Soil degradation, which includes soil erosion and loss of soil structure and nutrient content, has negative impact on crop production and threatens food security (Brevik, 2013b; Mishra et al., 2016). In recent times, environmental pollution and food safety are two of the most important emerging issues. Soil and water pollution have historically impacted on food safety, and have serious important threats to human health. Heavy metals in soil, which is toxic to humans, usually pass on to humans through crop uptake and thus compromise food safety. The increasing negative effects on food safety from water and soil pollution have put more people at risk of carcinogenic diseases (Lu et al., 2015). In this context, soil can have great influence on human health. When the soil degradation occurs, it generally has negative health effects on human health. It may directly affect health through drinking water or indirectly through food that accumulate the pollutants such as heavy metals, organic chemicals, or soil pathogens.

Heavy Metals

The heavy metals of greatest concern relating to human health are arsenic, lead, cadmium, chromium, copper, mercury, nickel, and zinc. Heavy metals occur in soil naturally due to the weathering of rocks. In addition, they are also introduced into the soil through human activities. Heavy metals are mostly the by-products of mining ores. They are present not only in the soils of mines and in the immediate surroundings of metal processing plants but also in far-off urban areas (Horváth et al., 2015). Heavy metals also lixiviate into soils from landfills that contain industrial and household wastes and from sewage sludge that comes from wastewater treatment plants. In recent times, E-wastes, or wastes associated with electronic appliances, are assuming larger source of metals like lead, strontium, mercury, cadmium, and nickel in the soil (Robinson, 2009). Urban soils are susceptible to significant accumulations of heavy metals from activities like coal burning, automobile exhaust, erosion of metal structures, and refuse incineration (Horváth et al., 2015). In the agricultural sector, the use of fertilizers, manures, and pesticides have all contributed to the accumulation of heavy metals in soils (Mishra et al., 2016). Arsenic has been used in pesticides, and the buildup of arsenic in orchard soils has become a major problem as it is known to persist for decades (Walsh et al., 1977).

Organic Chemicals

Organic chemicals are deposited into the soil, both naturally as well as anthropogenically. A wide and diverse type of organic chemicals that are released into the air and water consequently end up in the soil. Soil pollution with organic chemicals is a serious problem in all nations (Lu et al., 2015). The largest source of these organic chemicals is from agricultural sector. Pesticides have been playing an important role in the success of modern food production since the 1950s. The excessive and indiscriminate uses of herbicides, insecticides, and nematicides result in soil and water pollution. It is true and widely accepted that the use fertilizers and pesticides has greatly improved grain production. However, on the other hand, inefficient, incorrect, and excessive uses of pesticides have also posed considerable risks to human health. Inadequate management of pesticide application in food production constitutes a potential occupational hazard for farmers and environmental risks for agricultural ecosystems.

Soil pollution with organic chemicals is not only limited to farming areas, as soils in urban areas are, in fact, still more polluted with organic chemicals as a result of industrial activities, coal burning, motor vehicle emissions, waste incineration, sewage, and solid waste dumping (Mishra et al., 2016). Soil contamination of both agricultural and urban areas includes a complex mixture of organic chemicals, metals, microorganisms, and the growth caused by municipal and domestic septic system waste, farm animal waste, and other biowastes. In addition, more recently pharmaceutical waste derived from antibiotics, hormones, and antiparasitic drugs manufacturing factories are becoming rising sources of pollution (Albihn, 2001).

Polyhalogenated biphenyls, aromatic hydrocarbons, insecticides, herbicides, fossil fuels, and the by-products of fossil fuel combustion are some of the most common types of organic chemicals found in soil (Burgess, 2013). These organic chemicals are present in highly diluted form in the upper strata of soil. They form complex chemical mixtures by reactions due to microorganisms. Very scanty toxicological information is available regarding the health effects of these chemical mixtures (Brevik and Burgess, 2014). As they have very long half-lives, many such organic chemicals are referred to as "persistent organic pollutants." Several such organic chemicals resist decomposition in the environment and get concentrated in higher consumers by bioaccumulation as they move up the food chain.

Soil Pathogens

In general, soil is host of innumerable organisms. There is great biodiversity of soil organisms. The majority of these are useful and are not always harmful to humans. However, soil also serves as a home for many pathogenic organisms. More than 400 genera of bacteria have been identified together with as many as 10,000 species. These bacteria, with the exception of viruses, in most cases, are more abundant than any other organism in soils (Nieder et al., 2018). Around 300 soil fungi species, out of the vast numbers of more than 100,000 fungi species that occur in soils, are known to cause disease in humans (Bultman et al., 2005; Nieder et al., 2018). For instance, the soil fungus *Exserohilium rostratum* caused a fungal meningitis outbreak in the United States in 2012 (Brevik and Burgess, 2013). Most of the protozoans that are found in soil feed on organic matter, including bacteria and algae; however, a few of these cause human parasitic diseases like diarrhea and amebic dysentery (Brevik et al., 2018). Soil serves as a common reservoir for helminthes. Helminths are parasites that may inhabit the human intestines, lymph system, or other tissues. Helminthic diseases require a nonanimal intermediate host like humans for transmission. Billions of people are infected by helminths worldwide each year, with an estimated 130,000 deaths annually (Abdeltawabi et al., 2017). Helminth infections generally occur via ingestion of contaminated water and food, and in most cases it involves intestinal infection (Nieder et al., 2018). Soil is not a natural reservoir for viruses; however, viruses are known to survive and stay in a dormant state for years together in soil. Pathogenic viruses are generally introduced into soil via human septic or sewage waste. Viruses that cause communicable diseases like conjunctivitis, gastroenteritis, hepatitis, polio, aseptic meningitis, or smallpox have all been detected in soil (Bultman et al., 2005; Hamilton et al., 2007; Nieder et al., 2018).

Globally, inadequate and improper sewage sanitation is a common problem in approximately 40% of the world's population (WHO, 2017b). As a consequence, millions of people die each year from water-borne diseases (Massoud et al., 2009). We must take the advantage of nature and the amazing purifying powers of soil to address wastewater issues. However, to get the necessary benefits, the soil must be of good quality and have good structure. Well-designed, properly maintained, and functioning onsite sewage treatment systems become highly effective in reducing the risk of water-borne diseases, especially in areas with low population densities (Massoud et al., 2009). Efficient use of soils to deal with and solve several groundwater contamination issues is not a recent idea. It has been rediscovered with the ongoing research and developments in soil sciences. In many countries, food safety policies are not integrated with soil and water pollution management policies (Lu et al., 2015). Therefore, it is very essential that a comprehensive map of both soil and water pollution threats to food safety should be prepared together with implementation of integrated policies addressing soil and water pollution. This should be an ideal holistic approach in achieving food safety.

WATER AND GENETICALLY MODIFIED FOOD

Worldwide corporate companies in developed countries produce genetically engineered foods, often called "genetically modified organism" (GMO), by using biotechnology techniques. Instead of using the natural breeding methods that farmers have used for centuries to select for desirable traits, GMO crops and GMO animals are actually radically altered genetically to repel pests, withstand droughts, and grow faster. These changes are patented and help corporations to increase their control over the seed supply. However, they rarely have any interests of people or the environment in mind. The relevance of genetic manipulations in the food production may not be overlooked. In view of the present scenario of water use and availability for primary food production, in majorities of the countries in the world, it becomes a necessity to develop more water efficient crops and livestock.

Genetically modified crops have become increasingly popular since the last decade. Although it is a highly controversial topic, it may be viewed and if ethically used, it is an emerging highly promising technology. This genetic technology, if carefully regulated and tested, should have much-desired beneficial effects in terms of water conservation. The most logical approach in the genetic modification, for example, should be the selection of the traits to increase the rate of photosynthesis, depth of root structure, as well as a decrease the transpiration rate at which water is lost. This strategy should increase the nutrients yield of the crop and at the same time decrease the water requirement of the staple cereal crops. These changes have the great positive potential to lower pressures on the amount of global water resources required in food production. The enormous potential and the promising possibilities of genetic engineering and biotechnology research and development have the better future solutions to the current problems of water conservation, specifically in primary food production. Therefore, more precise and highly controlled and targeted genetic modifications should be very helpful in wheat, rice, and maize to increase their water efficiency and drought resistance. This approach will work more effectively by implementing genetically altered crops into the advanced and water efficient agricultural system.

Rice is well known as a very water-intensive food crop. The gene Arabidopsis HARDY is identified to have increased water efficiency of rice by increasing the rate of photosynthesis and decreasing the amount of water loss through transpiration (Karaba et al., 2007). Modification of this gene has also been demonstrated to increase the strength and amount of root structure. Plants with the HARDY gene have shown a 55% greater photosynthesis rate under normal conditions (Karaba et al., 2007).

Genetic modifications of the staple foods like the wheat plant are very much needed for water deficient countries. Wheat can be genetically modified to have deeper root structure by extending the vegetative growth period of the plant for later-flowering genotypes. Deeper root systems facilitate more water uptake, which means the plants require less irrigation and perform better under drought conditions. One example of drought tolerant wheat is SeriM82, also known as "stay-green wheat." This new variety has deeper root systems that lead to greater water uptake. It has been shown that this increased water uptake during drought periods gives 14.5% increase in yield compared to the traditional variety (Manschadi et al., 2006).

Although the FDA contends that there is no scientific evidence to demonstrate that consumption of GMO foods can cause chronic harm, in this context, the process of the FDA for evaluating the safety of these controversial GMO foods appears to be completely inadequate and unreliable. Recently, several media reports appeared supporting the safety of GMO foods. They all ignore growing concerns of the scientific community that say that GMO foods are far from safe. The scientific community criticizes the weak regulatory bodies that grant approval to GMO crops without safety concerns of humankind. The underlying truth of all the controversies is that there is no consensus on the safety of GMO foods. In fact, the evidence shows a serious potential for harm, particularly to our environment.

The GMO controversy does not end with GMO crops. GMO animals are now entering in the field. Recently, genetically engineered salmon has been approved by the FDA. It is feared that this GMO fish could present serious risks to consumer health, animal welfare, wild fish populations, fishing economies, and the environment.

Despite the approval of many genetically engineered foods by the FDA, questions persist about the safety of eating them. Safety concerns should be raised by strong public resistance and halting all sales of genetically engineered foods until the safety questions are scientifically addressed. Therefore, consumers should have a guarded approach in the consumption of GMO foods. When GMO food is used, it is mandatory that one should review the basic principles of food safety at every step of its production from raw material up until final consumption of the finished product.

CONCLUSION

Water plays a major and fundamental role in the safety of food production. Water is a critical resource for the food industry. In food production processes, water quality and its impact on products and operations are generally underestimated. Such underestimation often leads to mismanagement of water, equipment operation and maintenance issues, loss of income, food product quality, and safety. The agricultural sector is the largest consumer of water, compared to any other sector. The water footprint of any animal food product is larger than the water footprint of a plant product with equivalent nutritional value. In the world, 29% of the total water footprint of the agricultural sector is related to the production of animal products. The production of meat requires between 6 and 20 times more water as compared to cereals, vegetables, or fruits. The global meat production has almost doubled since 1980, and this trend is likely to continue given the projected doubling of meat production up to 2050. The shift from traditional extensive and mixed farming to industrial farming systems is likely to continue to meet this rising demand for animal products. Due to the larger dependence on concentrate feed in industrial systems, intensification of animal production systems will result in increasing blue and gray water footprints per unit of animal product. The pressure on the global freshwater resources will therefore increase both due to the increasing meat consumption together with the increasing blue and gray water footprint per unit of meat consumed. Therefore, the biggest opportunity and challenge for future water management for us is to store more green water in soil and plants, as well as to increase the storage as blue water.

The "water activity" is a key parameter in food storage and preservation. This important property is used to predict the stability and safety of food in relation to microbial growth, rates of deteriorative reactions, and chemical/physical changes in the food products. Food designers ingeniously use water activity values to formulate shelf-stable food. If a product is kept below a certain value of water activity, mold growth is inhibited. This results in a longer shelf-life.

The quality of water is the fundamental consideration in any food production process. The quality depends on the source of water and the actual application in food production. The pollution of water occurs at any stage in various ways. Physical, chemical, or biological agents cause significant health hazards in safe production of food. Contaminations of water by several pathological microorganisms create major challenges in food safety and jeopardize human and animal health. Water must be free from undesirable taste, odor, color, and impurities including pathogenic organisms that could be harmful to consumers and product quality, when water is used as an ingredient in food. The quality of water must adhere to the safe drinking water standard.

The quality of the water is greatly dependant of the source. It is the source of water that generally determines quality. The chemistry of waters from different sources may differ significantly. The pH and various impurities in water can drastically alter the effectiveness of water used in several processes in food production. Water hardness is the most important chemical property, which has profound direct effect on cleaning and sanitizing efficiency. Adequate treatment of the water is required to ensure that it meets drinking water standards and is safe to be used in food production, which is safe for human consumption. Various techniques and methods of water treatments should be used to purify water as per the guidelines of government authorities to improve food safety.

Water quality can be compromised by the presence of infectious agents, toxic chemicals, and radiological hazards. These hazards may arise due to poor quality of water used directly or indirectly in the food production. The majority of water-borne diseases by pathogenic microorganisms are transmitted through contaminated freshwater. Various forms of water-borne diarrheal diseases are the most prominent examples and affect mainly children in developing countries. An estimated 4.1% of the total DALYs global disease burden is due to such diseases. This also results in about 1.8 million human deaths annually. Therefore, great attention is necessary to learn various health hazards such as biological, chemical, or physical pollutants in water that jeopardize human health. Foods can become contaminated at any point along the production line. Control programs such as Hazard Analysis and Critical Control Points help in maintenance of food safety through the analysis and control of biological, chemical, and physical hazards from raw material production until the finished product. It is implemented in food industries to reduce food safety risks.

Soil has multiple roles in deciding water quality. Good quality of soil has better ability to remove all types of contaminants of water, whereas deteriorated soil becomes a source of heavy metals, organic chemicals, and soil pathogens in water contamination. Water quality controls should start at the source and should also include the review of incoming/used municipal water supplies. It is very essential to build a comprehensive map of both soil and water pollution threats to food safety together with implementation of integrated policies addressing soil and water pollution, which should be an ideal holistic approach in achieving food safety.

Use of genetically modified food is gaining much popularity recently. Even though a controversial topic, great care must be taken when GMO food is used. It is very essential that one should review the basic principles of food safety. Thus, there are several challenges in using water for the production of safe and secure food. One must keep a wide vision and rigid policies for use of proper water in food production for safety of human life.

References

Abdeltawabi, M.S., Seddik, N.E., Salem, H.K., 2017. World wide epidemiology of helminths infection. In: Rodrigo, L. (Ed.), Human Helminthiasis. IntechOpen, Rijeka. Available from: https://doi.org/10.5772/67273.

Albihn, A., 2001. Recycling biowaste-human and animal health problems. Acta Vet. Scand. 95 (Suppl.), 69–75. Available from: https://doi.org/10.1186/1751-0147-43-S1-S69.

Aquastat, 2014. <http://www.fao.org/nr/aquastat> (accessed 30.10.18.).

Brevik, E.C., 2013b. The potential impact of climate change on soil properties and processes and corresponding influence on food security. Agriculture 3, 398–417. Available from: https://doi.org/10.3390/agriculture3030398.

Brevik, E.C., Burgess, L.C., 2013. The 2012 fungal meningitis outbreak in the United States: connections between soils and human health. Soil Horizons 54. Available from: https://doi.org/10.2136/sh12-11-0030.

Brevik, E.C., Burgess, L.C., 2014. The influence of soils on human health. Nat. Educ. Knowl. 5 (12), 1.

Brevik, E.C., Pereg, L., Joshua, Steffan, J.J., Burgess, L.C., 2018. Soil ecosystem services and human health. Curr. Opin. Environ. Sci. Health 5, 87–92. <https://doi.org/10.1016/j.coesh.2018.07.003> (accessed 28.2.19.).

Bultman, M.W., Fisher, F.S., Pappagianis, D., 2005. The ecology of soil-borne human pathogens. In: Selinus, O. (Ed.), Essentials of Medical Geology. Elsevier, Amsterdam, pp. 481–511.

Burgess, L.C., 2013. Organic pollutants in soil. In: Brevik, E.C., Burgess, L.C. (Eds.), Soils and Human Health. CRC Press, Boca Raton, pp. 83–106.

Castilla, J., Barricarte, A., Aldaz, J., García Cenoz, M., Ferrer, T., Pelaz, C., et al., 2008. A large Legionnaires' disease outbreak in Pamplona, Spain: early detection, rapid control and no case fatality. Epidemiol. Infect. 136 (6), 823–832.

CFIA, 2017. Safe water for food processing. Canadian Food Inspection Agency, modified: 2013-01-08, < http://www.inspection.gc.ca/food/non-federally-registered/safe-food-production/safe-water/eng/1357705718160/1357706182889> (accessed 10.11.18.).

Codex, 2003. Codex Alimentarius Commission. 2003. "Hazard Analysis and Critical Control Point (HACCP) System and Guidelines for Its Application." Annex to CAC/RCP 1-1969 (Rev. 4-2003).

Dawson, D., 2005. Food borne protozoan parasites. Int. J. Food Microbiol. 103 (2), 207–227.

De, A., Bose, R., Kumar, A., Mozumdar, S., 2014. Targeted Delivery of Pesticides Using Biodegradable Polymeric Nanoparticles. Springer Briefs in Molecular Science. Springer, New Delhi, India. Available from: https://doi.org/10.1007/978-81-322-1689-6.

EU-OSHA, 2005. European Guidelines for Control and prevention of travel associated Legionnairesdisease. European Agency for Safety and Health at Work. <http://www.hpa.org.uk/webc/HPAwebFile/HPAweb_C/1274093149925> (accessed 11.11.18.).

FAO, 2002. Crops and Drops. Food and Agriculture Organization of the United Nations, Rome, Italy, <http://www.fao.org/docrep/005/Y3918E/y3918e02.htm#TopOfPage> (accessed on 6.11.2018).

FAO, 2003. Agriculture, Food and Water. Food and Agriculture Organization of the United Nations, Rome, Italy, <http://www.fao.org/docrep/006/Y4683E/y4683e07.htm#P0_0> (accessed 8.11.18.).

Fontana, A.J., 2000. Water Activity's Role in Food Safety and Quality. Second NSF International Conference on Food Safety, Savannah, GA, October 11-13, 2000.

Fontana, A.J., 2001. Water Activity's Role in Food Safety and Quality. INSIDE MICROBIOLOGY | Food Safety Magazine, February/March 2001. <https://www.foodsafetymagazine.com/magazine-archive1/february-march-2001/water-activitye28099s-role-in-food-safety-and-quality/> (accessed 10.9.18.).

Gorham, J.R., Zurek, L., 2006. Filth and other foreign objects in foods: a review of analytical methods and health significance. In: Hui, Y. (Ed.), Handbook of Food Science Technology and Engineering. Taylor and Francis, CRC Press, Boca Raton, FL.

Hamilton, A.J., et al., 2007. Wastewater irrigation: the state of play. Vadose Zone J. 6, 823–840. Available from: https://doi.org/10.2136/vzj2007.0026.

Hatami, H., 2013. Importance of water and water-borne diseases: on the occasion of the World Water Day (March 22, 2013). Int. J. Prev. Med. 4 (3), 243–245.

Helmke, M.F., Losco, R.L., 2013. Soil's influence on water quality and human health. In: Brevik, E.C., Burgess, L. C. (Eds.), Soils and Human Health. CRC Press, Boca Raton, FL, pp. 155–176.

Horváth, A., Szűcs, P., Bidló, A., 2015. Soil condition and pollution in urban soils: evaluation of the soil quality in a Hungarian town. J. Soils Sediments 15, 1825. <https://doi.org/10.1007/s11368-014-0991-4> (accessed on 28.2.19.).

ILSI, 2008. Considering Water Quality for Use in the Food Industry. ILSI Report, April 2008. International Life Sciences Institute; Europe Expert Group on Water Safety, Brussels, Belgium.

Karaba, A., Dixit, S., Greco, R., Aharoni, A., Trijatmiko, K.R., Marsch-Martinez, N., et al., 2007. Improvement of water use efficiency in rice by expression of HARDY, an Arabidopsis drought and salt tolerance gene. Proc. Natl. Acad. Sci. <http://www.pnas.org/content/104/39/15270.short> (accessed 28.11.18.).

Kirby, R.M., Bartram, J., Carr, R., 2003. Water in food production and processing: quantity and quality concerns. Food Control 14 (5), 283–299.

Koopmans, M., Duizer, E., 2004. Foodborne viruses: an emerging problem. Int. J. Food Microbiol. 90 (1), 23–41.

Kuchenmuller, T., Hird, S., Stein, C., Kramarz, P., Nanda, A., Havelaar, A.H., 2009. Estimating the global burden of foodborne diseases—a collaborative effort. Euro. Surveill. 14, 191–195. PMID: 19422776.

Lu, Y., Song, S., Wang, R., Liu, Z., Meng, J., Sweetman, A.J., et al., 2015. Impacts of soil and water pollution on food safety and health risks in China. Environ. Int. 77, 5–15.

Manschadi, A.M., Christopher, J., deVoil, P., Hammer, G.L., 2006. The role of root architectural traits in adaptation of wheat to water-limited environments. Funct. Plant Biol. 33, 823–837.

Massoud, M.A., Tarhini, A., Nasr, J.A., 2009. Decentralized approaches to wastewater treatment and management: applicability in developing countries. J. Environ. Manage. 90, 652–659. Available from: https://doi.org/10.1016/j.jenvman.2008.07.001.

Mead, P.S., Slutsker, L., Dietz, V., McCaig, L.F., Bresee, J.S., Shapiro, C., et al., 1999. Food-related illness and death in the United States. Emerg. Infect. Dis. 5, 607–625. Available from: https://doi.org/10.3201/eid0505.990502.

Mekonnen, M.M., Hoekstra, A.Y., 2010. The green, blue and grey water footprint of farm animals and animal products, Value of Water Research Report, Series No. 48, UNESCO-IHE, Delft, the Netherlands.

Mishra, R.K., Mohammad, N., Roychoudhury, N., 2016. Soil pollution: causes, effects and control. Van Sangyan, 3 (1), Tropical Forest Research Institute, Jabalpur, MP, India. <https://www.researchgate.net/publication/289281444> (accessed 28.2.19.).

Nieder, R., Benbi, D.K., Reichl, F.X., 2018. Soil as a transmitter of human pathogens. Soil Components and Human Health. Springer, Dordrecht, pp. 723–827.

Osmonics, 1997. Pure Water Handbook. Osmonics Inc., Minnetonka, MN. Available at <http://dwb4.unl.edu/Chem/CHEM869A/CHEM869AMats/PureWater/pwh-s.pdf>.

Renault, D., 2002. Value of Virtual Water in Food: Principles and Virtues. UNESCO-IHE Workshop on Virtual Water Trade, 12-13 December 2002, Delft, the Netherlands.

Robinson, B.H., 2009. E-waste: an assessment of global production and environmental impacts. Sci. Total Environ. 408, 183–191. Available from: https://doi.org/10.1016/j.scitotenv.2009.09.044.

Sandulachi, E., 2012. Water Activity Concept and Its role in Food Preservation. Available at <http://www.utm.md/meridian/2012/MI_4_2012/8_Art_Sandulachi_E_Water.pdf> (accessed 12.10.18.).

Sandulachi, E.I., Tataro, P.G., 2012. Water activity concept and its role in strawberries food. Chem. J. Mold. 7 (2), 103–115.

Scallan, E., Hoekstra, R.M., Angulo, F.J., Tauxe, R.V., Widdowson, M.A., Roy, S.L., et al., 2011. Foodborne illness acquired in the United States—major pathogens. Emerg. Infect. Dis. 17 (1), 7–15.

Schmidt, R.H., 1997. Basic Elements of Equipment Cleaning and Sanitizing in Food Processing and Handling Operations. FS14, (Reviewed March, 2009). University of Florida. <http://edis.ifas.ufl.ed> (accessed 15.10.18.).

Steinfeld, H., Gerber, P., Wassenaar, T., Castel, V., Rosales, M., de Haan, C., 2006. Livestock's Long Shadow: Environmental Issues and Options. Food and Agriculture Organization, Rome, Italy.

UNEP, 2016. A Snapshot of the World's Water Quality: Towards a Global Assessment. United Nations Environment Programme, Nairobi.

Veolia/IFPRI, 2015. The Murky Future of Global Water Quality: New Global Study Projects Rapid Deterioration in Water Quality. International Food Policy Research Institute, Washington DC/Chicago, Ill.

Walsh, L.M., Sumner, M.E., Keeney, D.R., 1977. Occurrence and distribution of arsenic in soils and plants. Environ. Health Perspect. 19, 67–71.

Washington, C.W., 1996. Medical microbiology. In: Baron, S. (Ed.), Legionella, Chapter 40, fourth ed University of Texas Medical Branch, Galveston (TX).

WHO, 2013. Water Sanitation Health, Burden of disease and cost-effectiveness estimates. <http://www.who.int/water_sanitation_health/diseases/burden/en/index.html> (accessed 30.01.13.).

WHO, 2015. WHO Estimates of the Global Burden of Foodborne Diseases: Foodborne Disease Burden Epidemiology Reference Group 2007-2015. World Health Organization, Geneva.

WHO, 2017a. Guidelines for Drinking-Water Quality: 4th edn, Incorporating the 1st Addendum. World Health Organization, Geneva.

WHO, 2017b. WHO Methods and Data Sources for Global Burden of Disease Estimates 2000-2015. World Health Organization, Geneva.

WWDR, 2017. World Water Development Report 2017: Wastewater: The Untapped Resource. Facts and Figures. (United Nations World Water Assessment Programme)/UN-Water. UNESCO, Paris.

WWDR, 2018. World Water Development Report 2018: Nature-Based Solutions for Water. (United Nations World Water Assessment Programme)/UN-Water. UNESCO, Paris.

Further Reading

Brevik, E.C., 2013a. Soils and human health: an overview. In: Brevik, E.C., Burgess, L.C. (Eds.), Soils and Human Health. CRC Press, Boca Raton, pp. 29–56.

Hakami, B.A., 2015. Impacts of soil and water pollution on food safety and health risks. Int. J. Civil Eng. Technol. 6 (11), 32–38.

Leake, J.R., Adam-Bradford, A., Rigby, J.E., 2009. Health benefits of "grow your own food" in urban areas: implications for contaminated land risk assessment and risk management. Environ. Health 8 (S1), 1–6. Available from: https://doi.org/10.1186/1476-069X-8-S1-S6.

Pimentel, D., Burgess, M., 2013. Soil erosion threatens food production. Agriculture 3, 443–463. Available from: https://doi.org/10.3390/agriculture3030443.

CHAPTER

10

Safety of Fresh Fruits and Vegetables

Charu Gupta and Dhan Prakash

Amity Institute for Herbal Research & Studies, Amity University, Noida, India

OUTLINE

Food Safety and Human Health
DOI: https://doi.org/10.1016/B978-0-12-816333-7.00010-2

INTRODUCTION

In today's health conscious society, vegetables and fruits play an important role and are an integral part of human diet. They provide essential vitamins, minerals, dietary fiber, and beneficial phytochemicals and offer reduced risk from several dangerous diseases. In many cases, they are associated with many foodborne diseases, resulting in safety threats (Linlin et al., 2016). Foodborne outbreaks related to the consumption of contaminated fresh fruits and vegetables occur frequently, despite the implementation of good agricultural and good manufacturing practices (GMP), control measures, and strict regulations to ensure food safety. Therefore, alternative and cost-effective technologies to prevent foodborne illnesses are necessary.

Fresh fruits and vegetables are a vibrant and important part of the human diet. Consumer confidence is eroded in particular by poor hygienic conditions in growing, handling, or transporting the produce (Oluoch, 2006). In various countries worldwide, fresh vegetables or fruit salads were contaminated by Escherichia *coli* O157:H7 through cow manure, irrigation water, flood water, or overload of organic material either at washing, growing, storage, or transport. Yeasts and molds invade and grow on grains, nuts, beans, and fruits in the field before harvest, during storage, and in processed foods (Pao and Petracek, 1997). In spoilage, fruit becomes too soft due to action of pectolytic enzymes deposited on them by microorganisms and they impart an unpleasant taste and smell due to rotting. Yellow fruit turns grayish due to the infection of mold or the juice becomes too sour due to fermentation (Thomas and O'Brien, 2000). The primary reasons for the variation of pathogens and fresh produce types involved in outbreaks are likely linked to various factors: the increased supply of fresh produce globally, higher consumption, aging population, and possible changes in climate (Tirado et al., 2010).

Degradation and inadequate food quality may result in rejection by consumers and reduction in sales, and food safety hazards may be concealed and go undetected until the product has been consumed. If detected, serious food safety hazards may result in market access exclusion and major economic loss and costs. Because food safety hazards directly affect public health and economies, achieving proper food safety must always take superiority over achieving high levels of other quality attributes. Furthermore, food safety and quality control processes should always be focused to prevent problems at the ground root level itself, not on simply curing them. Once product quality has been undermined, it is virtually impossible to restore.

Fresh fruits and vegetables are extremely perishable produce that can easily spoil or deteriorate during postharvest procedures of handling in the supply chain from the end of producer to the final consumer. Generally, spoilage and deterioration of fresh produce result in rapid decay and thus cause the product to be unsuitable for human consumption. Sometimes the postharvest losses of fresh product due to spoilage can be up to 50%, and even more in some delicate commodities. Therefore, reduction of such losses, particularly to avoid economically, would be of great importance for both producers and consumers alike. Fresh food products remain one of the leading causes of foodborne disease outbreaks, exceeding the classic carriers for pathogens like meat, dairy products, and seafood (Centers for Disease Control, 2017). Since 1990, there have been more than 400 outbreaks related to fresh produce. Sprouted seeds like alfalfa, cloves, and mung beans have been very often involved in foodborne disease outbreaks associated with *Salmonella*, shiga toxin-reducing *E. coli* (STEC), or *Listeria monocytogenes* (Callejón et al., 2015; Nuesch-Inderbinen and Stephan, 2016).

Therefore, strict food standards are necessary to maintain uniformity of product quality and safety, to gain market access, and to establish market presence. Further, to provide basic safety information to consumers about the product and to prevent economic fraud or market exclusion, guidelines should be followed. Standardization methods permit for correct food labeling forms and enhance the basis for consumer confidence. It is quite evident that maintaining the freshness of fruits and vegetables requires the efficient transport and storage of highly perishable horticultural produce. Their high rate of perishability is due to the fact that after harvest, fruits and vegetables maintain their physiological systems and continue with their metabolic activities. Respiration and transpiration lead to the consumption of main substrates like sugars and organic acids and loss of water accompanied by ripening and senescence, ultimately making the fresh produce unsuitable for marketing. In developed countries, access to cutting-edge postharvest technology is readily available to reduce losses and wastage while upholding food safety and quality norms. Traditionally, cold storage, like cellars, basements, caves, and ice houses, have been used to preserve fresh produce (Paull, 1999). In principle, fruit and vegetables should be cooled to remove heat: before processing, transporting, and storing (Dincer, 2017). Currently, techniques like refrigeration are more cost effective, sustainable, and consume less energy. They can be used as centralized systems that operate at a wide range of temperatures and respond effectively to the working temperature changes. At reduced storage temperature, the enzymatic, respiratory, and metabolic activities are decreased, which enhances the shelf life of produce. Processing fruit by drying method causes losses up to 88% of water to a level of 18%–25%, which is safe for storage and also results in increased intense flavor (Farkas, 2001). Canning process involves first washing of the fruits or vegetables, peeling, blanching, cutting into pieces or comminuting as pulp or juice followed by sterilization before being placed in the can. Synthetic chemical preservatives like benzoic acid are also used very frequently (The Schumacher Centre for Technology and Development, 2008).

Nowadays, there are several novel preservation technologies that have been developed for the safety of fruits and vegetables, like storage at modified or controlled temperature, preservation based on pressure modification methods such as high hydrostatic pressure (HHP), hyperbaric treatment, vacuum cooling, vacuum packaging, hypobaric storage, and focusing on their application to store fresh fruits and vegetables. In addition to these, some alternative preservation technologies, by using microbial metabolites, including bacteriocin (produced by lactic acid bacteria) that exert antagonistic activity against important foodborne pathogens, are also utilized. Some bacteriocins are already in practice in the food industry as natural antimicrobials to prevent the spoilage of various food products.

ORIGIN OF CONTAMINATION IN FRUITS AND VEGETABLES

The fresh produce production in open natural conditions is more susceptible to contamination from several sources. Various reviews on the source of contamination from soil, water, biological change, and activity of wild animals have been reported, which can introduce pathogens in humans (Warriner et al., 2009; Olaimat and Holley, 2012; Goodburn and Wallace, 2013; Martínez-Vaz et al., 2014; Nuesch-Inderbinen and Stephan, 2016). Under normal ideal conditions, soil, water, and biological amendments should be pathogen-free, thereby preventing contamination of pathogens. However, in fact, such type of pathogens can survive for longer time in the environment and become widely distributed to impart infection (Yang et al., 2012; Schwarz et al., 2014). Some pathogens like *Salmonella* can be even found in greenhouse operations that would have been considered as a controlled environment (Holvoet et al., 2014).

It is now accepted that pathogens are likely to be encountered, and tolerance limits for fecal indicator bacteria in water and soil have been proposed (Jongman and Korsten, 2017). However, testing in detecting contamination and in many instances mitigation strategies available should be adopted (Pagadala et al., 2015). Consequently, a philosophy has been postulated that contamination via water or manure amendments is very likely but it can be mitigated by allowing sufficient time to elapse before harvest, by which pathogens, if present, might have perished (Moynihan et al., 2013). Practically, this is the basis for 90–120 days rule, that when manure or biological amendment is applied onto land, at least 120 days should intervene before harvest (Xu et al., 2016). In another recommendation, it is advised not to irrigate crops before 2–7 days prior of harvest (Weller et al., 2015). In both the cases of manure improvements and irrigation, it is assumed that enteric pathogens will die-off in the environment or plant, resulting in reduced risk of infection. The waiting periods are largely based on laboratory trials under controlled conditions that monitored pathogens such as *E. coli* O157:H7 and *Salmonella* decrease over a time period of irrigation.

In several other studies it has been observed that a major proportion of pathogen populations died within 1–10 days of intervention (Astrom et al., 2006; Oliver et al., 2006; Liang et al., 2011; Erickson et al., 2014; Oladeinde et al., 2014; Généreux et al., 2015). However, in fact, the observed effect on pathogens is more related to nonculturable induced into a dormant state (Ayrapetyan et al., 2015). Another possible source of contamination at preharvest stage is from workers that might have been significant for transferring parasites such as norovirus and enteric protozoan (Bouwknegt et al., 2015; Jensen et al., 2017). Infection through contact surfaces like knives and reusable crates has also been identified with potential carrier of contamination between different batches of produce (Zilelidou et al., 2015). Further, a possible route of contamination into inner plant tissues is through hydro-cooling or vacuum cooling (Li et al., 2008). Under these processes, the product is packed into crates and placed in a sealed chamber before applying vacuum to draw air and remove moisture. The contamination on the surface of plants can also be affected through stomata and cut edges (Li et al., 2008). Once infected, the pathogens can be protected from stresses imposed during postharvest operations (Jablasone et al., 2005).

DETERIORATION, SPOILAGE, AND AFTER-HARVEST LOSSES

All fruits and vegetables are living parts of plants and contain around 65%–96% water. Their metabolic activity is continued even after harvesting and thus can change their characteristics. The effect depends on product handling, storage, and treatment, all of which

TABLE 10.1 Major Causes of Postharvest Losses for Different Groups of Fruits and Vegetables

Vegetables	Main Causes of Postharvest Losses and Poor Quality
Vegetables (Edible roots) include examples such as carrots, beets, onions, garlic, potatoes, sweet potatoes	Mechanical injury, improper curing, sprouting water loss, decay, chilling injury
Leafy vegetables examples such as lettuce, chard, spinach, cabbage, spring- onions	Mechanical injury, water loss and decay, relatively high respiration rates, loss of green color
Vegetables (Edible Flowers) examples such as artichokes, cauliflower, broccoli	Mechanical injury, water loss and decay, discoloration, abscission of florets
Immature fruit vegetables examples such as cucumbers, squash, okra, eggplant, peppers, snap beans	Bruising and other mechanical injury, water loss and decay, over-maturity at harvest, chilling injury
Mature fruit produce examples such as tomatoes, melons, bananas, mangoes, apples, grapes, cherries, peaches, apricots	Bruising and other mechanical injury, water loss and decay, over-ripeness at harvest, chilling injury

TABLE 10.2 Major Spoilage Factors and the Causes of Deterioration of Fresh Fruits and Vegetables

Types of Deterioration Factor	Causes
Chemical and biochemical factors include enzymatic, oxidation and nonenzymatic changes, and light oxidation.	Environment, handling and bruising; high oxygen concentration and availability; improper packaging, composition, heat; improper packaging.
Physical factors include bruising and crushing, wilting, texture and moisture change.	Improper handling and packaging; high relative humidity and improper packaging; environment and improper packaging; high relative humidity and improper packaging.
Biological and physiological factors include pests (e.g., insects, rodents, birds); spoilage microorganisms; respiration rate; ethylene production; growth and development; maturation, ripening, senescence; transpiration and water loss.	Inadequate good agricultural/manufacturing practices; inadequate hygiene and sanitation practices; excessive heat and high temperatures; environment conditions such as temperature and atmospheric pressure; time and environment, and improper packaging.

impact the shelf life of the product. The type and the nature of the plant produce strongly influences its vulnerability to different types of spoilage. The following, Table 10.1, gives an overview of the major causes of after-harvest losses for different important groups of fruits and vegetables (University of Maryland, 2002).

The food spoilage of fresh fruits and vegetables can occur because of biological, microbiological, physiological, biochemical, and/or physical factors. These factors arise due to the lack of proper training for product handlers, inadequate storage spaces, unequipped handling technologies, poor quality control, and adverse environmental conditions (Satin, 2000; Potter et al., 1995). Some of the major deterioration factors and their causes are discussed in Table 10.2.

CAUSES OF SPOILAGE IN FRUITS AND VEGETABLES

There are various causes of spoilage of fruits and vegetables that are associated with cultivation, harvesting, postharvest treatment methods, and processing protocols. The fresh produce is highly prone to damage if Good Agricultural Practices (GAP) and GMP are not properly implemented throughout the whole chain. If neglected, the safety and quality of final product might be adversely affected, which may pose a threat to the consumer's health.

Thus, it is necessary to first identify the potential hazards in the production environment to reduce the risk and to increase final produce safety. After the identification of the hazards, effective measures should be incorporated to control, reduce, or eliminate them.

Biological

The spoilage through biological methods in fruits and vegetables arises from the attack of various microorganisms that include bacteria, fungi (both yeasts and molds), protozoans, viruses, and helminths. The fruits and vegetables are more susceptible to microbial contamination if their outer protective peel is directly infested or injured by the pests that serve as a portal of entry for various microbes. Moreover, as microbes are ubiquitous, they exist in raw produce naturally and are capable of causing human disease. They may also gain entry through soil or surrounding contamination, and sometimes they are part of the fruit or vegetable natural microflora. In some cases, they might get introduced through improper food handling practices in agricultural production or during postharvest processing methods (CFSAN, 2004).

Thus, prevention of bacterial contamination is the most important control factor to enhance product safety because sometimes some bacteria have very low infective doses. Hence, it is essential to take assured measures to prevent the growth and reproduction of plant pathogens to prevent their growth to hazardous levels. The following, Table 10.3, lists the most important foodborne pathogens and their impact on human health.

Thus, the spoilage of fresh fruits and vegetables can be prevented by following these measures as described:

1. Keeping the bacteria from coming in contact with the surface of plant produce by following basic precautionary measures;
2. Preventing the growth and multiplication of bacteria on the plant produce by incorporating suitable physical parameters;
3. Ensuring GAP during handling of agricultural produce and their postharvest operations;
4. Ensuring GMP before final packing.

Similarly, spoilage of fruits and vegetables by parasites can cause serious health threats and diseases to humans. The following precautions should be taken to prevent and minimize the abundance of parasites on fresh fruits and vegetables at all stages of production:

1. Ensuring and preventing that no water contact or soil contaminated with human or animal feces should come in direct contact with fresh plant produce;

TABLE 10.3 Important Foodborne Pathogens and Their Effect on Human Health

Foodborne Pathogen	Effect on Human Health
Campylobacter spp.	Symptoms of *Campylobacter* infection, which usually occur within 2–10 days after the bacteria are ingested, include fever, abdominal cramps, and diarrhea (often bloody).
Clostridium botulinum	The toxin acts as a muscle paralyzer and leads to symptoms including double vision, muscle weakness, and eventually paralysis of the arms, legs, trunk, and respiratory muscles. Symptoms generally begin 18–36 h after ingestion of contaminated food.
Escherichia coli O157:H7	Infection causes severe bloody diarrhea and abdominal cramps.
Listeria monocytogenes	Common effects of *Listeriosis* are fever, muscle aches, and serious gastrointestinal symptoms. If infection spreads to the nervous system, symptoms such as headache, stiff neck, confusion, loss of balance, or convulsions can occur.
Salmonella spp.	*Salmonellosis* is an infection causing diarrhea, abdominal cramps, and fever within 8–72 h after ingestion of the contaminated food.
Shigella spp.	Common symptoms of *Shigellosis* include diarrhea, fever, and stomach cramps starting 1–2 days after exposure.
Staphylococcus aureus	The most common cause of contamination is improper hygiene during food handling.

2. Ensuring and preventing the contact of infected people in the form of product handlers with fresh plant produce;
3. Ensuring and prevention of contact between animals (pests) and fresh plant produce.

Chemical

Chemicals in the form of synthetic pesticides and fertilizers can pose a serious health hazard to the consumer if they are not used judiciously and can contaminate fresh fruit and vegetables in significant concentrations. Such contamination can occur either through naturally occurring substances or by synthetic chemicals that may be added or are already present during cultivation and/or postharvest treatment and processing. Table 10.4 highlights the common naturally occurring chemical hazards and their potential health risks for humans.

Thus, chemicals in any form (natural or artificial) can cause serious health conditions for consumers. It is important to make judicious use of chemical agents (e.g., agrochemicals, processing and treatment chemicals, packaging additives, pest control agents, and antibiotics, to name a few) and should prevent contamination during food product handling and processing by identifying potential risks and implementing adequate cultivation practices and their preventive measures.

Physical

Physical methods of contamination can be through introduction of foreign material into fresh fruits and vegetables at numerous points in the production chain. The presence of

TABLE 10.4 Naturally Occurring Chemical Hazards and Their Potential Health Risks for Humans

Types of Naturally Occurring Chemical Hazards	Health Risks for Humans
Allergens (e.g., weeds, peanuts)	Allergenic reactions
Fungal toxins (mycotoxins; e.g., aflatoxin)	Multiple poisonings (acute or chronic)
Phyto-haemagglutinin Alkaloids	Multiple poisonings (acute or chronic)
Agrochemicals (pesticides and fertilizers) contain toxic elements and compounds (e.g., lead, zinc, cadmium, mercury, arsenic, cyanide)	Multiple poisonings (acute or chronic)
Processing contaminants (e.g., lubricants, cleaning agents, sanitizers, coatings, paints, refrigerants and cooling agents, water/steam treatment chemicals, pest control chemicals)	Multiple poisonings (acute or chronic)
Persistent organic pollutants (POPs); examples are dioxins and polychlorinated biphenyls	Exposure to POPs may result in a wide variety of adverse effects in humans
Agents from packing material (e.g., plasticizers, vinyl chloride, adhesives, lead, tin)	Multiple poisonings (acute or chronic)

foreign material particles in fresh plant produce can cause serious injury and illness to the consumer. A majority of the food contamination arises due to poor handling practices such as harvesting, washing, sorting, and packing of plant products.

The following precautions should be taken to prevent the contamination of fresh plant produce:

- Correct identification of physical source of hazard along with the production chain step (both agriculture and postharvest processes);
- Ensuring proper implementation of practices and their countermeasures;
- Generating awareness and responsibility among the field workers.

Most of the foodborne diseases are associated with biological hazards (pathogenic bacteria) and the toxins generated as a result of their microbial activity. The highly perishable plant produce commodity can be preserved through refrigeration at low temperature to extend their shelf life. The storage of fresh plant produce at low temperature inhibits the growth of pathogenic bacteria. It is generally recommended to cool the products within 24 hours of harvesting.

Cooling Considerations

The plant harvesting should be preferably done at night or in the early morning hours to minimize exposure to high daytime temperatures. The harvested crop should be collected and shade dried under good ventilation conditions. Care must be taken to prevent cross-contamination by bird droppings if the plant produce is shade dried under trees. Drying under direct sunlight should be completely discouraged to prevent the spoilage of the plant produce. The produce should be promptly cooled after harvest. This results in the extension of the shelf life of products, the appearance is more attractive, and the products are of better quality. The amount of heat eliminated during the cooling step depends on the weight, specific heat, and initial and final temperature of the fresh plant produce. It should be carefully noted that both the cooling method and cooling medium should not contaminate the plant produce.

It is important to maintain sanitary conditions in the facility when using an air-based cooling system; for example, special attention should be given to the air source area. The air system should be properly maintained and the filters must be cleaned or changed regularly. The animals should be completely excluded from the surrounding areas, the compost storage spaces should be located far from air sources, and any other pathogen sources that could potentially contaminate the air of cooling systems should be completely eliminated. The water coolants using water or ice as the cooling mediums are the common medium for contamination of fruits and vegetables; therefore, their water should be replaced regularly depending on the amount used and produce conditions. It is essential that ice used in cooling should only be produced from chlorinated, potable water and stored under total sanitary conditions, so as to avoid cross contamination.

The water and ice used for cooling systems should be bacteria-free, which can be ensured through various microbiological tests (such as total coliforms, fecal coliforms, and *E. coli* tests) on water used in cooling and ice cooling systems. These are good indicators of bacteriological quality of water. Besides this, use of only chlorinated cold water and ice is recommended. As chlorine loses effectiveness when it reacts with organic compounds, its concentration should be monitored frequently in order to ensure its efficacy. A general recommendation is usually between 50 and 200 ppm chlorine concentrations are sufficient to destroy most vegetative cells of bacteria, but higher concentrations are needed to kill their spores. Furthermore, a water settling and filtration device should always be placed in the cooling water treatment system to remove organic material. Similarly, cooling equipment should be inspected and cleaned frequently. Thus, overall maintenance of equipment and use of appropriate sanitary procedures are critical to ensure the safety of the plant produce.

The foodborne pathogens can also be found in water handling systems such as dump tanks and flumes in which the water is recirculated, thereby contaminating the fresh plant produce (Sargent et al., 2000). Besides this, in some fruits and vegetables such as apples, celery, mangoes, and tomatoes, when such warm fruit or vegetables are placed in cold water, a differential pressure is generated that creates a suction effect, thereby resulting in infiltration of water into the fruit. Thus, more research is required to identify those plant produce that can bear cooling water infiltration. This problem to some extent can be solved through the use of good quality water for cooling. Hazard analysis and critical control point procedures should be stringently followed throughout the production process.

Showalter (1993) showed that in order to reduce potential produce contamination associated with water infiltration, the cooling/wash water temperature can be adjusted to 5°C and above the temperature of the flesh of the fruit. This could be an important precaution for washing systems; however, for cooling systems it interferes with the removal of field heat. Therefore, for such commodities, it was recommended to cool with air or other cooling methods or to combine hydro-cooling with an initial air cooling step to minimize the temperature differential between produce flesh and water temperature. The use of disinfectants, such as chlorine, in the cooling water could also help to reduce the risks associated with pathogen internalization.

A variety of other methods are commercially available to cool the fresh plant produce. However, it is important to know the principle of each cooling method so that the potential hazards associated with them can be identified and rectified. The other commonly used methods for reducing heat from fresh plant produce include room cooling, forced air cooling, hydro-cooling, package icing, and vacuum cooling.

In the method of room cooling, heat is transferred slowly from the mass of a product through convection, to the cold air being circulated around stacked containers of the fresh plant produce. Room cooling is used for a wide range of plant commodities but is usually a slow method and therefore not generally used. The other method of ceiling jet cooling is a slightly faster mode of room cooling where the ceiling jets direct cold air down directly over the stacked plant produce, thereby lowering the temperature and increasing their shelf life.

In the other method of forced air cooling as described by Droby (2006), the cooling air is pulled or pushed through the plant produce containers, providing greater air circulation around the plant produce and resulting in faster cooling. This method is commonly used on various crops such as grapes, berries, and other fruits.

Another method is hydro-cooling, a rapid cooling method that uses water showering down over the fresh plant produce as the cooling medium. This method is based on the principle that a pound of water can absorb more heat than a pound of air. Hydro-cooling can only be used for those plant produce and shipping containers that can tolerate wetting. Chlorine concentrations of 200 ppm (free chlorine) are generally used in hydro-coolers, and cooling water should be changed frequently. Hydro-cooling as described by Droby (2006) is used for those commodities that may be cooled in bulk or even in packed containers.

The other method of package icing is one of the oldest methods of plant produce cooling and is used on commodities that can tolerate contact with ice (e.g., root and stem vegetables, broccoli, brussels sprouts, etc.) where the direct contact of the plant produce with ice provides fast, initial conduction cooling. The method involves packing finely crushed or flaked ice over the packaged produce. An alternative process uses liquid ice (60% ice and 40% water) as the cooling medium. The advantage of liquid ice is that it gives a greater initial contact between the plant produce and the ice, and it can even be applied after the boxes have been palletized. It may be used to distribute ice around the plant produce in the shipping containers. The amount of ice added should be adjusted according to the initial plant produce temperature, produce weight, and the expected ambient temperatures during transit (Droby, 2006).

The recent method of cooling is vacuum cooling, in which the plant produce is placed in a strong, airtight steel chamber. Air is pumped out of the chamber to reduce the atmospheric pressure, causing the water in the plant produce to vaporize. Cooling occurs because the heat energy for vaporization comes from the produce. The cooling rate is related to the surface area to volume ratio of the produce. Thus, loose leafy vegetables cool faster than tight-headed cauliflower or celery. This method is used primarily for cooling leafy vegetables, celery, cauliflower, and to a limited extent, sweet corn, carrots, and sweet peppers. The only disadvantage of this method is that during cooling 1%

of the produce weight (primarily water) is lost for each 5°C–6°C drop in produce temperature (Holdsworth, 1985).

The other method, hydro-vacuum cooling, is a patented modification of vacuum cooling. It prevents the weight loss by providing a water shower at specific times during the cooling cycle. As with hydro-cooling, monitoring and maintaining water quality are also important during this process. Generally, vacuum coolers are portable and can be moved to different shipping points as the growing season progresses, but sometimes they may be large enough to hold an entire boxcar load of plant produce.

DEVELOPMENT OF BIOPRESERVATIVES FOR FRESH-HARVEST PRODUCE

Antisepsis and Preservation of Edible Flavor Vegetation

In ancient periods, antisepsis and fresh-keeping functions of edible flavor vegetation were applied for long times. Hoffman and Evans observed that mustard seeds, clove, and cassia have preventing fruits and vegetables from rotting. The addition of 0.5 g mustard seeds in 100 g of apple juice enhances its shelf life for 4 months. Edible flavor vegetation was found with antisepsis properties due to the presence of essential oil (Haisheng, 1993). They also found that essential oil in mustard seed, clove, cassia bark, *Elettaria cardamomum*, caraway seed, spices, and thyme all had more or less antisepsis function up to a certain extent (Gao et al., 2009).

Curry Leaf Essential Oil as a Biopreservative

Essential oil from aerial parts of Murraya can control and kill common food spoilage caused by molds. These were found as possible ideal raw materials of natural fresh-keeping agents in fruits. Vegetation like Zanthoxylum was widely used as food preservatives, anticorrosive with strong efficacy in controlling and killing function of fruit malignant disease-producing pathogens. Essential oil of citrus vegetation peel showed strong antagonistic effect against *Penicillium digitiatum* and *Penicillium italicum* causing orange blue and green molds (Gao et al., 2009).

Cinnamon Essential Oil as Biopreservative

The essential oil of *Cinnamomum* is a rich source of cinnamal, borneo camphor, 1,8-cineole, eugenyl methyl ether, linalool, and they exhibit good bactericidal property. Eugenyl methyl ether, cinnamal, and linalool may be used successfully in fruit fresh-keeping experiments (Gao et al., 2009).

Other Vegetation

In addition to these vegetation, several other flavor-containing plants had active ingredients with the potential of controlling and killing function to fruit malignant disease-producing pathogens. For example, *Ascitriodora, Eucalyptus*, Blue gum of *Eucalyptus* family Myrtaceae; Tsaoko, *Curcuma* of Amomum and *Curcuma* in Zingiberaceae were widely used as food anticorrosive and fresh-keeping agents. Further, Citronella, Lemongrass, Rue grass family Poaceae, *Acorusgramineus* Soland, *Anise calamus* in Araceae, perilla, Sweet basil in family Labiatae were also found to be helpful to increase the stability and shelf life of fresh fruits and vegetables (Gao et al., 2009).

Effect of Herbal Extracts on Shelf Life of Fruits and Vegetables

A wide range of medicinal and aromatic plants could be used as medicine and food preservatives. Their bioactive useful phytochemical ingredients extracted by decoction, alone and in combination with other natural drugs, were also used to treat fruits and vegetables to improve their shelf life and freshness.

Konjac (Amarphaluskonjac) *Extract as a Biopreservative*

Tuber of *Amarphaluskonjac* K. contained rich konjaku flour with konjakumannoside as main ingredient. *Konjakumannan* had no color, was completely innocuous, and had no peculiar smell. It had certain function to fruit fresh-keeping and antisepsis of fish and meat. The treatment involves the soaking of fresh strawberries in 0.05% *Konjakumannan* solution for 10 minutes followed by drying in air. The treated strawberries can be effectively stored for 1 week at room temperature, without permitting the growth of mildew and rot, even after 3 weeks of storage. On the other hand, untreated strawberries would lose their gloss and freshness after a period of storage for 2 days at room temperature and start to mildew and rot just after 3 days (Gao et al., 2009).

Lesser Galangal (Alpinia officinarum) *Extract as a Biopreservative*

The main antisepsis compound present in *Alpinia officinarum*is naphtha, comprises about 0.5%–1.5% and also includes 1,8-cineole, methyl cinnamate, clove oil, phenol pinene, cubebin, etc. Its main piquancy component is galangol and major yellow components are reported as galangin and kaempferide. Oranges are painted with its decoction on their surfaces and encased in wicker baskets. Such oranges can be stored under normal temperature for up to 95 days. It was reported that the rotting rate of treated oranges was only 7.8%, as compared to untreated oranges whose rotting rate was much higher of up to 37% (Yi, 2006).

Garlic (Allium sativum) *Oil Extract as a Biopreservative*

Allium sativum consists of garlic naphtha, in which the main ingredient was garlicin, which is produced by enzymatic activity of garlic. Allicin exhibit strong antifungal activity against anthracnose fungi, damping-off fungi, and *Rhizopusnigricans*. Preparation of garlic extraction includes soaking of fresh garlic pieces in cold water for about 12 hours, followed by boiling. In an experiment, 10% garlic extract was used for soaking hand-picked oranges for a period of about 10–15 minutes, dried in air, and packed boxes with some empty cases and hardboards. These boxes were stored in ventilated storehouses or common rooms. The shelf life of these oranges was prolonged for up to 70 days, which was more than 92.4% at room temperature (Gao, 1990a).

Inositol Hexakis-Phosphate (Phytic Acid) as a Biopreservative

Phytic acid a type of organic acid widely found in seeds and can be used as antisepsis and fresh-keeping of highly perishable fruits and vegetables and to prevent growth of fungal infection. Some melons and fruits, such as grapes, strawberries, Hami melons, bananas, and litchi could also be preserved by painting with these agents. Phytic acid was found to maintain their faint physiological activity, ideal air, water permeability, and also raises melon and fruit shine, enhances resistant ability from the invading pathogens and fungus, and inhibits enzyme activity. Edible fungi could be treated with phytic acid to prevent them from changing color, to solve the problem of residual SO_2, and enhance shelf life from 2–3 to 5–7 days. This method was also found effective in fresh cherries, which were difficult to store (Wu Fan, 2003). In addition, it has also been found that phytic acid has an antioxidant activity and efficiently prevents browning of fruits and vegetables.

Chinese Herbal Plant Extracts as a Biopreservative

Studies on Stemona, giant knotweed rhizome, galangal rhizome, and berberine extracts have shown promising potential to be effective for use as fresh-keeping agents. These extracts were blended with starch, konjaku, and lecithin to increase their natural semipermeability as film fresh-keeping agent. These constituents consist of starch, galangal rhizome, stemona, giant knotweed rhizome, konjaku, berberine, and lecithin. Apples treated with this mixture could be preserved for 6 months; tomatoes, mad apples, and cucumbers could be stored for 2 months at ambient temperature (Xuan, 2001; Qiulian et al., 2002).

In yet another composition of natural ingredients, Chinese prickly ash, cassia bark, clove, and starch oxide were studied. Starch, potassium permanganate ($KMnO_4$), borax, sodium hydroxide (NaOH), and water were mixed together, heated, stirred, and cooled, followed by addition of starch oxide to get the end product. Initially, the mixture was even painted on common packing paper to form a uniform and even Chinese herbal medicine painting coating. The layer was baked under 60–80 to form compound fresh-keeping paper, and Ya pears stored under these conditions at room temperature for 35 days could be kept in fresh condition (Song Xiaogang et al., 1996).

Apple-of-Peru (Nicandraphysalodes) *Seed Extracts as a Biopreservative*

Nicandra physaloides (L.) Gaertn seed extract have good film-forming property without any color or smell and can be easily removed by washing with water. It has great potential to use effectively for keeping fruit and vegetables fresh. Ya pears treated with extract (0.25%–0.5%) could be stored for 40 days in the open under room temperature without any adverse effect on their quality, while untreated pears become wizened and with poor quality. Similarly, using 0.5%–1% of the same extract to paint fruits and vegetables, like sweet peppers, cucumbers, and tomatoes could efficiently prolong their shelf life for more than 10 days.

The fresh-keeping mechanism of *Nicandra physaloides* seed extract might be based on the fact that its coating forms an even and colorless film on fruit surfaces which offers protection against pathogens. It may provide a small air-conditioning-like environment to the painted fruits. It efficiently controls fruit respiration metabolism, postpones cell becoming decrepit and pigment degradation, and extends fruit fresh-keeping period. Different varieties of fruits have different tolerance to CO_2, so different varieties of fruits need ideal treating concentration (Xiangqiu et al., 1997). After painting with the extract, the fruit surface gloss is enhanced as it imparts both glazing and waxing. Therefore, the extract is very suitable for fruits and vegetables fresh-keeping and to extend their shelf life.

Potential of Phosphoprotein Membrane as a Biopreservative

Phosphoproteins have high molecular weights, which are prevalent in animals and vegetation kingdom. Like other proteins, they also contain several hydrophilic groups, such as NH−, −O−, −COOH, −OH and −NH. Besides forming film, they also maintain appropriate air permeability, water permeability, and gas selectivity, thereby keeping fruits fresh for longer duration. After soaking fruits in phosphoprotein, a uniform film coating is formed on the surface of fruit. It is easy to adjust the thickness of this film among tens micron according to requirements of physiological development of fruit. The film coating could clearly control respiration intensity of fruits and vegetables (Wenqiang and Shufen, 2006). Hygienic toxicity experimental studies showed that in albino rat oral LD_{50} of phosphoprotein with high molecular weight was more than 10,000 mg/kg and hence can be concluded to be safe without any toxicity.

The experimental results showed that Jingguan apples can be stored up to 5 months, with an improvement of 95%; in other case of Guoguang apples can be stored for 6 months with improved self life up to 98%. Further, it was observed with better effects in the case of orange and banana storage under similar conditions. Studies on strawberries showed that treatment at the concentration of 1%−2% has prolonged their storage time for 2 days under normal temperature and for up to 15−20 days when stored at 4 −8 (Wenqiang and Shufen, 2006).

Recent Advancements in Biopreservatives for Fruits and Vegetables

Snow Fresh (MONSANTO Make)

Snow Fresh is a novel type of high effect multiple function agents made by the Monsanto chemical company (Missouri, United States) for keeping fruits and vegetables fresh for longer duration. It could defer oxidization and brown changing of fruits and vegetables. It had better effects on fresh-keeping properties of fruits and vegetables, semimanufactured goods, which are without peel and pit. Under experimental conditions, Snow Fresh was found effective to keep semimanufactured goods' color and shape for 5 days, and the effect was higher than that of sulfite, which is generally used commercially. Snow Fresh comprises four safe and avirulent components, sodium pyrophosphate, ascorbic acid, citric acid, and $CaCl_2$. Snow Fresh was different in its action from sulfite, which had blanching function. Fruits and vegetables treated with sulfite and stored for some time had a peculiar smell and rudimental sulfur contents. Fruits and vegetables treated with Snow Fresh are without such peculiar smell.

The application method of Snow Fresh is very simple. Under conditions of room temperature, a fixed amount of Snow Fresh (white powder) is added in water, mixed for about 1 minute, and 1%−3% of solution was compounded. The soaking time for fruits and vegetables in this solution was 0.5−3 minutes. It is always recommended to prepare a new solution of Snow Fresh every time just before being used in 1 day and it should be used in recommended doses to get optimal effect (Haisheng, 2002).

Semper Fresh (SEMPER Make)

Semper Fresh is another type of agent used for fresh-keeping of fruits and vegetables made by SEMPER Genetic Engineering Company in England. It comprises several features, such as it has no color or smell, and is nonvirulent, ecofriendly, and without any adverse side effects. The edible Semper Fresh has wide application in fruit and vegetable fresh-keeping processes and also succeeded in the case of flower fresh-keeping. It was a compound formulation, containing vegetable oil and sugars as its main active ingredients mixed with some other ingredients like sugar ester, cellulose, edible oil, etc.

The possible mechanism of fresh-keeping might be that it could control respiration and prevent water evaporating in order to make them rest and slow down the aging speed of fruits and vegetables during storage. It can successfully be used at the concentration of 0.8%–1.0% for melons, strawberries, and green leafy vegetables. In the case of apples, pears, oranges, and bananas, the optimum concentrations were 1.0%, 0.8%, 1.0%–1.5%, and 1.2%, respectively. Further studies on fresh-keeping experiments in the case of strawberries, cherries, and apricots by using Semper Fresh were performed at Shanxi Academy of Agricultural Science and similar results were observed. The Semper Fresh process could be effectively used for the storage in both cool and dry places for a longer duration. The blended solution could be successfully stored for more than 5 days. In other studies, a different composition of agent, water, and Semper Fresh was dissolved in water and stored overnight for further use. This ensured the powder saturation was complete. Some selected fresh fruits and vegetables were soaked in this Semper Fresh solution for about 30 seconds and dried in air. In general, approximately 1 kg Semper Fresh powder is sufficient to treat about 28–35 tonnes of apples (Haisheng, 1990; Fan, 2003).

Chitosan as Biopreservatives

The process of arthropod shell extraction was generally mentioned as chitosan. Its main ingredient was derivative of deacetylated chitin, a type of cation consisting of high molecular polysaccharide (Dongxing and Shengyin, 2003). The extraction process is safe and innoxious and can be washed away by treatment with normal water and degraded by microflora. This process does not generate any type of rudimental toxic substance.

The semipermeable film coating on the outer surface of fruit could regulate physiological metabolism and control various types of microorganisms. It has been found with higher efficacy in the case of grapes fresh-keeping potential under normal temperature conditions. During the treatments, extraction containing Ca^{2+} was found with best efficacy on fresh-keeping of grapes by reducing rotting rate and loss in fresh weight. The product was found effective and stable, inhibiting function under high temperature and high pressure also (Guohui et al., 2004).

Mineral Halite Extract as Biopreservatives

Halite extraction, a compound obtained from mineral in rock layer by extraction process, is a white powder, comprising metallic salts of Ca, P, Mg, Na, and Mn. It is used as a fresh-keeping agent to enhance the storage period of fruits and vegetables. About 10 g of halite extraction is dissolved in 75 L water and used for preservation. Under experimental conditions, it was observed that this agent had significant effects for the protection of fruits and vegetables from rotting and fresh-keeping, particularly in the case of strawberries, arbutus, and mushrooms. In addition to protection, it also has another advantage of decomposing chemical fertilizers and agricultural chemicals on the outer surface of fruits and vegetables and destroys parasite eggs (Fan, 2003).

Limitations of Coating Films in Food Preservation

The film coating method for fresh-keeping of fruits and vegetables has some limitations also. They include its own structural limitations; polysaccharides in particular have defects in a number of features of film coating, like amount of moisture retention and antibacterial activity. Therefore, it is essential to add functional additives and active ingredients to modify the characteristics of polysaccharides to make them proficient for antisepsis.

It is important to take care at every step of application of coating reagents, for factors like film thickness, concentration, preserving time, and apparatus can influence the final results. It is difficult to control the thickness of film and concentrations of reagents due to the roughness of surface that varies with breeds and batches. In contrast, film coating is also not suitable for all type of produce. Out of a range of different produce, some are capable of self-protection due the presence of a layer of fuzz on their outer surfaces. It may be possible that the fuzz can be destroyed by film coating process and limiting storage time (Gao et al., 2009). In the following Table 10.5, a brief account of some important food preservation techniques to enhance the shelf life of the products is mentioned.

STRATEGIES FOR SAFETY OF FRUITS AND VEGETABLES

The strategies for safety of fruits and vegetables are summarized in Fig. 10.1.

Controlled Atmosphere Storage

The method of controlled atmosphere storage (CAS) is one of the most recent and successful methods developed by the postharvest sectors. The method generally comprises increasing the concentration of carbon dioxide and correspondingly decreasing the level of oxygen gas. This modification in the internal gas atmosphere of the plant produce reduces the metabolic activity of fruits and vegetables, thereby delaying their senescence and increasing their keeping quality. Thus this method helps to increase the seasonal availability of the plant produce and maintains their physicochemical and functional quality, which can indirectly reduce the cost to the consumer. Besides this, the method also helps to reduce storage disorders such as chilling injury and indirectly aids in reducing food waste. Overall, the total economic, social, and environmental impact is lowered (del Carmen Alamar et al., 2018). This method can also be used as an alternative to postharvest chemicals, which are expensive (Li et al., 2015). Care should be taken to optimize the gas concentrations according to each commodity. It is generally recommended to store most

TABLE 10.5 Various Food Preservation Techniques to Increase Shelf Life of the Products

Method	Effect on Microbial Growth or Survival
Refrigeration	Low or reduced temperature retards the microbial growth
Freezing	Causes reduction of water activity to prevent microbial growth, slowing of oxidation reactions
Drying, curing, and conserving	Delays or prevents microbial growth
Vacuum and oxygen-free modified atmosphere packaging	Inhibits strict aerobes and delays growth of facultative anaerobes
Carbon dioxide-enriched and/or modified atmosphere packaging	Specific inhibition of some microorganisms
Addition of weak acids; for example, sodium lactate	Reduction of the intracellular pH of microorganisms
Lactic fermentation	Reduction of pH value in situ by microbial action and by lactic and acetic acids formed (e.g., ethanol, bacteriocins)
Sugar preservation	Creates too high osmotic pressure for most microbial survival
Ethanol preservation	Produces toxic inhibition of microbes; can be combined with sugar preservation
Emulsification	Compartmentalization and nutrient limitation within the aqueous droplets in water-in-oil emulsion foods
Addition of preservatives such as nitrite or sulfite ions	Inhibition of specific groups of microorganisms
Pasteurization and appertization	Delivery of heat sufficient to inactivate target microorganisms to the desired extent
Food irradiation -Radurization, radicidation, and radappertization	Disrupts cellular RNA/DNA
Application of high hydrostatic pressure (Pascalization)	Pressure-inactivation of vegetative bacteria, yeasts, and molds
Pulsed electric field processing treatment	Microbial inactivation

fruits and vegetables under low-oxygen concentrations that should be close to the anaerobic compensation point (ACP). It is generally believed that oxygen levels above the ACP quickly increase respiration rate, and when below, fermentation adversely affects the fruit metabolism (Bessemans et al., 2016).

Nowadays, CAS technique has evolved to more recent dynamic controlled atmosphere (DCA) storage through the development of more accurate control systems. The objective of DCA storage is to maintain the lowest possible oxygen level, simultaneously adapting the gas concentrations dynamically by sensing the changing physiological response of the plant produce. If during storage the system detects low-oxygen stress, it increases the oxygen level until the plant produce response is achieved back to the optimal threshold level

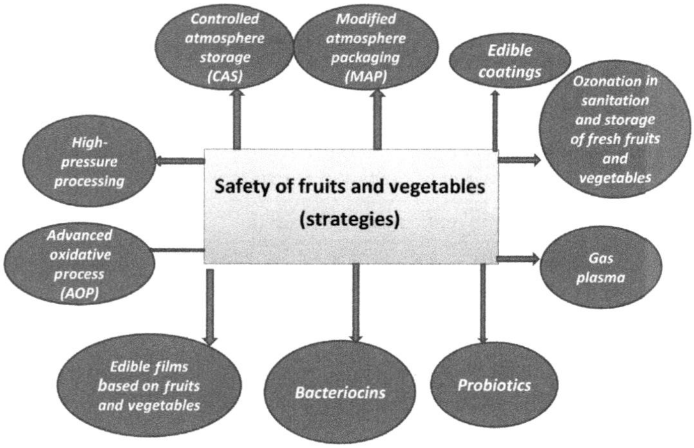

FIGURE 10.1 Strategies for safety of fruits and vegetables.

(Bessemans et al., 2016). This method is more efficient because it uses existing CAS technology that is improved by controlling parameters in near real time, thereby extending the plant produce storage life longer than the traditional CAS. It can also reduce the effect of storage disorders (e.g., superficial scald in apples, pears, etc.).

Modified Atmosphere Packaging

Modified atmosphere packaging (MAP) alters the atmosphere within the package according to the interaction that occurs between the plant product respiration rate and the transfer of gases through the package (Oliveira et al., 2015). The gas diffusion that occurs through the package depends on the film characteristics such as permeability, area, and thickness and the temperature of the surrounding environment (Beaudry, 2007). An equilibrium modified atmosphere (EMA) can be established in the package when the packaging technology is adapted to the plant produce respiration rate. This leads to a reduction in the respiration rate and metabolic processes, and correspondingly improved shelf life of the plant produce. The most commonly used gases in MAP are oxygen, carbon dioxide, and nitrogen.

Based on gaseous transmission rates, there are two types of MAP: passive and active. The passive uses the natural permeability and thickness of the packaging film to establish the desired atmosphere for the product as a result of its respiration (Somboonkaew and Terry, 2010).

The most commonly used polymers are polyamide, polypropylene, polyethylene, low-density polyethylene, linear low-density polyethylene, polystyrene, polyester, polyethylene terephthalate, ethylene vinyl alcohol, and polyvinylchloride (Zhang et al., 2015; Ghidelli and Pérez-Gago, 2018).

This technique has been successfully used to preserve the whole and fresh-cut products such as artichokes (Giménez et al., 2003), lettuce (Posada-Izquierdo et al., 2014), and strawberries (Siro et al., 2006). These packaging films can also be micro-perforated so as to enable gas exchange between inside and outside of the packaging material.

A variety of new structural polymers are now available to improve packaging materials that are biobased, biodegradable, and nonpetroleum (e.g., polylactic acid made from corn or other starch or sugar sources) (Mistriotis et al., 2016), polylactide aliphatic copolymer, and polymers derived from high proportions of recycled plastics (Wilson et al., 2017). Nanotechnology has recently been used to enhance packaging functionality by contributing toward antimicrobial, structural, and barrier properties to the packaging material (Eleftheriadou et al., 2017).

In addition to this, oxygen scavengers in the form of active inserts (lower oxygen concentrations within the sealed packs, thereby slowing deterioration caused by oxidation) are also incorporated during packaging of the horticultural produce. The synthetic antioxidants such as sulfites (potassium sulfite) are also used. Natural antioxidants (tocopherols, lecithin, organic acids, and plant extracts) are currently being explored to reduce the oxidation of the fresh produce and delay denaturation of proteins (Cruz et al., 2012; Pereira de Abreu et al., 2011).

The carbon dioxide scavengers, also called chemical absorbers (e.g., calcium hydroxide, sodium carbonate, calcium oxide) and physical absorbers (e.g., zeolite and activated carbon), are other attractive alternatives to delay senescence and reduce browning and mold-spoilage. This is particularly important for climacteric products, which produce high concentrations of carbon dioxide and affect their organoleptic characteristics.

Another strategy used commonly is to remove the ethylene from the package of the plant produce. Ethylene scrubbers (e.g., potassium permanganate pellets and clay mineral-coated strips) can also slow down senescence and reduce decay by neutralizing the effect of the plant hormone. Other alternatives are the use of carbon dioxide emitters that increase the carbon dioxide concentration within the package, thereby helping to achieve the optimal gas mixture for each product (Yahia, 2009).

Recently, smart or intelligent packaging has been used to fit packaging materials with sensors that are able to monitor the quality, microbiological growth, and temperature of the plant produce (Jedermann et al., 2017). The important components of intelligent packaging are radio frequency identification sensors; time-temperature-ripeness indicators; and biosensors (Brecht et al., 2016).

Edible Coatings

The other strategy to preserve fresh plant produce is through the use of edible coatings. These coatings create physical barriers on the fruit/vegetable surface and provide protection against moisture loss and control oxygen and carbon dioxide concentrations, in a similar way to MAP, and change the internal atmosphere of the plant produce (Dhall, 2013). The ideal edible coating extends the shelf life of the produce without causing complete anaerobic conditions and also reduces decay and water loss (Dhall, 2013). They can also act as effective antimicrobial agents. The edible coatings are generally prepared from

"Generally Regarded as Safe" (GRAS) materials. The chief characteristics of edible coating are that they should be water resistant, should cover the product completely, be able to reduce water-vapor permeability, generate the optimal atmosphere, improve the produce appearance, melt over 40ºC without decomposition, dry with high-efficiency performance, should have low viscosity, be easily emulsifiable, be economical, be translucent, and should not interfere with the plant produce quality. The composition of edible coating is also based on natural compounds such as aloe vera gel (Valverde et al., 2005), alginate-based edible coatings (Falagán et al., 2016), shellac (Chitravathi et al., 2016), or silk fibroin (Marelli et al., 2016).

The main advantages of using edible coatings are that they extend the shelf life of perishable products, maintain their initial attractive appearance such as color and gloss, and delay decaying. The correct formulation should not affect flavor or appearance. To maintain safety within the packaging, an application of solutions such as natural antimicrobials, like cinnamon or vanillin (Falagán et al., 2016), and essential oils are also used within the edible coating materials. Besides the natural materials, the films can also be coated with inhibitors such as titanium dioxide (TiO_2) to inactivate pathogens like *Escherichia coli* (Othman et al., 2014).

Edible films are made of different polymers such as pectin, proteins, oils, etc. They are generally applied to fresh fruits and vegetables to improve appearance and to prevent moisture loss. They also can serve as a carrier for antimicrobial compounds such as organic acids (Beuchat and Golden, 1989), methyl jasmonate (Buta and Moline, 1998), and bacteriocins onto the produce surface. However, more research is needed to determine the effectiveness of films in controlling microbial growth.

Edible Films as Biopreservatives of Fresh Produce

The food packaging materials should be natural and safe to use without any chemical side effects. Earlier petroleum-derived polymers were widely used, but the rising concern with their nonrenewable and/or nonbiodegradable nature paved the way for the development of other natural and herbal alternatives, such as polysaccharides and polypeptides. The use of food-grade biomacromolecules provides edible packaging material with suitable physical-mechanical properties and unique sensory and nutritional properties. Ideally, edible films and coatings should be transparent and flavorless and should not interfere with food sensory properties. However, under certain cases, specific sensory properties may be desirable for specific applications, such as sushi wraps, pouches to be melted on cooking, films between crust and toppings of pizzas, and film snacks.

McHugh et al. (1996) produced the first edible films based on fruit purees (McHugh et al., 2006). Since then, several studies have been carried out on the development of films made up of fruits and vegetables. This finding focuses on combinations of film-forming hydrocolloids such as starch, pectin, cellulose, and hemicellulose with fruits and vegetable purees (Azeredo et al., 2009). Moreover, fruits and vegetables are the sources of both nutrients and antioxidants that may be consumed in the form of edible films prepared from them (Deng and Zhao, 2011; Espitia et al., 2014).

Bacteriocins

A well-known group of nonpathogenic bacteria with antagonistic action against pathogens consists of lactic acid bacteria (LAB), which is a very heterogeneous group of gram-positive bacteria, categorized by the production of lactic acid as the main fermentation produce. LAB has been traditionally used in the production of different dairy products. Some of the strains are producers of potential antagonistic products like hydrogen peroxide, short-chain fatty acids, antimicrobial peptides, and bacteriocins. The production and assimilation of bacteriocins in pre- and postharvest processing of fruits and vegetables could be an effective ecological approach to ameliorate the quality and safety of fresh products in horticultural.

Nowadays, nisin, a bacteriocin produced by *Lactococcus lactis*, is considered under the category of GRAS by the U.S. FDA (Cano-Garrido et al., 2015) and is presently being used as a preservative in the food industry to prevent the growth of *L. monocytogenes*. The commercialization of nisin since the 1950s triggered research interest to isolate new bacteriocins from different sources, and by the 1990s, there was a variety of bacteriocins with diverse activity spectra; some of them are still in the process of seeking approval for use as a food additive. Bacteriocins are ribosomally synthesized peptides that when secreted act selectively on other bacteria by permeabilizing its membrane leading to cell death (Jordan et al., 2014). Bacteriocins are classified as potential antimicrobial agents, comparable to the roles played by defensins (produced by mammals) and thionines (produced by plants) (Karpiński and Szkaradkiewicz, 2013). Among the large number of bacteriocins considered so far, LAB are the most frequent producers, including some strains that are capable to synthesize up to three bacteriocins having different characteristics (Snyder and Worobo, 2014). While some bacteriocins are highly target specific, some others are known to effective in both gram-positive and gram-negative bacteria (Perez et al., 2016). The main reason behind such accuracies, along with the regulatory mechanism consisting of their production and processing, have not been yet fully explained but may vary differently according to the surface features of target bacterial and bacteriocin structural characteristics (Gabrielsen et al., 2014; Perez et al., 2016). Some bacteriocins are synthesized as precursors required in posttranslational processing like glycosylation or hydrolysis in specific signaling sequences (Kaškonienė et al., 2017).

The efficacy pathway of bacteriocins is one of the mechanisms about bacterial antagonism that has received more attention, mainly due to their similarity in function with antibiotics. Concisely, bacteriocins can cause the formation of pores in the membrane of sensitive bacteria, which alters cell permeability and decompensates the electrochemical homeostasis. Once inside the target cells, these molecules can bind nucleic acids to prevent gene expression and to interrupt cell biosynthesis (Snyder and Worobo, 2014).

Bacteriocins as an Innovative Biopreservative

Bacteriocins are a type of antimicrobial peptides synthesized ribosomes produced by bacteria that can inhibit or kill bacterial strains closely related or nonrelated to produced bacteria, without any harm to bacteria themselves through specific immunity proteins. Bacteriocins become one of the important weapons against microorganisms due to the specific characteristics, wide range of diversity in structure and function, natural resources, and stability to heat. In several recent studies, purified and identified bacteriocins were found suitable for application in food technology for extended food preservation time, in treating pathogen disease, in cancer therapy, and to maintain human health. Some chief examples of bacteriocins used in food safety are colicins and microcins. Additionally, co-culture methods have proved to enhance the antimicrobial activity of LAB and to increase the production of bacteriocins in comparison with monocultures in a strain-specific manner (Maldonado-Barragán et al., 2013). Among the LAB bacteriocins successfully induced through co-culture strategies so far are lactacin B, kimchicin G7, paracin 1.7, and plantaricins A, NC8, and MG (González-Pérez et al., 2018). The activity of bacteriocins in the case of fruits and vegetables is determined by the type of food product. The bacteriocin enterocin 416K1 (produced by *Enterococcus casseliflavus* IM 416K1) was able to completely control *L. monocytogenes* from contaminated processed apple and grapes within 8 hours posttreatment; however, no considerable inhibition of this pathogen was observed for processed pineapple and melon fruits (Anacarso et al., 2011). Likewise, enterocin AS-48, produced by *Enterococcus faecalis* A-48-32, inhibited *L. monocytogenes* growth in whole raspberries and sliced strawberries and blackberries stored at low temperatures (Molinos et al., 2008).

Bacteriocins have many advantages that allow them to be used in the food industry, such as they are resistant to surfactants, active in a wide pH range, and are often thermostable (Alvarez-Sieiro et al., 2016) and also their sensitivity toward digestive proteases. Bacteriocins in purified form may be used directly to the food matrix as food additive or applied as coatings using a carrier matrix; alternatively, if bacteriocins are produced naturally by food-grade bacteria, the producer strain may be considered GRAS category and may be inoculated for the production of fermented food (O'Bryan et al., 2018).

Considering that fresh fruits and vegetables are natural reservoirs of LAB, the screening of epiphytic bacteriocinogenic strains isolated from their surfaces would provide novel bacteriocin-producing strains adapted to the same environmental conditions used for the growth and storage of fresh horticultural products. The incorporation of bacteriocins-based technologies to avoid the proliferation of pathogens in these types of fresh foods may be achieved through either one of the following strategies or their combination: application of prebiotic oligosaccharides on the surface of fruits and vegetables, for inducing the production of bacteriocin by epiphytic LAB; inoculation of these products with GRAS bacteriocin-producing LAB; or application of active antimicrobial films or coatings containing commercially available bacteriocins. Thus, the use of bacteriocins on fresh horticultural products may be an ecological alternative to combat pathogens and ameliorate the incidence of foodborne diseases (González-Pérez et al., 2018).

Probiotics

Probiotics produce many antibacterial metabolites, such as bacteriocins, short-chain fatty acids, and hydrogen peroxide, that are responsible for inhibiting gastrointestinal microorganisms or pathogens. Bacteriocins are one of the important types of probiotics (Dobson et al., 2012). Some common examples of probiotics are LAB, nonpathogenic *E. coli*, bacilli, and yeasts. *E. coli* strain H22 produces the bacteriocins, namely colicinIb, E1, andmicrocin C7 (Cursino et al., 2006). They possess the ability to inhibit the growth of both pathogenic or nonpathogenic bacteria, such as *Enterobacter*, *Escherichia*, *Klebsiella*, *Morganella*, *Salmonella*, *Shigella*, and *Yersinia*. In a study by Cursino et al. (2006) carried out on a germ-free mouse model, *E. coli* strain H22 after a 6-day oral inoculation exhibited the bacteriostatic activity against *Shigellaflexneri* 4 to undetectable levels in feces.

Other Methods for Biopreservation of Fruits and Vegetables

The other traditional methods for biopreservation of fruits and vegetables are irradiation with ultraviolet (UV) rays, ozone treatment, and chlorine dioxide. They have been well-known for their germicidal activity for a long time and are used for decontaminating fruits and vegetable surfaces. An advanced oxidative process (AOP) or gas plasmas are recent developments in biopreservation of fresh plant produce.

Radiation Exposure

Sterilization with radiation is commonly known as the cold-sterilization process, and it is one of most important methods for nonthermal treatment of foods. This method was previously used to retard ripening, for sprout inhibition, and for killing insects in fresh plant produce (Ramos et al., 2013). Sterilization with radiation exposure is effective against all major food spoilage microorganisms and contaminants without causing any increase in temperature.

Cursino et al. (2006) reported that an irradiation dose of 1 kGy can cause a 5 log CFU reduction of *E. coli* O157:H7 populations on leafy greens including both those present in the leaves and incorporated into biofilms. The inhibition of *E. coli* O157:H7 to irradiation can also be achieved by dipping green leafy vegetables in sanitizer solutions such as chlorine, peracetic acid, or quaternary ammonium salt (Moosekian et al., 2014). However, only vegetative bacterial cells are sensitive to irradiation, but the viruses, endospores, and enteric protozoa are resistant and can tolerate even high doses of radiation.

In addition to green leafy vegetables, the fruits can also be subjected to irradiation to reduce the load of pathogens. For example, when a 1.5-kGy dose of radiation is applied to the cantaloupes that are inoculated with *Salmonella*, it could be decreased by 3.6 log CFU (Palekar et al., 2015). Similarly, irradiation up to a dose of 1 kGy can decrease pathogens such as *E. coli* O157:H7 on spinach or lettuce that is also approved by the US FDA (Shayanfar et al., 2017). Likewise, application of 1 kGy irradiation dose to strawberries supported a 4 log CFU reduction on STEC levels (Shayanfar et al., 2017). There are several examples of pathogen reduction through irradiation that have previously been reviewed by Pinela and Ferreira (2017).

UV Exposure

Out of all three types of UV rays A, B, and C, the UV-C type at 254 nm wavelength has the most potent germicidal activity and hence is recommended for surface disinfection of plant produce and to control microbial growth (Gayan et al., 2014; Nigro and Ippolito, 2016). Nigro and Ippolito (2016) reported an enhancement of bioactive compounds after exposure with UV rays. The other advantage is that there are no residues in the fruit or vegetable produce after UV-C treatment.

Another upcoming technology to improve food safety is cold plasma. It functions by applying an electric current to normal air or a gas that generates reactive gaseous species with antimicrobial activity. It involves no chemical use and therefore leaves no residues.

In a study, UV-C was used to reduce *E. coli* O157:H7 on apple surfaces by 3 log CFU and *Salmonella* on tomatoes by 2 log CFU (Yaun et al., 2004). However, pathogen reduction on uneven surfaces such as cantaloupes and berries is restricted to 1 log CFU with comparable doses (12 kJ/m2) (Adhikari et al., 2015). The only limitation of UV is that the light is coherent and cannot penetrate shaded areas on the surfaces of plant produce.

Similarly, pulsed light can generate a wide spectrum range varying between 200–1100 nm with a high-intensity flash. This can reduce the surface bacteria up to 1–3 log CFU reduction, although again limited by cells being located in shaded areas (Kramer et al., 2017). This limitation of shading can be overcome through the use of light emitting diode (LEDs), which are amenable to constructing novel reactors that deliver UV at multiple angles (Chen et al., 2017). Other advantages of LEDs are that they can be used in a wide range of different wavelengths, thereby providing a synergistic antimicrobial activity (Kim et al., 2017).

Ozonation

Ozone, a potential oxidant, is very effective in controling bacteria, molds, protozoa, and viruses (Rodoni et al., 2010). It was declared under GRAS category of preservatives in 1997 by the FDA (Habibi Najafi and Haddad Khodaparast, 2009). This application of ozone resulted in increased focus on research for its use in the food industry. It can be successfully used for numerous purposes like disinfecting surfaces of equipment, water, fresh fruits, and vegetables (Habibi Najafi and Haddad Khodaparast, 2009). The efficacy as disinfectants depends according to the type of surface to be treated, nature of product, and specific characteristics of contaminated microorganisms. Moreover, the vulnerability of microorganisms to ozone also differs depending on temperature, moisture, pH, physiology of tissue, and type of the preservatives or additives used (Karaca and Velioglu, 2007).

In recent times, there has been increasing interest in the utilization of ozone treatments during processing and storage of fresh fruits and vegetables (Tzortzakis et al., 2007). Microbial adulteration of fruit and vegetables can happen at several stages from farm to table. The proliferation of microorganisms generally occurs during growth in field, harvest, postharvest, transportation, storage, processing, and/or during human consumption. In some studies, it was observed that ozone treatment has a very high beneficial effect in enhancing storage periods of fresh commodities like broccoli, cucumber, apples, grapes, oranges, pears, raspberries, and strawberries by decreasing microbial contamination and via ethylene oxidation (Beuchat, 1992). Usages of ozone on apples resulted in reduction of weight loss and spoilage. Similarly, treatment of onions with ozone during storage resulted in a considerable reduction in molds and microbial counts without any adverse effect on their chemical composition and sensory quality (Song et al., 2000).

Continuous exposure with ozone at 0.3 ppm (v/v) restricted the growth of fungi, microbes, and spores during storage in Elegant Lady peaches (Palou et al., 2002). It was also observed that it reduces growth of gray mold in Thompson seedless table grapes. In the case of whole and fresh-cut Thomas tomatoes, storage for a longer duration and treatment with higher doses (7 μL/L) of ozone efficiently reduced the population of bacteria and fungus (Aguayo et al., 2006). Further, investigations on the induction of antioxidant accumulation in white table grapes (var. Superior) after treatment with ozone at different concentrations (3.88 and 1.67 g/h) for 1, 3, and 5 hours during storage at 22°C showed good results (Gonzales-Barrio et al., 2006). Furthermore, in another study, fresh-cut salad, washed with ozonized water and packed under ozone, exhibited extended storage shelf life (Zambuchini et al., 2006).

In some other investigations to observe the effect of ozone interaction on orange juice, color degradation was carried out. Ozonization of organic dyes causes loss of color due to the oxidation of chromophores by the attack on conjugated double bonds. Similarly, the oxidation of chromophore of conjugated double bonds and aromatic rings in the case of carotenoids was found responsible for the color change of orange juice. In carotenoid, pigments are responsible for their yellow, orange, or red color in orange juices. The ozone and its hydroxyl radicals (OH−) generated during the ozonization process in aqueous medium may open aromatic rings and lead to a partial oxidation of these substances to organic acids, aldehydes, and ketones (Tiwari et al., 2008). Ozone has been observed to be a more effective biocide as compared to other constituents due to its strong reactivity and high oxidation power. Conversely, it is essential to define the safety parameters of exposure to ozone in a direction to prevent damage of food quality and human health.

Use of Combination of Preservation Methods

Several studies analyzed the use of combined preservation methods with ozone treatments including hydrostatic pressure, UV, and H_2O_2. The application of pressure aids the penetration of the sanitizers into the inaccessible cracks and crevices of foods, thus enhancing microbial decontamination without compromising quality. The main advantage of applying hydrostatic pressure includes uniform transmission of pressure, regardless of the size and shape of sample. The use of ozone in combination with initiators such as UV or H_2O_2 can result in advanced oxidation processes that are highly effective against the most resistant microorganisms (Rodoni et al., 2010).

High-Pressure Processing

High-pressure processing (HPP) and HHP are processing techniques that utilize high pressure ranging from 100 to 700 MPa, with or without external heat. In order of commercial terms, HHP has been mainly used in nonthermal pasteurization of juices, purees, meat, and seafood (Tadapaneni et al., 2014). Processing of whole or semiprocessed fresh fruits and vegetables through these techniques is not so common, although commercial products like avocado halves are available. The advantage of HHP, particularly in relation to avocados, is to inactivate enzymes and reduce microbial contamination, leading to enhanced shelf life of produce up to 60 days (Woolf et al., 2013). Under HHP processing, pressure applied at 600 MPa for 3 minutes reduces coliform count level of fresh basil to <1 log CFU without effecting the leaf integrity (Koutchma and Warriner, 2017). Therefore, HHP is an effective process to reduce contamination in fresh produce with the added benefit of increasing shelf life by reducing endogenous enzyme activity (Georget et al., 2015). Yet, this process has limited commercial application due to high cost, limited capacity, batch system, and compatibility with product packing under MAP. In case of leafy vegetables, there are also some limitations of texture loss caused by plant cell disruption and browning due to chlorophyll degradation (Seifert and Zude-Sasse, 2016; Murray et al., 2017).

Gas-Phase Treatments

In addition to water-based wash systems, gas-phase treatments are also interesting as alternative or in combination with synergistic efficacy (Shynkaryk et al., 2015). The advantage of using such type of treatment and the issues related to cross-contamination are minimized with higher penetration into subsubsurface of fresh fruits and vegetables (Goodburn and Wallace, 2013). Fumigation involves use of ethylene oxide and some other potential gases, but it has limited application due to carcinogenic residues (Bononi et al., 2014). Consequently, the main gas-phase processes used for direct food contact are acetic acid, chlorine dioxide, hydrogen peroxide, and ozone (Netramai et al., 2016). On commercial level, chlorine dioxide and ozone are more feasible and cost effective in addition to negligible or no effects on organoleptic characters of fresh food produce (Shynkaryk et al., 2015). The most important advantage of ozone application is its availability and absence of any disinfection byproducts during process. Chlorine dioxide is very effective, but a main concern over its use is persistency of chlorite residues (Smith et al., 2015). Ozone has added advantage to degrade ethylene, resulting in delaying of ripening, but prolonged exposure can lead to a decrease in antioxidants. As a decontamination agent, it has a limited efficacy due to the low concentration (3 ppm) that can be used during storage process and also lacks penetration into packed produce (Tzortzakis and Chrysargyris, 2017).

Ozone gas treatment process is being incorporated into vacuum cooling process, where it was introduced into the chamber that facilitated uptake of this antimicrobial gas into inner leaf structures (Shynkaryk et al., 2016). In another study by Yesil et al. (2017), spinach was inoculated with *E. coli* O157:H7, followed by introduction of ozone into a chamber at a rate of 1.5 g/h for 30 minutes. This treatment supported >3 log CFU reduction in *E. coli* O157:H7 levels on spinach, without any detrimental effect on the leafy green (Yesil et al., 2017). Although ozone has been found effective antimicrobial gas for the decontamination of fresh fruits and vegetables produce, it is rarely used at industrial level. Main reasons behind this are difficulty in containing ozone gas and extensive corrosion of metal surfaces like fans, fittings, and condensers (Coelho et al., 2015).

As compared to ozone, chlorine dioxide is less corrosive and is generated onsite by mixing sodium chlorite with an acid, either organic or inorganic to generate gas. In an experiment, orange fruits treated with 0.5 ppm chlorine dioxide for 14 minutes caused a reduction of 5 log CFU *Salmonella* (Bhagat et al., 2011). In another study, *Salmonella* on cantaloupes was reduced by 3 log CFU on exposure to 5 ppm chlorine dioxide for 10 minutes, thereby indicating that uneven surfaces could be treated with efficacy via gas treatment (Mahmoud et al., 2008). Chlorine dioxide treatment has limitations due to its strong bleaching action that negatively affects sensory quality of fresh products and also has risk of generating toxic byproducts like chlorite (Kaur et al., 2015). To overcome such limitations, the main focus has been to find novel delivery methods based on slow release of gas for extended time. In an experiment, it was observed that spraying spinach with sodium chlorite, a precursor of chlorine dioxide, followed by hydrochloric acid vapor inactivated >5 log CFU of *Salmonella* and *L. monocytogenes* (Hwang et al., 2017). It was postulated that the inert chlorite might have penetrated into the inner leaf tissues, which was converted to chlorine dioxide by reaction with HCl. In another study, a high reduction of pathogens (>3 log CFU) was observed but there was no comment on any change in quality of product or risk of chlorite residues. Slow-releasing chlorine dioxide pads have been also developed for use in sealed packaging or containers. In this study, the chlorite is held within one layer of the pad, and acid (tartaric acid in current case) diffuses slowly through chlorite layer to generate chlorine dioxide gas (Bai et al., 2016). In another study, the pads were placed within a clam shell pack having inoculated tomatoes, chlorine dioxide maintained at 3.5 ppm for 10 days supported a 3 log CFU reduction of *E. coli* (Sun et al., 2017).

The introduction of ozone into the packs has also been studied in tomatoes inoculated with *E. coli* O157:H7, *L. monocytogenes*, or *Salmonella*. The headspace of packing was filled with 4000 ppm ozone that showed 2–3 log CFU reduction of pathogen (Fan et al., 2012). In this case, ozone would be depleted within minutes and thereby would not show residual antimicrobial effect as observed for chlorine dioxide gas (Sun et al., 2017).

Advanced Oxidative Process

AOP is the generation of hydroxyl radicals by decomposition of hydrogen peroxide and/or ozone. These hydroxyl radicals have oxidation potential much higher than that of hydrogen peroxide or ozone, but they have short half-life; thereby, it is required to be generated at them at point of application (Assalin et al., 2010).

Hydroxyl radicals can be generated by several methods like UV-C mediated decomposition of hydrogen peroxide and/or ozone. It is also possible to generate radicals by reacting ozone and hydrogen peroxide in solution followed by their application in the form of fine mist. The formation of hydroxyl radicals is promoted by Fe ions (Fenton reaction) at an operating temperature of 48°C and is influenced by the rate of formation and hence stability (Zhang et al., 2014). Higher concentration of hydroxyl radicals is formed by the use of a combination of hydrogen peroxide, ozone under UV-C at 48°C (Chen et al., 2015). Combination of UV and hydrogen peroxide was found to inactivate *E. coli* and *Salmonella* inoculated on the surface or subsurface of a variety of fresh produce like lettuce, cauliflower, and onion (Hadjok et al., 2008).

An AOP-based process has been found successful to inactivate *Listeria* and *E. coli* in case of mushrooms without any negative effect on sensory characters (Guan et al., 2013; Murray et al., 2015). The AOP process has been found to be an effective tool; but the major constraint was its use on commercial scale related to throughput given treatment times required of 30 seconds. In case of leafy vegetables, contact with all the fresh surface layers is another challenge. In spite of all these limitations, it is expected that the treatment will find commercial utility to decontaminate whole produce, where material transfer issues would not be significant (Murray et al., 2017).

Gas Plasma

As compared to AOP process, gas plasma process comprises a mixture of ions, electrons, radicals, and UV photons (Pignata et al., 2017). Under this process to generate gas plasmas, high voltage is passed through a gas phase comprising oxygen, helium, hydrogen, and argon (Perni et al., 2008). The antimicrobial effects of gas plasmas have been known since the 19th century; however, it requires application of high-voltage generators, excessive heat generation, and use of noxious working gases like hydrogen peroxide or peracetic acid-based mixture, and requirement of low pressure for treatment has restricted its commercial application. In recent advances, atmospheric gas plasmas have been developed that can be operated at normal atmospheric conditions of pressure and temperature around 50°C (Mir et al., 2016; Murray et al., 2017).

Problems in Natural Products Applications

The essential oils have been industrialized as pesticides, but they are still not industrialized as antibacterial agents and they are barely used in preservatives for fruits and vegetables. The studies of mechanism and applied technology are not thorough yet. Many studies of these fields are in the test period and the mature products are absence.

The reports about natural products for fresh-keeping of the bacteriostatic test are all carried through in medium. The results showed that the bacteriostatic concentration should be increased in the production application. While the essential oils are all hydrophobic compounds, they can influence the flavor of fruits and vegetables. The studies were mostly processed in the conditions that environment in variety and single microorganisms exist. Lacking in the studies are the combination and application of natural extractives on antisepsis characteristic.

CONCLUSION

In the majority of cases of fresh fruits and vegetables, the high level of contamination is brought into processing and subsequently disseminated to various batches of products. Therefore, after thorough study of available literature, it may be concluded that in order to reduce the postharvest losses of the fruits and vegetable produce, initiation of awareness about possible sources of contamination is extremely necessary. The chain of such activities should begin from farmers followed by salespersons and finally toward the consumers. Special emphasis should be directed toward the sources of possible contamination and their remediation. In order to reduce the risk of contamination, GAP and GMP should be adopted at all stages of harvest, processing, and also by consumers. Storage, processing, and packaging under controlled conditions are some of the latest and important tools to minimize losses and maintain good quality of fruits and vegetables along with the supply chain to reduce wastage and to maintain fresh produce availability year-round. In addition to these, there are several other available promising processes based on bacteriocins, chlorine dioxide, irradiation, ozone, and AOP. The use of UV radiation, pulsed light, and gas plasma might have limited utility due to their limited efficacy to control contamination on a commercial scale. Further, a combination of different interventions, which include postharvest washing, may also be applied to control foodborne pathogens. It is possible that application of multiple processing techniques can have synergistic antimicrobial effect without significant detrimental effects on produce sensory quality. Finally, it may be concluded that these technologies might have potential to extend the shelf life and achieve a better preservation of quality of fresh fruits and vegetables.

References

Adhikari, A., Syamaladevi, R.M., Killinger, K., Sablani, S.S., 2015. Ultraviolet-c light inactivation of *Escherichia colio*157: h7 and *Listeria monocytogenes* on organic fruit surfaces. Int. J. Food Microbiol. 210, 136–142.

Aguayo, E., Escalona, V.H., Artes, F., 2006. Effect of cyclic exposure to ozone gas on physicochemical, sensorial and microbial quality of whole and sliced tomatoes. Postharvest Biol. Technol. 39, 169–177.

Alvarez-Sieiro, P., Montalbán-López, M., Mu, D., Kuipers, O.P., 2016. Bacteriocins of lactic acid bacteria: extending the family. Appl. Microbiol. Biotechnol. 100 (7), 2939–2951.

Anacarso, I., De Niederhaeusern, S., Iseppi, R., Sabia, C., Bondi, M., et al., 2011. Anti-listerial activity of chitosan and Enterocin 416K1 in artificially contaminated RTE products. Food. Control. 22 (12), 2076–2080.

Assalin, M.R., De Moraes, S.G., Queiroz, S.C., Ferracini, V.L., Duran, N., 2010. Studies on degradation of glyphosate by several oxidative chemical processes: ozonation, photolysis and heterogeneous photocatalysis. J. Environ. Sci. Health, Part B 45, 89–94.

Astrom, J., Carlander, A., Sahlen, K., Stenstrom, T.A., 2006. Fecal indicator and pathogen reduction in vegetation microcosms. Water, Air, Soil Pollut. 176, 375–387.

Ayrapetyan, M., Williams, T.C., Baxter, R., Oliver, J.D., 2015. Viable but non-culturable and persister cells coexist stochastically and are induced by human serum. Infect. Immunity 83, 4194–4203.

Azeredo, H.M.C., Mattoso, L.H.C., Wood, D., Williams, T.G., Avena-Bustillos, R.J., McHugh, T.H., 2009. Nanocomposite edible films from mango puree reinforced with cellulose nanofibers. J. Food Sci. 74 (5), N31–N35.

Bai, Z., Cristancho, D.E., Rachford, A.A., Reder, A.L., Williamson, A., Grzesiak, A.L., 2016. Controlled release of antimicrobial clo2 gas from a twolayer polymeric film system. J. Agric. Food. Chem. 64, 8647–8652.

Beaudry, R., 2007. MAP as a basis for active packaging. In: Wilson, C.L. (Ed.), Intelligent and Active Packaging for Fruits and Vegetables. Taylor and Francis Group LLC, Boca Raton, FL, pp. 31–56.

Bessemans, N., Verboven, P., Verlinden, B.E., Nicolaï, B.M., 2016. Postharvest Biol. Technol. 115, 91.

Beuchat, L.R., 1992. Surface disinfection of raw produce. Dairy Food Environ. Sanit. 12 (1), 69.

Beuchat, L.R., Golden, D.A., 1989. Antimicrobials occurring naturally in foods. Food Technol. 43, 135.

Bhagat, A., Mahmoud, B.S., Linton, R.H., 2011. Effect of chlorine dioxide gas on *Salmonella enterica*inoculated on navel orange surfaces and its impact on the quality attributes of treated oranges. Foodborne Pathog Dis. 8, 77–85.

Bononi, M., Quaglia, G., Tateo, F., 2014. Identification of ethylene oxide in herbs, spices and other dried vegetables imported into Italy. Food Addit. Contam. Part A Chem. Anal. Control Expo Risk Assess. 31, 271–275.

Bouwknegt, M., Verhaelen, K., Rzeżutka, A., Kozyra, I., Maunula, L., et al., 2015. Quantitative farm-to-fork risk assessment model for norovirus and hepatitis A virus in European leafy green vegetable and berry fruit supply chains. Int. J. Food Microbiol. 198, 50–58.

Brecht, J.K., Loaza, F.E., Nunes, M.C.N., Emond, J.P., Uysal, I., Badia, F., et al., 2016. Reducing strawberry waste and losses in the postharvest supply chain via intelligent distribution management. Acta Hortic. 1120, 253–260.

Buta, J.G., Moline, H.E., 1998. Methyl jasmonate extends shelf life and reduces microbial contamination of fresh-cut celery and peppers. J. Agric. Food Chem. 46, 1253.

Callejón, R.M., Rodríguez-Naranjo, M.I., Ubeda, C., Hornedo-Ortega, R., Garcia-Parrilla, M.C., Troncoso, A.M., 2015. Reported foodborne outbreaks due to fresh produce in the United States and European Union: trends and causes. Foodborne Pathog. Dis. 12, 32–38.

Cano-Garrido, O., Seras-Franzoso, J., Garcia-Fruitós, E., 2015. Lactic acid bacteria: reviewing the potential of a promising delivery live vector for biomedical purposes. Microb. Cell Fact. 14 (1), 137.

CDC (Center of Disease Control), 2017. Foodborne Illness and Outbreaks. <https://www.cdc.gov/foodsafety/outbreaks/multistate-outbreaks/outbreaks-list.html.> (accessed August, 2017).

CFSAN, 2004. Produce Safety From Production to Consumption: 2004 Action Plan to Minimize Foodborne Illness Associated with Fresh Produce Consumption. Center for Food Safety and Applied Nutrition (CFSAN), US Department of Health and Human Services, Food and Drug Administration, October 2004.

Chen, Y., Peng, R., Shen, T., Tong, S., Ma, C., 2015. A promising ozone-based advanced oxidation process for effective generation of hydroxyl radicals in acidic solution. Sep. Purif. Technol. 151, 269–275.

Chen, J., Loeb, S., Kim, J.H., 2017. LED revolution: fundamentals and prospects for UV disinfection applications. Environ. Sci. Water Res. Technol. 3, 188–202.

Chitravathi, K., Chauhan, O.P., Raju, P.S., 2016. Shelf life extension of green chillies (*Capsicum annuum* L.) using shellac-based surface coating in combination with modified atmosphere packaging. J. Food Sci. Technol. 53 (8), 3320–3328.

Coelho, C.C.D., Freitas-Silva, O., Campos, R.D., Bezerra, V.S., Cabral, L.M.C., 2015. Ozonation as post-harvest technology in conservation of fruits and vegetables: a review. Rev. Bras. Eng. Agr. Amb. 19, 369–375.

Cruz, R.S., Camilloto, G.P., dos Santos Pires, A.C., 2012. Oxygen scavengers: an approach on food preservation. In: Eissa, A.A. (Ed.), Structure and Function of Food Engineering. InTech, Rijeka, Croatia, pp. 21–42.

Cursino, L., Smajs, D., Smarda, J., Nardi, R.M., Nicoli, J.R., Chartone-Souza, E., et al., 2006. Exoproducts of the *Escherichia coli* strain H22 inhibiting some enteric pathogens both *in vitro* and *in vivo*. J. Appl.Microbiol. 100, 821–829.

del Carmen Alamar, M., Falagán, N., Aktas, E., Terry, L.A., 2018. Minimising food waste: a call for multidisciplinary research. J. Sci. Food Agric. 98 (1), 8–11.

Deng, Q., Zhao, Y., 2011. Physicochemical, nutritional, and antimicrobial properties of wine grape (cv. Merlot) pomace extract-based films. J. Food Sci. 76 (3), E309–E317.

Dhall, R.K., 2013. Advances in edible coatings for fresh fruits and vegetables: a review. Crit. Rev. Food. Sci. Nutr. 53 (5), 435–450.

Dincer, I., 2017. Refrigeration Systems and Applications, third ed. John Wiley & Sons Ltd, Chichester, UK, p. 752.

Dobson, A., Cotter, P.D., Ross, R.P., Hill, C., 2012. Bacteriocin production: a probiotic trait? Appl.Environ. Microbiol. 78, 1–6.

Dongxing, Z., Shengyin, L., 2003. Summarization of applying studies on preservatives for fruits and vegetables. Shan Xi J. Agric. Sci. 1, 30–34.

Droby, S., 2006. Improving quality and safety of fresh fruits and vegetables after harvest by the use of biocontrol agents and natural materials. Acta Hort. 709, 45–52.

Eleftheriadou, M., Pyrgiotakis, G., Demokritou, P., 2017. Nanotechnology to the rescue: using nano-enabled approaches in microbiological food safety and quality. Curr. Opin. Biotechnol. 44, 87–93.

Erickson, M.C., Habteselassie, Liao, J., Webb, C.C., Mantripragada, V., Davey, L.E., et al., 2014. Examination of factors for use as potential predictors of human enteric pathogen survival in soil. J. Appl. Microbiol. 116, 335–349.

Espitia, P.J.P., Du, W.-X., Avena-Bustillos, R.J., Soares, N.F.F., McHugh, T.H., 2014. Edible films from pectin: physical-mechanical and antimicrobial properties—a review. Food Hydrocoll. 35, 287–296.

Falagán, N., Artés, F., Aguayo, E., 2016. Natural additives to preserve quality and improve nutritional value of fresh-cut nectarine. Food Sci. Technol. Int. 22 (5), 429–439.

Fan, W., 2003. Application of natural fresh-keeping agents for fruits and vegetables in abroad. Shan Dong Food Sci. Technol. 4, 28–30.

Fan, X., Sokorai, K.J., Engemann, J., Gurtler, J.B., Liu, Y., 2012. Inactivation of *Listeria innocua, Salmonella Typhimurium,* and *Escherichia colio*157: h7 on surface and stem scar areas of tomatoes using in-package ozonation. J. Food. Prot. 75, 1611–1618.

Farkas, J., 2001. Physical methods of food preservation. In: Doyle, M.P., Beuchat, L.R., Montville, T.J. (Eds.), Food Microbiology: Fundamentals and Frontiers, second ed. American Society for Microbiology, New York, pp. 567–591.

Gabrielsen, C., Brede, D.A., Nes, I.F., Diep, D.B., 2014. Circular bacteriocins: Biosynthesis and mode of action. Appl. Environ. Microbiol. 80 (22), 6854–6862.

Gao, H., Shi, P., Zhao, Y., 2009. IFIP-International Federation for Information Processing, Volume 295. In: Li, D., Chunjiang, Z. (Eds.), Computer and Computing Technologies in Agriculture II, Volume 3. Springer, Boston, pp. 1655–1669.

Gayan, E., Condon, S., Alvarez, I., 2014. Biological aspects in food preservation by ultraviolet light: a review. Food Bioprocess Technol. 7, 1–20.

Généreux, M., Breton, M.J., Fairbrother, J.M., Fravalo, P., Côté, C., 2015. Persistence of indicator and pathogenic microorganisms in broccoli following manure spreading and irrigation with fecally contaminated water: field experiment. J. Food Prot. 78, 1776–1784.

Georget, E., et al., 2015. Inactivation of microorganisms by high isostatic pressure processing in complex matrices: a review. Innovat. Food Sci. Emerging Technol. 27, 1–14.

Ghidelli, C., Pérez-Gago, M.B., 2018. Recent advances in modified atmosphere packaging and edible coatings to maintain quality of fresh-cut fruits and vegetables. Crit. Rev. Food Sci. Nutr. 58 (4), 662–679.

Giménez, M., Olarte, C., Sanz, S., Lomas, C., Echávarri, J.F., Ayala, F., 2003. Relation between spoilage and microbiological quality in minimally processed artichoke packaged with different films. Food Microbiol. 20 (2), 231–242.

Gonzales-Barrio, R., Beltran, D., Cantos, E., Gil, M.I., Espın, J.C., Tomas-Barberan, F.A., 2006. Comparison of ozone and UV-C treatments on the postharvest stilbenoid monomer, dimer, and trimer Induction in Var. Superior White Table Grapes. J. Agric. Food Chem. 54, 4222–4228.

González-Pérez, C.J., Aispuro-Hernández, E., Vargas-Arispuro, I., Martínez-Téllez, M.A., 2018. Induction of bacteriocins from lactic acid bacteria; a strategy to improve the safety of fresh fruits and vegetables. Agric. Res. Technol. Open Access J. 14 (4), 555927.

Goodburn, C., Wallace, C.A., 2013. The microbiological efficacy of decontamination methodologies for fresh produce: a review. Food Control. 32, 418–427.

Guan, W., Fan, X., Yan, R., 2013. Effect of combination of ultraviolet light and hydrogen peroxide on inactivation of *Escherichia coli* O157:H7, native microbial loads, and quality of button mushrooms. Food Control 34, 554–559.

Guohui, Q., Yan, N., Baoguo, L., 2004. Effect of free-pollutant fresh-keeping agents on Taihang chestnut (*Castaneamollissima* Bl.). J. Agri. Univ. Hebei 27 (1), 53–56.

Habibi Najafi, M.B., Haddad Khodaparast, M.H., 2009. Efficacy of ozone to reduce microbial populations in date fruits. Food Control 20, 27–30.

Hadjok, C., Mittal, G.S., Warriner, K., 2008. Inactivation of human pathogens and spoilage bacteria on the surface and internalized within fresh produce by using a combination of ultraviolet light and hydrogen peroxide. J. Appl. Microbiol. 104, 1014–1024.

Haisheng, G., 1990. Introductions about several natural fresh-keeping agents for fruits and vegetables. World Agric. 4, 44–46.

Haisheng, G., 1993. Edible spice plant and application of it. World Agric. 5, 23–26.

Haisheng, G., 2002. Production and application of natural fresh-keeping agents for fruits and vegetables. Farmers Consultant 8, 19.

Holdsworth, S.D., 1985. The Preservation of Fruit and Vegetable Food Products, first ed. The Macmillian Press Ltd., London.

Holvoet, K., Sampers, I., Seynnaeve, M., Jacxsens, L., Uyttendaele, M., 2014. Agricultural and management practices and bacterial contamination in greenhouse versus open field lettuce production. Int. J. Environ. Res. Public Health 12, 32–63.

Hwang, C.A., Huang, L., Wu, V.C., 2017. *In situ* generation of chlorine dioxide for surface decontamination of produce. J. Food. Prot. 80, 567–572.

Jablasone, J., Warriner, K., Griffiths, M., 2005. Interactions of Escherichia colio157: h7, *Salmonella typhimurium* and *Listeria monocytogenes* plants cultivated in a gnotobiotic system. Int. J. Food Microbiol. 99, 7–18.

Jedermann, R., Nicometo, M., Uysal, I., Lang, W., 2017. Reducing food losses by intelligent food logistics. Philosophical transactions Series A, Mathematical, physical, and engineering sciences. Philos. Trans. Royal Soc. A 372, 20130302.

Jensen, D.A., Danyluk, M.D., Harris, L.J., Schaffner, D.W., 2017. Quantifying bacterial cross-contamination rates between fresh-cut produce and hands. J. Food Prot. 80, 213–219.

Jongman, M., Korsten, L., 2017. Assessment of irrigation water quality and microbiological safety of leafy greens in different production systems. J. Food Safety 37, 12.

Jordan, K., Dalmasso, M., Zentek, J., Mader, A., Bruggeman, G., et al., 2014. Microbes versus microbes: control of pathogens in the food chain. J. Sci. Food Agric. 94 (15), 3079–3089.

Karaca, H., Velioglu, S.Y., 2007. Ozone applications in fruit and vegetable processing. Food Rev. Int. 23, 91–106.

Karpiński, T.M., Szkaradkiewicz, A.K., 2013. Characteristic of bacteriocines and their application. Polish J. Microbiol. 62 (3), 223–235.

Kaškonienė, V., Stankevičius, M., Bimbiraitė-Survilienė, K., Naujokaitytė, G., Šernienė, L., et al., 2017. Current state of purification, isolation and analysis of bacteriocins produced by lactic acid bacteria. Appl. Environ. Microbiol. 101 (4), 1323–1335.

Kaur, S., Smith, D.J., Morgan, M.T., 2015. Chloroxyanion residue quantification in cantaloupes treated with chlorine dioxide gas. J. Food. Prot. 78, 1708–1718.

Kim, D.K., Kim, S.J., Kang, D.H., 2017. Bactericidal effect of 266 to 279nm wavelength uvc-leds for inactivation of gram positive and gram negative foodborne pathogenic bacteria and yeasts. Food Res. Int. (Ottawa, Ont.) 97, 280–287.

Koutchma, T., Warriner, K., 2017. Regulatory aspects of high pressure processed foods in North America, Europe, Asia, New Zealand and Australia. In: Houska, M., Silva, F. (Eds.), High Pressure Processing of Fruit and Vegetable Products. CRC Press, New York.

Kramer, B., Wunderlich, J., Muranyi, P., 2017. Recent findings in pulsed light disinfection. J. Appl. Microbiol. 122, 830–856.

Li, H., Tajkarimi, M., Osburn, B.I., 2008. Impact of vacuum cooling on Escherichia colio157: h7 infiltration into lettuce tissue. Appl. Environ. Microbiol. 74, 3138–3142.

Li, W., Wang, K., Chen, L., Johnson, J.A., Wang, S.J., 2015. Tolerance of Sitophilus zeamais (Coleoptera: Curculionidae) to heated controlled atmosphere treatments. J. Stored Prod. Res. 62, 52–57.

Liang, Z.B., He, Z.L., Powell, C.A., Stoffella, P.J., 2011. Survival of *Escherichia coli* in soil with modified microbial community composition. Soil Biol. Biochem. 43, 1591–1599.

Linlin, L., Min, Z., Benu, A., Zhongxue, G., 2016. Recent advances in pressure modification-based preservation technologies applied to fresh fruits and vegetables. Food Rev. Int. 33 (5), 538–559.

Mahmoud, B.S., Vaidya, N.A., Corvalan, C.M., Linton, R.H., 2008. Inactivation kinetics of inoculated *Escherichia colio*157: h7, *Listeria monocytogenes* and *Salmonella* Poona on whole cantaloupe by chlorine dioxide gas. Food Microbiol. 25, 857–865.

Maldonado-Barragán, A., Caballero-Guerrero, B., Lucena-Padrós, H., Ruiz-Barba, J.L., 2013. Induction of bacteriocin production by coculture is widespread among plantaricin-producing *Lactobacillus plantarum*strains with different regulatory operons. Food Microbiol. 33 (1), 40–47.

Marelli, B., Brenckle, M.A., Kaplan, D.L., Omenetto, F.G., 2016. Silk fibroin as edible coating for perishable food preservation. Sci. Rep. 6, 25263.

Martínez-Vaz, B.M., Fink, R.C., Diez-Gonzalez, F., Sadowsky, M.J., 2014. Enteric pathogen-plant interactions: molecular connections leading to colonization and growth and implications for food safety. Microbes Environ. 29, 123–135.

McHugh, T.H., Huxsoll, C.C., Krochta, J.M., 1996. Permeability properties of fruit puree edible films. J. Food Sci 61 (1), 88–91.

McHugh, T.H., de Bord, M.D., Olsen, C.W., 2006. Edible film for wrapping around foods, has specific dry film thickness, moisture content and water activity. U.S. Patent US 2006051464 A1.

Mir, S.A., Shah, M.A., Mir, M.M., 2016. Understanding the role of plasma technology in food industry. Food Bioprocess Technol. 9, 734–750.

Mistriotis, A., Briassoulis, D., Giannoulis, A., D'Aquino, S., 2016. Design of biodegradable bio-based equilibrium modified atmosphere packaging (EMAP) for fresh fruits and vegetables by using micro-perforated poly-lactic acid (PLA) films. Postharvest Biol. Technol. 111, 380–389.

Molinos, A.C., Abriouel, H., López, R.L., Omar, N.B., Valdivia, E., et al., 2008. Inhibition of *Bacillus cereus* and *Bacillus weihenstephanensis* in raw vegetables by application of washing solutions containing enterocin AS-48 alone and in combination with other antimicrobials. Food Microbiol. 25 (6), 762–770.

Moosekian, S.R., Jeong, S., Ryser, E.T., 2014. Inactivation of sanitizer injured *Escherichia coli* O157:H7 on baby spinach using X-ray irradiation. Food Control. 36, 243–247.

Moynihan, E.L., Richards, K.G., Ritz, K., Tyrrel, S.F., Brennan, F.P., 2013. Impact of soil type, biology and temperature on the survival of non-toxigenic *Escherichia coli* O157. Biol. Environ. Proc. Royal Irish Acad. 113B, 41–46.

Murray, K., Wu, F., Aktar, R., Namvar, A., Warriner, K., 2015. Comparative study on the efficacy of bacteriophages, sanitizers, and UV light treatments to control *Listeria monocytogenes* on sliced mushrooms (*Agaricusbisporus*). J. Food Prot. 78, 1147–1153.

Murray, K., Wu, F., Shi, J., Xue, S.J., Warriner, K., 2017. Challenges in the microbiological food safety of fresh produce: limitations of post-harvest washing and the need for alternative interventions. Food Qual. Safety 1 (4), 289–301.

Netramai, S., Kijchavengkul, T., Sakulchuthathip, V., Rubino, M., 2016. Antimicrobial efficacy of gaseous chlorine dioxide against *Salmonella entericatyphimurium* on grape tomato (*Lycopersiconesculentum*). Int. J. Food Sci. Technol. 51, 2225–2232.

Nigro, F., Ippolito, A., 2016. UV-C light to reduce decay and improve quality of stored fruit and vegetables: a short review. Acta Horti. 1144, 293–298.

Nuesch-Inderbinen, M., Stephan, R., 2016. Fresh fruit and vegetables as vehicles of bacterial foodborne disease: a review and analysis of outbreaks registered by proMED-mail associated with fresh produce. J. Food Safety Food Qual. Archiv Für Lebensmittelhygiene 67, 32–39.

O'Bryan, C.A., Koo, O.K., Sostrin, M.L., Ricke, S.C., Crandall, P.G., et al., 2018. Characteristics of bacteriocins and use as food antimicrobials in the United States. Food and Feed Safety Systems and Analysis. Elsevier, Netherlands, pp. 273–286.

Oladeinde, A., Bohrmann, T., Wong, K., Purucker, S.T., Bradshaw, K., Brown, R., et al., 2014. Decay of fecal indicator bacterial populations and bovine-associated source-tracking markers in freshly deposited cow pats. Appl. Environ. Microbiol. 80, 110–118.

Olaimat, A.N., Holley, R.A., 2012. Factors influencing the microbial safety of fresh produce: a review. Food Microbiol. 32, 1–19.

Oliveira, M., Abadias, M., Usall, J., Torres, R., Teixidó, N., Viñas, I., 2015. Application of modified atmosphere packaging as a safety approach to fresh-cut fruits and vegetables–A review. Trends Food Sci. Technol. 46 (1), 13–26.

Oliver, D.M., Haygarth, P.M., Clegg, C.D., Heathwaite, A.L., 2006. Differential *E.coli* die-off patterns associated with agricultural matrices. Environ. Sci. Technol. 40, 5710–5716.

Oluoch, F., 2006. Lack of clean water kills more people than War: UNDP's Human Development Report 2006. The East African, November 13–19, 2006.

Othman, S.H., Abd Salam, N.R., Zainal, N., Basha, R.K., Talib, R.A., 2014. Antimicrobial activity of TiO_2 nanoparticle-coated film for potential food packaging applications. Int. J. Photoenergy 2014, 1–6.

Pagadala, S., Marine, S.C., Micallef, S., et al., 2015. Assessment of region, farming system, irrigation source and sampling time as food safety risk factors for tomatoes. Int. J. Food Microbiol. 196, 98–108.

Palekar, M.P., Taylor, T.M., Maxim, J.E., Castillo, A., 2015. Reduction of *Salmonella enterica*serotype Poona and background microbiota on fresh cut cantaloupe by electron beam irradiation. Int. J. Food Microbiol. 202, 66–72.

Palou, L., Crisosto, C.H., Smilanick, J.L., Adaskaveg, J.E., Zoffoli, J.P., 2002. Effects of continuous 0.3 ppm ozone exposure on decay development and physiological responses of peaches and table grapes in cold storage. Postharvest Biol. Technol. 24, 39–48.

Pao, S., Petracek, P.D., 1997. Shelf life extension of peeled oranges by citric acid treatment. J. Food Microbiol. 14, 485–491.

Paull, R.E., 1999. Effect of temperature and relative humidity on fresh commodity quality. Postharvest. Biol. Technol. 15 (3), 263–277.

Pereira de Abreu, D.A., PaseiroLosada, P., Maroto, J., Cruz, J.M., 2011. Natural antioxidant active packaging film and its effect on lipid damage in frozen blue shark (*Prionaceglauca*) Innov. Food Sci. Emerg. Technol. 12 (1), 50–55.

Perez, R.H., Ishibashi, N., Inoue, T., Himeno, K., Masuda, Y., et al., 2016. Functional analysis of genes involved in the biosynthesis of Enterocin NKR- 5-3B, a novel circular bacteriocin. J. Bacteriol. 198 (2), 291–300.

Perni, S., Shama, G., Kong, M.G., 2008. Cold atmospheric plasma disinfection of cut fruit surfaces contaminated with migrating microorganisms. J. Food Prot. 71, 1619–1625.

Pignata, C., D'Angelo, D., Fea, E., Gilli, G., 2017. A review on microbiological decontamination of fresh produce with non-thermal plasma. J. Appl. Microbiol. 122, 1438–1455.

Pinela, J., Ferreira, I.C., 2017. Non-thermal physical technologies to decontaminate and extend the shelf-life of fruits and vegetables: trends aiming at quality and safety. Crit. Rev. Food Sci. Nutr. 57, 2095–2111.

Posada-Izquierdo, G.D., Pérez-Rodríguez, F., López-Gálvez, F., Allende, A., Gil, M.I., Zurera, G., 2014. Modeling growth of *Escherichia coli* O157:H7 in fresh-cut lettuce treated with neutral electrolyzed water and under modified atmosphere packaging. Int. J. Food Microbiol. 177, 1–8.

Potter, H., Potter, N.N., Hotchkiss, J.H., 1995. Food science, fifth ed. Chapman & Hall, New York.

Qiulian, K., Mingrong, Z., Lili, Z., 2002. Inhibition on *Aspergillus niger* and *Penicillium expansun* of Chinese medicinal herb. Storage Process 3, 17–18.

Ramos, B., Miller, F.A., Brandao, T.R.S., Teixeira, P., Silva, C.L.M., 2013. Fresh fruits and vegetables-an overview on applied methodologies to improve its quality and safety. Innovative Food Sci. Emerg. Technol. 20, 1–15.

Rodoni, L., Casadei, N., Concellon, A., Chaves, A.R., Vicente, A.R., 2010. Effect of short-term ozone treatments on tomato (*Solanum lycopersicum* L.) fruit quality and cell wall degradation. J. Agric. Food Chem. 58, 594–599.

Sargent, S.A., Ritenour, M.A., Brecht, J.K., 2000. Handling, Cooling, and Sanitation Techniques for Maintaining Postharvest Quality. University of Florida. Available from: <http://edis.ifas.ufl.edu/CV115>.

Satin, M., 2000. Trends in Post-Production Technology. FAO, Rome Italy.

Schwarz, K., Sidhu, J.P., Pritchard, D., Li, Y., Toze, S., 2014. Decay of *Salmonella enterica, Escherichia coli* and bacteriophage ms2 on the phyllosphere and stored grains of wheat (*Triticum aestivum*). Lett. Appl. Microbiol. 58, 16–24.

Seifert, B., Zude-Sasse, M., 2016. High hydrostatic pressure effects on spectral-optical variables of the chlorophyll pool in climacteric fruit. Lwt-Food Sci. Technol. 73, 303–310.

Shayanfar, S., Mena, K.D., Pillai, S.D., 2017. Quantifying the reduction in potential infection risks from non-O157 Shiga toxin producing *Escherichia coli* in strawberries by low dose electron beam processing. Food Control. 72, 324–327.

Showalter, R.K., 1993. Postharvest water intake and decay of tomatoes. Hort. Technol. 3, 97–98.

Shynkaryk, M.V., Pyatkovskyy, T., Mohamed, H.M., Yousef, A.E., Sastry, S.K., 2015. Physics of fresh produce safety: role of diffusion andtissue reaction in sanitization of leafy green vegetables with liquid and gaseous ozone-based sanitizers. J. Food. Prot. 78, 2108–2116.

Shynkaryk, M.V., Pyatkovskyy, T.I., Yousef, A.E., Sastry, S.K., 2016. Gaseous ozone treatment of baby spinach within the existing production chain for inactivation of *Escherichia coli*O157:H7. J. Food Eng. 191, 10–18.

Siro, I., Devlieghere, F., Jacxsens, L., Uyttendaele, M., Debevere, J., 2006. The microbial safety of strawberry and raspberry fruits packaged in high-oxygen and equilibrium-modified atmospheres compared to air storage. Int. J. Food Sci. Technol. 41 (1), 93–103.

Smith, D.J., Ernst, W., Herges, G.R., 2015. Chloroxyanion residues in cantaloupe and tomatoes after chlorine dioxide gas sanitation. J. Agric. Food Chem. 63, 9640–9649.

Snyder, A.B., Worobo, R.W., 2014. Chemical and genetic characterization of bacteriocins: antimicrobial peptides for food safety. J. Sci. Food. Agric. 94 (1), 28–44.

Somboonkaew, N., Terry, L.A., 2010. Physiological and biochemical profiles of imported litchi fruit under modified atmosphere packaging. Postharvest Biol. Technol. 56 (3), 246–253.

Song, J., Fan, L., Hildebrand, P.D., Forney, C.F., 2000. Biological effects of corona discharge on onions in a commercial storage facility. Hort. Technol. 10 (3), 608–612.

Sun, X.X., Baldwin, E., Plotto, A., Narciso, J., Ference, C., Ritenour, M., et al., 2017. Controlled-release of chlorine dioxide in a perforated packaging system to extend the storage life and improve the safety of grape tomatoes. J. Visualized Exp. 6, 122.

Tadapaneni, R.K., Daryaei, H., Krishnamurthy, K., Edirisinghe, I., Burton-Freeman, B.M., 2014. High-pressure processing of berry and other fruit products: implications for bioactive compounds and food safety. J. Agric. Food. Chem. 62, 3877–3885.

The Schumacher Centre for Technology and Development. *Fruit Juice Processing*. E-mail: inforserv@practicalaction.org.uk 27 August, 2008.

Thomas, C., O'Brien, D., 2000. Evaluation of the impact of short term temperature abuse on the microbiology and shelf life of a model ready to eat vegetable combination product. J. Food Microbiol. 59, 47–57.

Tirado, M.C., Clarke, R., Jaykus, L.A., McQuatters-Gollop, A., Franke, J.M., 2010. Climate change and food safety: a review. Food Res. Int. 43, 1745–1765.

Tiwari, B.K., Muthukumarappan, K., O'Donnell, C.P., Cullen, P.J., 2008. Modelling colour degradation of orange juice by ozone treatment using response surface methodology. J. Food Eng. 88, 553–560.

Tzortzakis, N., Chrysargyris, A., 2017. Postharvest ozone application for the preservation of fruits and vegetables. Food Rev. Int. 33, 270–315.

Tzortzakis, N., Borland, A., Singleton, I., Barnes, J., 2007. Impact of atmospheric ozone-enrichment on quality-related attributes of tomato fruit. Postharvest Biol. Technol. 45, 317–325.

University of Maryland 2002. *Improving the safety and quality of fresh fruits and vegetables.*

Valverde, J.M., Valero, D., Martínez-Romero, D., Guillén, F., Castillo, S., Serrano, M., 2005. Novel edible coating based on aloe vera gel to maintain table grape quality and safety. J. Agric. Food Chem. 53 (20), 7807–7813.

Warriner, K., Huber, A., Namvar, A., Fan, W., Dunfield, K., 2009. Recent advances in the microbial safety of fresh fruits and vegetables. In: Taylor, S.L. (Ed.), Advances in Food and Nutrition Research, vol. 57. Elsevier Academic Press Inc, San Diego, CA, pp. 155–208.

Weller, D., Wiedmann, M., Strawn, L.K., 2015. Spatial and temporal factors associated with an increased prevalence of *Listeria monocytogenes* in spinach fields in New York state. Appl. Environ. Microbiol. 81, 6059–6069.

Wenqiang, G., Shufen, L., 2006. Research advances in application of natural plant extracts to post-harvest preservation of fruits and vegetables. Trans. Chinese Soc. Agric. Eng. 22 (7), 200–204.

Wilson, M.D., Stanley, R.A., Eyles, A., Ross, T., 2017. Innovative processes and technologies for modified atmosphere packaging of fresh and fresh-cut fruits and vegetables. Crit. Rev. Food Sci. Nutr. 11, 1–12.

Woolf, A.B., Wibisono, R., Farr, J., Hallett, I., Richter, L., Oey, I., et al., 2013. Effect of high pressure processing on avocado slices. Innovative Food Sci. Emerging Technol. 18, 65–73.

Xiangqiu, Z., Jianmei, W., Zhenchuan, M., 1997. The Colloid Characteristc of Seeds of Saloides (L). Gaertn and Its Application on Food Industry. Colloquium Collection of Essays About International Food Process and Packing Technology. Engineering Industry Publishing House, Beijing.

Xiaogang, S., Min, C., Yahong, W., 1996. Studies on the fresh-keeping effect of several Chinese medicinal herb and compound preservative paper for juicy pear. Food Sci. 17 (2), 67–69.

Xu, C., Wang, D., Huber, A., Weese, S.J., Warriner, K., 2016. Persistence of *Clostridium difficile* in wastewater treatment-derived biosolids during land application or windrow composting. J. Appl. Microbiol. 120, 312–320.

Xuan, Z., 2001. Advance in using natural Chinese herb as fruits and vegetables preservatives. Xinjiang Agric. Sci. 41 (Special Issue), 110–111.

Yahia, E.M., 2009. In: Yahia, E.M. (Ed.), Modified and Controlled Atmospheres for the Storage, Transportation, and Packaging of Horticultural Commodities. Taylor and Francis Group LLC, Boca Raton, FL, 608 pp.

Yang, Y., Luo, Y., Millner, P., Turner, E., Feng, H., 2012. Assessment of *Escherichia colio*157: h7 transference from soil to iceberg lettuce via a contaminated field coring harvesting knife. Int. J. Food Microbiol. 153, 345–350.

Yaun, B.R., Sumner, S.S., Eifert, J.D., Marcy, J.E., 2004. Inhibition of pathogens on fresh produce by ultraviolet energy. Int. J. Food Microbiol. 90, 1–8.

Yesil, M., Kasler, D.R., Huang, E., Yousef, A.E., 2017. Efficacy of gaseous ozone application during vacuum cooling against *Escherichia coli o*157: h7 on spinach leaves as influenced by bacterium population size. J. Food. Prot. 80, 1066–1071.

Zambuchini, B., Giosia, L., Sturba, M., 2006. Study of the ozone effect on the shelf-life of minimally processed salad (Review). Postharvest Biol. Technol. 45, 636–643.

Zhang, Y., Zhou, L., Zhang, Y., Tan, C., 2014. Inactivation of *Bacillus subtilis* spores using various combinations of ultraviolet treatment with addition of hydrogen peroxide. Photochem. Photobiol. 90, 609–614.

Zhang, M., Meng, X., Bhandari, B., Fang, Z., Chen, H., 2015. Recent application of modified atmosphere packaging (MAP) in fresh and fresh cut foods. Food Rev. Int. 31 (2), 172–193.

Zilelidou, E.A., Tsourou, V., Poimenidou, S., Loukou, A., Skandamis, P.N., 2015. Modeling transfer of *Escherichia colio*157: h7 and *Listeria monocytogenes* during preparation of fresh-cut salads: impact of cutting and shredding practices. Food Microbiol. 45, 254–265.

Further Reading

Castillo, A., Mckenzie, K.S., Lucia, L.M., Acuff, G.R., 2003. Ozone treatment for reduction of *Escherichia coli* O157: H7 and *Salmonella* serotype Typhimurium on beef carcass surfaces. J. Food Prot. 66 (5), 775–779.

FDA, 2005. US Food and Drug Administration (US FDA). Centre for Food Safety and Applied Nutrition (CFSAN). Foodborne Pathogenic Microorganisms and Natural Toxins Handbook.

Liangfeng, Z., Biyao, L., 1991. Studies on natural fresh-keeping agents for fruits. Lett. China Food Addit. 4, 32–35.

Ming, Y., 2006. Studies on natural fresh-keeping agents for fruits and vegetables. China Net Appl. Food Addit. 2006.1.10.

CHAPTER

11

Utility of Nanomaterials in Food Safety

Ravindra Pratap Singh

Department of Biotechnology, Indira Gandhi National Tribal University, Amarkantak, India

OUTLINE

INTRODUCTION

The recent utility of nanomaterials in food science which involves processing, packaging, storage, transportation, and a few significant functionalities has raised the cause of concern of food safety and human health. The nanostructured materials or nanomaterials often used or applied within the food trade for a large variety of advantages and rising to

Food Safety and Human Health
DOI: https://doi.org/10.1016/B978-0-12-816333-7.00011-4

accumulate not solely in human bodies; however, additionally within the environment to cause serious issues or threats to human health and food safety in terms of biohazards (Cockburn et al., 2012). Therefore, food safety, human health issues, food regulative policies bearing on producing, processing, good packaging, and overwhelming nanofood product should be focusing and supply basic understanding relating to the utility of nanomaterials within the food packaging and process industries (Martirosyan and Schneider, 2014). Recent innovations as well as economical globalization have changed the people's eating habits due to the use of a variety of diverse chemicals in our food, namely additives, agricultural chemicals, contaminants, toxicants, nanomaterials, and other ingredients to improve human health and prevent lifestyle diseases. The food safety concerns are increasing rapidly (Zeng et al., 2014). Nanomaterial use in food is the main cause of concern due to their small size (diameters of ≤ 100 nm), shapes, and unique properties. When the size of nanomaterial decreases, their surface area increases with increase in surface coverage. Several nanomaterials containing food products are known, but the toxicity and safety of nanomaterials has not been thoroughly investigated (McGovern, 2010). The current uses of nanomaterials in the food industry have to be studied in this book chapter. Fig. 11.1 shows a variety of nanomaterials utilized in food trade.

Application of nanotechnologies within the food business has begun. New tools and techniques, that is, nanobiosensors were developed for the nanobiosensing to find type of

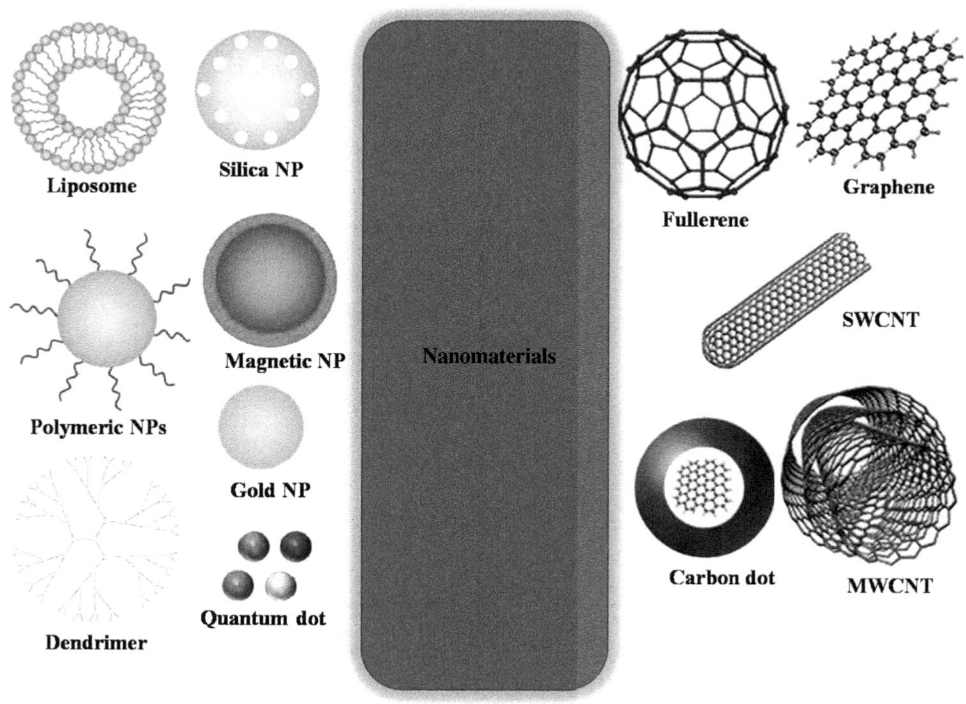

FIGURE 11.1 A few varieties of nanomaterials utilized in food trade.

analytes of interest within the food trade. Consumers' demand for healthy foods has encouraged investigators to utilize nanomaterials in the food and nutrition. The nanomaterials are used to coat food packaging, and nanosieves are used to filter food microbes. The food products containing manmade nanomaterials are titanium dioxide and silica nanoparticles which are used as food additives and antimicrobial activity of chitosan, silver nanoparticles, and photocatalytic titanium dioxide used in the food industry. The nanoencapsulation technologies are used for the development of manmade colorants, preservatives, and aroma. Containers and wrapping materials for the food contain nanomaterials that have improved the packaging properties, to check gas exchange against temperature and moisture. Nanomaterials containing food showed antibacterial activity. Containers and wrapping materials containing nanosensors, referred to as active/smart/intelligent food packaging are used to detect the condition of food when packaged (Sekhon, 2014; Wang et al., 2013).

The book chapter highlights wide discussion pertaining to utility of nanomaterials in food safety management and security. It covers a colossal background of the present literature supporting the proposed topic in numerous domains that is extremely helpful to researchers concerned in this topic of an interdisciplinary approach and additionally throws light on the recent trends and developments to create essential work to scientists, technologists, and graduate and postgraduate students.

NANOMATERIALS IN FOOD PROCESSING

Various artificial and natural polymers-based mostly encapsulating delivery systems are developed for the improved bioavailability and preservation of the active food parts. The importance of nanomaterials in food process are often evaluated by considering its role within the improvement of food product in terms of food texture, food look, food taste, organic process worth of the food, and food time period. The nanomaterial in nanofood not only enhances taste and texture but also provides consistency. Fig. 11.2 presents the utility of nanomaterials in food processing in the food industry.

Pradhan et al. (2015) reported use of technology in food process, packaging, and preservation trade. Nanomaterials have increased the time period of various types of foodstuffs to ascertain wastage of food. Bratovcic et al. (2015) reported the nanocarriers that are used to hold food additives in food products. Nanomaterials are employed in the formation of nanoencapsulation, nanoemulsions, nanobiopolymer matrices, or nano polymers utilized in food packaging. Nanosensors are accustomed to detect pollutants, toxicants, fungal toxins, and microbial pathogens in food. Ubbink and Kruger (2006) reported that nanoparticles have higher properties for encapsulation and distribution potency than ancient encapsulation systems. Nanoencapsulations mask odors or tastes, management interactions of active ingredients with the food matrix, management the discharge of the active agents, guarantee accessibility at a target time and specific rate, and shield them from wetness, heat, chemical, or biological degradation throughout process, storage, and utilization, and additionally exhibit compatibility with alternative compounds within the system. Lamprecht et al. (2004) reported the nanodelivery systems for the economical delivery of active compounds in vivo. Nanomaterials might give to enhance not only the food quality

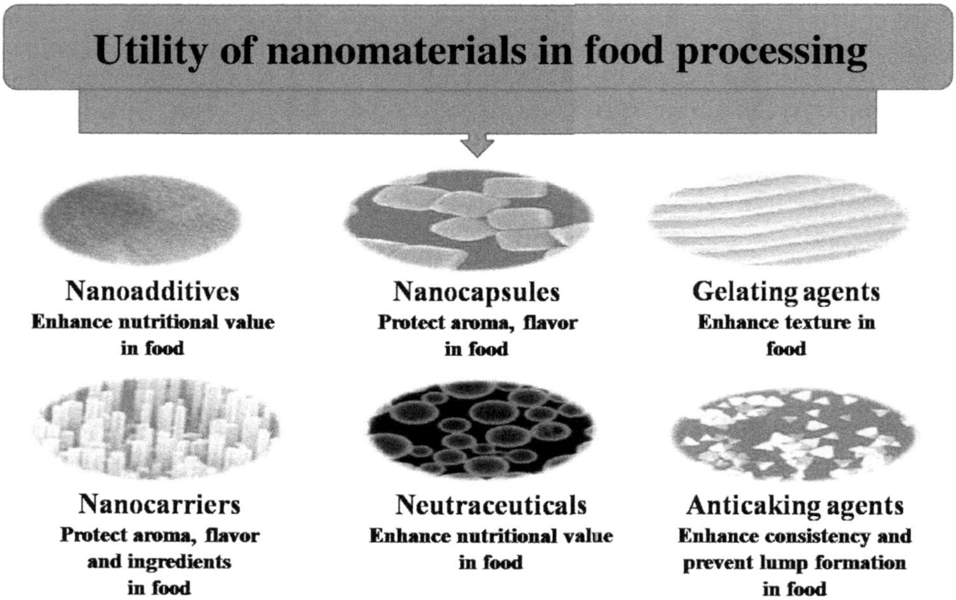

FIGURE 11.2 Utility of nanomaterials in food processing in the food industry.

but also provide taste and texture of food. Nanoencapsulation has enhanced the food flavor distribution and retention. Zhang et al. (2014) reported to utilize the nanoencapsulation of anthocyanins for not only the look but also for the design of multifunctional nanocarriers. Yang et al. (2015) reported that rutin, a typical dietary flavonoid with medical specialty activities, however, has poor solubility and restricted application within the food trade. However, encapsulated ferritin nanocages have increased the solubility, thermal stability of ferritin cornered rutin when put next with the free rutin. Ozturk et al. (2015) reported the use of natural biopolymers for the formation of nanoemulsion-based vitamin E delivery systems to deliver natural lipid bioactive food ingredients that enhance water-dispersion and bioavailability.

Dekkers et al. (2011) reported SiO_2 in foodstuff and showed its health risks. They have also demonstrated that TiO_2 and silicon oxide (SiO_2) showed coloring agents in food product with long shelf-life. However, SiO_2 as nanomaterials in foodstuff are as flavoring agents. The bulk of bioactive compounds, particularly carbohydrates, lipids, proteins, enzymes, hormones, and vitamins are very prone at acidic condition in abdomen and small intestines so that if encapsulated bioactive compound has developed then it is not solely resistant to the adverse conditions. However, it additionally allows assimilating in food. Koo et al. (2005) reported that nanoparticles based mostly edible capsules not only enhance drug delivery but also facilitate maximum intake of vitamins and trace elements. Langer and Peppas (2003) reported in their critical review that the various techniques that are utilized to nanoencapsulate the active ingredients for the delivering of nutrients like saccharide, supermolecules, and antioxidants. Weiss et al. (2006) reported the purposeful materials in food technology. The useful food with bioactive part once nanoencapsulated, then the food product extends its shelf-life. The nanocoatings of edible foodstuffs do not seem to be solely checked wetness and gas but additionally provide coloring, flavoring, texturing, antioxidant activity, and enzymatic activity. Excluding these, it might extend the time period of foodstuffs. Sari et al. (2015) reported nanoemulsion-based encapsulated curcumin and demonstrated more active and stable with antioxidant activity when compared to normal curcumin.

NANOMATERIALS IN FOOD PACKAGING

The nanomaterials in food packaging have been performing smartly. Fig. 11.3 shows utility of nanomaterials in food packaging in food trade.

AgNPs or TiO_2 as antimicrobials are utilized in food packaging to ascertain spoilage of foods. The clay nanoparticles employed in food packaging as air-tight packaging material are accountable for obstruction oxygen, carbon dioxide, and wet to forestall spoilage of food because of microbes, for instance, a clear film Durethan containing nanoparticles of clay derived from combination of polyamide 6 and polyamide 66. The clay nanoparticles exhibit strength and toughness to avoid abrasion, crackness. Because of these distinctive properties of Durethan it's utilized in medical as packaging film and also in food packaging. The nanocomposites primarily based bottles minimize the discharge of carbon dioxide and will increase the period of time of an effervescent drink. AgNPs embedded plastics kill microorganisms to forestall food from spoilage. The nanosensors are the sensors that are embedded in plastic packaging to not solely

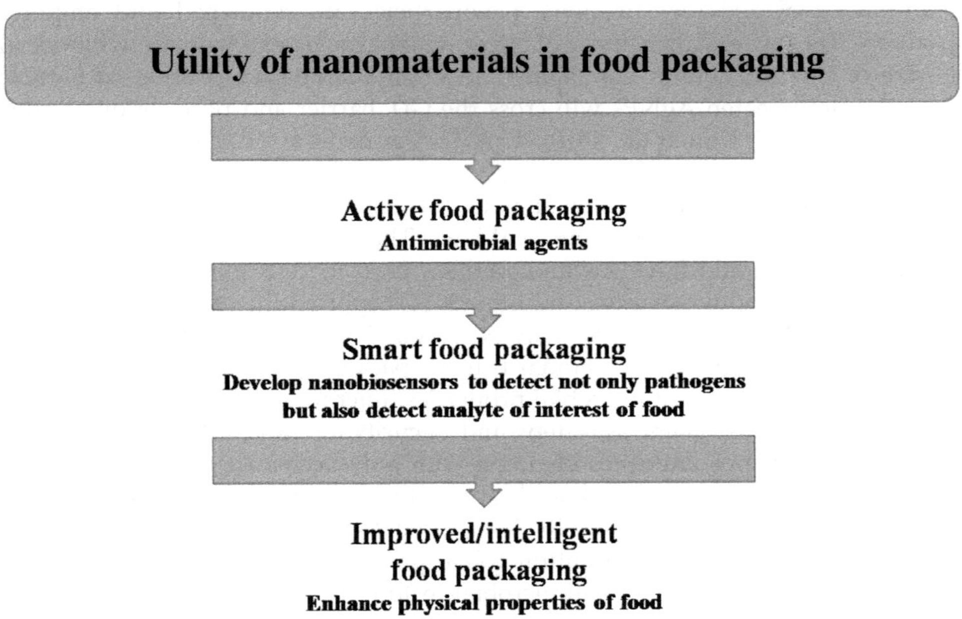

FIGURE 11.3 Utility of nanomaterials in food packaging in food trade.

find gases/microbes given off by food once it spoils, however, additionally find the physicochemical changes like color to alert us that food package has terminated. Plastic films containing silicate nanoparticles enable food to remain fresh longer that scale back the flow of oxygen into the package and therefore the leaky of wet out of the package. Nanobiosensors are able to find microorganisms like salmonella on the surface of food at an inside and outdoors packaging plant. The frequent food packaging testing can be potential by nanosensors among less time and low price instead of analyzed in an exceeding laboratory. The point-of-food packaging testing has utilized nanosensors to ascertain the standard of food to not solely scale back the prospect of contaminated food, however, additionally find pesticides on fruit and vegetables. The packaging of food has to be compelled to keep the merchandise safe and secure. That's why food safety measures are obligatory for us. We have a tendency to not need covering natural ingredients; environmental gases and water droplets are governed by natural forces. Although, within the food packaging, quality and porousness of gases are completely obstruction or forestalled or prohibited for the fresh fruits and vegetables packaging, otherwise internal respiration begins to spoil each. However, just in case of packaging of effervescent beverages, the flow of carbon dioxide and O_2 forestall decarbonation and oxidation and stable the product. In order that water content, carbon dioxide, and O_2 concentration varies, relying upon the food matrix likewise as what food packaging materials are used for higher packaging. So we will say that food packaging is an extremely complicated and sophisticated method. It is centered and overcome by mistreatment type of varied eco-friendly materials and nanomaterials. The key nanomaterials employed in food products are silver, silica, and titanium oxide. These nanomaterials are the most widely used and common in food products (Frohlich and Roblegg, 2012).

The antimicrobial activities of silver nanoparticles are renowned and employed in food products. The bactericide activity of silver nanoparticles will increase with decreasing particle size or increase within the expanse to mass quantitative relation as particle size decreases. But orally eaten AgNPs will cross the GIT barrier and pose serious health problems (Lok et al., 2007; Kim et al., 2010). Van der Zande et al. (2012) reported that in rats exposed to silver nanoparticles for 28 days and recommended that the silver nanoparticles evoked no acute hepatotoxicity. Couch et al. (2016) reported in their critical review as food technology, projected uses, safety issues, and laws. They illustrated that a fascinating packaging food material should have gas and wetness porosity and mirrored its strength and biodegradability. Mihindukulasuriya and Lim (2014) reported the technology development in food packaging. Nanomaterial primarily based good, active, and intelligent food packagings have many benefits over standard food packaging like improved mechanical strength, barrier properties, antimicrobial properties, microorganism detection by nanobiosensor, and alerting customers as safety and security of foodstuff. Pinto et al. (2013) reported the copper nanocomposites blending with polysaccharide and showed antibacterial activity. They have demonstrated application of nanocomposites for food packaging to forestall food spoilage. Galvez et al. (2007) reported food biopreservation using essential oils and bacteriocins. They showed antimicrobial properties and their utility in polymeric matrices once encapsulated act as antimicrobial in food packaging. The food process procedure needs high temperatures and pressures in order that chemical compounds could not face up to as a result of their sensitivity to temperature and pressure.

Schirmer et al. (2009) reported a completely unique packaging technique using carbon dioxide head and organic acids which enhance the time period to preserve contemporary fish salmon. Soares et al. (2009) reported smart packaging for dairy product using nanoparticles that showed antimicrobial activity. An antimicrobial food packaging may be a quite active/smart/intelligent food packaging type. The Ag, Cu, TiO_2, and ZnO are reported to act as bactericide activity. Tan et al. (2013) reported an antimicrobial agent using quaternized chitosan which demonstrated an antimicrobial activity. The appliance of nanoparticles is not restricted to food packaging, but shelf-life of food packaging might be increased using nanocomposite and nanolaminates. However, addition of nanoparticles into food packaging enhances not only the food quality but also prolongs the time period of the food. Apart from these, polymeric nanocomposites in food packaging provide an additional mechanical as well as thermostability (Duncan, 2011). Sorrentino et al. (2007) reported the potential views of bionano composites for food packaging applications. The inorganic or organic fillers are accustomed created polymer composites, once incorporation of nanoparticles in polymer composite become polymer nanocomposites that are well-known to resist food packaging material. The clay, silicate nanoplatelets, silicon dioxide (SiO_2) nanoparticles, chitin or chitosan are well-known inert nanofillers, if they are incorporated in polymeric matrix and showed fire retardant activity. Othman (2014) reported the bionanocomposite in food packaging using biopolymer and nanofiller. Rhim and Ng (2007) reported the natural biopolymer primarily based nanocomposite films for packaging applications. They have demonstrated that antimicrobial nanocomposite films are ready by impregnating the fillers into the polymers that showed its structural integrity and barrier properties. Table 11.1 shows utility and potential result of few normally well-known nanomaterials in food packaging.

TABLE 11.1 Some Commonly Known Nanomaterials in Food Packaging

Utility of Nanomaterials	In Food Packaging	Potential Effects	References
AgNPs	Asparagus, orange juice, poultry meat, fresh-cut melon, beef meat	Prevent the growth of yeasts and molds as an antimicrobial agent in response to *E. coli* and *S. aureus*	An et al. (2008)
			Fernandez et al. (2009)
			Fernandez et al. (2010a)
			Fernandez et al. (2010b)
			Emamifar et al. (2010)
Au nanorod	*E. coli* O157:H7	–	Fu et al. (2008)
ZnO NPs	Orange juice, egg albumen, etc.	Reduces *Lactobacillus plantarum*, *Salmonella*, yeast and mold counts	Emamifar et al. (2011)
			Jin and Gurtler (2011)
TiO_2 NPs	Chinese jujube, strawberry	Reduces browning, ripening, senescence and spoilage	Li et al. (2009)
AgO NPs	Apple slice	Retards microbial spoilage	Zhou et al. (2011)

NANOMATERIALS IN SMART/ACTIVE/INTELLIGENT FOOD

The nanomaterials utility in food business are rising and increasing within the domain of food process, food packaging, food safety, food shelf-life, food product, purposeful food, and food-borne pathogens. Fig. 11.4 shows the application of nanomaterials in active/smart/intelligent foodstuffs.

In the food business, processed food and food products should be safe and contamination free for the customers with improved useful properties. There are some insights on food questions of safety alongside food regulative rule on nanoprocessed food product which would be matter of our issues. The patron issues pertaining to food quality and health advantages are essential aspects. The demand of nanomaterials has been accrued within the food business. Nanomaterials based mostly on packagings extend the food period to examine gas and wetness exchange. Smart food packagings have nanosensors (tiny chips, i.e., pursuit device or electronic barcode) and antimicrobial activators to find food spoilage and accountable microbes and ready to extend food period of time (Brody et al., 2008). Smart foods are client minded food and individualize their food to dynamical color, flavor, concentration, texture, and nutrients on demand by using nanoencapsulation technique. Nanofood packaging needs nanosensor or nanobiosensor to find

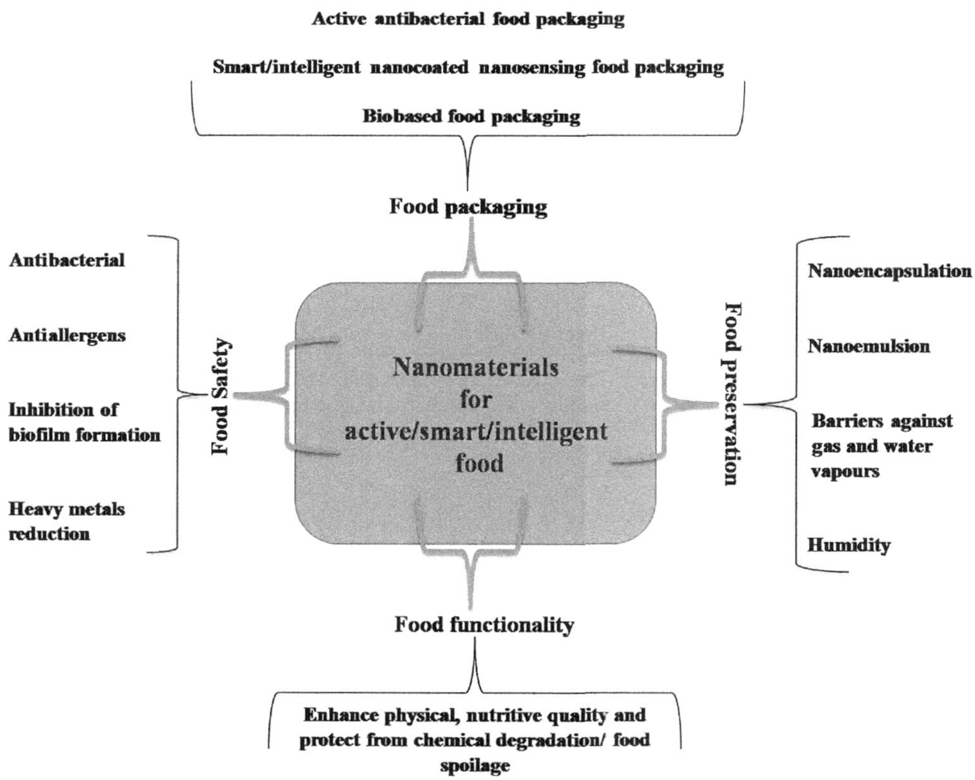

FIGURE 11.4 Application of nanomaterials in active/smart/intelligent foodstuffs.

food contamination, food adulteration, toxicants detection, pathogen detection, for stable food packaging as well as food storage, and pesticide detection, pursuit likewise astracing whole protection, texture foil, flavor foil, and microorganism elimination (Kirtiraj et al., 2018).

Nowadays R&D on smart food packaging nanofoods and its observation may be a major focus within the food business. Nanofoods will reply to environmental conditions not solely repair it; however, additionally alert a client regarding food contaminations because of the pathogens. Nanomaterials based on bottom-up approach in food packaging to create nanobioindustrial product (i.e., nanofood product) with safety packaging ready to find spoilage or microbes, contaminates, toxins, and pesticide (Janjarasskul and Suppakul, 2018). Gupta et al. (2016) reported nanomaterials on the nanometer scale (1−100 nm) to utilize the method and its nanofood product due to high surface to volume aspect ratio and physiochemical properties. Ezhilarasi et al. (2013) reported the nanomaterials utility in the food sector (i.e., from food processing to food packaging) which have been classified as nanostructured food components and nanosensing of food.

NANOMATERIALS UTILITY IN POLYMER NANOCOMPOSITES

Nanocomposites became a central domain of scientific and technical activity to assist humanity. The advances in nanostructured composites and their trends, organic process applications in numerous fields together with in water treatment, green energy generation, anticorrosive, arduous coatings, antiballistic, optoelectronic devices, solar cells, biosensors, and nanodevices, have been reportable by many investigators enormously (Singh, 2017). The nanocomposites are new polymers having mechanical and thermal properties with high strength and stability to check gas in and out and acts as a gas barrier. In food packaging, they were used as a high barrier to examine O_2, CO_2, and moisture content and become used as a property food packaging material (Dean and Yu, 2005).

Nanocomposite materials are mixtures of two or a lot of elements, generally consisting of a couple of matrixes containing one or additional fillers created from particles, sheets, and fibers with dimensions but 100 nm, having a better surface to volume quantitative relation. The vital property of nanocomposites is that they're less porous than regular plastics, creating them as ideal to use within the packaging of foods and drinks, vacuum packs, and to safeguard medical instruments, film, and different products from outside contamination. of these developments are attainable because of their distinctive properties of nanostructures composites at nanoscale level which might amendment dramatically by a reduction in size, shape, dimension, and exhibit newer properties together with reactivity, electrical conduction, insulating behavior, elasticity, strength, and color (Singh, 2011, 2016; Singh et al., 2014).

Filler is incorporated in nanocomposites a minimum of one dimension <100 nm and shown higher mechanical, thermal properties in term of additional heat resistant, and high barrier materials, which are beneficial for the food process, food transportation, as well as food storage (Thostenson et al., 2005). Ramanathan et al. (2008) reported graphene nanoplates, which showed heat resistant along with gas barrier and was utilized in smart food packaging.

Multifunctional nanocomposite is an exciting and quick evolving field that has higher surface assimilation capability, property, and stability. Therefore, they need vast potential for significant metal detection, pollutants, toxicants, and adulterants from the contaminated water. The nanocomposites having additives moreover as filler counterparts showed substantial property enhancements like mechanical properties, diminished porosity to gases, water and hydrocarbons, thermal stability, flame retardancy and reduced smoke emissions, chemical resistance, surface look, electrical conduction, and optical property (Zhu et al., 2001; Bourbigot et al., 2002).

The gas barrier property enhancements shown by nanoclay as nanofiller in nanocompotes are established. Such glorious barrier characteristics have resulted in food packaging applications for processed meats, cheese, confectionery, cereals, fruit crush, dairy farm products, brews, and effervescent drink bottles. They need the ability to reinforce the period of time of the many foods. Honeywell encompasses a combined active/passive oxygen barrier system in nature. Artificial polyamide-6 with nanoclay particles incorporation might show an oxygen scavenging species as a manmade combined active/passive system. Nanoclay polymer composites are presently used in food packaging materials that do not need refrigeration and are capable of maintaining food freshness for a while. It's a glorious vaporous barrier. It's going to additionally be attainable to develop films for artificial intestines in future. A nanoclay filler material in nanocomposites is based mostly for fuel tanks for cars, which might be reduced to solvent outflow and additionally reduced the price. The presence of nanofiller in nanocomposites might have vital effects on the transparency and haze characteristics once targeted to form films for use in car windows. Nanoclay might enhance transparency and cut back haze. Nano changed polymers are used as coating chemical compound transparency materials that enhance toughness and hardness while not impeding light-weight transmission (Garces et al., 2000; Caroline, 2002; Singh, 2018).

The vast nanocomposites utility in food packaging has high barrier properties to prevent O_2 and CO_2 mobility, and moisture to prolong time period of processed foods and maintain the freshness of the taste, texture, and flavor of nanofoods. Best effects are reported as smart compatibility between filler and polymer, and enhancements can be expected for higher food process technologies (Karger-Kocsis and Zhang, 2005). Furthermore, biopolymer nanocomposites are used in food packaging derived from plants, animals, and microbial products, mainly polysaccharides, proteins, and polyhydroxybutyrate (Wu et al., 2002). Seashells are natural nanocomposites (carbonate and aragonite) derived from mineralized collagen fiber, that is, hydroxyapatite (Ca_5 $(PO_4)_3OH$) plates with high strength and toughness (Fratzl et al., 2004).

Starch is a carbohydrate utilized in food packaging with high access and low cost value (Charles et al., 2003). Once inorganic materials and artificial polymers (Avella et al., 2005; Cyras et al., 2008) are mixed with starch, then it improves water resistance. Starch-clay could be perishable nanocomposites used in food packaging (Yoon and Deng, 2006) as a barrier in bottling. Starch/ZnO-carboxymethyl cellulose metallic element nanocomposite was used to check water vapor porosity. Polylactic acid (PLA) is utilized in food packaging having biocompatible, perishable, with smart mechanical and optical properties (Valerinia et al., 2018). Sinclair (1996) reported the mixture of PLA and montmorillonite bedded silicate as a nanocomposite which was utilized in food packaging as a smart food. Lee et al. (2005) reported film-forming proteins, for example, casein, whey, collagen, albumin, soybean, and zein in food packaging. Whey protein is used as film-forming material. Sothornvit and Krochta (2005) reported the whey protein films act as a gas barrier and also showed antimicrobial properties if blended with TiO_2 to form a nanocomposite. Similarly, soy protein has shown thermoplastic and perishable properties and if blended with nanomaterial to form nanocomposite films which prevent gas permeableness, and prolong durability. Zein is also protein derived from maize, a hydrophobic protein forming protein film utilized in the food industry. Abbaspour et al. (2015) have addressed complexities within the food packaging and observe bacterium. Nanomaterials have a diameter within the range of 1–100 nm and showed numerous structures of nanomaterials and its utility within the various fields of the food industry. Ghanbarzadeh et al. (2014) reported nanostructured biopolymeric material utilized in food packaging which act as a barrier to stop gases to avoid spoilage of food. Abdollahi et al. (2012) reported a unique active bio-nanocomposite using rosemary oil, nanoclay, and chitosan. They utilized perishable polymers strengthened with nanofillers in food packaging, which are eco-friendly; however; we could not rule out its toxicity concerning the consumption of this nanofood product. Azeredo et al. (2011) reported in their critical review that nanomaterials migrate inside the material body and showed impact of toxicity and immunogenic effects. Moreover, Klaine (2009) reported the issues for environmental analysis and its fate along with impact of nanoparticles for the worldwide researchers to seek out the ecofriendly resolution. Hannon et al. (2015) reported prospects and challenges of nanoparticles in food contact materials. They have classified food packaging as smart, active, and intelligent. Intelligent food packaging materials have unleashed nanoparticles and act as antibacterial drug agents. Echegoyen and Nerin (2013) reported nanoparticle unleashed from nanosilver antimicrobial food containers. Mills (2005) reported nanosized titanium dioxide (TiO_2) as an intelligent ink to check oxygen in food packaging.

Nowadays, food packaging trade has addressed the various forms of nanostructures and desired nanomaterials, nanoclay as nanocomposite because of their affordable, method ability, accessibility, and nice performance. Arora and Padua (2010) reported in their critical review on nanocomposites in food packaging. They demonstrated the utility of graphene nanosheets, and carbon nanotubes in food packaging. Stormer et al. (2017) in their critical review illustrated that polymeric films over bottles (glass and plastic) using nanocoating technique have the ability to dam out oxygen and aromas. They're in nice demand in material packaging. Duncan (2011) reported an intelligent kind of nanocoating film primarily based nanofood which might not solely check food contamination throughout production and storage but also prevents CO_2, wetness, and O_2. Ariyarathna et al. (2017) reported maximum utility of inorganic nanomaterial in food packaging. Yu et al. (2017) reported polyvinyl alcohol/chitosan polymer-based perishable films employing a nanocomposite of silicon dioxide in place that forestalls the permeableness of oxygen and wetness considerably. Swaroop and Shukla (2018) reported nanomagnesium chemical compound strengthened PLA biopolymer biofilms in food packaging to defend from microorganism biofilms.

NANOMATERIALS AS ANTIMICROBIAL AGENTS

The treatment of infection with drugs/antibiotics while not harming the host cells was established. In another sense, infections caused by pathogens which may be prevented, treated through antibacterial activity of compounds known as antibiotics/nanomaterials. These are natural, semisynthetic or artificial in nature that kill or inhibit the growth of infectious agents (Woon and Fisher, 2016). Dasgupta et al. (2015) reported in their critical review on technology in agrofood from field to plate. They highlighted the employment of nanoparticles having antimicrobial properties shield food products from food-borne unwellness outbreaks (i.e., spoiled packaged food). An energetic food-packaging has the power of a passive barrier to forestall oxygen or water vapors that enhance food stability. The active or intelligent food packaging contains active molecules and has the power to soak up or distribute the specified constituents into or out of the encircling surroundings of the packaged food. In food packaging, numerous desired bioactive constituents are often value-added by encapsulation with nanomaterials. Ghaani et al. (2016) reported within their critical review on an outline of the intelligent food packaging technologies. Fang et al. (2017) reported smart food packaging in the meat business. They illustrated that antimicrobial active food packaging makes sure of the food quality by physical look and sensory levels. Hu et al. (2017) reported organic compound based nanodelivery systems for polyphenols. The molecule-based nanoparticles in food not solely improve the bioavailability of bioactive polyphenolics, like resveratrol, epigallocatechin-3-gallate, and curcumin, however, additionally enhance the solubility of those polyphenols and therefore forestall their degradation within the gastrointestinal surroundings. Food process and food storage needs encapsulation of nutraceuticals and useful antimicrobial ingredients for not solely the food preservation, however, additionally for the bioavailability of bioactive ingredients. There are numerous encapsulation techniques that are accustomed for turn out nano systems, specifically nanoemulsion, coacervation, extrusion technique, spray cooling, and spray drying (Silva et al., 2012; De Conto et al., 2013; Gibbs et al., 2010; Murugesan and Orsat, 2012).

Nanoencapsulated nanomaterials are utilized in the consumption of food. Dekkers et al. (2011) reported nanosilica (SiO_2) was used as a fragrance carrier within the form of food product. Mozafari et al. (2006) reported within their critical review that lipid-based nanoencapsuled antioxidants in food which enhance not only the activity but also increasing bioavailability and solubility of antioxidant contents. Additionally, they mention that nanoencapsulated foods are target orientating, site-specific with economical absorption. Flores-Lopez et al. (2015) reported nanosized edible coating of fruit and vegetables not only enhances the effective preservation but also extends storage time period of food to prevent microorganism in spoilage of food. However, Shi et al. (2013) reported the chitosan/nanosilica coating on longan fruit underneath close temperature and recommended that this nanocoating is critical to forestall food spoilage. Medeiros et al. (2014) reported alginate or lysozyme nanolaminate coating on fresh foods prolongs food preservation and storage. Yang et al. (2010) reported impact of nanopackaging on strawberries throughout storage at 40°C to enhance food preservation quality. They demonstrated that the nanopackaging-based technique used polyethylene and nanopowders like silver, kaolin, anatase TiO_2, and mineral TiO_2 to preserve fruits like strawberries.

A lot of analysis is to be administrated by investigators on nanoencapsulation utilizing numerous nanomaterials; however, the precise delivery of nanoencapsulated foods also as their safety could not be studied very well. During this context, future researches on nanoencapsulated foods have to be compelled to be administrated bearing on short-term/long-term toxicity (acute/chronic) (Elgadir et al., 2015; Jovanovic, 2015). Additionally to the current, Johnston (2010) reported fabricated clay minerals blended calcium silicate (CS) with silver ions to form nanostructured CS-Ag and showed antimicrobial activity useful in food packaging. Fujishima et al. (2000) reported titanium dioxide surface coater and used as a photocatalytic disinfecting agent whereas Chaweng kijwanich and Hayata (2008) reported TiO_2 to form film in food packaging and showed to inactivate Escherichia *coli* in vitro. Qi et al. (2004) reported an antibacterial drug activity potential of chitosan-capsulated nanoparticles and result in increased membrane permeableness and run of the intracellular material. Sarwar et al. (2018) reported fabrication of polyvinyl alcohol PVA/nanocellulose/Ag nanocomposite films and showed antimicrobial potentiality used in food packaging. Valerinia et al. (2018) reported aluminum-doped zinc oxide coatings on polylactic acid films and showed sturdy antibacterial drug activity potential used in the active food packaging. Lu et al. (2018) reported nanoemulsions using volatile oil and showed antimicrobial activity used in food packaging or even in the food system. Metal and metal oxide NPs based nanocomposites are also used in active food packagings and known to act as antimicrobial activity. Liau et al. (1997) reported AgNPs metal nanoparticles toxic to food pathogens as a result by degrading lipopolysaccharide. However, Sondi and Salopek-Sondi (2004) reported AgNPs penetrability potential due to Ag^+ as an antibacterial nature which not solely harms the microorganism cell but additionally harm its DNA, and unleash antimicrobial silver ions that bind to S, O, N containing electron donor groups and as a result inhibited ATP formation and DNA replication (Morones et al., 2005). Karimi et al. (2018) reported Ag^+ potentiality to cause living substance shrinkage rupture of cell walls and breakage of peptidoglycan within the plasma membrane, ribosomes, and DNA harm successively inhibits DNA synthesis and necrobiosis.

Pathkoti et al. (2017) reported the microorganism toxicity of metal-containing NPs that causes the nucleotide depletion and ROS production because of that oxidative cellular harm probably occurred. Suppakul et al. (2003) reported nanomaterial-based smart technologies related to food packaging free from microbes. Ahmed et al. (2017) reported in their comprehensive review on active food packaging technologies to muscle foods with high antimicrobial result that can be achieved onto food packaging films coated with numerous antimicrobial parts. Arfat et al. (2017) reported designed bionanocomposite films supported fish skin, gelatin, and Ag/Cu NPs and showed antibacterial activity response potential against Listeria *monocytogenes* and *Salmonella enterica*.

NANOMATERIALS IN FOOD PATHOGENS DETECTION

Nanomaterials for the utilization within the construction of nanosensors and nanobiosensors supply the high level of sensitivity with different attributes as Table 11.2 shows the utility and impact of a few most typical nanomaterials pertaining to detection of

TABLE 11.2 Utility and Impact of a Few Most Commonly Used Nanomaterials for the Detection of Food-Borne Pathogens

Nanomaterials-Based Nanodevices	Detection of Food-Borne Pathogens	References
Fe_2O_4 NPs, SWCNTs	*E. coli* O157:H7, *S. aureus, S. epidermidis, E. coli*	Zhao et al. (2004)
QD	*S. enterica* serotype Typhi, *E. coli* O157:H7, *L. monocytogenes*	Hahn et al. (2008), Yang and Li (2005), Tully et al. (2006), Wang and Irudayaraj (2008)
CdSe/ZnS QDs	Salmonella	Kim et al. (2015a,b)
RuBpy doped silica	*E. coli* O157:H7	Su and Li (2004), Yang and Li (2006)
Au-encapsulated SiO_2 NPs	*E. coli*, Salmonella, Listeria	Weidemaier et al. (2015)
ImmunoFe_2O_4 nanoparticle	*Cronobacter sakazakii*	Shukla et al. (2016)
QDs and C NPs	Vibrio, Salmonella	Duan et al. (2015a,b)
Aptamer conjugated AuNPs	*S. typhimurium*	Oh et al. (2017), De Souza Reboucas et al. (2012)
Rare-earth doped NPs, magnetic NPs	*S. aureus*, Vibrio, Salmonella	Wu et al. (2014)
Magnetic bead/QD	*E. coli* 0157:H7	Yang and Li (2006)
Graphene oxide	Salmonella	Duan et al. (2014)
Magnetic NP clusters	Salmonella	Lee et al. (2014)
AuNPs	*S. enterica*	Vikesland and Wigginton (2010)
Graphene oxide, QDs	*E. coli*	Morales-Narvaez et al. (2013)
AgNPs on PVA particles	*E. coli*, Listeria, Salmonella, *S. aureus* SERS	Sundaram et al. (2013)
Au/silicon nanorod	*S. enterica* serotype Typhi; respiratory syncytial virus	Dungchai et al. (2008)
Fe_2O_4 NPs, QDs	Salmonella	Wen et al. (2013)
AuNP MWCNT	*S. typhimurium*	Dong et al. (2013), Wang and Alocilja (2015)
Nanoporous Al_2O_3	*E. coli* O157:H7 and *S. aureus*	Tian et al. (2016), Joung et al. (2013)
Fe_2O_4NPs; AuNPs	*E. coli* O157:H7	Wang and Alocilja (2015)
AuNPs	Salmonella	Zong et al. (2016)
Fe_2O_4, AuNPs, CdS NPs	*E. coli* O157:H7	Wang and Alocilja (2015)
MWCNTs; CdS, PbS, and CuS NPs	*E. coli* O157:H7, Campylobacter and Salmonella	Viswanathan et al. (2012)
SWCNT	*E. coli* K-12 and *S. aureus*	Yamada et al. (2016)
Superparamagnetic NPs	Salmonella	Ozalp et al. (2015)
AuNPs	*E. coli* O157:H7	Guo et al. (2012)
Superparamagnetic NPs	*S. aureus*	Issadore et al. (2013)

causal agents of food spoilage. In the food industry, nanobiosensors are potentially immense for the detection of desired analyte of interest, especially microbes in foodstuffs, food constituents detection as a result warning the consumers and distributors to not use such a type of food (Helmke and Minerick, 2006; Cheng et al., 2006). Bouwmeester et al. (2009) reported nanobiosensor to measure not only response changes in wetness in food storage rooms but also to detect microbes in food contamination. Subramanian (2006) have developed self-assembled monolayer-based SPR immunosensor for the detection of *E. coli* O157:H7. Tan et al. (2011) reported microfluidic electrical phenomenon immunosensor via antibody-immobilized nanoporous aluminum oxide membrane using dimethylsiloxane for speedy detection of *E. coli* O157:H7 and Staphylococcus *aureus*.

Nanomaterials employed in nanobiosensors are ready to detect pesticides, pathogens, and toxins to enhance food quality tracking-tracing-monitoring chain (Palchetti and Mascini, 2008; Liu et al., 2008; Inbaraj and Chen, 2015). Nachay (2007) reported nanobiosensors supported carbon nanotubes to detect microbes and toxicants in food and beverages. Wang et al. (2009) reported SWNT-paper sensing element linked with enzyme-linked-immunosorbent assay to detect water-borne toxins. Garcia et al. (2006) reported the electronic nose and electronic tongue as the nanosensors to detect wine type food aroma or gases. Kanazawa and Cho (2009) reported the quartz crystal balance based electrical nose to detect chemical odorants. Nanobiosensors are sensitive bioanalytical devices using nanomaterials and biological entities to detect food-borne pathogens that spoil the foodstuffs. Surface increased Raman scattering technique primarily based nanobiosensing using silver nano colloids to detect microorganism or microorganism pathogens. Besides nanosilver colloids, AgNPs, AuNPs, graphene oxide, magnetic beads, and carbon nanotubes (CNTs) were accustomed to detect food-borne microorganism pathogens (Thakur and Ragavan, 2013; Li and Church, 2014; Jarvis and Goodacre, 2004; Kahraman et al., 2008; Baranwal et al., 2016; Zuo et al., 2013; Holzinger et al., 2014). Bhattacharya et al. (2007) reported MEMs based technology for fast estimation of food pathogens in foodstuffs. Chen and Durst (2006) reported the concurrent estimation of *E. coli* O157:H7, *Salmonella species.*, and *L. monocytogenes* using immunosorbent assay based G-liposomal nanovesicles in pure and mixed cultures. Further, DeCory et al. (2005) reported immunomagnetic bead immunoliposome light assay to estimate *E. coli* O157:H7 using sulfur hodamine B. Shukla et al. (2016) reported immunoliposome based immunomagnetic separation assay to estimate *Cronobacter sakazakii*. Tominaga (2018) reported estimation of Klebsiella *pneumoniae, Klebsiella oxytoca, Raoultella ornithinolytica* using lateral-flow check strip immunoassays. Thakur et al. (2018) reported estimation of *E. coli* using graphene-based field-effect transistor device. Recently, Shukla et al. (2018) reported electrochemical sensing platform supported graphene oxide gold NPs to estimate *C. sakazakii*, a microorganism that is harmful to children when found in infant formula powder. Moreover, Song et al. (2018) reported estimation of *Cronobacter species* in pulverized child formula using immunoliposome-based immunomagnetic separation assay. Oh et al. (2017) reported gold nanoparticle aptamer-based LSPR sensing chips to estimate Salmonella *typhimurium* in pork meat.

NANOMATERIALS FOR PROTECTION FROM FOOD ALLERGENS

Nanomaterials are utilized in nanobioanalytical devices to estimate food allergens. Kumar et al. (2012) reported biosensors to estimate food-borne pathogens and allergens. Pilolli et al. (2013) reported nanobiosensor to estimate and manage food-allergens. Many nanomaterials are accountable to cause allergic pneumonic inflammation in humans. Vogelbruch et al. (2000) reported aluminum evoked granulomas when inaccurate intradermic hyposensitization injections of aluminum adsorbate depot preparations. The therapy of allergies used hydrated aluminum oxide as an adjuvant showed facet effects, like swelling, erythema, and edema. Localized surface plasmon resonance-based label-free biosensing strategies are highlighted. Lee et al. (2018) reported LSPR aptasensor modified gold nanorods to estimate Ochratoxin-A in fruit crush samples (food plant toxin inflicting allergy). Zhang et al. (2018) reported magnetic NPs based aptamer light assay to estimate allergens in food. Brotons-Cantoa et al. (2018) reported nanoparticle as oral vehicles for therapy and evaluated the positive effects of poly(anhydride) nanoparticles against peanut allergies. Di Felice et al. (2015) reported nanoparticles adjuvants in medicine and demonstrated in allergen therapy.

NANOMATERIALS IN HEAVY METALS DETECTION IN FOOD

He et al. (2015) reported metal oxide nanomaterials in nanomedicine and showed that a nanomaterial as metals in food could cause harmful effects. Metals from nanofood products once used then accumulated within the body and gave adverse effects to the health. Metal and metal oxide nanomaterials, like ZnO, Ag (nanosilver), and CuO are accountable to reinforce the living thing to reactive oxygen species level and finally cause lipid peroxidation and DNA injury because of aerobic stress. McShan et al. (2014) reported molecular toxicity mechanism of nanosilver. Additionally, Karlsson et al. (2013) reported cytomembrane injury and supermolecule interaction elicited by copper- containing nanoparticles which give the good proof for the metal distribution method. Karatapanis et al. (2011) reported silica-modified magnetite NPs functionalized with cetylpyridinium bromide and ascertained that ion surfactant act as adsorbents. Amin et al. (2014) reported magnetic iron-ore nanoparticles as nanomaterials that have large potential for the rectification of pollutants, contaminants from numerous sources. Zhang et al. (2011) reported Fe at Fe_2O_3 core/shells, nanowires, and nanonecklaces as antimicrobial nanofibrous membranes to filter chromium in contaminated solutions. Lin et al. (2017) reported aminated magnetic iron oxide NPs act as adsorbents to get rid of toxic metal ions. Cai et al. (2017) reported extremely active MgO nanoparticles synthesis via sol-gel method for synchronal microorganism inactivation and significant metal removal from contaminated water samples. Lingamdinne et al. (2017) reported biogenic magnetic iron oxide nanoparticles to remove toxic heavy metals without losing their stability. Simpson et al. (2018) reported a thermal method within the presence of H_3PO_4 and glycerin to get toxic metal ions from solution.

NANOMATERIALS AND BIOFILM AS THREAT TO FOOD SAFETY

Biofilms are microorganism cells that aggregate along on a surface and might be created of one sort of cell. The positioning for biofilm formation is natural materials, metals, plastics, etc., and needs wetness, nutrients, and a surface. Biofilms are controlled along by extracellular compound substances. Food-borne diseases have continually been a threat to human health. They are thought of a nascent public health concern throughout the globe. Several outbreaks are reported to be related to biofilm (Anselmo, and Mitragotri, 2016; Joo and Otto, 2012). It has been established that biofilm became a retardant in food industries (dairy, fish process, poultry, meat, and ready-to-eat foods), immune to antimicrobial agents and cleanup. Biofilms on surfaces are a retardant in a variety of food industries because of EPS that are an integral part of biofilms to supply structural support and stability. Biofilms are a serious concern in food industries and caused economic issues, also as exposed human health risk (Flemming et al., 2016; Characklis and Marshall, 1990). Throughout food production, microorganisms attach to surfaces and develop internally within the product. A variety of studies have evaluated numerous strategies to stop and eradicate biofilm formation, together with Clean-in-place, chemical-based management (sodium hypochlorite, peroxide, ozone, peracetic acid), and enzymes. Biofilm could be a protection growth pattern of bacterium, that is, concerned with food hygiene, that may well be to blame for food spoilage or food contamination because of morbific or nonpathogenic bacterium, and having risk to human health. The biofilm cells are also additionally immune to medical aid methods; its formation could be a terribly advanced process within the food trade. Therefore, the importance of biofilms in food safety management is obligatory to develop biofilm-free food-processing trade (Hall-Stoodley et al., 2004). Biofilm could be a tightly packed bunch of microorganism cells kept on with the substrates and manufacture a compound extra-cellular matrix that could not penetrate simply. In another sense, a biofilm could be a skinny layer, tightly packed microorganisms encapsulated inside an aqueous matrix of proteins, nucleic acids, and polysaccharides. The abundance of wetness and nutrients is liable to be biofilm; it will have an effect on any trade like paper and textile producing unit, cooling devices, potable system, health care or medical devices, and food process. Biofilm causes persistent low-level of food contamination because of human pathogens and it will impair food safety (Cucarella et al., 2001; Natan and Banin, 2017).

The food-borne diseases related to contemporary fruits and vegetables could represent biofilms containing bacterium because of environmental factors, temperature changes, pH changes, desiccation, light, etc., that may well be main culprits for food-borne malady outbreaks. Prepackaged salads are a frequent supply of food-borne malady. The North American nation government agency reported outbreaks of food-borne diseases caused by lettuce, spinach, basil, cabbage, onions, and parsley. The triple-water wash treatments and disinfectants to wash vegetables could cut back infective agent levels are insufficient to make sure microbiological safety. Foodstuff contamination in food trade that occurred sporadically by the biofilm formation and cells discharged are morbific, and the merchandise causes a food-borne malady occurrence (Brooks and Steve, 2008; Wei et al., 2003).

The most of the compounds and procedures are known as inhibitors of biofilm formation as a biocides and disinfectants in clinical usage; however, not in food process. Indeed, during this direction the analysis work is insufficient for distinctive methods that check the formation of biofilms on food and food process product. Generally, biofilms play a helpful role in nature for organisms within the food system; however, food-borne pathogens cause a big threat to food safety. The power of biofilm microbes improves microorganism colony and is immune to cleanup so food safety strategies are not effective so to cut back such risks to the food trade. Additional analysis is required not solely to know biofilm formation in morbific organisms but additionally to work out effective techniques for inactivating biofilms on foods. Vetter and Schlievert (2005) reported that glycerine monolaurate inhibits *Bacillus anthracis* and supported that glycerine monolaurate (GML) is safe declared by the U.S. Food and Drug Administration. In addition to that, Schlievert and Peterson (2012) reported glycerine monolaurate as a medicine shown to inhibit the biofilm formation of three totally different strains of *S. aureus*. Zhang et al. (2011) reported nanofibrous membranes of polyacrylonitrile nanofibers and shown to stop biofilm formation. Shahrokh and Emtiazi (2009) reported nanosilver and found that nanosilver particles not solely increased microorganism metabolism but additionally stop biofilm formation. Saleem et al. (2017) reported NiO nanoparticles synthesized from leaf extract of *Eucalyptus globulus* plants and shown that nickel oxide nanoparticles used as a medicine, antitumor agents, and antibiofilm activity. Ahmed et al. (2016) reported biofilm repressing result of antiseptic conjugated gold nanoparticles against klebsiella respiratory disease. Ranmadugala et al. (2017) reported superparamagnetic iron oxide against *B. subtilis* and determined vital reduction within the total biomass of the microorganism biofilm while not loss of cell viabilities and recommended that iron oxide nanoparticles may well be utilized in industries against the expansion of microorganism biofilm. Moreover, Thuptimdang et al. (2017) reported AgNPs against *Pseudomonas putida* biofilm recommended to change the biofilm body.

NANOMATERIALS VIS-A-VIS FOOD SAFETY ISSUES

Locally and globally, food safety issues are a public health concern. Food should be protected from physical, chemical, and biological contamination during food processing, handling, and its distribution. Within the food business food process, safeties, and security are vital parameters that directly and indirectly enhance nutraceutical price and time period (Pal, 2017; Wesley et al., 2014).

Nowadays, food safety could be a major cause of concern. We know that food-borne pathogens, toxicants, adulterants, and contaminants are main threats to human health. Table 11.3 presents food nanosensors for the detection of a good type of food analytes (i.e., gasses and water vapors, biogenic amines, significant metal, antibiotics, pesticides, biomolecules like allergenic proteins microorganism toxins, oligonucleotides, unhealthful microorganisms, vitamins, antioxidants, etc.). The traditional strategies to detect food pathogens and their toxins are labor intensive and time overwhelming. Nanomaterials

TABLE 11.3 Food Nanosensors for the Detection of a Wide Variety of Food Analytes

Nanomaterials	Analyte of Interest (in Foodstuffs)	References
AgNPs	Onion: organosulfur compounds	Sachdev et al. (2016)
AgNPs, PVA nanofibers	Shrimp meat: biogenic amines	Marega et al. (2015)
Nanoporous TiO_2 film	Pork meat: trimethylamine	Xiao-wei et al. (2016)
AuNPs	Milk: neomycin (antibiotic)	Ling et al. (2016)
Dye-doped silica NPs and QDs	Beverages, urine, serum: glucose	Zhai et al. (2016)
C QDs, MnO_2 nanosheets	Fruits/vegetables; juices: ascorbic acid	Liu et al. (2016)
ZnO NPs, CNTs	Milk powder: cholesterol	Hayat et al. (2015)
Au nanofingers	Drinking water, apple skin rinse: pesticides	Kim et al. (2015)
AgNPs	Paprika extract: Sudan III	Jahn et al. (2015)
AgNPs	Water, apple, carrot: carbendazim (pesticide)	Patel et al. (2015)
ZnSe QDs, AgNPs	Raw milk and egg: melamine	Cao et al. (2014)
Cu nanoclusters	Soy sauce and vinegar: kojic acid (antioxidant)	Gao et al. (2014)
Ag nanoclusters	Chili powder: Sudan I–IV (colorant)	Chen et al. (2014)
Dye-doped silica NPs	Chicken meat extract: enrofloxacin (antibiotic)	Huang et al. (2013)
CeO_2 NPs	Tea and mushrooms: various antioxidants	Sharpe et al. (2013)
Nano-TiO_2	Green, herbal, and black tea: tea catechins	Apak et al. (2012)
Graphene QDs and AuNPs	Root vegetables: cyanide	Wang and Alocilja (2015)
Triangular Ag nanoplates	Dried kelp: iodide	Hou et al. (2014)
Ag nanoplates	Water, tomato juice, rice: copper ion	Chaiyo et al. (2015)
Nanostructured Au surface	Solution: peanut allergen	Gezer et al. (2016)
Graphene sheet, Au, AgNPs	Buffer: bacterial DNA	Duan et al. (2015)
Lanth-doped NPs, Graphene oxide	Milk: bacterial enterotoxins	Huang et al. (2015)
AuNPs	Raw milk, egg powder, banana, meat, cheese: bacterial RNA	Liu et al. (2014)
AuNPs	Milk: bacterial DNA	Fu et al. (2013)
AuNPs	Meat balls: swine DNA	Ali et al. (2012)
Silica NPs	Fish, shrimp, rice, tobacco: Cd^{2+}, Cu^{2+}, Hg^{2+}	Afkhami et al. (2013)
Cobalt nitroprusside NPs	Dry fruits, wine, sugar, water: sulfite	Devaramani and Malingappa (2012)

(Continued)

TABLE 11.3 (Continued)

Nanomaterials	Analyte of Interest (in Foodstuffs)	References
Platinum NPs	Sausage: nitrite	Saber-Tehrani et al. (2013)
Platinum NPs and CNTs	Chili powder; chili, tomato, and strawberry sauces: Sudan I (colorant)	Elyasi et al. (2013)
Graphene/mesoporous TiO_2	Soft drinks, sausage: azo dyes	Gan et al. (2013)
AuNPs/rGO nanocomposite	Food contact materials: melamine	Chen et al. (2015)
Peptide-functionalized nanoporous membrane	Buffer: paraoxon (pesticide)	Liebes-Peer et al. (2014)
AgNPs and rGO	Grape juice, wine: ochratoxin A	Yola et al. (2016)
Au-Fe_3O_4 NPs	Cereal: aflatoxin B1	Chauhan et al. (2015)
GO-chitosan composite	Buffer: Salmonella	Singh et al. (2013)
MWCNTs; CdS NP	Beef: *E. coli* O157:H7	Abdalhai et al. (2015)
Polypyrrole/TiO_2 film	Mango, egg, fish: ethanol, H_2S, and trimethylamine TiO_2 absorption to gases	Cui et al. (2016)
MWCNTs	Buffer: putrescine, maleic anhydride	Tanguy et al. (2015)
MIP nanofilm	Red yeast rice: lovastatin (statin drug)	Eren et al. (2015)
Au-carbon nanocomposite	Rice, wheat, rice vinegar: citrinin (toxin)	Fang et al. (2016)
Cantilever	Buffer: Kanamycin (antibiotic)	Bai et al. (2014)
Cantilever	Buffer: oxytetracycline (antibiotic)	Hou et al. (2013)
Superparamagnetic NPs	Milk: staphylococcal enterotoxins	Orlov et al. (2013)
Cantilever; AuNPs	Buffer: listeria	Sharma and Mutharasan (2013)
Cantilever	Shrimp: *Vibrio cholerae*	Khemthongcharoen et al. (2015)

under flagship of technology are addressing food issues of safety associated with microbic contaminants and additionally improved the poison detection, shelf-life, and food packaging. Nanomaterials based nanosensors and nanobiosensors have been developed to detect food microbes (Inbaraj and Chen, 2015).

Many researchers have mentioned regarding food safety issues and regulative problems concerning nanomaterials for the food packaging and human health (Bradley et al., 2011). Savolainen et al. (2010) reported nanomaterials and its occupational health hazard and safety. Cushen et al. (2014) reported Ag and CuNPs blended polyethylene nanocomposites model to check the migration of these nanoparticles in food packaging. Mahler et al. (2012) reported vinylbenzene NPs on iron absorption. Moreover, the regulative authorities should develop gold standards for the food product to confirm food quality, health and safety concern, and environmental laws.

CHALLENGES, PERSPECTIVES, AND HEALTH RISKS

There are rising developments toward the utility of nanomaterials in food business as Fig. 11.5 shows in delineate presentation relating utility and impact of nanomaterials.

The safety and security of smart nanofoods is also attainable by the detection of pesticides, herbicides, harmful microorganism pathogens, toxins, pollutants, adulterants, and toxicants that facilitate within the management of food quality. CNTs are carbon nanomaterials used in food packaging to detect toxicants, contaminants, microbes, and food-borne pathogens. They are capable to rework into active food material for the long run and additionally referred to as intelligent packaging materials. But in its enhancing application as nanomaterials, there are also potential risks and toxicity problems that should be demonstrated and targeted. The impact of CNTs because of its dynamical properties on human, animals, and environment could not be ignoring. The inhalations of ultrafine particles or nanoparticles that are present within the surroundings by numerous human activities are increasing daily and cause health risks via crossing biological barriers or blood brain barrier and at last enter into numerous organs, tissues, and cells.

In food process utilizing nanomaterials, bioaccumulation of nanomaterials could not be ruled out, as example, nanosilver in nanopackaging food. The utility of nanomaterials as nanocatalysts, nanopesticides, and nanoherbicides might cause an unknown human health risk, so health risk assessment approaches should be strictly established. The challenges during this regard the use of nanomaterials for the betterment of foodstuffs. The end users must be educated to occupational health risks, safety, and environmental impacts of the nanomaterials.

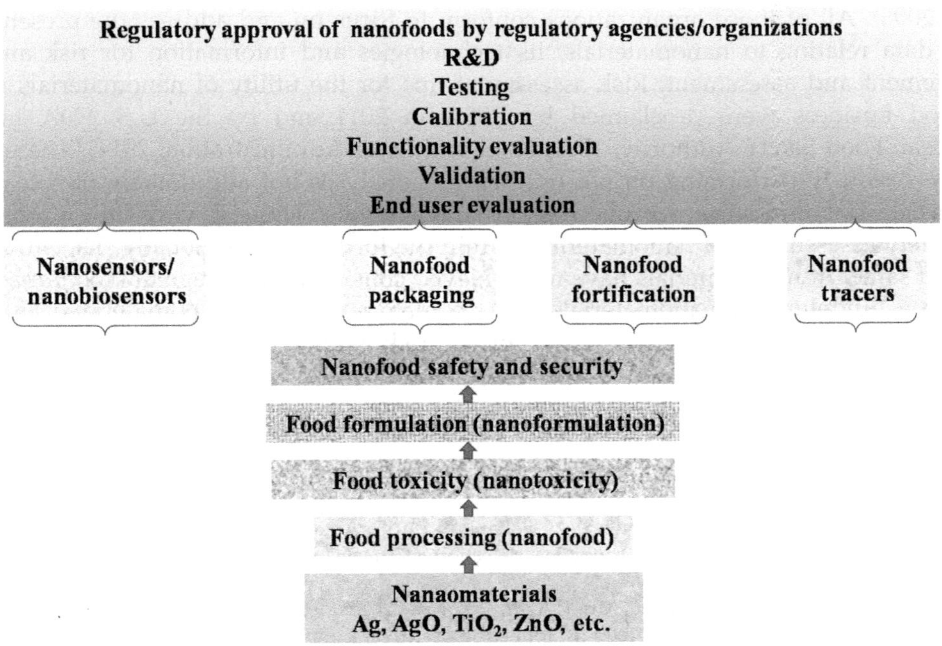

FIGURE 11.5 Diagrammatical presentations pertaining to utility and impact of nanomaterials.

CONCLUSION

Nanomaterials provide large opportunities for food packaging and profit to each customer and business. However, we are still within the early stages and want additional analysis and development. The food safety and food security of smart nanofoods are necessary to arrange information based mostly on data to forestall outbreaks of food-borne diseases to reinforce decision-making terms and conditions for the food process and food completion. Therefore, the utility of nanomaterials can herald the long run nanofoods to enhance not solely food safety, but additionally food dependableness, food time period, and food security. The challenge of world population is the problem relating food, human health, and food safety that affects end users. The recognition and challenges of the nanomaterials utility in food sector is increasing. The food packaging, food preservation, and food safety are renowned domain using nanomaterials that shield the food from supermolecule, moisture, gases, flavors, texture, and odors. Nanomaterials supply nice vehicle systems that open new methods with several challenges and opportunities to boost technological problems with the buyer issues relating transparency of issues of safety and environmental impact in food systems at the side of testing of nanofoods that are a unit necessary to distribute to the market to finish users (consumers). The history of nanomaterials utility within the food business is not long; however, speedy advances in their development have prompted their food safety concern domestically and globally. The Food and Agriculture Organization of the world organization, World Health Organization, and the U.S. Food and Drug Administration Nanotechnology Task Force are accountable organizations to understand the tremendous advantages of nanomaterials within the food business that were acknowledged and devise restrictive policies to manage the utilization of nanomaterials (FDA, 2006). All of those organizations conform to focus on and address the present food safety data relating to nanomaterials, its technologies and information for risk analysis, management and assessment. Risk assessment tips for the utility of nanomaterials within the food business were proclaimed by EFSA in 2011 and by the U.S. FDA in 2012 (European Food Safety Authority, 2011; Food and Drug Administration, 2014). These agencies are seriously performing on not only hazard analyses but additionally of research of the particle size and surface properties of nanomaterials. There is very little progress in food safety assessments of nanomaterials within the food business globally. The studies on the food safety of nanomaterials have not achieved considerably to regulate food quality, to develop a guideline for nanomaterials, and regulate security of nanomaterials. Thus, it's expected to produce the scientific basis for the risk assessments that are needed for the event and safe use of nanomaterials within the food business.

References

Abbaspour, A., Norouz-Sarvestani, F., Noori, A., Soltani, N., 2015. Aptamer-conjugated silver nanoparticles for electrochemical dual-aptamer-based sandwich detection of *Staphylococcus aureus*. Biosens. Bioelectron. 68, 149–155.

Abdalhai, M.H., Fernandes, A.M., Xia, X., Musa, A., Ji, J., Sun, X., 2015. Electrochemical genosensor to detect pathogenic bacteria (*Escherichia coli* O157:H7) as applied in real food samples (fresh beef) to improve food safety and quality control. J. Agric. Food Chem. 63, 5017–5025.

Abdollahi, M., Rezaei, M., Farzi, G., 2012. A novel active bio-nanocomposite film incorporating rosemary essential oil and nanoclay into chitosan. J. Food Eng. 111 (2), 343–350.

Afkhami, A., Soltani-Felehgari, F., Madrakian, T., Ghaedi, H., Rezaeivala, M., 2013. Fabrication and application of a new modified electrochemical sensor using nano-silica and a newly synthesized Schiff base for simultaneous determination of Cd^{2+}, Cu^{2+} and Hg^{2+} ions in water and some foodstuff samples. Anal. Chim. Acta 771, 21–30.

Ahmed, A., Khan, A.K., Anwar, A., Ali, S.A., Shah, M.R., 2016. Biofilm inhibitory effect of chlorhexidine conjugated gold nanoparticles against *Klebsiella pneumonia*. Microb. Pathog. 98, 50–56.

Ahmed, I., Lin, H., Zou, L., Brody, A.L., Li, Z., Qazi, I.M., et al., 2017. A comprehensive review on the application of active packaging technologies to muscle foods. Food Control 82, 163–178.

Ali, M.E., Hashim, U., Mustafa, S., Man, Y.B.C., Islam, K.N., 2012. Gold nanoparticle sensor for the visual detection of pork adulteration in meatball formulation. J. Nanomater. 103607.

Amin, M.T., Alazba, A.A., Manzoor, U., 2014. A review of removal of pollutants from water/wastewater using different types of nanomaterials. Adv. Mater. Sci. Eng. 2014, 1–24.

An, J., Zhang, M., Wang, S., Tang, J., 2008. Physical, chemical and microbiological changes in stored green asparagus spears as affected by coating of silver nanoparticles-PVP. LWT Food Sci. Technol. 41 (6), 1100–1107.

Anselmo, A.C., Mitragotri, S., 2016. Nanoparticles in the clinic. Bioeng. Transl. Med. 1, 10–29.

Apak, R., Cekic, S.D., Cetinkaya, A., Filik, H., Hayvali, M., Kilic, E., 2012. Selective determination of catechin among phenolic antioxidants with the use of a novel optical fiber reflectance sensor based on indophenol dye formation on nano-sized TiO_2. J. Agric. Food Chem. 60, 2769–2777.

Arfat, Y.A., Ahmed, J., Hiremath, N., Auras, R., Joseph, A., 2017. Thermo mechanical, rheological, structural and antimicrobial properties of bionanocomposite films based on fish skin gelatin and silver-copper nanoparticles. Food Hydrocoll. 62, 191–202.

Ariyarathna, I.R., Rajakaruna, R.M.P.I., Nedra Karunaratne, D., 2017. The rise of inorganic nanomaterial implementation in food applications. Food Control 77, 251–259.

Arora, A., Padua, G.W., 2010. Review: nanocomposites in food packaging. J. Food Sci. 75 (1), R43–R49.

Avella, M., De Vlieger, J.J., Errico, M.E., Fischer, S., Vacca, P., Volpe, M.G., 2005. Biodegradable starch/clay nanocomposite films for food packaging applications. Food Chem. 93, 467–474.

Azeredo, H.M.C., Mattoso, L.H.C., McHugh, T.H., 2011. Nanocomposites in food packaging review. In: Reddy, B. (Ed.), Advances in Diverse Industrial Applications of Nanocomposites. InTech, p. 550.

Bai, X., Hou, H., Zhang, B., Tang, J., 2014. Label-free detection of kanamycin using aptamer-based cantilever array sensor. Biosens. Bioelectron. 56, 112–116.

Baranwal, A., Mahato, K., Srivastava, A., Maurya, P.K., Chandra, P., 2016. Phytofabricated metallic nanoparticles and their clinical applications. RSC Adv. 6 (107), 105996–106010.

Bhattacharya, S., Jang, J., Yang, L., Akin, D., Bashir, R., 2007. Biomems and nanotechnology based approaches for rapid detection of biological entities. J. Rapid Methods Autom. Microbiol. 15, 1–32.

Bourbigot, S., Devaux, E., Flambard, X., 2002. Flammability of polyamide-6/clay hybrid nanocomposite textiles. Polym. Degrad. Stab. 75, 397–402.

Bouwmeester, H., Dekkers, S., Noordam, M.Y., Hagens, W.I., Bulder, A.S., Heer, C., et al., 2009. Review of health safety aspects of nanotechnologies in food production. Reg. Toxicol. Pharmacol. 53, 52–62.

Bradley, E.L., Castle, L., Chaudhry, Q., 2011. Applications of nanomaterials in food packaging with a consideration of opportunities for developing countries. Trends Food Sci. Technol. 22, 603–610.

Bratovcic, A., Odobasic, A., Catic, S., Sestan, I., 2015. Application of polymer nanocomposite materials in food packaging. Croat. J. Food Sci. Technol. 7, 86–94.

Brody, A.L., Bugusu, B., Han, J.H., Sand, C.K., McHugh, T.H., 2008. Innovative food packaging solutions. J. Food. Sci. 73, R107–R116.

Brooks, J.D., Steve, H.F., 2008. Biofilms in the food industry: problems and potential solutions. Int. J. Food Sci. Technol. 43 (12), 2163–2176.

Brotons-Cantoa, A., Gamazo, C., Martın-Arbella, N., Abdul karim, M., Matıas, J., Gumbleton, M., et al., 2018. Evaluation of nanoparticles as oral vehicles for immunotherapy against experimental peanut allergy. Int. J. Biol. Macromol. 110, 328–335.

Cai, Y., Li, C., Wu, D., Wang, W., Tan, F., Wang, X., et al., 2017. Highly active MgO nanoparticles for simultaneous bacterial inactivation and heavy metal removal from aqueous solution. Chem. Eng. J. 312, 158–166.

Cao, X., Shen, F., Zhang, M., Sun, C., 2014. Rapid and highly-sensitive melamine sensing based on the efficient inner filter effect of Ag nanoparticles on the fluorescence of eco-friendly ZnSe quantum dots. Sens. Actuators B 202, 1175–1182.

Caroline, E., 2002. Auto applications drive commercialization of nanocomposites. Plast. Add. Comp. 4 (11), 30–33.

Chaiyo, S., Siangproh, W., Apilux, A., Chailapakul, O., 2015. Highly selective and sensitive paper-based colorimetric sensor using thiosulfate catalytic etching of silver nanoplates for trace determination of copper ions. Anal. Chim. Acta 866, 75–83.

Characklis, W.G., Marshall, K.C., 1990. Biofilms. John Wiley & Sons, Inc., New York, NY.

Charles, A., Kao, H.M., Huang, T.C., 2003. Physical investigations of surface membrane–water relationship of intact and gelatinized wheat–starch systems. Carbohydr. Res. 338 (22), 2403–2408.

Chauhan, R., Singh, J., Solanki, P.R., Basu, T., O'Kennedy, R., Malhotra, B.D., 2015. Electrochemical piezoelectric reusable immunosensor for aflatoxin B1 detection. Biochem. Eng. J. 103, 103–113.

Chaweng kijwanich, C., Hayata, Y., 2008. Development of TiO_2 powder-coated food packaging film and its ability to inactivate Escherichia coli in vitro and in actual tests. Int. J. Food Microbiol. 123, 288–292.

Chen, C.S., Durst, R.A., 2006. Simultaneous detection of *Escherichia coli* O157:H7, *Salmonella* spp. and *Listeria monocytogenes* with an array-based immunosorbent assay using universal protein G-liposomal nanovesicles. Talanta 69, 232–238.

Chen, N.Y., Li, H.F., Gao, Z.F., Qu, F., Li, N.B., Luo, H.Q., 2014. Utilizing polyethyleneimine-capped silver nanoclusters as a new fluorescence probe for Sudan I-IV sensing in ethanol based on fluorescence resonance energy transfer. Sens. Actuators B 193, 730–736.

Chen, N., Cheng, Y., Li, C., Zhang, C., Zhao, K., Xian, Y., 2015. Determination of melamine in food contact materials using an electrode modified with gold nanoparticles and reduced grapheme oxide. Microchim Acta 182, 1967–1975.

Cheng, Q., Li, C., Pavlinek, V., Saha, P., Wang, H., 2006. Surface-modified antibacterial TiO_2/Ag^+ nanoparticles: preparation and properties. Appl. Surf. Sci. 252, 4154–4160.

Cockburn, A., Bradford, R., Buck, N., Constable, A., Edwards, G., Haber, B., et al., 2012. Approaches to the safety assessment of engineered nanomaterials (ENM) in food. Food. Chem. Toxicol. 50 (6), 2224–2242.

Couch, L.M., Wien, M., Brown, J.L., Davidson, P., 2016. Food nanotechnology: proposed uses, safety concerns and regulations. Agro. Food Ind. Hi Tech 27, 36–39.

Cucarella, C., Solano, C., Valle, J., Amorena, B., Lasa, I., Penades, J.R., 2001. Bap, a *Staphylococcus aureus* surface protein involved in biofilm formation. J. Bacteriol. 183, 2888–2896.

Cui, S., Yang, L., Wang, J., Wang, X., 2016. Fabrication of a sensitive gas sensor based on PPy/TiO_2 nanocomposites films by layer-bylayer self-assembly and its application in food storage. Sens. Actuators B 233, 337–346.

Cushen, M., Kerry, J., Morris, M., Cruz-Romero, M., Cummins, E., 2014. Evaluasian and simulation of silver and copper nanoparticle migration from polyethylene nanocomposites to food and an associated exposure assessment. J. Agric. Food Chem. 62, 1403–1411.

Cyras, V.P., Manfredi, L.B., Ton-That, M.T., Vazquez, A., 2008. Physical and mechanical properties of thermoplastic starch/montmorillonite nanocomposite films. Carbohydr. Polym. 73 (1), 55–63.

Dasgupta, N., Ranjan, S., Mundekkad, D., Ramalingam, C., Shanker, R., Kumar, A., 2015. Nanotechnology in agro-food: from field to plate. Food Res. Int. 69, 381–400.

De Conto, L.C., Grosso, C.R.F., Goncalves, L.A.G., 2013. Chemometry as applied to the production of omega-3 microcapsules by complex coacervation with soy protein isolate and gum Arabic. LWT Food Sci. Technol. 53 (1), 218–224.

De Souza Reboucas, J., Esparza, I., Ferrer, M., Sanz, M.L., Irache, J.M., Gamazo, C., 2012. Nanoparticulate adjuvants and delivery systems for allergen immunotherapy. J. Biomed. Biotechnol. 474605, 13.

Dean, K., Yu, L., 2005. Biodegradable Polymers for Industrial Application. CRC Press, Boca Raton, FL, pp. 289–309.

DeCory, T.R., Durst, R.A., Zimmerman, S.J., Garringer, L.A., Paluca, G., DeCory, H.H., et al., 2005. Development of an immunomagnetic bead immunoliposome fluorescence assay for rapid detection of *Escherichia coli* O157: H7 in aqueous samples and comparison of the assay with a standard microbiological method. Appl. Environ. Microbiol. 71, 1856–1864.

Dekkers, S., Krystek, P., Peters, R.J., Lankveld, D.X., Bokkers, B.G., van Hoeven Arentzen, P.H., et al., 2011. Presence and risks of nanosilica in food products. Nanotoxicology 5, 393–405.

Devaramani, S., Malingappa, P., 2012. Synthesis and characterization of cobalt nitroprusside nano particles: application to sulfite sensing in food and water samples. Electrochim. Acta 85, 579–587.

Di Felice, G., Barletta, B., Bonura, A., Butteroni, C., Corinti, S., Colombo, P., 2015. Nanoparticles adjuvants in allergology: new challenges and pitfalls. Curr. Pharmaceut. Des. 21, 4229–4239.

Dong, J., Zhao, H., Xu, M., Ma, Q., Ai, S., 2013. A label-free electrochemical impedance immunosensor based on AuNPs/PAMAMMWCNT-Chi nanocomposite modified glassy carbon electrode for detection of *Salmonella typhimurium* in milk. Food Chem. 141, 1980–1986.

Duan, Y.F., Ning, Y., Song, Y., Deng, L., 2014. Fluorescent aptasensor for the determination of *Salmonella typhimurium* based on a graphene oxide platform. Microchim. Acta 181, 647–653.

Duan, B., Zhou, J., Fang, Z., Wang, C., Wang, X., Hemond, H.F., et al., 2015a. Surface enhanced Raman scattering by graphene-nanosheet-gapped plasmonic nanoparticle arrays for multiplexed DNA detection. Nanoscale 7, 12606–12613.

Duan, N., Wu, S., Dai, S., Miao, T., Chen, J., Wang, Z., 2015b. Simultaneous detection of pathogenic bacteria using an aptamer based biosensor and dual fluorescence resonance energy transfer from quantum dots to carbon nanoparticles. Microchim. Acta 182, 917–923.

Duncan, T.V., 2011. Applications of nanotechnology in food packaging and food safety: barrier materials, antimicrobials and sensors. J. Colloid Interface Sci. 363, 1–24.

Dungchai, W., Siangproh, W., Chaicumpa, W., Tongtawe, P., Chailapakul, O.S., 2008. Typhi determination using voltammetric amplification of nanoparticles: a highly sensitive strategy for metalloimmunoassay based on a copper-enhanced gold label. Talanta 2, 727–732.

Echegoyen, Y., Nerin, C., 2013. Nanoparticle release from nanosilver antimicrobial food containers. Food Chem. Toxicol. 62, 16–22.

Elgadir, M.A., Uddin, M.S., Ferdosh, S., Adam, A., Khan, A.J., Saraker, M.Z.I., 2015. Impact of chitosan composites and chitosan nanoparticle composites on various drug delivery systems: a review. J. Food Drug Anal. 23 (4), 619–629.

Elyasi, M., Khalilzadeh, M.A., Karimi-Maleh, H., 2013. High sensitive voltammetric sensor based on Pt/CNTs nanocomposite modified ionic liquid carbon paste electrode for determination of Sudan I in food samples. Food Chem. 141, 4311–4317.

Emamifar, A., Kadivar, M., Shahedi, M., Soleimanian-Zad, S., 2010. Evaluation of nanocomposite packaging containing Ag and ZnO on shelf life of fresh orange juice. Innovative Food Sci. Emerg. Technol. 11 (4), 742–748.

Emamifar, A., Kadivar, M., Shahedi, M., Soleimanian-Zad, S., 2011. Effect of nanocomposite packaging containing Ag and ZnO on inactivation of *Lactobacillus plantarum* in orange juice. Food Control 22 (3–4), 408–413.

Eren, T., Atar, N., Yola, M.L., Karimi-Maleh, H., 2015. A sensitive molecularly imprinted polymer based quartz crystal microbalance nanosensor for selective determination of lovastatin in red yeast rice. Food Chem. 185, 430–436.

European Food Safety Authority, 2011. Guidance on the risk assessment of the application of nanoscience and nanotechnologies in the food and feed chain. EFSA J. 9, 2140.

Ezhilarasi, P.N., Karthik, P., Chhanwal, N., Anandharamakrishnan, C., 2013. Nanoencapsulation techniques for food bioactive components: a review. Food Bioprocess Technol. 6, 628–647.

Fang, G., Liu, G., Yang, Y., Wang, S., 2016. Quartz crystal microbalance sensor based on molecularly imprinted polymer membrane and three-dimensional Au nanoparticles@mesoporous carbon CMK-3 functional composite for ultrasensitive and specific determination of citrinin. Sens. Actuators B 230, 272–280.

Fang, Z., Zhao, Y., Warner, R.D., Johnson, S.K., 2017. Active and intelligent packaging in meat industry. Trends Food Sci. Technol. 61, 60–71.

Food and Drug Administration (FDA), 2006. Nanotechnology: a report of the USA food and drug administration nanotechnology task force.

Fernandez, A., Soriano, E., Lopez-Carballo, G., Picouet, P., Lloret, E., Gavara, R., 2009. Preservation of aseptic conditions in absorbent pads by using silver nanotechnology. Food Res. Int. 42 (8), 1105–1112.

Fernandez, A., Picouet, P., Lloret, E., 2010a. Cellulose-silver nanoparticle hybrid materials to control spoilage-related microflora in absorbent pads located in trays of fresh-cut melon. Int J Food. Microbiol. 142 (1–2), 222–228.

Fernandez, A., Picouet, P., Lloret, E., 2010b. Reduction of the spoilagerelated microflora in absorbent pads by silver nanotechnology during modified atmosphere packaging of beef meat. J. Food Protect. 73 (12), 2263–2269.

Flemming, H.C., Wingender, J., Szewzyk, U., Steinberg, P., Rice, S.A., Kjelleberg, S., 2016. Biofilms: an emergent form of bacterial life. Nat. Rev. Microbiol. 14, 563–575.

Flores-Lopez, M.L., Cerqueira, M.A., de Rodrıguez, D.J., Vicente, A.A., 2015. Perspectives on utilization of edible coatings and nano-laminate coatings for extension of post harvest storage of fruits and vegetables. Food Eng. Rev. 5, 1–14.

Food and Drug Administration, 2014. Guidance for industry: assessing the effects of significant manufacturing process changes, including emerging technologies, on the safety and regulatory status of food ingredients and food contact substances, including food ingredients that are color additives; availability. Fed. Regist. 2014 (79), 36533–36534.

Fratzl, P., Gupta, H.S., Paschalis, E.P., Roschger, P., 2004. Structure and mechanical quality of the collagen–mineral nano-composite in bone. J. Mater. Chem. 14, 2115–2123.

Frohlich, E., Roblegg, E., 2012. Models for oral uptake of nanoparticles in consumer products. Toxicology 291 (1–3), 10–17.

Fu, J., Park, B., Siragusa, G., Jones, L., Tripp, R., Zhao, Y., 2008. An Au/Si hetero-nanorod-based biosensor for Salmonella detection. Nanotechnology. 15, 155502.

Fu, Z., Zhou, X., Xing, D., 2013. Rapid colorimetric gene-sensing of food pathogenic bacteria using biomodification-free gold nanoparticle. Sens. Actuators B 182, 633–641.

Fujishima, A., Rao, T.N., Tryk, D.A., 2000. Titanium dioxide photocatalysis. J Photochem. Photobiol. C Photochem. Rev. 1, 1–21.

Galvez, A., Abriouel, H., Lopez, R.L., Omar, N.B., 2007. Bacteriocin based strategies for food biopreservation. Int. J. Food Microbiol. 120, 51–70.

Gan, T., Sun, J.Y., Zhu, H.J., Zhu, J.J., Liu, D.G., 2013. Synthesis and characterization of graphene and ordered mesoporous TiO_2 as electrocatalyst for the determination of azo colorants. J. Solid State Electron. 17, 2193–2201.

Gao, Z., Su, R., Qi, W., Wang, L., He, Z., 2014. Copper nanocluster-based fluorescent sensors for sensitive and selective detection of kojic acid in food stuff. Sens. Actuators B 195, 359–364.

Garces, J.M., Moll, D.J., Bicerano, J., Fibiger, R., McLeod, D.G., 2000. Polymeric nanocomposites for automotive applications. Adv Mater. 12 (23), 1835–1839.

Garcia, M., Aleixandre, M., Gutiérrez, J., Horrillo, M.C., 2006. Electronic nose for wine discrimination. Sens. Actuators B 113, 911–916.

Gezer, P.G., Liu, G.L., Kokini, J.L., 2016. Development of a biodegradable sensor platform from gold coated zein nanophotonic films to detect peanut allergen, Ara h1, using surface enhanced Raman spectroscopy. Talanta 150, 224–232.

Ghaani, M., Cozzolino, C.A., Castelli, G., Farris, S., 2016. An overview of the intelligent packaging technologies in the food sector. Trends Food Sci. Technol. 51, 1–11.

Ghanbarzadeh, B., Oleyaei, S.A., Almasi, H., 2014. Nano-structured materials utilized in biopolymer based plastics for food packaging applications. Crit. Rev. Food Sci. Nutr. 7, 37–41.

Gibbs, B.F., Kermash, S., Alli, I., Mulligan, C.N., 2010. Encapsulation in the food industry: a review. Int. J. Food Sci. Nutr. 199, 213–224.

Guo, X., Lin, C.S., Chen, S.H., Ye, R., Wu, V.C.H., 2012. A piezoelectric immunosensor for specific capture and enrichment of viable pathogens by quartz crystal microbalance sensor, followed by detection with antibody-functionalized gold nanoparticles. Biosens. Bioelectron. 38, 177–183.

Gupta, A., Eral, H.B., Hatton, T.A., Doyle, P.S., 2016. Nanoemulsions: formation, properties and applications. Soft Matter. 12, 2826–2841.

Hahn, M.A., Keng, P.C., Krauss, T.D., 2008. Flow cytometric analysis to detect pathogens in bacterial cell mixtures using semiconductor quantum dots. Anal. Chem. 3, 864–872.

Hall-Stoodley, L., Costerton, J.W., Stoodley, P., 2004. Bacterial biofilms: from the natural environment to infectious diseases. Nat. Rev. Microbiol. 2, 95–108.

Hannon, J.C., Kerry, J., Cruz-Romero, M., Morris, M., Cummins, E., 2015. Advances and challenges for the use of engineered nanoparticles in food contact materials. Trends Food Sci. Technol. 43, 43–62.

Hayat, A., Haider, W., Raza, Y., Marty, J.L., 2015. Colorimetric cholesterol sensor based on peroxidase like activity of zinc oxide nanoparticles incorporated carbon nanotubes. Talanta 143, 157–161.

He, X., Aker, W.G., Huang, M.J., Watts, J.D., Hwang, H.M., 2015. Metal oxide nanomaterials in nanomedicine: applications in photodynamic therapy and potential toxicity. Curr. Top. Med. Chem. 15, 1887–1900.

Helmke, B.P., Minerick, A.R., 2006. Designing a nano-interface in a microfluidic chip to probe living cells: challenges and perspectives. U.S.A. Proc. Natl. Acad. Sci. 103, 6419–6424.

Holzinger, M., Le Goff, A., Cosnier, S., 2014. Nanomaterials for biosensing applications: a review. Front. Chem. 2, 63.

Hou, H., Bai, X., Xing, C., Gu, N., Zhang, B., Tang, J., 2013. Aptamer-based cantilever array sensors for oxytetracycline detection. Anal. Chem. 85, 2010–2014.

Hou, X., Chen, S., Tang, J., Xiong, Y., Long, Y., 2014. Silver nanoplates-based colorimetric iodide recognition and sensing using sodium thiosulfate as a sensitizer. Anal. Chim. Acta 825, 57–62.

Hu, B., Liu, X., Zhang, C., Zeng, X., 2017. Food macromolecule based nanodelivery systems for enhancing the bioavailability of polyphenols. J. Food Drug Anal. 25, 3–15.

Huang, X., Aguilar, Z.P., Li, H., Lai, W., Wei, H., Xu, H., et al., 2013. Fluorescent Ru(phen)(3)(2 +)-doped silica nanoparticles-based ICTS sensor for quantitative detection of enrofloxacin residues in chicken meat. Anal. Chem. 85, 5120–5128.

Huang, Y., Zhang, H., Chen, X., Wang, X., Duan, N., Wu, S., et al., 2015. A multicolor time-resolved fluorescence aptasensor for the simultaneous detection of multiplex *Staphylococcus aureus* enterotoxins in the milk. Biosens. Bioelectron. 74, 170–176.

Inbaraj, B.S., Chen, B.H., 2015. Nanomaterial-based sensors for detection of foodborne bacterial pathogens and toxins as well as pork adulteration in meat products. J. Food Drug Anal. 24, 15–28.

Issadore, D., Chung, H.J., Chung, J., Budin, G., Weissleder, R., Lee, H., 2013. mHall chip for sensitive detection of bacteria. Adv. Healthcare Mater. 2, 1224–1228.

Jahn, M., Patze, S., Bocklitz, T., Weber, K., Cialla-May, D., Popp, J., 2015. Towards SERS based applications in food analytics: lipophilic sensor layers for the detection of Sudan III in food matrices. Anal. Chim. Acta 860, 43–50.

Janjarasskul, T., Suppakul, P., 2018. Active and intelligent packaging: the indication of quality and safety. Crit. Rev. Food Sci. Nutr. 58 (5), 808–831.

Jarvis, R.M., Goodacre, R., 2004. Discrimination of bacteria using surface enhanced Raman spectroscopy. Anal. Chem. 76, 40–47.

Jin, T., Gurtler, J.B., 2011. Inactivation of Salmonella in liquid egg albumen by antimicrobial bottle coatings infused with allyl isothiocyanate, nisin and zinc oxide nanoparticles. J. Appl. Microbiol. 110 (3), 704–712.

Johnston, C.T., 2010. Probing the nanoscale architecture of clay minerals. Clay Miner. 45 (3), 245–279.

Joo, H.S., Otto, M., 2012. Molecular basis of in vivo biofilm formation by bacterial pathogens. Chem. Biol. 19, 1503–1513.

Joung, C.K., Kim, H.N., Lim, M.C., Jeon, T.J., Kim, H.Y., Kim, Y.R., 2013. A nanoporous membrane-based impedimetric immunosensor for label-free detection of pathogenic bacteria in whole milk. Biosens. Bioelectron. 44, 210–215.

Jovanovic, B., 2015. Critical review of public health regulations of titanium dioxide, a human food additive. Integr. Environ. Assess. Manag. 11, 10–20.

Kahraman, M., Yazıcı, M.M., Sahin, F., Culha, M., 2008. Convective assembly of bacteria for surface-enhanced Raman scattering. Langmuir 24, 894–901.

Kanazawa, K., Cho, N.J., 2009. Quartz crystal microbalance as a sensor to characterize macromolecular assembly dynamics. J. Sens. 6, 1–17.

Karatapanis, A.E., Fiamegos, Y., Stalikas, C.D., 2011. Silica-modified magnetic nanoparticles functionalized with cetylpyridinium bromide for the preconcentration of metals after complexation with 8-hydroxy-quinoline. Talanta 84, 834–839.

Karger-Kocsis, J., Zhang, Z., 2005. In: Balta Calleja, J.F., Michler, G. (Eds.), Mechanical Properties of Polymers Based on Nanostructures and Morphology. CRC Press, New York, pp. 547–596.

Karimi, M., Sadeghi, R., Kokini, J., 2018. Human exposure to nanoparticles through trophic transfer and the biosafety concerns that nanoparticle-contaminated foods pose to consumers. Trends Food Sci. Technol. 75, 129–145.

Karlsson, H.L., Cronholm, P., Hedberg, Y., Tornberg, M., Battice, L.D., Svedhem, S., et al., 2013. Cell membrane damage and protein interaction induced by copper containing nanoparticles-importance of the metal release process. Toxicology 313, 59–69.

Khemthongcharoen, N., Wonglumsom, W., Suppat, A., Jaruwongrungsee, K., Tuantranont, A., Promptmas, C., 2015. Piezoresistive microcantilever-based DNA sensor for sensitive detection of pathogenic Vibrio cholerae O1 in food sample. Biosens. Bioelectron. 63, 347–353.

Kim, Y.S., Song, M.Y., Park, J.D., et al., 2010. Subchronic oral toxicity of silver nanoparticles. Part Fibre Toxicol. 7, 20.

Kim, G., Moon, J.H., Moh, C.Y., Lim, J.G., 2015a. A microfluidic nanobiosensor for the detection of pathogenic Salmonella. Biosens. Bioelectron. 67, 243–247.

Kim, A., Barcelo, S.J., Li, Z., 2015b. SERS-based pesticide detection by using nanofinger sensors. Nanotechnology 26, 011502.

Kirtiraj, K.G., Suman, S., Lee, Y.S., 2018. Oxygen scavenging films in food packaging. Environ. Chem. Lett. 16 (2), 523–538.

Klaine, S.J., 2009. Considerations for research on the environmental fate and effects of nanoparticles. Environ. Toxicol. 28 (9), 1787–1788.

Koo, O.M., Rubinstein, I., Onyuksel, H., 2005. Role of nanotechnology in targeted drug delivery and imaging: a concise review. Nanomed. Nanotechnol. Biol. Med. 1, 193–212.

Kumar, S., Dilbaghi, N., Barnela, M., Bhanjana, G., Kumar, R., 2012. Biosensors as novel platforms for detection of food pathogens and allergens. BioNano Sci. 2, 196–217.

Lamprecht, A., Saumet, J.L., Roux, J., Benoit, J.P., 2004. Lipid nanocarriers as drug delivery system for ibuprofen in pain treatment. Int. J. Pharm. 278, 407–414.

Langer, R., Peppas, N.A., 2003. Advances in biomaterials, drug delivery, and bionanotechnology. AIChE J. 49, 2990–3006.

Lee, C., Scheufele, D.A., Lewenstein, B.V., 2005. Public attitudes toward emerging technologies:examining the interactive effects of cognitions and affect on public support for nanotechnology. Sci. Commun. 27 (2), 240–267.

Lee, W., Kwon, D., Chung, B., Jung, G.Y., Au, A., Folch, A., et al., 2014. Ultrarapid detection of pathogenic bacteria using a 3D immunomagnetic flow assay. Anal. Chem. 86, 6683–6688.

Lee, B., Park, J.H., Byun, J.Y., Kim, J.H., Kim, M.G., 2018. An optical fiber based LSPR aptasensor for simple and rapid in-situ detection of ochratoxin A. Biosens. Bioelectron. 102, 504–509.

Li, Y.S., Church, J.S., 2014. Raman spectroscopy in the analysis of food and pharmaceutical nanomaterials. J. Food Drug Anal. 22 (1), 28–48.

Li, H., Li, F., Wang, L., Sheng, J., Xin, Z., Zhao, L., et al., 2009. Effect of nano-packing on preservation quality of Chinese jujube (*Ziziphus jujuba* Mill. var. inermis (Bunge) Rehd). Food Chem. 114 (2), 547–552.

Liau, S.Y., Read, D.C., Pugh, W.J., Furr, J.R., Russell, A.D., 1997. Interaction of silver nitrate with readily identifiable groups: relationship to the antibacterial action of silver ions. Lett. Appl. Microbiol. 25, 279–283.

Liebes-Peer, Y., Rapaport, H., Ashkenasy, N., 2014. Amplification of single molecule translocation signal using b-strand peptide functionalized nanopores. ACS Nano 8, 6822–6832.

Lin, S., Xu, M., Zhang, W., Hu, X., Lin, K., 2017. Quantitative effects of amination degree on the magnetic iron oxide nanoparticles (MIONPs) using as adsorbents to remove aqueous heavy metal ions. J. Hazard Mater. 335, 47–55.

Ling, K., Jiang, H., Zhang, L., Li, Y., Yang, L., Qiu, C., et al., 2016. A self assembling RNA aptamer-based nanoparticle sensor for fluorometric detection of Neomycin B in milk. Anal. Bioanal. Chem. 408, 3593–3600.

Lingamdinne, L.P., Chang, Y.Y., Yang, J.K., Singh, J., Choi, E.H., Shiratani, M., et al., 2017. Biogenic reductive preparation of magnetic inverse spinel iron oxide nanoparticles for the adsorption removal of heavy metals. Chem. Eng. J. 307, 74–84.

Liu, S., Yuan, L., Yue, X., Zheng, Z., Tang, Z., 2008. Recent advances in nanosensors for organophosphate pesticide detection. Adv. Powder. Technol. 19, 419–441.

Liu, H., Zhan, F., Liu, F., Zhu, M., Zhou, X., Xing, D., 2014. Visual and sensitive detection of viable pathogenic bacteria by sensing of RNA markers in gold nanoparticles based paper platform. Biosens. Bioelectron. 62, 38–46.

Liu, J., Chen, Y., Wang, W., Feng, J., Liang, M., Ma, S., et al., 2016. Switch-On' fluorescent sensing of ascorbic acid in food samples based on carbon quantum dots-MnO_2 probe. J. Agric. Food Chem. 64, 371–380.

Lok, C.N., Ho, C.M., Chen, R., et al., 2007. Silver nanoparticles: partial oxidation and antibacterial activities. J. Biol. Inorg. Chem. 12 (4), 527–534.

Lu, W.C., Huang, D.W., Wang, C.C.R., Yeh, C.H., Tsai, J.C., Huang, Y.T., et al., 2018. Preparation, characterization, and antimicrobial activity of nanoemulsions incorporating citral essential oil. J. Food Drug Anal. 26, 82–89.

Mahler, G.J., Esch, M.B., Tako, E., Southard, T.L., Archer, S.D., Glahn, R.P., et al., 2012. Oral exposure to polystyrene nanoparticles affects iron absorption. Nat. Nanotechnol. 7, 264–271.

Marega, C., Maculan, J., Rizzi, G.A., Saini, R., Cavaliere, E., Gavioli, L., et al., 2015. Polyvinyl alcohol electrospun nanofibers containing Ag nanoparticles used as sensors for the detection of biogenic amines. Nanotechnology 26, 075501.

Martirosyan, A., Schneider, Y.J., 2014. Engineered nanomaterials in food: implications for food safety and consumer health. Int. J. Environ. Res. Public Health 11 (6), 5720–5750.

McGovern, C., 2010. Commoditization of nanomaterials. Nanotechnol. Perceptions 6 (3), 155–178.

McShan, D., Ray, P.C., Yu, H., 2014. Molecular toxicity mechanism of nanosilver. J. Food Drug Anal. 22, 116–127.

Medeiros, B.S., Souza, M.P., Pinheiro, A.C., Bourbon, A.I., Cerqueira, M.A., Vicente, A.A., et al., 2014. Physical characterization of an alginate/lysozyme nano-laminate coating and its evaluation on "Coalho" cheese shelf life. Food Bioprocess Technol. 7, 1088–1098.

Mihindukulasuriya, S.D.F., Lim, L.T., 2014. Nanotechnology development in food packaging: a review. Trends Food Sci. Technol. 40, 149–167.

Mills, A., 2005. Oxygen indicators and intelligent inks for packaging food. Chem. Soc. Rev. 34, 1003–1011.

Morales-Narvaez, E., Hassan, A.R., Merkoci, A., 2013. Graphene oxide as a pathogen-revealing agent: sensing with a digital-like response. Angew. Chem. Int. Ed. 52, 13779–13783.

Morones, J.R., Elechiguerra, J.L., Camacho, A., Holt, K., Kouri, J.B., Ramirez, J.T., 2005. The bactericidal effect of silver nanoparticles. Nanotechnology 16, 2346–2353.

Mozafari, M.R., Flanagan, J., Matia-Merino, L., Singh, H., 2006. Recent trends in the lipid-based nanoencapsulation of antioxidants and their role in food. J. Sci. Food Agric. 86 (13), 2038–2045.

Murugesan, R., Orsat, V., 2012. Spray drying for the production of nutraceutical ingredients—a review. Food Bioprocess Technol. 5 (1), 3–14.

Nachay, K., 2007. Analyzing nanotechnology. Food Technol. 1, 34–36.

Natan, M., Banin, E., 2017. From nano to micro: using nanotechnology to combat microorganisms and their multidrug resistance. FEMS Microbiol. Rev. 41, 302–322.

Oh, S.Y., Heo, N.S., Shukla, S., Cho, H.Y., Ezhil Vilian, A.T., Kim, J., et al., 2017. Development of gold nanoparticle aptamer-based LSPR sensing chips for the rapid detection of *Salmonella typhimurium* in pork meat. Sci. Rep. 7, 10130.

Orlov, A.V., Khodakova, J.A., Nikitin, M.P., Shepelyakovskaya, A.O., Brovko, F.A., Laman, A.G., et al., 2013. Magneticimmunoassay for detection of staphylococcal toxins incomplex media. Anal. Chem. 85, 1154–1163.

Othman, S.H., 2014. Bio-nanocomposite materials for food packaging applications: types of biopolymer and nano-sized filler. Agric. Agric. Sci. Procedia 2, 296–303.

Ozalp, V.C., Bayramoglu, G., Erdem, Z., Arica, M.Y., 2015. Pathogen detection in complex samples by quartz crystal microbalance sensor coupled to aptamer functionalized core-shell type magnetic separation. Anal. Chim. Acta 853, 533–540.

Ozturk, A.B., Argin, S., Ozilgen, M., McClements, D.J., 2015. Formation and stabilization of nanoemulsion-based vitamin E delivery systems using natural biopolymers: whey protein isolate and gum. Food Chem. 188, 256–263.

Pal, M., 2017. Nanotechnology: a new approach in food packaging. J. Food Microbiol. Saf. Hyg. 2, 121.

Palchetti, I., Mascini, M., 2008. Electroanalytical biosensors and their potential for food pathogen and toxin detection. Anal. Bioanal. Chem. 391, 455–471.

Patel, G.M., Rohit, J.V., Singhal, R.K., Kailasa, S.K., 2015. Recognition of carbendazim fungicide in environmental samples by using 4-aminobenzenethiol functionalized silver nanoparticles as a colorimetric sensor. Sens. Actuators B 206, 684–691.

Pathkoti, K., Manubolu, M., Hwang, H.M., 2017. Nanostructures: current uses and future applications in food science. J. Food Drug Anal. 25 (2), 245–253.

Pilolli, R., Monaci, L., Visconti, A., 2013. Advances in biosensor development based on integrating nanotechnology and applied to food-allergen management. Trends Anal. Chem. 47, 12–26.

Pinto, R.J.B., Daina, S., Sadocco, P., Neto, C.P., Trindade, T., 2013. Antibacterial activity of nanocomposites of copper and cellulose. BioMed Res. Int. 6, 280512.

Pradhan, N., Singh, S., Ojha, N., Srivastava, A., Barla, A., Rai, V., et al., 2015. Facets of nanotechnology as seen in food processing, packaging, and preservation industry. BioMed Res. Int. 365672.

Qi, L.F., Xu, Z.R., Jiang, X., Hu, C., Zou, X., 2004. Preparation and antibacterial activity of chitosan nanoparticles. Carbohydr. Res. 339, 2693–2700.

Ramanathan, T., Abdala, A.A., Stankovich, S., Dikin, D.A., Herrera-Alonso, M., Piner, R.D., et al., 2008. Functionalized graphene sheets for polymer nanocomposites. Nat. Nanotechnol. 3 (6), 327–331.

Ranmadugala, D., Ebrahiminezhad, A., Manley-Harris, M., Ghasemid, Y., Berenjian, A., 2017. The effect of iron oxide nanoparticles on *Bacillus subtilis* biofilm, growth and viability. Process. Biochem. 62, 231–240.

Rhim, J.W., Ng, P.K., 2007. Natural biopolymer-based nanocomposite films for packaging applications. Crit. Rev. Food Sci. Nutr. 47, 411–433.

Saber-Tehrani, M., Pourhabib, A., Husain, S.W., Arvand, M., 2013. A simple and efficient electrochemical sensor for nitrite determination in food samples based on Pt nanoparticles distributed poly (2-aminothiophenol) modified electrode. Food Anal. Methods 6, 1300–1307.

Sachdev, D., Kumar, V., Maheshwari, P.H., Pasricha, R., Deepthi, Baghel, N., 2016. Silver based nanomaterial, as a selective colorimetric sensor for visual detection of post harvest spoilage in onion. Sens. Actuators B 228, 471–479.

Saleem, S., Ahmed, B., Khan, M.S., Al-Shaeri, S., Musarrat, J., 2017. Inhibition of growth and biofilm formation of clinical bacterial isolates by NiO nanoparticles synthesized from *Eucalyptus globulus* plants. Microb. Pathog. 111, 375–387.

Sari, P., Mann, B., Kumar, R., Singh, R.R.B., Sharma, R., Bhardwaj, M., et al., 2015. Preparation and characterization of nanoemulsion encapsulating curcumin. Food Hydrocol. 43, 540–546.

Sarwar, M.S., Niazi, M.B.K., Jahan, Z., Ahmad, T., Hussain, A., 2018. Preparation and characterization of PVA/nanocellulose/Ag nanocomposite films for antimicrobial food packaging. Carbohydr. Polym. 184, 453–464.

Savolainen, K., Pylkkanen, L., Norppa, H., Falck, G., Lindberg, H., Tuomi, T., et al., 2010. Nanotechnologies, engineered nanomaterials and occupational health and safety – a review. Saf. Sci. 6, 1–7.

Schirmer, B.C., Heiberg, R., Eie, T., Møretrø, T., Maugesten, T., Carlehøg, M., 2009. A novel packaging method with a dissolving CO_2 head space combined with organic acids prolongs the shelf life of fresh salmon. Int. J. Food Microbiol. 133, 154–160.

Schlievert, P.M., Peterson, M.L., 2012. Glycerol monolaurate antibacterial activity in broth and biofilm cultures. PLoS One 7, 40350.

Sekhon, B.S., 2014. Nanotechnology in agri-food production: an overview. Nanotechnol. Sci. Appl. 7, 31–53.

Shahrokh, S., Emtiazi, G., 2009. Toxicity and unusual biological behavior of nanosilver on Gram positive and negative bacteria assayed by microtiter-plate. Eur. J. Biol. Sci. 1, 28–31.

Sharma, H., Mutharasan, R., 2013. hlyA gene-based sensitive detection of *Listeria monocytogenes* using a novel cantilever sensor. Anal. Chem. 85, 3222–3228.

Sharpe, E., Frasco, T., Andreescu, D., Andreescu, S., 2013. Portable ceria nanoparticle-based assay for rapid detection of food antioxidants (NanoCerac). Analyst 138, 249–262.

Shi, S., Wang, W., Liu, L., Wu, S., Wei, Y., Li, W., 2013. Effect of chitosan/nano-silica coating on the physicochemical characteristics of longan fruit under ambient temperature. J. Food Eng. 118, 125–131.

Shukla, S., Lee, G., Song, X., Park, S., Kim, M., 2016. Immuno liposome based immunomagnetic concentration and separation assay for rapid detection of *Cronobacter sakazakii*. Biosens. Bioelectron. 77, 986–994.

Shukla, S., Haldorai, Y., Bajpai, V.K., Rengaraj, A., Hwang, S.K., Song, X., et al., 2018. Electrochemical coupled immunosensing platform based on graphene oxide/gold nanocomposite for sensitive detection of *Cronobacter sakazakii* in powdered infant formula. Biosens. Bioelectron. 30, 139–149.

Silva, H.D., Cerqueira, M.A., Vicente, A.A., 2012. Nanoemulsions for food applications: development and characterization. Food Bioprocess Technol. 5 (3), 854–867.

Simpson, A., Pandey, R.R., Chusuei, C.C., Ghosh, K., Patel, R., Wanekaya, A.K., 2018. Fabrication characterization and potential applications of carbon nanoparticles in the detection of heavy metal ions in aqueous media. Carbon 127, 122–130.

Sinclair, R.G., 1996. The case for polylactic acid as a commodity packaging plastic. J. Mass Spec-Pure Appl. Chem. 33 (5), 585–597.

Singh, R.P., 2011. Prospects of nanobiomaterials for biosensing. Int. J. Electrochem. 30. Publisher SAGE-Hindawi journal collection. 2011, 2011, Review article ID 125487.

Singh, R.P., 2016. Nanobiosensors: potentiality towards bioanalysis. J. Bioanal. Biomed. 8, e143. Available from: https://doi.org/10.4172/1948-593X.1000e143.

Singh, R.P., 2017. Application of nanomaterials toward development of nanobiosensors and their utility in agriculture. In: Prasad, R., Kumar, M., Kumar, V. (Eds.), Nanotechnology: An Agricultural Paradigm. Springer Publisher, Singapur, pp. 293–303.

Singh, R.P., 2018. Nanocomposites: recent trends, developments and applications. In: Aliofkhazraei, M. (Ed.), Advances in Nanostructured Composites: Vol 1: Carbon Nanotube and Graphene Composites, first ed. CRC Press, Chapter 2.

Singh, A., Sinsinbar, G., Choudhary, M., Kumar, V., Pasricha, R., Verma, H.N., et al., 2013. Graphene oxide-chitosan nanocomposite based electrochemical DNA biosensor for detection of typhoid. Sens. Actuators B 185, 675–684.

Singh, R.P., Choi, J.W., Tiwari, A., Pandey, A.C., 2014. Functional nanomaterials for multifarious nanomedicine. In: Tiwari, A., Turner, A.P.F. (Eds.), Biosensors Nanotechnology. John Wiley & Sons, Inc., Hoboken, NJ, pp. 141–197. , Chapter 6.

Soares, N.F.F., Silva, C.A.S., Santiago-Silva, P., Espitia, P.J.P., Gonçalves, M.P.J.C., Lopez, M.J.G., et al., 2009. Active and intelligent packaging for milk and milk products. In: Coimbra, J.S.R., Teixeira, J.A. (Eds.), Engineering Aspects of Milk and Dairy Products. CRC Press, New York, NY, pp. 155–174.

Sondi, I., Salopek-Sondi, B., 2004. Silver nanoparticles as antimicrobial agent: a case study on *E. coli* as a model for gram-negative bacteria. J. Colloid Interface Sci. 275, 177–182.

Song, X., Shukla, S., Kim, M., 2018. Detection of Cronobacter species in powdered infant formula using immunoliposome-based immunomagnetic concentration and separation assay. Food Microbiol. 72, 23–30.

Sorrentino, A., Gorrasi, G., Vittoria, V., 2007. Potential perspectives of bionanocomposites for food packaging applications. Trends Food Sci. Technol. 18, 84–95.

Sothornvit, R., Krochta, J.M., 2005. Plasticizers in edible films and coatings. In: Han, J.H. (Ed.), Innovations in Food Packaging. Elsevier Publishers, New York, pp. 40–55.

Stormer, A., Bott, J., Kemmer, D., Franz, R., 2017. Critical review of the migration potential of nanoparticles in food contact plastics. Trends Food Sci. Technol. 63, 39–50.

Su, S.L., Li, Y., 2004. Quantum dot biolabeling coupled with immunomagnetic separation for detection of *Escherichia coli* O157:H7. Anal. Chem. 76 (16), 4806–4810.

Subramanian, A., 2006. A mixed self-assembled monolayer-based surface Plasmon immunosensor for detection of *E. coli* O157H7. Biosens. Bioelectron. 7, 998–1006.

Sundaram, J., Park, B., Kwon, Y., Lawrence, K.C., 2013. Surface enhanced Raman scattering (SERS) with biopolymer encapsulated silver nanosubstrates for rapid detection of food borne pathogens. Int. J. Food Microbiol. 167, 67–73.

Suppakul, P., Miltz, J., Sonneveld, K., Bigger, S.W., 2003. Active packaging technologies with an emphasis on antimicrobial packaging and its applications. J. Food Sci. 68, 408–420.

Swaroop, C., Shukla, M., 2018. Nano-magnesium oxide reinforced polylactic acid biofilms for food packaging applications. Int. J. Biol. Macromol. 113, 729–736.

Tan, F., Leung, P.H.M., Liud, Z., Zhang, Y., Xiao, L., Ye, W., et al., 2011. Microfluidic impedance immunosensor for *E. coli* O157:H7 and *Staphylococcus aureus* detection via antibody-immobilized nanoporous membrane. Sens. Actuators B. Chem. 159, 328–335.

Tan, H., Ma, R., Lin, C., Liu, Z., Tang, T., 2013. Quaternized chitosan as an antimicrobial agent: antimicrobial activity, mechanism of action and biomedical applications in orthopedics. Int. J. Mol. Sci. 14, 1854–1869.

Tanguy, N.R., Fiddes, L.K., Yan, N., 2015. Enhanced radio frequency biosensor for food quality detection using functionalized carbon nanofillers. ACS Appl. Mater. Interfaces 7, 11939–11947.

Thakur, M.S., Ragavan, K.V., 2013. Biosensors in food processing. J. Food Sci. Technol. 50 (4), 625e41.

Thakur, B., Zhou, G., Chang, J., Pu, H., Jin, B., Sui, X., et al., 2018. Rapid detection of single *E. coli* bacteria using a graphene-based field-effect transistor device. Biosens. Bioelectron. 110, 16–22.

Thostenson, E.T., Li, C.Y., Chou, T.W., 2005. Nanocomposites in context. Compos. Sci. Technol. 65, 491–516.

Thuptimdang, P., Limpiyakorn, T., Khan, E., 2017. Dependence of toxicity of silver nanoparticles on *Pseudomonas putida* biofilm structure. Chemosphere 188, 199–207.

Tian, F., Lyu, J., Shi, J., Tan, F., Yang, M., 2016. A polymeric microfluidic device integrated with nanoporous alumina membranes for simultaneous detection of multiple foodborne pathogens. Sens. Actuators B, Chem. 225, 312–318.

Tominaga, T., 2018. Rapid detection of *Klebsiella pneumoniae, Klebsiella oxytoca, Raoultella ornithinolytica* and other related bacteria in food by lateral-flow test strip immunoassays. J. Microbiol. Methods 147, 43–49.

Tully, E., Hearty, S., Leonard, P., O'Kennedy, R., 2006. The development of rapid fluorescence-based immunoassays, using quantum dot-labelled antibodies for the detection of *L. monocytogenes* cell surface proteins. Int. J. Biol. Macromol. 1-3, 127–134.

Ubbink, J., Kruger, J., 2006. Physical approaches for the delivery of active ingredients in foods. Trends Food Sci. Technol. 17, 244–254.

Valerinia, D., Tammaroa, L., Di Benedettoa, F., Vigliottab, G., Capodiecia, L., Terzia, R., et al., 2018. Aluminum-doped zinc oxide coatings on polylactic acid films for antimicrobial food packaging. Thin Solid Films 645, 187–192.

Van der Zande, M., Vandebriel, R.J., Van Doren, E., et al., 2012. Distribution, elimination, and toxicity of silver nanoparticles and silver ions in rats after 28-day oral exposure. ACS Nano 6 (8), 7427–7442.

Vetter, S.M., Schlievert, P.M., 2005. Glycerol monolaurate inhibits virulence factor production in *Bacillus anthracis*. Antimicrob. Agents Chemother. 49, 1302–1305.

Vikesland, P.J., Wigginton, K.R., 2010. Nanomaterial enabled biosensors for pathogen monitoring e a review. Environ. Sci. Technol. 10, 3656–3669.

Viswanathan, S., Rani, C., Ho, J.A., 2012. Electrochemical immunosensor for multiplexed detection of food-borne pathogens using nanocrystal bioconjugates and MWCNT screen-printed electrode. Talanta 94, 315–319.

Vogelbruch, M., Nuss, B., Korner, M., Kapp, A., Kiehl, P., Bohm, W., 2000. Aluminium induced granulomas after inaccurate intradermal hyposensitization injections of aluminium adsorbed depot preparations. Allergy 55, 883–887.

Wang, Y., Alocilja, E.C., 2015. Gold nanoparticle-labeled biosensor for rapid and sensitive detection of bacterial pathogens. J. Biol. Eng. 9, 16.

Wang, C., Irudayaraj, J., 2008. Gold nanorod probes for the detection of multiple pathogens. Small 12, 2204–2208.

Wang, L., Chen, W., Xu, D., Shim, B.S., Zhu, Y., Sun, F., et al., 2009. Simple, rapid, sensitive, and versatile SWNT-paper sensor for environmental toxin detection competitive with ELISA. Nano Lett. 9, 4147–4152.

Wang, H., Du, L.J., Song, Z.M., Chen, X.X., 2013. Progress in the characterization and safety evaluation of engineered inorganic nanomaterials in food. Nanomedicine (Lond) 8 (12), 2007–2025.

Wei, J., Ravn, D.B., Gram, L., Kingshott, P., 2003. Stainless steel modified with poly (ethylene glycol) can prevent protein adsorption but not bacterial adhesion. Colloids Surf. B: Biointerfaces 32, 275–291.

Weidemaier, K., Carruthers, E., Curry, A., Kuroda, M., Fallows, E., Thomas, J., et al., 2015. Real-time pathogen monitoring during enrichment: a novel nanotechnology-based approach to food safety testing. Int. J. Food Microbiol. 198, 19–27.

Weiss, J., Takhistov, P., McClements, J., 2006. Functional materials in food nanotechnology. J. Food Sci. 71, R107–R116.

Wen, C.Y., Hu, J., Zhang, Z.L., Tian, Z.Q., Ou, G.P., Liao, Y.L., et al., 2013. One-step sensitive detection of *Salmonella typhimurium* by coupling magnetic capture and fluorescence identification with functional nanospheres. Anal. Chem. 85, 1223–1230.

Wesley, S.J., Raja, P., Sunder Raj, A.A., Tiroutchelvamae, D., 2014. Review on nanotechnology applications in food packaging and safety. Int. J. Eng. Res. 3, 645–651.

Woon, S.A., Fisher, D., 2016. Antimicrobial agents-optimising the ecological balance. Woon Fisher BMC Med. 14 (114), 1–9.

Wu, Y., Weller, C.L., Hamouz, F., Cuppett, S.L., Schnepf, M., 2002. Development and application of multicomponent edible coatings and films: a review. Adv. Food Nutr. Res. 44, 347–394.

Wu, S., Duan, N., Shi, Z., Fang, C., Wang, Z., 2014. Simultaneous aptasensor for multiplex pathogenic bacteria detection based on multicolor upconversion nanoparticles labels. Anal. Chem. 86, 3100–3107.

Xiao-wei, H., Zhi-hua, L., Xiao-bo, Z., Ji-yong, S., Han-ping, M., Jiewen, Z., et al., 2016. Detection of meat-borne trimethylamine based on nanoporous colorimetric sensor arrays. Food Chem. 197, 930–936.

Yamada, K., Choi, W., Lee, I., Cho, B.K., Jun, S., 2016. Rapid detection of multiple foodborne pathogens using a nanoparticle functionalized multi-junction biosensor. Biosens. Bioelectron. 77, 137–143.

Yang, L., Li, Y., 2005. Quantum dots as fluorescent labels for quantitative detection of *S. typhimurium* in chicken carcass wash water. J. Food Protect. 6, 1241–1245.

Yang, L., Li, Y., 2006. Quantum dot bioconjugates for simultaneous detection of *Escherichia coli* O157:H7 and *Salmonella typhimurium*. Analyst 131, 394–401.

Yang, F.M., Li, H.M., Li, F., Xin, Z.H., Zhao, L.Y., Zheng, Y.H., et al., 2010. Effect of nano-packaging on preservation quality of fresh strawberry (*Fragaria ananassa* Duch. cv Fengxiang) during storage at 4 C. J. Food Sci. 75 (3), 236–240.

Yang, R., Zhou, Z., Sun, G., Gao, Y., Xu, J., Strappe, P., et al., 2015. Synthesis of homogeneous protein-stabilized rutin nanodispersions by reversible assembly of soybean (Glycine max) seed ferritin. RSC Adv. 5, 31533–31540.

Yola, M.L., Gupta, V.K., Atar, N., 2016. New molecular imprinted voltammetric sensor for determination of ochratoxin A. Mater. Sci. Eng. C 61, 368–375.

Yoon, S., Deng, Y., 2006. Clay-starch composites and their application in papermaking. J. Appl. Polym. Sci. 100 (2), 1032–1038.

Yu, Z., Lib, B., Chud, J., Zhang, P., 2017. Silica in situ enhanced PVA/chitosan biodegradable films for food packages. Carbohydr. Polym. 184, 214–220.

Zeng, S., Baillargeat, D., Ho, H.P., Yong, K.T., 2014. Nanomaterials enhanced surface plasmon resonance for biological and chemical sensing applications. Chem. Soc. Rev. 43 (10), 3426–3452.

Zhai, H., Feng, T., Dong, L.Y., Wang, L.Y., Wang, X., Liu, H., et al., 2016. Development of dual-emission ratiometric probe-based on fluorescent silica nanoparticle and CdTe quantum dots for determination of glucose in beverages and human body fluids. Food Chem. 204, 444–452.

Zhang, L., Luo, J., Menkhaus, T.J., Varadaraju, H., Sun, Y., Fong, H., 2011. Antimicrobial nano-fibrous membranes developed from electrospun polyacrylonitrile nanofibers. J. Membr. Sci. 369, 499–505.

Zhang, T., Lv, C., Chen, L., Bai, G., Zhao, G., Xu, C., 2014. Encapsulation of anthocyanin molecules within a ferritin nanocage increases their stability and cell uptake efficiency. Food Res. Int. 62, 183–192.

Zhang, Y., Wu, Q., Sun, M., Zhang, J., Mo, S., Wang, J., et al., 2018. Magnetic-assisted aptamer-based fluorescent assay for allergen detection in food matrix. Sens. Actuator B Chem. 263, 43–49.

Zhao, X., Hilliard, L., Mechery, S., Wang, Y., Bagwe, R., Jin, S., 2004. A rapid bioassay for single bacterial cell quantitation using bioconjugated nanoparticles. Proc. Natl. Acad. Sci. USA 42, 15027–15032.

Zhou, L., Lv, S., He, G., He, Q., Shi, B.I., 2011. Effect of PE/AG_2O nanopackaging on the quality of apple slices. J. Food Qual. 34 (3), 171–176.

Zhu, J., Morgan, A.B., Lamelas, F.J., Wilkie, C.A., 2001. Fire properties of polystyrene-clay nanocomposites. Chem. Mat. 13, 3774–3780.

Zong, Y., Liu, F., Zhang, Y., Zhan, T., He, Y., Hun, X., 2016. Signal amplification technology based on entropy-driven molecular switch for ultrasensitive electrochemical determination of DNA and *Salmonella typhimurium*. Sens. Actuators B 225, 420–427.

Zuo, P., Li, X., Dominguez, D.C., Ye, B.C., 2013. A PDMS/paper/glass hybrid microfluidic biochip integrated with aptamerfunctionalized graphene oxide nano-biosensors for onestep multiplexed pathogen detection. Lab Chip 13, 3921–3928.

Further Reading

Chaudhry, Q., Scotter, M., Blackburn, J., Ross, R., Boxall, A., Castle, L., et al., 2008. Applications and implications of nanotechnologies for the food sector. Food Addit. Contam. Part A Chem. Anal. Control Expo. Risk Assess. 25 (3), 241–258.

Li, X.Y., Chen, X.G., Sun, Z.W., Park, H.J., Cha, D.S., 2011. Preparation of alginate/chitosan/carboxymethyl chitosan complex microcapsules and application in Lactobacillus casei ATCC 393. Carbohydr. Polym. 83 (4), 1479–1485.

Morgan, A.B., Harris, R.H., Kashiwagi, T., Chyall, L.J., Gilman, J.W., 2002. Flammability of polystyrene layered silicate (clay) nanocomposites: carbonaceous char formation. Fire Mater. 26, 247–253.

Wang, L., Zheng, J., Yang, S., Wu, C., Liu, C., Xiao, Y., et al., 2015a. Two-photon sensing and imaging of endogenous biological cyanide in plant tissues using graphene quantum dot/gold nanoparticle conjugate. ACS Appl. Mater. Interfaces 7, 19509–19515.

Wang, Y., Fewins, P.A., Alocilja, E.C., 2015b. Electrochemical immunosensor using nanoparticle-based signal enhancement for *Escherichia coli* O157:H7 detection. IEEE Sens. J. 15, 4692–4699.

Yang, M., Peng, Z., Ning, Y., Chen, Y., Zhou, Q., Deng, L., 2013. Highly specific and cost-efficient detection of Salmonella paratyphi combining aptamers with single-walled carbon nanotubes. Sensors 13, 6865–6881.

CHAPTER

12

International Laws and Food-Borne Illness

Tek Chand Bhalla, Monika, Sheetal and Savitri

Department of Biotechnology, Himachal Pradesh University, Shimla, India

OUTLINE

Food Safety and Human Health
DOI: https://doi.org/10.1016/B978-0-12-816333-7.00012-6

INTRODUCTION

Food is an essential component for the very existence and sustenance of life, and it is of plant or animal origin. It contains nutrients like carbohydrates, protein, fat, vitamins, and minerals, which provide nutritional support for the growth and maintenance of an organism. On the other hand, it is also a major cause for the ill health of people around the globe, as it acts as a common transmission route of diseases due to the presence of microorganisms and contaminants (Odeyemi and Sani, 2016). Food supports the growth of microorganisms, and sometimes it contains antinutritional components and toxins that a plant or animal produce for self-defense from which these foods are obtained (Adams and Moss, 2008). Microorganisms causing food-borne illness are found in a wide range of foods with various virulence factors and may illicit a diverse range of adverse responses that may be acute, chronic, or intermittent. Some microbial pathogens are invasive and may cause generalized infection; other pathogens produce toxins that cause severe damage in susceptible tissues and organs. The complications normally require medical care and frequently result sometimes in hospitalizations. There may be a risk of mortality, as not all the patients recover fully and may suffer from residual symptoms throughout life (Forsythe, 2002).

Food-borne diseases are defined as any disease infectious or toxic in nature caused by ingestion of contaminated food. They pose potential economical and health loss to society (Adams and Moss, 2003). They are the major contributors to the estimated 1.5 billion annual episodes of diarrhea in children under the age of 5 years, and up to 70% of such episodes are due to ingestion of contaminated foods. Hence, a child is caught up in malnutrition, infection, and many of them do not survive under these circumstances. The chronic sequelae (secondary complications) following food-borne infections are also the cause of concern recognized together with the variability of human response. The secondary complications may be more serious and result in dreadful chronic disorders or even death (Lindsay, 1997).

The globalizations of trades and travels have increased the emergence and reemergence of certain food-borne pathogens (Tauxe et al., 2010). This may be due to weak public health infrastructure, economic problems, changing health policies, poverty, uncontrolled population displacement and urbanization, ineffective disease control programs, resistant microbial strains, and few new diseases or cases identified as a result of increased knowledge or new, improved methods of identification and diagnosis (NRC, 1998).

The common consumers are usually not much aware that there is a possible problem with the food, and if contaminated food is ingested, they become ill. Therefore it is difficult to trace which food was the actual cause of food illness because of late onset of disease symptoms. Sources of food-borne pathogens are classified on the basis of severity of illness according to ICMSF (1996a,b) (International Commission on Microbiological Specifications for Food). Despite the increasing awareness and understanding of the food-associated contaminations, food-borne illnesses remain a major hurdle and are the main causes of reduced economic productivity.

In addition to human suffering, control and cure of food-borne illness can also be costly affairs (Buzby and Roberts, 1997). The cost of human illness due to only six bacterial pathogens is US$ 9.3–12.9 billion annually; of these costs, US$ 2.9–6.7 billion are attributable to the food-borne bacteria *Salmonella* serovars, *Campylobacter jejuni*, *Escherichia coli* O 157:H7, *Listeria monocytogenes*, *Staphylococcus aureus*, and *Clostridium perfringens*. The cost of food-borne salmonellosis is high, an estimated US$ 2.3 billion annually by FoodNet (Frenzen et al., 1999). In England and Wales, the medical costs and value of lives lost due to just five food-borne illnesses were estimated to be £300–700 million per year (Roberts, 1996). In 1999 the cost of food poisoning in Australia was estimated to be $2.6 billion per year by ANZFA (The Australia New Zealand Food Authority) and in Canada is US$ 1.3 billion per year (Todd, 1989). The cost of food-borne illnesses in developing countries is more and determined to be billions of US dollars. The World Health Organization (WHO, 2015) estimates that almost 2 million children die every year from diarrhea worldwide.

The impact of food losses due to contamination is also considerable and affects trade as well as tourism. Worldwide losses of grain and nongrain food are estimated to be at least 10% and as high as 50% of production. The cost benefits in preventing food-borne illness through the ensured food safety and laws have been estimated (Crutchfield et al., 1999). Thus there is considerable need, in both developed and developing nations, for more consistent steps to be taken to mitigate substantially the risks posed by the microbial food-borne pathogens. A significant proportion of deaths are caused by the contaminated foods. The food-borne illnesses are a major challenge of the people in administration, public health, medicine, food trade, environment management, food production, and processing, and thus coordinated and collaborative efforts are required to reduce the food-borne illnesses. Therefore in this chapter, various food-borne illnesses and the international laws involved in consumer safe food production and trade are discussed.

FOOD-BORNE ILLNESS

Bacterial Food-Borne Illness

The world is greatly suffering from food-borne illnesses and economic losses to mankind. People from the developing nations are more affected and the rate of mortality and morbidity is higher as compared to developed nations (Glavin, 2003). Various studies conducted at different part of the world have shown that the causative agent in most of the food-borne outbreaks belongs to bacterial counterpart (66% of all food-borne illnesses) among all pathogens (Addis and Sisay, 2015). Bacterial pathogens causing food-borne illnesses include *Bacillus cereus*, *Clostridium botulinum*, *C. perfringens*, *S. aureus*, *E. coli*, *C. jejuni*, *Brucella melitensis*, *L. monocytogenes*, *Vibrio cholerae*, *Vibrio parahaemolyticus*, *Shigella* spp., *Salmonella typhi*, and *S. paratyphi* (Rao et al., 1989). Pathogenesis caused by some of these bacteria is discussed next.

Staphylococcus aureus

S. aureus was first isolated and named by Sir Alexander Oguston in 1882. Cell division occurs in more than one plane, which is why they form irregular colonies like a bunch of grapes (Adams and Moss, 2008). *S. aureus* are gram-positive cocci found as single, chains, and tetrads, oxidase-negative facultative anaerobes and catalase-positive. They are mesophilic and grow in temperature ranges between 7°C and 48°C, with 37°C being the optimum growth temperature. This organism produces endotoxin, which is responsible for cases of food poisoning. Production of endotoxin depends on the temperature; a narrow range of temperature (35°C–40°C) is optimal for endotoxin production (Quinn et al., 2001; Adams and Moss, 2008). Food poisoning is caused by the strains that produce endotoxin (Walderhaug, 2007).

EPIDEMIOLOGY:

Half of the *S. aureus* bacterial species are known to infect humans and animals, and these are very much adapted to their host. In their host, *S. aureus* reside near body openings and moist perineal areas like the nose and throat, and their number may reaches to 10,00,000 pre square centimeter under favorable conditions. Contamination of food mostly happens due to the infected food handlers, who carry bacteria in carbuncles and boils on their arms and hands (Quinn et al., 2001). Studies carried out in the United Kingdom and United States have revealed that cold cooked meat, poultry products, salted meats (ham), and corned beef are the most common vehicles for transmission of *S. aureus*. Other processed, packaged, or canned food also provides a safe and competition free environment to *S. aureus* (Adams and Moss, 2008).

Various types of food products (e.g., meat and meat products, dairy and milk products, poultry and egg products, bakery products like cream-filled pastries, cakes, and sandwiches) are associated with growth and transmission of *S. aureus* (Tamarapu et al., 2001; le Loir et al., 2003). Industrially prepared food is distributed in large and distant areas, and food-borne disease can have grave consequences in this case. In 2000 there was an outbreak of food-borne illness in Japan; over 13,000 cases were reported as a result of staphylococcal contamination of milk at a dairy-food production plant (Murray, 2005).

SYMPTOMS:

Incubation period of food poisoning is typically short, about 2–4 hours, and symptoms of intoxication include vomiting, nausea, retching (to make the sound and movement of vomiting), abdominal cramps, hypersalivation, prostration, and diarrhea (Balaban and Rasooly, 2000). The illness is self-limiting and lasts 24–48 hours; however, it can be severe in case of immune-compromised, elderly, and infants (Argudín et al., 2010). Dehydration, masked pallor (an unhealthy pale appearance), and collapse occur in severe cases that may require treatment by intravenous infusion. In case of a shorter incubation period, the patient may have ingested contaminated food with preformed toxin. Role of enterotoxin to produce gastrointestinal illness is unknown (Adams and Moss, 2008).

PATHOGENESIS:

The pathogen present in contaminated food starts producing enterotoxin when stored at room temperature. *S. aureus* produce eleven different enterotoxins designated SEA to SEJ (SEA, SEB, SEC_1, SEC_2, SED, SEE, SEF, SEG, SHE, SEI, and SEJ). Enterotoxins are strictly neurotoxic in nature and act on receptors in the gut, stimulating the vomiting center in the brain. Most potent enterotoxin produced by *S. aureus* is SEA followed by SED. Enterotoxin is a small, single-chain polypeptide with molecular weight ranging from 26 to 30 kDa. Enterotoxins of *S. aureus* are superantigen as they are able to stimulate a higher amount of T cells (Adams and Moss, 2008; Anderson and Pritchard, 2008).

Enterotoxins of *S. aureus* are gut proteases and heat stable, with type SEB being the most heat resistant. Heat and proteases stability is explained on the bases of their compact structure due to single disulfide loop near the center of the peptide chain molecule. Enterotoxin also inhibits water and sodium absorption in the small intestine (Quinn et al., 2001).

Clostridium perfringens

C. perfringens is an anaerobic, spore-forming, gram-positive, rod-shaped ($1 \times 3-9\ \mu m$) and catalase negative bacteria. Depending on the type of enterotoxin production, they are categorized in five different types, A–E. This toxin-producing bacterium was formerly called *welchii* as isolated by American bacteriologist Welch in 1892. *C. perfringens* A is mainly responsible for food-borne illness because it produce only α toxin, which has lecithinase (phospholipase C) activity. It shows growth over a wide temperature range starting from 12°C to 50°C, and optimum growth occurs at 43°C–47°C. But at 41°C temperature, growth is very rapid with generation time of only 7.1 minutes (Labbe and Nolan, 1981; Adams and Moss, 2008; RIVM Report, 2011).

Strain of *C. perfringens* producing enterotoxin A is widely distributed in the environment as compared to other strains producing toxins B, C, D, and E as they are obligate parasites. Type A enterotoxin producing *C. perfringens* is found in soil, water, dust, sediments, human intestines, and on raw and cooked food (Adams and Moss, 2008).

EPIDEMIOLOGY:

Western countries are heavily burdened by food-borne illness due to enterotoxin A of *C. perfringens*. Between years 2000 and 2008 there were 1.0 million cases of *C. perfringens*–borne illness, which were 10% of total food-borne illnesses in the United States (Scallan et al., 2011). In the Netherlands, 160,000 cases of food-borne illness are reported annually. Different studies have confirmed the presence of 2%–6% of nonhemolytic and heat-resistant strains in general population. Around 20%–30% of healthy hospital personnel and their families have been found to carry these organisms in their feces, and the carrier rate of victims after 2 weeks may be 50% or as high as 88% (Jay, 2000).

Most of the outbreaks investigated show inadequate reheating or cooling as the main reason for onset of disease. Preservation of food after its preparation is a very important step but often is ignored at private households and obscure restaurants. Meat-containing dishes like stews and soups are most likely implicated in outbreaks. It is confirmed from the report of nVWA (nieuwe Voedsel en Waren Autoriteit) that food commodity groups spice, herbs, and prepared food are responsible for the highest risk of contamination with *C. perfringens* (RIVM Report, 2011).

SYMPTOMS:

Symptoms appear in 8–24 hours, which include diarrhea, acute abdominal pain, and vomiting, and patient recovers within 24 hours. Illness starts after 4–8 hours, and the typical symptoms of consumption of type atoxin of *C. perfringens* include pain in the lower abdominal area and diarrhea. In some of the cases, fever and vomiting are also reported (Robinson et al., 2000).

PATHOGENESIS:

Spores produced by vegetative cells of *C. perfringens* survive in food due to inadequate cooking, which produce toxin. *C. perfringens* produces 12 different toxins, illness is caused by α and θ toxins produced by type A strains. Toxins released in the intestine by sporulating vegetative cells result in accumulation of excessive fluid that causes diarrhea (Center for Disease Control and Prevention, 2011).

C. perfringens enterotoxin (CPE) binds with receptors present on the epithelial cells of the intestinal tract. Toxin accumulates in the plasma membrane and forms small complexes which combine with membrane proteins to make large complexes. These complexes along with CPE-induced membrane permeability alterations inhibit the formation of macromolecules in intestinal epithelial cells that cause pores in the cell membrane, and eventually the cell dies (Radostits et al., 2007). Protein chain of CPE is 35 kDa with isoelectric point of 4.3. It reverses the flow of Na^{+}, $C1^{-}$, and water across the epithelium of the alimentary canal from absorption to secretion (Adams and Moss, 2008).

Clostridium botulinum

C. botulinum is a motile, gram-positive, anaerobic, spore-bearing, and rod-shaped or slightly curved 2–10 μm bacteria, which is widely distributed in soil, sediments of lakes, ponds, and decaying vegetations. Flagella are peritrichous in position and form central or subterminal oval spores. Serologically, *C. botulinum* is classified into seven different serotypes (A–G). They produce eight different toxins recognized as A, B, C_1, C_2, D, E, F, and G; all are neurotoxins except C_2. Among all outbreaks of botulism, most of them are associated with fish and seafood products. Botulism in animals is predominantly associated with types C and D and rarely with types A and B. On the basis of physiological diversity within species, *C. botulinum* are divided into four groups (Table 12.1). Members of group I are proteolytic, members of group II are nonproteolytic and can grow and produce toxin at refrigerated temperature (i.e., 3°C). Group III members produce toxin types C and D, and group IV produce serological type G toxin, and some nontoxigenic clostridia are also placed in this group (Hall et al., 1985; Adams and Moss, 2008).

TABLE 12.1 Properties of Different Groups of *C. botulinum*

Group	Type of Toxin	Pathogenicity	Heat Resistance	Psychotrophicity	Proteolytic	Saccharolytic	Lipolytic
I	A, B, or F	Humans	+	−	+	+	+
II	B, E, or F	Humans	−	+(3°C)	−	+	+
III	C_1, C_2, or D	Animal and birds	±	−	−	+	+
IV	G	Humans	No data	15°C–12°C	+	−	−

Source: *Adams, M.R., Moss, M.O., 2008. Food Microbiology, third ed. The Royal Society of Chemistry, Cambridge, pp. 158–309.*

EPIDEMIOLOGY:

Botulism is widely distributed in many parts of the world, as *C. botulinum* is a soil saprophyte. Food storage and management habits vary in different parts of the globe, which decides the fate of botulinum toxin exposure to the human population. Outbreaks associated with consumption of toxin-containing food are common in Europe and the northern United States, while outbreaks in cattle on pastures have been reported mainly from the United States, Australia, and South Africa (Radostits et al., 2007).

Geographic distribution of ideological agents of botulism differs considerably. In an investigation carried out in the United States, the type A strain was found predominantly in neutral and alkaline soil of the west, whereas types B and C strains were present in damp or moist soil all over the United States. However, type B strain was not found in the southern US region. The type C strain was found in soils of the Gulf Coast, while type D was present in alkaline soil of western parts of the United States. The hay made at a time of mouse plague has been reported to be a source of this illness in Australia. An incident took place in California where 427 out of 444 lactating Holstein cows died after feeding on feed contaminated with botulinum type C toxin from the carcass of cat. Although spores of *C. botulinum* are present throughout the world, most of the outbreaks of botulism have been recorded at north of the Tropic of Cancer, except in Argentina. The geographical prevalence of the disease is linked with some important observations such as home canning of fruits and vegetables in most tropical countries (Jay, 2000; Moeller et al., 2003; Radostits et al., 2007).

SYMPTOMS:

Incubation period of *C. botulinum* is 12–48 hours after the consumption of food containing enterotoxin. The most common features include vomiting, thirst, dryness of mouth, urine retention, constipation, ocular paresis (blurred vision), and difficulty in speaking (dysphonia), breathing, and swallowing (dysphagia). Death occurs due to respiratory paralysis or cardiac failure after 1–7 days. Botulism is clinically recognized as a lower motor neuron disease resulting in progressive flaccid paralysis (Labbe and Nolan, 1981).

PATHOGENESIS:

C. botulinum strain produces highly potent neurotoxin during growth, which causes neuroparalytic disease called botulism in humans and animals without any histological lesions. Botulism may be fatal due to respiratory and cardiac muscle paralysis unless it is properly treated and timely (Jay, 2000). Botulism toxin is absorbed when ingested with contaminated food in the stomach and anterior small intestine or the wound and carried through the blood stream to receptor and enters the nerve cell. Enterotoxin acts on cholinergic nerves of the peripheral nervous system, causing paralytic attack (Hirsh et al., 2004).

Experiments conducted with animals have shown that after ingestion of toxin-containing food, toxin makes its way through the upper part of the small intestine and reaches the blood stream via the lymphatic system. At the nerve muscle junction, it binds with nerve endings and blocks the release of neurotransmitter acetylcholine. Botulinum toxin is a most potent known toxin and is lethal in very small quantities. Example, for an adult human, only 10^{-8} g is enough to be fatal. Protein chain of *C. botulinum* protoxin is

150 kDa, produced during logarithmic growth phase as complexes and released from the cell after lyses. This protoxin is activated by proteolytic cleavage of 150 kDa chain linked with disulfide linkage, into 50 (light) and 100 kDa (heavy) protein chains by host proteolytic enzyme trypsin. Light chain is responsible for cell penetration, and heavy chain plays a significant role in specific binding to neuronal cells (Adams and Moss, 2008).

Salmonella

The *Salmonella* are small, gram-negative, nonspore-forming rods ranging between 0.5 and 3 μm, facultative anaerobe, catalase positive, oxidase negative, and indistinguishable from the *E. coli* under compound microscope. They are generally motile with peritrichous flagella. Humans and animals are the primary host of salmonella. They grow at a vast range of temperature ranging from 5°C to 47°C with optimum growth at 37°C. Salmonella food poisoning is categorized as enteritis and systemic disease. In the last several decades, major changes have occurred in the taxonomy of *Salmonella* (Le Minor and Popoff, 1987; Adams and Moss, 2008).

EPIDEMIOLOGY:

The *Salmonella* species mainly inhabit the alimentary canal of humans, farm animals, reptiles, insects, and birds, yet they are also found in the other parts of the body at different stages of life (Kalpelmecher, 1993). After excretion from the intestine with feces, vegetative cells of *Salmonella* may be propagated by other organisms like birds and insects to distant places (Jay, 2000).

Epidemiologically, *Salmonella* can be divided into three groups. The first group includes those species that infect only humans and are causative agents of diarrhea, typhoid, and paratyphoid fevers. These are *S. typhi*, *S. paratyphi A*, and *S. paratyphi C*. Severity of the illness called enteritis is very high caused by this group. The second group comprises the host-adapted serotypes, few of which are human pathogens and may be present in contaminated food, and include *S. dublin* (cattle), *S. gallirinum* (poultry), *S. abortus-ovis* (sheep), *S. choleraesuis* (swine), and *S. abortusequi* (equine). The host-adapted serotypes cause systemic disease, and they are more invasive due to resistance for phagocytic killing (Acha and Szyfres, 2001; Adams and Moss, 2008).

The third group consists of unadapted serotypes, which means that they are nonhost specific. There are pathogenic to humans and other animals. The epidemiology of *Salmonella* is very complex, which makes it difficult to control. Humans and animals are the main reservoir of this notorious pathogen, and therefore complete eradication is not possible (Quinn et al., 2001).

SYMPTOMS:

Symptoms of salmonellosis (enteritis and systemic disease caused by *Salmonella*) start between 6 to 48 hours after consuming contaminated food. The clinical manifestations are diarrhea (watery or greenish colored with foul-smelling stool) that lasts for a few days, nausea, vomiting, abdominal pain, headache, and chills (mild fever), and the disease is self-limiting (Bean and Griffins, 1990). The organism does not produce any toxin, but the symptoms of illness are due to the infection and invasiveness of this pathogen. Symptoms can be severe, especially in young children and elderly persons. Osteomyelitis, sickle-cell anemia, and reactive arthritis due to salmonella infection are much more common in the general population (Jay, 2000; Radostits et al., 2007).

PATHOGENESIS:

Ingestion of the contaminated food is the main way through which the pathogen enters into the host body. *Salmonella* may execute several mechanisms to skip acidic pH of the stomach, and reaches to the small intestine. Microflora of the intestine provide protection against colonization of pathogen to some extent, but administration of antibiotics kills normal microflora and helps the colonization of pathogen. In the intestine, salmonella uses mannose-resistant fimbriae to adhere to epithelial cells, and they are then engulfed by epithelial cells using receptor-mediated endocytosis. This is a prerequisite for pathogenicity, and it is responsible for nonphagocytic entry of pathogen into the host cell (Bryan et al., 1971; Adams and Moss, 2008). The ruffles assist uptake of the bacterium in membrane-bound vesicles, which often coalesce. The organism replicates in these vesicles and is subsequently released from the cells (Walderhaug, 2007). Advances in molecular techniques have revealed that a signaling molecule is synthesized on a 35–40 kb region of bacterial chromosome called "pathogenicity island," which is responsible for the perversion of host cell and leads to uptake of bacterium (Adams and Moss, 2008).

Shigella

Shigella is primary parasite of human and other primates, as they are fragile organisms not able to survive outside their natural habitat (i.e., intestine of humans or other primates). They belong to the enterobacteriaceae family. Cells are nonmotile, nonsporeforming, gram-negative, catalase-positive, oxidase-negative, facultative anaerobe, and rod--shaped (Jay, 2000; Adams and Moss, 2008). This organism is typically adapted to mesophilic range of temperature (10°C–45°C) and heat sensitive, and grows well in a pH range of 6–8. Different biochemical tests are employed to distinguish between four known human pathogenic species such as *Sh. dysenteriae*, *Sh. flexneri*, *Sh. Boydii*, and *Sh. sonnei* (Richmond, 1990; Adams and Moss, 2008).

EPIDEMIOLOGY:

Main route of transmission is contaminated food and water with feces or person-to-person contact. Often 10–100 viable cells of *Shigella* are enough to cause infection in a healthy individual. The cases of shigellosis are low in developed countries as compared to developing nations because in developed countries only fecal–oral spread is responsible, and in developing nations both fecal–oral route and contamination of food and water are responsible (Radostits et al., 2007; Addis and Sisay, 2015). Annual report of Laboratory-based Enteric Disease Surveillance 2016 (CDC) has shown 12,597 cases of *Shigella* infection have been reported from all 52 states of United States. *Shigella sonnei* was the largest contributor (80.5%) followed by *Shigella flexneri* (12.6%), *Shigella boydii* (0.2%), and *Shigella dysenteriae* (0.1%) (NCEZID, 2018).

In Indian perspective, *Shigella* is mainly responsible for diarrheal illnesses, *S. flexneri* being the main culprit followed by *S. sonnei* and *S. dysenteriae.* However, it is eradicated from northern and eastern parts of the country. Reports of outbreaks of shigellosis from various parts of India surface from time to time, but exact data of morbidity and mortality is lacking. *Shigella* is developing and exhibiting antibiotic resistance. Few cases have been reported for the development of antibiotic resistance in *Shigella*, which is an issue of serious concern (Taneja and Mewara, 2016).

SYMPTOMS:

Clinical symptoms of illness caused by *Shigella* include abdominal cramps and pain, watery loose stool, fever, tenesmus, and in severe cases, bloody diarrhea. Diarrhea caused by a different pathogenic species is mostly self-limiting (Kuo et al., 2008). More severe symptoms or acuteness of illness are prevalent in immune-compromised patients and children. Complications include peritonitis, toxic megacolon, and septicemia. Untreated cases of *Shigella* dysentery may become severe and cause anorexia, hemolytic-uremic syndrome, dilation of large intestine, kidney damage, weight loss, and seizures (Sur et al., 2004). Bacteremia is also reported in HIV-infected patients and malnourished children (Miller et al., 2005).

PATHOGENESIS:

Infection caused by *Shigella* is invasive in nature. Invasiveness of pathogen depends on large plasmid encoding pathogenic factors. After reaching the intestine, bacterial cells adhere to enterocytes with the help of outer membrane adhesins. These are subsequently engulfed by intestinal enterocytes; inside, the cytoplasm bacterial cells escape from the phagosome and multiply, infecting the adjacent enterocytes and underlying connective tissue. In response to this, the colon gets inflamed and produces abscess, ulcerations, and results in diarrhea, sometimes blood along with watery stool (Adams and Moss, 2008).

Species of *Shigella* produce enterotoxin known as shiga toxin with different biological activities. Shiga toxin (Stx) is among the most potent bacterial toxins known. Toxin is reported to be mostly formed by *Shigella dysenteriae* and some *E. coli* strains. Shiga toxin (Stx) consists of two protein subunits A and B joined noncovalently. Subunit A of shiga toxin (Stx) inactivate 60s subunit of eukaryotic ribosome and halt protein synthesis in the cell. Subunit B is a pentamer that binds to cellular receptor globotriaosylceramide, Gb3 of endothelial cells, and it acts as a strong cytotoxin (Adams and Moss, 2008; Melton-Celsa, 2014).

Escherichia coli

E. coli is gram-negative, catalase-positive, oxidase-negative, fermentative, nonspore-forming rods and the most studied microorganism. This was first described by German bacteriologist Theodor Escerich. *E. coli* belong to the family enterobactereriae. It is closely related to *Shigella* spp. but can be biochemically differentiated from other members of enterobactereriae (Bryan et al., 1971; Quinn et al., 2001). *E. coli* is a versatile bacterium and is an important component of the normal microflora of humans and other warm-blooded animals. It has been used for a long time as a laboratory workhorse for cloning and expression purposes. Besides this, *E. coli* is a deadly pathogen for humans and other animals. It causes severe intestinal and extraintestinal illness through its several virulence factors (Kaper et al., 2004).

EPIDEMIOLOGY:

E. coli produces three types of general clinical syndromes resulting from infection of different pathotypes. They are sepsis/meningitis, urinary tract infections, and enteric/diarrheal diseases. There are six well-characterized intestinal pathogenic *E. coli*, namely enteropathogenic *E. coli* (EPEC), enterotoxigenic *E. coli* (ETEC), enteroaggregative *E. coli* (EAEC), enterohemorrhagic *E. coli* (EHEC), diffusely adherent *E. coli* (DAEC), and enteroinvasive *E. coli* (EIEC). Among these, the sixth enteroinvasive *E. coli* (EIEC) is the true intracellular pathogen (Nataro and Kaper, 1998). Enterohemorrhagic *E. coli* (EHEC) strain produces verocytotoxin that is responsible for major food-borne diseases in humans. This toxin resembles shiga toxin and is responsible for producing diarrhea in its host. Animals are identified as main reservoirs of *E. coli* pathotypes (Buchanan and Doyle, 1997). Enterotoxigenic *E. coli* (ETEC) is responsible for infantile diarrhea in most developing counties, and leads to acute dehydration. Enteroinvasive *E. coli* (EIEC) invades epithelial cells of the colon like *Shigella* but does not produce toxin. Enteropathogenic *E. coli* (EPEC) attaches to the enterocytes, effaces them, and produces lesions leading to loss of microvilli and diarrhea (Adams and Moss, 2008).

SYMPTOMS:

Onset of symptoms, in the case of enterotoxigenic *E. coli* (ETEC), starts 12–36 hours after ingestion of organism along with food or water. The disease is self-limiting and persists for 2–3 days. Symptoms include abdominal pain along with vomiting, severe cholera-like syndrome (watery stool), and loss of body fluids, leading to dehydration. Enterohemorrhagic *E. coli* (EHEC) produce acute bloody diarrhea that remains for 5–10 days. Stomach pain, acute watery diarrhea initially present for 1–3 days, and then progresses to bloody diarrhea that becomes a life-threating condition in elderly people. Enteroinvasive *E. coli* (EIEC) produces symptoms of invasive bacillary dysentery causing ulceration and inflammation in the colon. Fever, malaise, abdominal pain, stool containing blood and mucus, followed by watery diarrhea are the clinical symptoms. Symptoms of enteropathogenic *E. coli* (EPEC) appear after 12–36 hours, with infection including vomiting and diarrhea, stool containing mucus without blood, and malaise. Infection is self-limiting, persists for 5–7 days, but in case of infants it can persists more than 2 weeks (Kaper et al., 2004; Adams and Moss, 2008).

PATHOGENESIS:

Pathogenic *E. coli* use specific factors for adherence to different sites in the intestine (small intestine) where usually nonpathogenic *E. coli* does not inhabit. They produce distinct morphological structures known as pili or fimbriae (Cassels and Wolf, 1995). Other than pili, pathogenic *E. coli* uses adhesins, which are outer membrane proteins, such as intimin of enterohemorrhagic *E. coli* (EHEC) and uropathogenic *E. coli* (UPEC) (Tieng et al., 2002). Different strains of enterotoxigenic *E. coli* (ETEC) secret heat-labile enterotoxin (LT), heat-stable enterotoxin a (STa), and heat-stable enterotoxin b (STb). These enterotoxins affect a number of important eukaryotic processes, such as increasing the concentration of intracellular messengers (cyclic AMP, GMP, and Ca^{+}) leading to ion secretion in the intestinal lumen. Shiga-like toxin produced by enterohemorrhagic *E. coli* (EHEC) disrupts ribosomal RNA, which halts protein synthesis and intoxicated endothelial or epithelial cells die (Melton-Celsa and O'Brien, 1998). Enteropathogenic *E. coli* (EPEC) and enterohemorrhagic *E. coli* (EHEC) produce Map protein having two different biological activities; first, it targets mitochondria and disrupt their membrane potential, and second it stimulates Cdc42-dependent filopodia formation (Kenny et al., 2002).

Shiga-like toxin produced by enterohemorrhagic *E. coli* (EHEC) is differentiated in two forms: Stx 1 and Stx 2. Stx 1 resembles closely shiga toxin and is composed of two subunits, A (M_r 32 kDa) and B (M_r 7.7 kDa). Stx 2 also has two subunits A and B with molecular weight of 35 and 10.7 kDa, respectively (Adams and Moss, 2008).

Listeria monocytogenes

This organism is widespread in the environment. It is a gram-positive catalase-positive, oxidase-negative, nonspore-forming, facultative intracellular parasite and facultative anaerobe. Shape of bacteria varies from coccoid to rods of 0.4–0.5 μm × 0.5–2.0 μm in size. This organism is motile and peritrichous flagellated and moves by a characteristic tumbling motion. Optimum growth temperature is 30°C–35°C, but it can also survive over a wide range of temperatures from 0°C to 42°C. *L. monocytogenes* can easily tolerate salt concentration up to 10% and can survive for almost 1 year in 6.0 pH and 16% salt concentration (Adams and Moss, 2008).

EPIDEMIOLOGY:

L. monocytogenes can be isolated from soil, sewage, decaying vegetation, and fresh and salt water and survive more than 8 weeks in the environment (Hoelzer et al., 2013). Food vehicles with which *L. monocytogenes* is associated mostly are milk and other dairy products, ground beef or meat and their products, poultry products, and raw fruits and vegetables. In most of the recent listeriosis investigations, *L. monocytogenes* was associated with the food vehicles (stone fruits, apple, caramel, ice cream, celery, cantaloupe, and mung bean sprouts) that were not reported in earlier investigations. Advancement in molecular typing has made it possible to precisely track the pathogenic strains responsible for outbreaks in a particular area (Buchanan et al., 2017).

In a study conducted in the European Union, sporadic cases and outbreaks of listeriosis were found to be increase, with 1763 confirmed cases of listeriosis in 27 states. Five member states of the EU reported seven confirmed outbreaks of listeriosis, and the food vehicles identified include shellfish, mollusks, and crustaceans (EFSA, 2015). From 1998 to 2008, the United States has implemented regulatory initiatives on industries processing of ready-to-eat red meat and poultry, which reduced outbreaks of listeriosis (Cartwright et al., 2013). Outbreaks of listeriosis were reported from the United States as a result of the consumption of ice cream in March 2015. All of the patients were hospitalized, and there were two deaths (Pouillot et al., 2016). In 2011 PulseNet reported a multistate (28 states) listeriosis outbreak in the United States infecting 147 people, with 33 deaths and one miscarriage. The US Food and Drug Administration (FDA) identified *Listeria* from environment and food products; it was found that most of the infected individuals had consumed cantaloupes.

It is a fatal proven disease in the west, and from last two decades it has surfaced in India. In September 2011, 500 students of the Premier Institute in India became infected with an outbreak of listeriosis after consuming contaminated food (Tirumalai, 2013). India has poor networking and reporting systems on the disease outbreaks. Very limited information is available on food-borne listeriosis outbreaks, but incidence of listeriosis outbreaks including animals can be traced back to the 1930s (Janakiraman, 2008). A study carried out by Chugh (2008) has summarized increasing sporadic cases of human listeriosis and reported that it is a growing food-borne disease in India (Tirumalai, 2013).

SYMPTOMS:

Incubation period varies between 1 and 90 days, and typical symptoms of disease arise in a few weeks. Symptoms vary in pregnant women, elderly, or the very young and the immunocompromised considerably from flu-like illness to meningoencephalitis and meningitis. In pregnant women, flu-like symptoms arise initially, which are further associated with fever, gastrointestinal discomforts, and headache. A transplacental fetal infection results in premature-labor, miscarriage, and stillbirth. Early onset of listeriosis in newborns results in in utero (in the uterus) infection characterized by septicemia, pneumonia, and abscesses. Meningitis mostly occurs as late onset of disease in newborns (Adams and Moss, 2008).

PATHOGENESIS:

After entering in the intestine, *L. monocytogenes* penetrates in the endothelial cells or crosses the Peyer's patches. Human gastrointestinal cells have receptors for internalin A and B produced by the pathogen, which facilitates internalization of bacteria (Bonazzi et al., 2009). Internalin is a bacterial surface protein of 800 amino acids encoded by chromosomal gene *inlA*. Internalized bacteria are entrapped in phagosome, where it produces listeriolysin O (58 kDa hemolysin), which breaks the membrane of phagosome and releases itself prior to fusion of phagosome with lysosome. Pathogen multiplies intracellularly and reaches to mesenteric lymph nodes and then spreads in different parts of the body including the liver, placenta, and the central nervous system through the blood (Adams and Moss, 2008; Smith et al., 2008).

Food-Borne Bacterial Diseases Outbreaks

Occurrence of a large number of cases of a disease than normally expected in a defined season and geographical area for a short or long period of time is called disease outbreak. In the United States, a federal agency known as The Centers for Disease Control and Prevention (CDC) with the help of FoodNet, a reporting system, keeps track of disease outbreaks and related pathogens (Vemula et al., 2012). Analysis of data of the last 10 years from CDC showed that the frequency of food-borne disease outbreak due to *Salmonella* is very high. In the last 10 years, 35 different strains of *Salmonella* have caused 67 outbreaks, and a total 6214 people were affected, 1409 were hospitalized, and 16 people died. Similarly, four different strains of *E. coli* have killed 5 people, 843 people suffered, 324 were hospitalized in 24 outbreaks. *Cyclospora cayetanensis* caused 2242 cases of food-borne illness; out of them, 109 were hospitalized in 5 outbreaks. In two outbreaks of

TABLE 12.2 CDC[a] Data on Food-Borne Disease Outbreaks in the Last 10 Years (2009–18)

S. no.	Pathogen	Number of Pathogenic Strains	Number of Outbreaks	Number of Cases	Hospitalizations	Deaths
1.	*Salmonella*	35	67	6214	1409	16
2.	*Escherichia coli*	4	24	843	324	5
3.	*Listeria monocytogenes*	1	14	310	293	63
4.	*Cyclospora cayetanensis*	1	5	2242	109	0
5.	*Vibrio parahaemolyticus*	1	2	130	15	0

[a]CDC, *Centers for Disease Control and Prevention, United States.*

V. parahaemolyticus, 130 people were affected and 15 were hospitalized. *L. monocytogenes* has proven to be the most virulent bacteria, causing 14 major outbreaks in the last 10 years. The cases of listeriosis reported were less as compared to *Salmonella* and *E. coli*, as only 310 cases were reported—out of them, 293 were hospitalized; however, 63 deaths were reported with this food-borne disease (Table 12.2).

All of these outbreaks were associated with commercial food products produced by top US brands, and few of them were imported from European countries. These food products include beef, ground beef, deli ham, other meat products, milk products, cheese, fruits, vegetables, salads, frozen food products, nuts, and sprouts.

Food-borne disease outbreaks occur in both developed and developing countries due to pathogens and their toxins (Käferstein and Abdussalam, 1999). According to the WHO, in developed countries about 30% of individuals suffer from food- and water-borne illnesses every year (WHO, 2006). Another study reported that about 1.8 million people die due to diarrheal disease outbreaks, mostly in developing nations (Yadav and Rekhi, 2015). Four lakh children in India below age 5 years die each year due to diarrheal outbreaks. In India, reporting of food-borne diseases and their categorization is not done separately in the Health Information of India. In the official document of the Government of India for 2004, about 95,75,112 cases of diarrhea were reported; these cases might include the cases of food-borne illnesses, but they were categorized under diarrheal cases (Health Information of India, 2004). In Indian context, studies on food safety and food-borne diseases are much less (Parvathy et al., 2005). WHO estimates that only 1% of cases of food-borne illnesses were reported in developing countries (Bhat and Rao, 1987). Most of the outbreaks in India remained unreported and unrecognized until a major health and economic disaster took place (Kohli and Garg, 2015).

In 1998 at the high altitude of the western Himalayas, 78 out of 103 soldiers of an army unit suffered from food-borne illness after consuming contaminated food with *Salmonella enteritidis* (Singh et al., 1998). A study conducted on bacterial food-borne illness outbreaks in India for a period of 1980–2009 showed 24 outbreaks, in which 1130 individuals suffered. Important pathogenic agents involved in the outbreaks were *S. aureus*, *Vibrio*, *Salmonella*, *E. coli*, and *Yersinia enterocolitica*. *Y. enterocolitica* belongs to Enterobacteriaceae, gram-negative, facultative anaerobic, catalase-positive, and oxidase-negative responsible for gastroenteritis (Vemula et al., 2012). One hundred and fifty individuals were affected by the outbreak at Kharar town (Punjab) reported in 2009 after consuming *kheer* contaminated with *S. enteritidis* (Dikid et al., 2009). In the state of Madhya Pradesh (India), more than 100 children and adults were affected after consuming a snack *tikki* made up of potato balls fried in vegetable oil. Clinical and food samples were isolated and investigated in which *S. aureus* was found to be the pathogen of this outbreak (Nema et al., 2007). Food-borne disease outbreaks in India from the years 1983 to 2009 are summarized in Table 12.3.

TABLE 12.3 Data on Bacterial Food-Borne Illness From 1983 to 2009 in India

S. no.	Year	Pathogen	Associated Food	No. of Affected Individuals	References
1.	1983	*Vibrio parahaemolyticus*	Fish	34	Lalitha et al. (1983)
2.	1985	*Salmonella bornum*	Chicken	No data	Chaudhary et al. (1998)
3.	1985	*Salmonella weltevreden*	Stale rice	4	Aggarwal et al. (1985)
4.	1987	*Staphylococcus aureus*	Sweet meat	31	Mandokhot et al. (1987)
5.	1990	*Vibrio fluvialis*	Vegetarian meal	14	Thekdi et al. (1990)
6.	1993	*Staphylococcus aureus*	Meat	42	Nayar et al. (1993)
7.	1995	*Salmonella paratyphi A var durazoo (2, 12: a:-)*	Vegetarian meal	33	Fule et al. (1996)
8.	1997	*Yersinia enterocolitica*	Buttermilk	48	Abraham et al. (1997)
9.	1998	*Salmonella enteritidis*	Fowl meat	78	Singh et al. (1998)
10.	2007	*Staphylococcus aureus*	Potato *tikki*	100	Nema et al. (2007)
11.	2009	*Salmonella enteritidis*	*kheer*	150	Dikid et al. (2009)
12.	2009	*Salmonella wein*	Chicken and poultry product	10	Antony et al. (2009a)
13.	2009	*Salmonella weltevreden*	Vegetarian meal	34	Antony et al. (2009b)

Control and Prevention of Food-Borne Bacterial Diseases

As microbes are ubiquitous in nature and the very common organisms are found in the environment, so their association with our food is not surprising. Illness caused by bacterial contamination can be prevented, and the efforts are being made to keep food safe and wholesome by legislative, agricultural, industrial, and public health authorities. These approaches will be very fruitful, if food handlers at final stages also have food safety education (Bredbenner et al., 2013). Toxin-producing pathogens and intoxication of already produced toxins in the food can be avoided by proper cooking at high temperature (Bryan et al., 1971). The data of food-borne diseases and their outbreaks should be maintained to investigate and identify the risk factors like infectious agent, relation with host, environmental factors, and the food as vector for that particular pathogen. It will further help in preventing the disease outbreaks (WHO, 2005). Other important measures for prevention of bacterial food-borne illness are improvement in personal hygiene, adequate cooking, heat processing, and refrigeration of food. The following control measures can be executed to eradicate food-borne bacterial disease throughout the world:

1. Educating food handlers for proper personal hygiene.
2. Prohibiting individuals with skin lesions or other abscess from handling food.
3. Keeping food at proper refrigerated temperature.
4. Avoiding the chances of cross-contamination.
5. Proper disposal of leftout or contaminated food to avoid contamination of other food articles.
6. Adequate washing and handling of raw food articles.
7. Boiling of home canned vegetables for at least 3 minutes prior to serving in order to destroy botulinum toxin.
8. Exposure of food to a temperature of 80°C for 30 minutes or boiling for 10 minutes to destroy preformed toxin I food.
9. Proper refrigeration of left-out food (WHO, 2008).

AGENTS OF NONBACTERIAL FOOD-BORNE ILLNESS

Food-borne illnesses are also caused by nonbacterial agents like toxic metabolites of plants, called plant toxins, algal toxins, nematodes, helminthes, dinoflagellates, toxigenic fungi, food-borne viruses, and cyanobacteria (Adams and Moss, 2008). These are in themselves a specialized area so they cannot be explained in details, but only some important agents are included here, keeping in view the scope of the present discussion.

Natural Plant Toxins

Many plants used by humans and animals as a food source produce toxigenic metabolites as naturally occurring constituents. Some plants produce these metabolites as their defense molecules from parasites or grazing animals. Although the risk of potential toxicity due to natural food toxins is very low, still there is always the possibility of undetected contamination and idiosyncratic response (Dolan et al., 2010). Some important toxins and their related plants are discussed next.

Lectins

Lectins are glycoproteins present in leguminous plants and other grains (e.g., kidney beans, soybeans, lentils, and black beans) (Shibamoto and Bjeldanes, 1993). Lectins are also called hemagglutinins because these agglutinate red blood cells and also interfere with nutrient absorption by the intestine. Ricin is a lectin produced by the castor oil plant (*Ricinus communis*) and is very poisonous as a very small dose (oral LD50 values in rat and mice were 20–30 mg/kg body weight) of purified ricin can kill a human, and due to this property, ricin is used in bioterrorism (Omaye, 2004). 3D structure of ricin is shown in Fig. 12.1.

The lectin found in significant amount in leguminous plants such as red kidney beans and fava beans is called phytohemagglutinin (PHA). PHA can agglutinate red blood cells, can affect membrane permeability of a cell, and can induce mitosis (Banwell et al., 1983). Symptoms of PHA intoxications in humans include diarrhea, vomiting, and nausea. In low doses, intoxication is self-limiting, and recovery occurs in 4–5 hours (FDA, 2009).

Lathyrus

Lathyrus, also called α,γ-diaminobutyric acid (Fig. 12.2A), is produced in pulse *Lathyrus sativa*, also called grass pea. Long-term consumption of this pulse results in a very serious health condition known as lathyrism (Adams and Moss, 2008). Lathyrism is an upper motor neuron degeneration caused by prolonged dependence on grass pea (*L. sativa*). This legume is used as a rich source of calories and protein in many parts of the Indian subcontinent, China, South America, and Ethiopia. Manifestation of lathyrism appears as weakening hind limbs successively and ultimately leads to hind limb paralysis (Spencer, 1999).

Solanine

Solanine is a green-colored pigmented glycoalkaloid and acts as natural pesticide, also known as α-solanine (Fig. 12.2B). α-Solanine is naturally produced in the plants of the

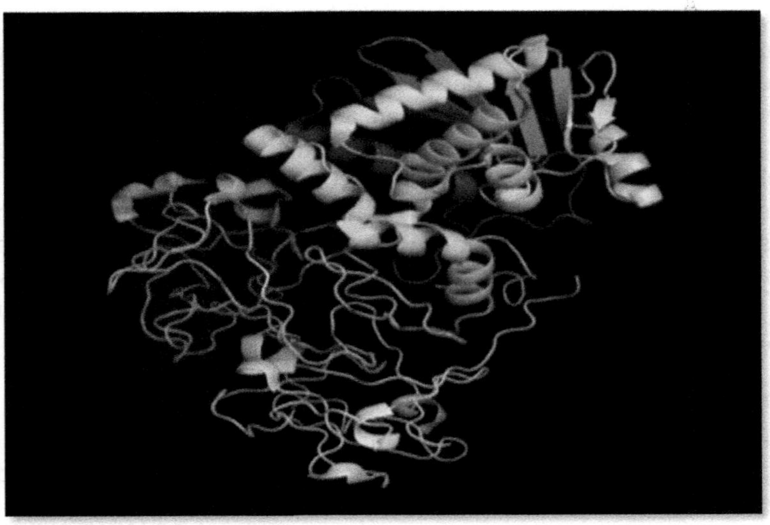

FIGURE 12.1 Three-dimensional structure of ricin. PDB ID: 2aai (Rutenber et al., 1991).

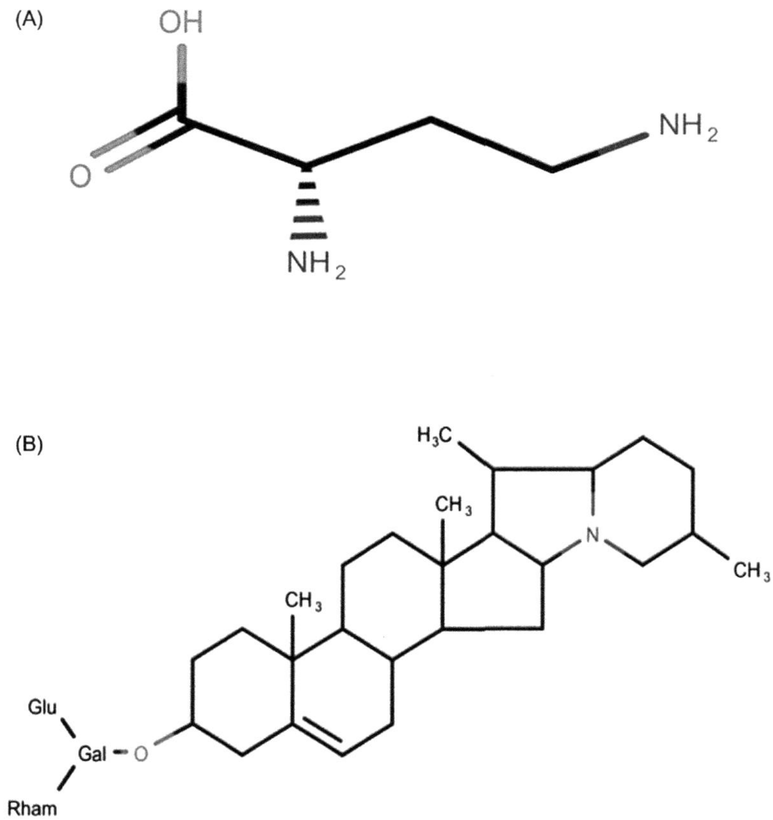

FIGURE 12.2 Chemical structure of various toxins: (A) α,γ-diaminobutyric acid (*Lathyrus*), (B) α-solanine.

solanaceae family and other plants such as potatoes, tomatoes, apples, bell peppers, cherries, and sugar beets. It is found in high concentrations at areal and the green part of the plant but also is found in potato tubers at low concentration (Shibamoto and Bjeldanes, 1993). Synthesis of α-solanine in potato tubers is stimulated by sunlight exposure, mechanical injury, and tuber aging. High concentration of α-solanine is unsafe for human consumption (Dalvi and Bowie, 1983; Jones, 1995) as it leads to inhibition of acetylcholinesterase, cell membrane disruption. Manifestations include itchiness in the neck region, increased sensitivity called hyperesthesia, drowsiness, stomach pain, labored breathing, nausea, vomiting, and diarrhea (Shibamoto and Bjeldanes, 1993). A 1–5 mg/kg dose of α-solanine is highly toxic for humans, and 3–6 mg/kg doses leads to death (Tice, 1998).

Pyrrolizidine Alkaloids

Pyrrolizidine alkaloids (PAs) are found in plants of the families of boraginaceae, compositae, apocynaceae, fabaceae, and asteraceae. Intoxication of PA compounds (Fig. 12.3) mostly happens due to accidental mixing of seeds of plants containing PAs with cereals.

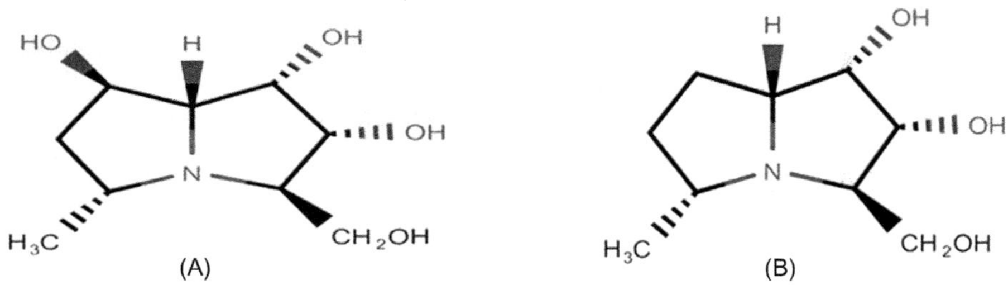

FIGURE 12.3 Chemical structure of polyhydroxylated pyrrolizidines: (A) hyacinthacine B_3 and (B) hyacinthacine A_3 (Asano et al., 1999).

PAs compounds are also present in honey and the milk of goats and cows. PAs containing necine rings are proven carcinogenic, mutagenic, and hepatotoxic in nature. In the liver, PAs are converted enzymatically into pyrrols, which are alkylating in nature. These pyrrols, when reach the lungs, cause pulmonary hypertension by making pulmonary vasculature thick (Deshpande, 2002; Prakash et al., 1999).

FOOD-BORNE ILLNESS DUE TO AGRICULTURAL PESTICIDES AND INSECTICIDES

Pesticides and insecticide are used to control pests and insects damaging crops in fields and storage houses and as of today have become part of our agricultural practices. Excessive use of these agrochemicals are showing a deleterious effect on human health and also polluting soil, water, and the environment. Pesticide intoxication is a big and unrecognized health issue in the developing nations around the globe. According to a study, about 250,000–370,000 people die every year due to ingestion of pesticides (Gunnell and Eddleston, 2003; Gunnell et al., 2007). Study conducted during 2010–14 has shown that total average annual pesticide use (kg/ha) in Japan during 2010–14 was highest (18.94), followed by China (10.45), Mexico (7.87), Brazil (6.166), Germany (5.123), France (4.859), United Kingdom (4.034), United States (3.886), and India (0.261) (Zhang, 2018). Humans and animal are exposed to pesticides in many ways, including drinking water, food, inhalation, and oral and dermal contact (EFSA, 2008). Exposure to these pesticide mixtures may have deleterious long-lasting impacts on human health in the long term. In studies carried out in Western countries, it was observed that pesticides cause and increase intensity of neurodevelopmental abnormalities, cancer, and chronic degenerative diseases (Hernández et al., 2013).

When pesticides are applied to crops in fields or during storage, their residues accumulate in food and potable water (Laetz et al., 2009). When food supplies containing pesticide residue or contaminated water are used by humans, they get exposed to the mixture of pesticides. This mixture of two or more pesticides in vivo involves in induction or inhibition of enzymes involved in detoxification (Hernández et al., 2013). Earlier studies conducted on the bioaccumulation of pesticides have revealed the bioaccumulation of dichlorodiphenyltrichloroethane in the fish muscles reached up to 57 ppm on the coast of California and 3 ppm in the Baltic Sea (Jensen et al., 1972). The bioaccumulation of pesticide residues in the organisms occurs through several routes (e.g., food chains, direct uptake from water, and suspended materials, etc.) (Miles and Harris, 1971). Biomonitoring studies have revealed bioaccumulation of detectable concentrations of pesticide residues in the bodies of adults and children (Zeliger, 2011).

Heavy Metals

Heavy metals are elements having high density as compared to water and include metalloids that are toxic at very low level. Toxicity of the heavy metals is related to their heaviness (Fergusson, 1990; Duffus, 2002). Heavy metals contamination in our environment occurs due to natural and anthropogenic activities (He et al., 2005). Most prominent sources of heavy metal contamination in the environment are metal-based industries, mining, foundries, and smelters (Fergusson, 1990; Duffus, 2002; He et al., 2005). Heavy metals are known to affect structure and function of cellular organelles (Wang and Shi, 2001) and enter our body through consumption of food and water contaminated with these heavy metals. These may induce carcinogenicity and toxicity in humans and exposed animals (Tchounwou et al., 2004). A few important heavy metals are transmitted with food and water; their toxic effects are discussed in following sections.

Arsenic

Food and water are the largest source of arsenic poisoning for most of the individuals, and average intake in developing countries is apparently 50 μg/day. Other means of arsenic poisoning are usually much smaller as compared to diet (NRC, 2001). It is estimated that millions of people are exposed to arsenic poisoning around the globe, and the situation is worst in developing nations like India, Uruguay, Bangladesh, Chili, Taiwan, and Mexico (Tchounwou et al., 1999).

High-level exposures of arsenic in humans cause cancer and systemic health problems (Tchounwou et al., 2003). Various clinicopathological developments were reported from parts of Argentina, Bangladesh, Chile, China, Finland, Hungary, Inner Mongolia, Mexico, Taiwan, Thailand, and West Bengal (India), where human populations were exposed to higher concentrations of arsenic in food and water. Clinical manifestations from arsenic poisoning comprise cardiovascular and peripheral vascular disease, portal fibrosis, neurologic and neurobehavioral disorders, hearing loss, diabetes, developmental anomalies, hematologic disorders, and carcinoma (Tchounwou et al., 2004; Centeno et al., 2005).

Chromium

Chromium is present in Earth's crust with oxidation states ranging from chromium (II) to chromium (VI) (Jacobs and Testa, 2005). Nonoccupational exposure of humans to chromium happens mostly due to contaminated food and water (Langard and Vigander, 1983). Generally, fresh food contaminated with heavy metals contains levels of chromium ranging from 10 to 1300 μg/kg, but the workers working in chromium industries are exposed to double concentration.

Chromium VI compounds are extremely toxic to humans and animals; clinical and pathological symptoms include anemia, damage to male reproductive system, low sperm count, and irritation and ulcers in the stomach and small intestine. A few individuals develop hypersensitivity toward chromium VI and chromium III compounds; allergic reaction includes swelling of skin and redness. Reports of increase in stomach tumors by drinking chromium VI contaminated water have been observed in humans and animals. Extremely high doses of chromium VI in humans resulted in cardiovascular, gastrointestinal, hepatic, renal, and neurological effects leading to death (ATSDR, 2008).

Mercury

Mercury is a transition metal found in nature in three forms: elemental, organic, and inorganic (Clarkson et al., 2003), and all the forms of are toxic. Methylmercury is the most toxic form of mercury poisoning formed in the environment from inorganic mercury by methylation caused by microorganisms (Dopp et al., 2004). Exposure of humans to all forms of mercury occurs during dental care, through environmental pollution, food contamination, and industrial and agricultural work (Bhan and Sarkar, 2005). Mercury once enters in water is methylated by algae and bacteria and then it enters into fish, shellfish, and finally the consumption of seafood contaminated with mercury leads to toxic effects in humans (Sanfeliu et al., 2003).

Studies have revealed that mercury and other toxic metals affect cellular organelles and adversely impair their biological functions (Zalups and Koropatnick, 2000). Mercury has a property of bioaccumulation, and it also increases the production of reactive oxygen species (ROS) by causing defect in oxidative phosphorylation and electron transport at the ubiquinone-cytochrome B5 reductase level. Mercury induces the premature shedding of electrons to molecular oxygen, which is responsible for an increase in the generation of ROS in mitochondria through normal metabolism of eukaryotic cells. All this play a major role in the mediation of metal-induced cellular responses and carcinogenesis (Crespo-Lopez et al., 2009).

FOOD-BORNE VIRAL INFECTIONS

Viruses are very small entities with diameter ranging between 25 and 300 nm and therefore are not visible under light microscope and can only be viewed using electron microscopes. They possess only one type of nucleic acid (DNA or RNA) inside a protein coat of capsid without any other cellular structure, and thus viruses are obligate parasites of plants, animals, and humans and are very specific to host and cannot multiply other than in susceptible host cells by using its cellular metabolism and machinery they hijack for their replication (Adams and Moss, 2008). Food-borne viral infections are of worldwide concern as around 125 million cases of viral food-borne diseases have been reported globally in the year 2010. Norovirus has alone killed 35,000 people worldwide (Kirk et al., 2015). People living in developing nations such as Africa, Southeast Asian countries, and parts of the Eastern Mediterranean are suffering more from food-borne viral infections (Iturriza-Gomara and O'Brien, 2010).

Poliomyelitis is a highly infectious disease and can be traced back to ancient Egyptian paintings and carvings. Poliovirus is an enterovirus belonging to Picornaviridae family. Clinical manifestations range from mild cases of respiratory illness, gastroenteritis, and mild to severe forms of paralysis (Mehndiratta et al., 2014). Infection of poliomyelitis remains asymptomatic for 3–5 days; after that, mild illness (abortive poliomyelitis), aseptic meningitis (nonparalytic poliomyelitis), and paralytic poliomyelitis take place. This disease has been associated with crippling deformities affecting thousands of lives throughout the world (McQuillen and McQuillen, 2005).

TABLE 12.4 Important Food-Borne Viral Infection and Their Causative Agents

S. no.	Illness	Virus	Genome	Family	Transmission Route	Sources of Contamination
1.	Poliomyelitis	Poliovirus types 1–3.	ssRNA	Picornaviridae	Fecal–oral	Fecal contaminated food and water
2.	Gastroenteritis	Sapovirus	ssRNA	Caliciviridae	Fecal–oral, person-to-person	Fecal contaminated food and water, by food handler
3.	Gastroenteritis	Astrovirus	ssRNA	Astroviridae	Fecal–oral, person-to-person	Fecal contaminated food and water, by food handler
4.	Hepatitis	Hepatitis A virus	ssRNA	Picornaviridae	Fecal–oral, contaminated water	Fecal contaminated food and water, by food handler
5.	Hepatitis	Hepatitis E virus	ssRNA	Hepeviridae	Contaminated water and fecal–oral route	Contaminated pork, fecal contamination water
6.	Gastroenteritis	Aichi virus	ssRNA	Picornaviridae	Fecal–oral, Contaminated water	Fecal contaminated food and water, by food handler
7.	Gastroenteritis	Rotavirus	dsRNA	Reoviridae	Fecal–oral, person-to-person	Fecal contaminated food and water, By food handler
8.	Gastroenteritis	Adenovirus	dsDNA	Adenoviridae	Fecal–oral, contaminated water	Fecal contaminated food and water, By food handler
9.	Shellfish associated gastroenteritis	Parvovirus	ssDNA	Parvoviridae	Fecal–oral, person-to-person	Fecal contaminated food and water, By food handler
10.	Gastroenteritis, Neonatal necrotizing enterocolitis	Human enteric coronavirus (HECV)	ssRNA	Coronaviridae	Person-to-person, prepared food contaminated with infected body fluid	Contamination of food and water with infected body fluids

Most outbreaks of hepatitis A and E occur due to ingestion of fecal contaminated water and food. Hepatitis A and E viruses multiply in endothelial cells of gut and then are carried via blood to the liver. Incubation period ranges from 2 to 6 weeks. Symptoms include fever, malaise, anorexia, nausea, and vomiting. Pale yellow-colored pigment is released into body tissues and urine due to liver damage. This situation is called jaundice. Identification of the source is difficult because of the long incubation period (Adams and Moss, 2008). Some important food-borne viral infections and their causative viral agents are listed in Table 12.4.

FUNGI AND MYCOTOXINS

Fungi are the diverse group of organisms and a very important part of our ecosystem. These are utilized by man in fermentation processes, production, certain medically important compounds, and beverages since antiquity (Buckley, 2008; Benedict et al., 2016). Fungi are also harmful to human and animals. About 300 known pathogenic fungi have been reported to cause illness in a healthy human, ranging from allergic reaction to invasive infection (Hawksworth, 2001). Many of the fungi synthesize some secondary metabolites during their growth in different foods. These metabolites are highly toxic and carcinogenic in nature and are called mycotoxins (Marroquín-Cardona et al., 2014). Mycotoxins have caused many dreadful epidemics in the past of man. During 1942–48 alimentary toxin aleukia became responsible for claiming 100,000 lives in Russia (Smith and Moss, 1985; Joffe, 1978), and stachybotryotoxicosis killed thousands of horses during 1930 in USSR (Moreau, 1979).

Mycotoxins are very diverse chemical compounds which elicit symptoms in the host that ultimately result in deformities or death. Mycotoxins-induced illness includes immunosuppression, skin necrosis, leucopoenia, mutation, and cancer (Pitt, 2000). Most important mycotoxins produced by various fungi are aflatoxins, fumonisins, ochratoxin A, zearalenone, and trichothecenes, listed in Table 12.5 and discussed in the following sections.

Aflatoxins

Aflatoxins (Fig. 12.4A) are produced by some species of *Aspergillus* (e.g., *A. flavus*, *A. parasiticus*, and *A. nomius*). *A. flavus* is widely distributed in the environment and is a most reported food-borne fungus (Stoloff, 1977). Naturally produced aflatoxins are of four types, B_1, B_2, G_1, and G_2. B and G refer to blue and green light emission by these compounds under ultraviolet light. Aflatoxins produced by these fungi are known to induce hepatocellular carcinoma in human and animals (Wu, 2013). These mycotoxins induce acute aflatoxicosis; manifestation includes abdominal pain, pulmonary edema, vomiting, and fatty infiltration and necrosis of the liver. Most cases of acute aflatoxicosis in humans were reported from developing nations (Shank et al., 1971). During the 1970s, about 97 fatal cases of aflatoxicosis were reported from western India due to consumption of heavily molded corn (Krishnamachari et al., 1975; Bhat and Krishnamachari, 1977).

Exposure of aflatoxins is also reported to cause childhood stunting, a condition in which the height of child in respect to age is much less, by WHO growth reference (Ricci et al., 2006). Aflatoxins are also reported to be responsible for the immune system dysfunction. Studies carried out to confirm the relation between aflatoxin exposure and

TABLE 12.5 Different Mycotoxins of Public Health Concern

S. no.	Mycotoxin	Producing Fungi	Associated Food
1.	Aflatoxin	*Aspergillus flavus* *A. parasiticus*	Maize, peanuts, tree nuts (Hazelnuts), copra (dried coconut kernels), spices, cottonseed
2.	Fumonisins	*Fusarium verticillioides* *F. proliferatum* *A. niger*	Maize
3.	Ochratoxin A	*Penicillium verrucosum* *A. ochraceus* *A. carbonarius* *A. niger*	Maize, wheat, barley, oats, dried meats and fruits, coffee, wine
4.	Zearalenone	*Fusarium. graminearum, F. cerealis, F. equiseti, F. crookwellense, d F. semitectum*, and *F. culmorum*	Maize, wheat, barley, oats,
5.	Trichothecenes	*F. graminearum* *F. culmorum*	Maize, wheat, barley, oats

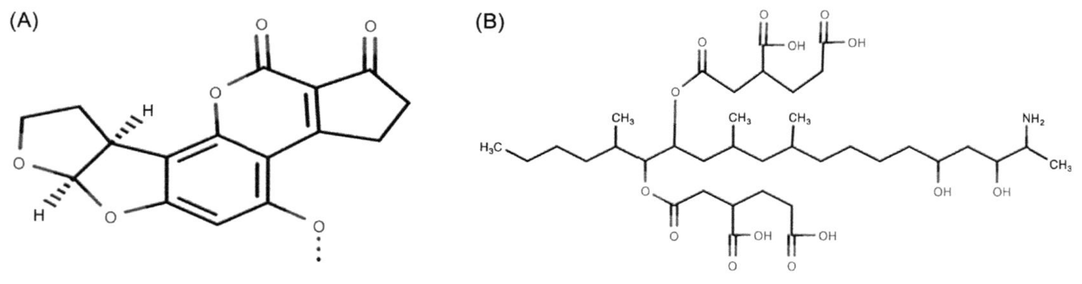

FIGURE 12.4 Chemical structure of various toxins: (A) aflatoxin and (B) fumonisin.

immune dysfunctions have revealed the increase of impaired markers of human immunity (Jiang et al., 2005; Turner et al., 2003).

Fumonisins

In late 1980s, scientists were studying a disease in horses called equine leucoencephalomalacia and discovered a toxin produced by fungi responsible for this disease. They named it fumonisins toxin (Fig. 12.4B), as it was produced by *Fusarium verticillioides*, *F. proliferatum*, *A. niger*, and some related fungal species (Bezuidenhout et al., 1988; Marasas et al., 1988). *F. verticillioides* is a plant pathogen and grows universally on maize and produces ear rot. Maize is the main staple food of the Republic of Transkei in Southern Africa (southeastern region), where the frequency of human esophageal cancer is very high. Fumonisins have been implicated with esophageal cancer in the Republic of Transkei (Marasas et al., 1988).

Fumonisins are of three types, fumonisin FB_1, FB_2, and FB_3. FB2 and FB_3 are the cocontaminators (Wu, 2013). Fumonisins resembles sphingosin; it consists of 20 carbon aliphatic chains with two ester-linked hydrophilic side chains. Toxicity of fumonisins can be explained on the basis of its competition with sphingosin in sphingolipid metabolism (Riley et al., 1996).

Fumonisins are also responsible for neural tube defects. In this, fumonisins interfere with formation of the neural tube of the developing embryo. It causes nerve damage, which results in partial leg paralysis, anencephaly (undeveloped brain), and stillborn birth (Marasas et al., 2004).

Ochratoxin A

Ochratoxin A is a metabolite that was first reported in *Aspergillus ochraceus* and later also found in *Penicillium viridicatum* (Van der Merwe et al., 1965; Van Walbeek et al., 1969). Later studies revealed that ochratoxin A is mainly produced by *Aspergillus carbonarius* (Varga et al., 1996).

Ochratoxin A is an acute nephrotoxin (Fig. 12.5A), which is lethal to dogs, pigs, mice, and trout. Fatal doses of ochratoxin A cause necrosis in renal tubules and periportal

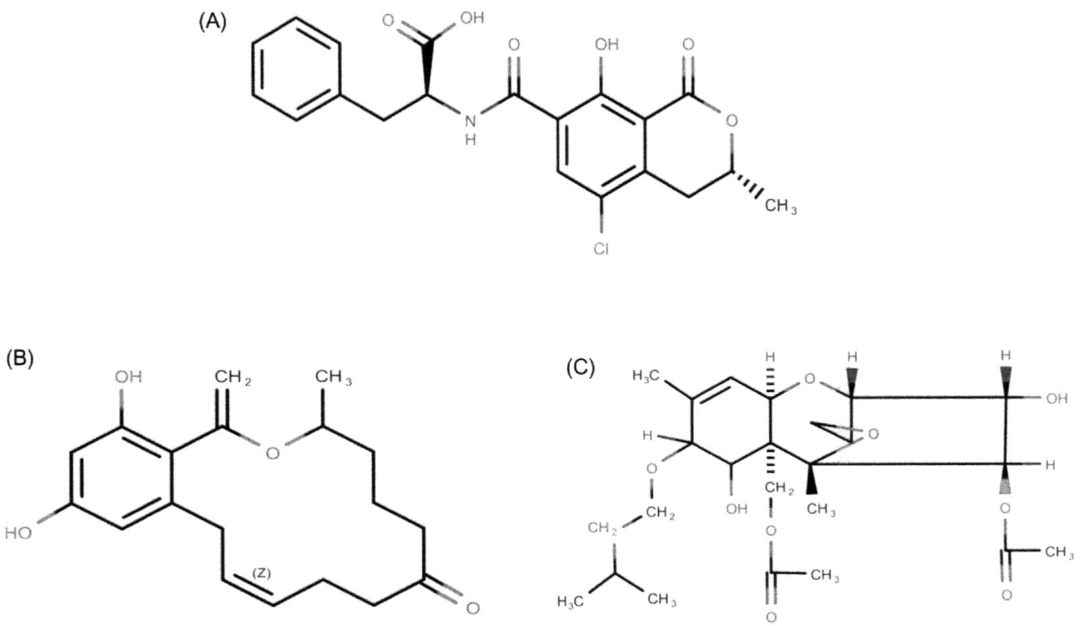

FIGURE 12.5 Chemical structure of various toxins: (A) ochratoxin A, (B) zearalenone, and (C) trichothecenes toxin (TCT).

liver cells. Other pathological effects include immunosuppression, damage to the embryo, and induction of cancer (Scott, 1977). Exposure to humans occurs due to feeding on pork or meat, as ochratoxin A is fat soluble so it accumulates in the fat of affected animals. Another source of exposure to humans of ochratoxin A is bread made from toxin-containing wheat or barley.

Zearalenone

Zearalenone was previously known as F-2 toxin. It is mainly produced by *Fusarium* species (e.g., *F. graminearum*, *F. cerealis*, *F. equiseti*, *F. crookwellense*, *F. semitectum*, and *F. culmorum*). These fungi are pathogens of cereal crops worldwide and produce zearalenone as a side product of cell metabolism (Bennett and Klich, 2003). Chemically, zearalenone (Fig. 12.5B) is resorcyclic acid lactone described as 6-(10-hydroxy-6-oxo-trans-1-undecenyl)-B-resorcyclic acid lactone (Cheeke, 1998). This toxin produces estrogenic effects on farm animals (Schwarzer, 2009). Clinical symptoms include increased uterine size and secretions, swelling of the vulva, hyperplasia of mammary glands and secretion, increased incidences of pseudopregnancy, decreased libido, infertility, stillbirth, and complications of vaginal and rectal prolapses (Gupta et al., 2018). In humans, studies reported that zearalenone acts as a ligand for pregnane X receptor (hPXR), which activate transcription factors involved in the expression of many hepatic drug-metabolizing enzymes, including cytochrome P450 enzymes (Ding et al., 2006). In vitro studies carried out with different human cell lines revealed the effect of zearalenone on human estrogen and androgen receptors. Zearalenone found agonist in MCF-7 cells (breast cancer cell line) for estrogen receptor alpha (hERα) and antagonist for androgen receptor (hAR) in PALM cell line (Gupta et al., 2018).

Trichothecenes

Trichothecenes toxins (TCT) are produced by several fungal genera; however, most of them have been isolated from *Fusarium* spp. TCT have been found to contaminate wheat, barley, corn, rice, rye, oats, and other crops. The effect of TCT has been extensively studied on poultry and farm animals as TCT contaminate feed to a large extent (Leeson et al., 1995).

Epoxides are found in all TCT at the C_{12} and C_{13} positions (Fig. 12.5C), which is responsible for its toxic activity. TCT affect cell division in the body, where cells are actively dividing such as the skin, gastrointestinal tract, lymphoid, and erythroid cells. In these cells, TCT inhibit the protein synthesis, resulting in acute necrosis in mucosal lining and skin. More fatal consequences are depressed immune function and reduced bone marrow (Schwarzer, 2009).

HELMINTHES (PARASITIC WORMS)

During 1990, the World Bank conducted its first investigation on the Global Burden of Diseases (GBD) and developed a standardized indicator called "Disability Adjusted Life Years" (DALY) to assess the burden of 107 diseases and injuries around the globe. Disease caused by parasitic worms was not taken into consideration in this study, so they are also called neglected tropical diseases (NTD) (Ndimubanzi et al., 2010). Parasitic worms are transmitted through food and water and parasitize on humans and animals. They do not grow and multiply in food, but they contaminate water with their eggs, which further grow and pass through different stages of development and form cysts. These cysts, when swallowed by specific hosts, grow and produce disease (Adams and Moss, 2008).

Nematodes (Roundworms)

Nematodes causing harmful effects to humans include *Ascaris lumbricoides* (roundworm), *Ancylostoma duodenale* (Ancylostoma) and *Necator americanus* (new world hookworms), *Trichuris trichiura* (whipworm), *Trichinella spiralis, Enterobius vermicularis* (pinworm), and *Strongyloides stercoralis* (threadworm). People living in the slums and rural areas of developing countries are at high risk of getting infected with these nematodes (Crompton, 1999). The reason behind this is poor hygiene, lack of education, overcrowding, and poor health services (Conway et al., 1995). Climatic conditions also favor the growth and propagation of these nematodes (Kappus et al., 1994).

Ancylostoma duodenale, Necator americanus, A. lumbricoides, T. trichiura, E. vermicularis, and *S. stercoralis* are the nematodes which are specific to humans as they have no intermediate host or reservoir. Occasionally other animals like pigs also get infected with human specific nematodes but the life cycle of nematodes does not complete in them. Except *Enterobius vermicularis,* all of the human specific nematodes require a period of development in soil to become infective to humans (Booth and Bundy, 1992). Life cycle of human gastrointestinal nematodes is generalized in Fig. 12.6.

Symptoms of the disease are not produced by adult worms in the case of *Trichinella spiralis,* but thousands of larvae produced by female worms are responsible for the symptoms. Symptoms include abdominal pain, nausea, and vomiting that appear when larvae burrow in the intestinal wall and reach to the specific muscle tissues. Muscle pain and fever start when these larvae invade the muscle and finally encyst (Adams and Moss, 2008).

Trematoda and Cestoda

In the context of food-borne illness, two helminthes are most important, which are discussed in this chapter. Trematoda, which include *Fasciola hepatica* are called liver fluke, and the cestoda which include the genus *Taenia.* The life cycle (Fig. 12.7) of these organisms is very complex and includes different intermediate hosts at different stages of development; humans, sheep, and cattle are definitive hosts in cases of trematoda (Adams and Moss, 2008). Infection of liver fluke is a food-borne trematodiases, and its transmission takes place through contaminated undercooked aquatic food. Infection of liver fluke is most prevalent in Southeast Asia and Latin America. International travel, human migration, and food trade pose the threat of spreading infection to the other parts of the world. Clinical symptoms of liver fluke infections are unspecific and appear as per the location of the parasite in the body of the host (Fürst et al., 2012).

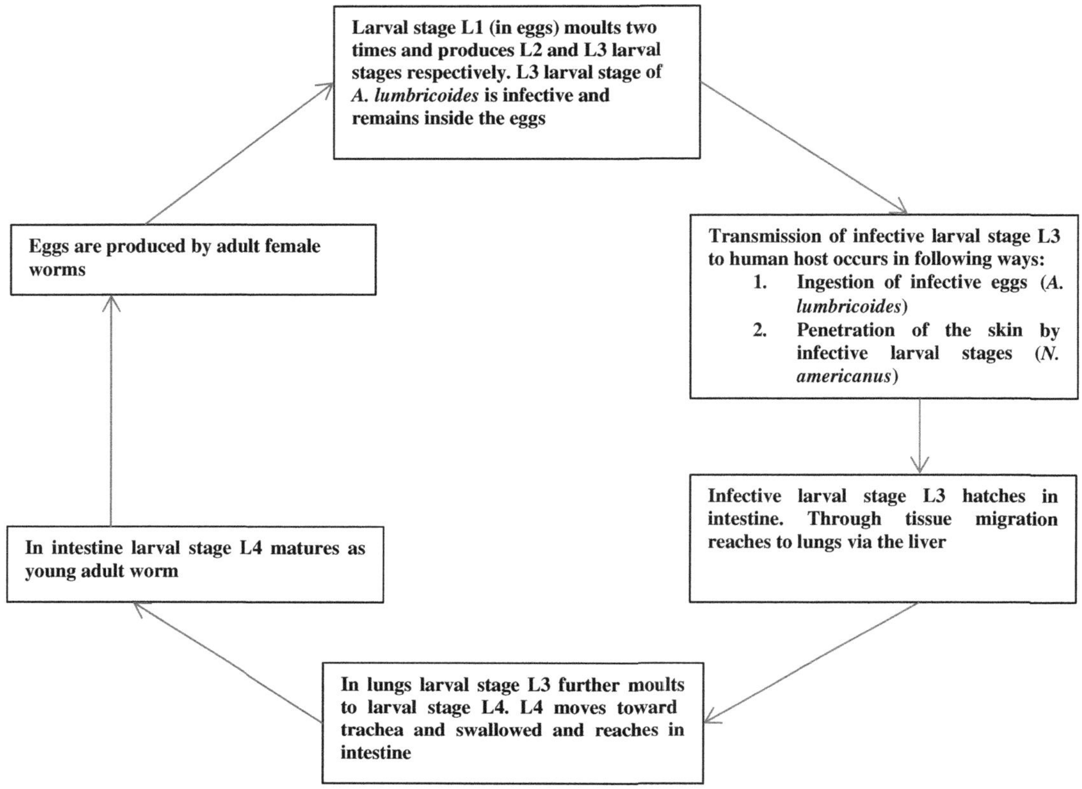

FIGURE 12.6 Life cycle of human specific nematodes (*A. lumbricoides*) (Stepek et al., 2006).

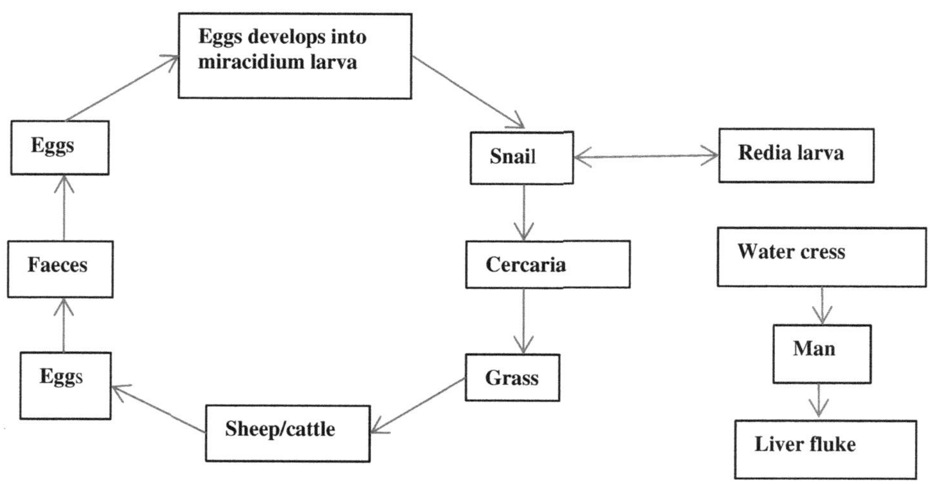

FIGURE 12.7 Life cycle of *Fasciola hepatica*.

About 56 million people worldwide have been infected by these parasites. *F. hepatica* and *F. gigantica* are responsible for infecting 2.6 million people around the globe alone (FuÈrst et al., 2012). It is NTD, and most of the affected population belong to developing countries (Mas-Coma et al., 2014). Patients infected with liver fluke remain asymptomatic for a long time; acute infection symptoms include abdominal pain in the liver region, and fever and chronic symptoms are biliary colic, cholecystitis, and cholangitis (Marcos et al., 2008).

Cestoda include *Taenia solium*, *Taenia saginata*, and *Taenia asiatica*, which are a long, ribbon-like flat worms. *Taenia solium* infect humans as its primary host after consumption of undercooked pork infected with larva of organism called cysticerci (García et al., 2003). Its infection (taeniasis) resulting in neurocysticercosis is prevalent in developing countries where people live in close proximity with pigs and rear them as a source of food and livelihood (Ndimubanzi et al., 2010).

Life cycle (Fig. 12.8) of *Taenia solium* is complex and includes two hosts: humans are primary and pigs are the intermediate host. Infection in humans and pigs starts with ingestion of eggs or gravid proglottids by fecal–oral route or by contaminated food or water (Murray and Lopez, 1997; Chimelli et al., 1998). In case of humans, after the ingestion, eggs hatch into oncospheres in the intestine of the host (Hotez and Brown, 2009; Cruz et al., 1999). These oncospheres than invade intestinal walls and migrate to striated muscles, brain, liver, and other tissues (Varma and Gaur, 2002). Cysticerci also attach to the wall of the small intestine by their scolex, and infection of brain tissue results in neurocysticercosis (Edwards and Krishna, 2004; Hotez et al., 2008).

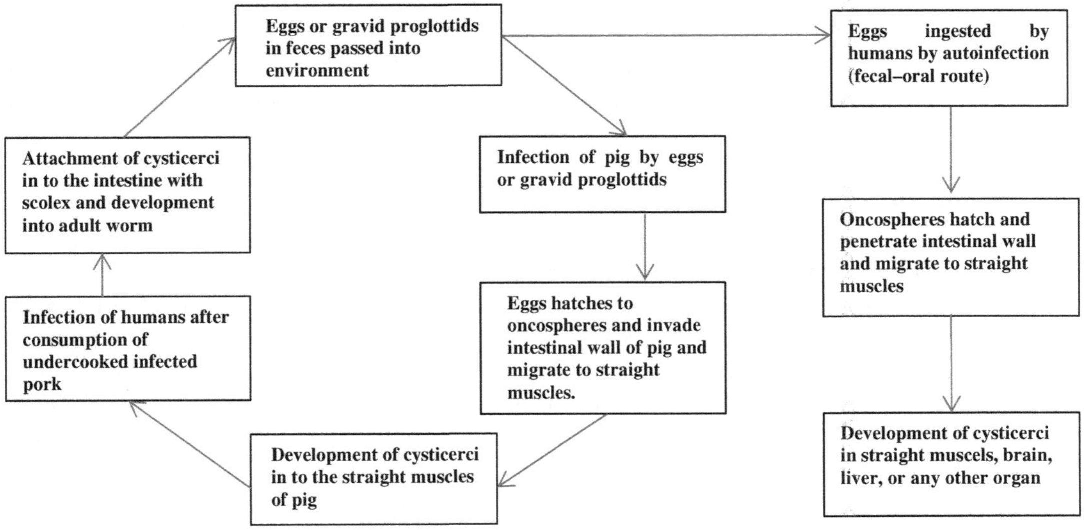

FIGURE 12.8 Life cycle of *Taenia solium*.

PROTOZOA

Protozoa are unicellular eukaryotic organisms, which are known to infect humans and produce environmentally stable cyst or oocyst. Cyst helps these organisms to survive in a harsh environment and infect other intermediate hosts through water or food (Vitaliano et al., 2015). There are several intestinal protozoa that cause diarrhea, but *Entamoeba histolytica, Giardia intestinalis,* and *Cryptosporidium* species are the most dreadful culprits of diarrhea (Thompson and Ash, 2016). In a recent study, it was estimated that different species of protozoa cause 45.3% of food-borne illnesses, of which *E. histolytica* is the major causative agent with 24.7% followed by *G. intestinalis* with 11.2% and *Cryptosporidium* species with 2.2% contribution (Berhe et al., 2018).

Entamoeba histolytica

E. histolytica is a parasitic pathogen of man which causes amebiasis. Amebiasis is a serious health problem mainly persistent in developing nations of tropical and subtropical areas including Central and South America, South and West Africa, Mexico, India, and Pakistan (Khan, 2017). Infection of *E. histolytica* is quite common in children and adults in the tropical and subtropical areas. Amebiasis claims 50 million lives, and 450 million individuals suffer from it every year (Ohnishi et al., 2004; Ravdin and Petri, 1995). In a recent study, it was reported that out of a total 45.3% of protozoan born disease, *E. histolytica* is major etiological agent with 24.7% contribution (Berhe et al., 2018). There are six species of *Entamoeba* (*E. histolytica, E. dispar, E. moshkovskii, E. polecki, E. coli,* and *E. hartmanni*), which harbor human intestines (Fotedar et al., 2007). *E. histolytica* is the most hostile among all because it dissolves intestinal tissue (Harold, 1975).

Life cycle of this parasite consists of two stages: one is the infective stage represented by cyst and the other is vegetative phase represented by trophozoites (Tanyuksel and Petri, 2003). Infection starts with ingestion of cyst with infected food or water, which exist in the intestine and form trophozoit. This trophozoite further feeds on food and red blood cells, and also damages the intestinal cells. Trophozoit form cysts during precyst phase and are excreted out with feces (Espinosa-Cantellano and Martinez-Palomo, 2000; Tanyuksel and Petri, 2003).

Giardia lamblia

Giardia lamblia is one of the protozoan intestinal parasites which cause diarrhea in both adult and children. It is also a major concern of public health in most of the developing and developed nations (Azian et al., 2007; Ayeh-Kumi et al., 2009). Life cycle of *Giardia lamblia* (Fig. 12.9) also includes infective cyst stage and feeding trophozoite stage. Some of the trophozoites in the intestine form cysts and excrete out with feces (Adam, 2001). *Giardia* is not invasive, and it is not confirmed how the diarrheal symptoms appears. Along with diarrhea, abdominal cramps, pain, and vomiting also exist (Adams and Moss, 2008).

Food-borne illnesses when compared with other diseases are proven more severe and cause huge economic loss to mankind in terms of suffering, expenses of medicine and

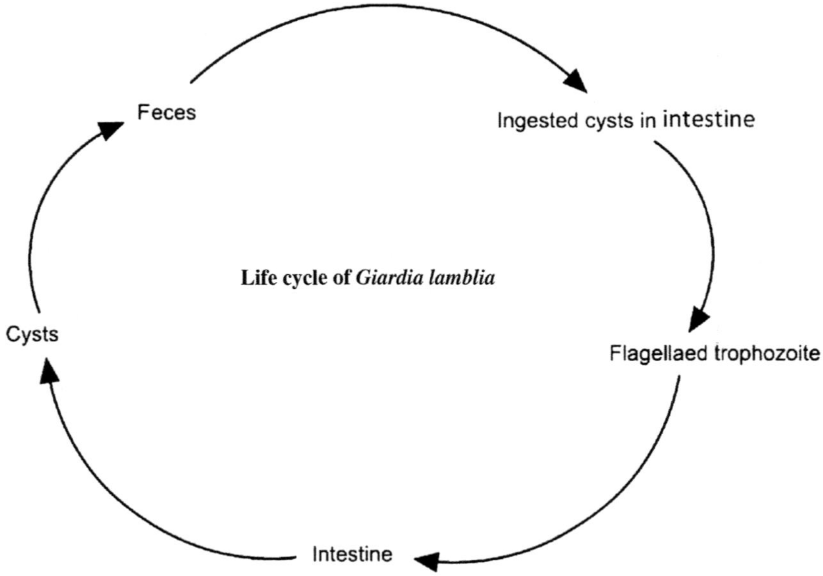

FIGURE 12.9 Life cycle of *Giardia lamblia*.

hospitalization, and food loss. Food gets contaminated with harmful pathogens, chemicals, and toxins by cross-contamination, adulteration, anthropogenic activities, and different environmental factors, and consumption of such unhygienic and unsafe food leads to food-borne illnesses their outbreaks. The rapid globalization and ease of transportation of food across the countries are often the cause of spread of food-borne pathogens from one part to the world to another. To deal with food-borne illnesses, it is necessary rather mandatory to regulate the process of food from "farm to fork" with stringent hygienic practices. There are agencies in every country vis-à-vis international agencies, which have framed regulations and laws in this concern that need to be strictly implemented by the regulatory bodies and observed by the food producers, processors, and traders to reduce the spread of food-borne illnesses. Some international regulations/laws made by several important agencies are discussed in the proceeding sections of this chapter.

INTERNATIONAL LAWS

Need of International Food Laws

Food-borne illnesses are of immense importance as global public health issue, and therefore, these have been recognized as a priority area by the WHO. In addition, the rise in international trade in food has augmented the risk of transmission of food contaminants from one country to another, and the need to estimate the risk that infectious agents pose to human health has become crucial. Global emergence and reemergence of food-borne pathogens have made microbiological safety of food an important issue (Odeyemi and Sani, 2016). Globally, more than 250 sources of food-borne diseases have been identified. The increase in the incidence of food-borne diseases has led to the imposition of several food quality regulations in different countries (Scallan et al., 2011). Although the globalization and liberalization of world food trade offer many benefits and opportunities, these present new risks of microbial pathogen from the original point of processing and packaging to the consumers' locations thousands of miles away because of the global nature of food production, manufacturing, and marketing to spread. Most of the countries have introduced food control systems to ensure that foods are safe for human consumption so that people are protected against unsafe, adulterated, or poor-quality food (Gauthier and Mahabir, 2012). There are four major challenges that need to be addressed to safeguard the health of consumers (Brundtland, 2001):

1. To set up consumer confidence from the farm to the table by assessing and upgrading existing food safety measures.
2. To ascertain reasonable food safety levels and serve all countries to achieve those standards.
3. To elaborate intercontinental standards for premarket compliance of genetically modified (GM) food.
4. To ensure the safety and benefits of new products for consumers.

The developing and developed countries in their endeavor to ensure global food safety need to closely interact and participate to establish food safety laws and develop sustainable integrated food safety systems for the mitigation of health risks all over the world.

What Are Food Laws?

These are the legislations and food control services to promote a good quality safe food supply to the consumers and also to protect them from adulterated, spoiled, and contaminated foods at international and national levels that refer to food safety laws, food inspection laws, and export and import rules for food (FAO, 2005). Food laws mainly contain a basic food act and regulations, sometimes food standards, lists of food additives, chemical tolerances, and others are also included in the basic food laws. The act itself sets out principles, while regulations contain various categories of products coming under the sovereignty. Food standards are components of the enforcement structure and are meant to implement the basic food laws as part of the regulation or separate enactments. The basic food laws should include the purposes and scope of the law, definitions of basic concepts, inspection and analytical procedures, enforcement, regulations for additives, pesticides, contaminants, and penalties. The proper implementation of food laws ensures fair trade practices, growth of the food industry, and also protects the honest manufacturer and dealer against unfair competition (Lasztity, 2009).

International Food Laws and Regulation

A large number of countries have enacted food laws and regulations to ensure that the food is safe, that is, up to the desired quality, and consumers receive adequate and precise information about the foodstuff they are buying in the market. However, the differences among countries need and description often make it unfavorable to trade food internationally (FAO/WTO, 2017). Trade can be more cumbersome, when two nations define the same product unlike and have nonidentical guidelines and norms to check that product. The following international agencies and agreements help to develop international standards in the production, processing, and preservation of food to be exported and imported with the objective of protecting public health and safeguard fair practices in the food trade, thereby facilitating international trade:

- World Health Organization (WHO)
- Food and Agriculture Organization (FAO)
- Codex Alimentarius Commission (CAC)
- Sanitary and Phytosanitary (SPS) agreement
- WTO/SPS agreements—Office International des Epizooties (OIE), International Plant Protection Convention (IPPC)
- Joint FAO/WHO Expert Committee on Food Additives (JECFA)
- Joint FAO/WHO Meeting on Pesticide Residues (JMPR)
- Joint FAO/WHO Expert Committee on Microbiological Risk Assessment (JEMRA)

World Health Organization

The WHO is an authoritative organization of the United Nations that is mainly involved in the public health matters all over the world. It was established in 1948 and its headquarters is in Geneva, Switzerland. WHO works to promote the accessibility of safe, healthy, and nutritious food for everyone, and its members have recognized food safety, food- borne diseases, GM food, and food additives as a worldwide challenge. From time to time, WHO frames and promotes various guidelines to meet these challenges, and its main roles in food safety and security are following (Forsythe, 2002; FAO, 2005; http://www.who.int):

- To monitor the use of antimicrobials in food of animal origin and minimize antimicrobial resistance associated with the use of antimicrobials.
- To define safe exposure levels, which form the basis for food safety standards to ensure fair trade practices by the development of scientific risk assessments.
- To prevent, detect, and manage food-borne risks by generating data on food-borne outbreaks and supporting administration of adequate infrastructures (e.g., laboratories).
- To improve food safety and security from "farm to consumer" by educational programs based on scientific research directed to train food handlers and the customers in order to reduce the food-borne illness.
- To provide scientific guidance on the evaluation of foods obtained from genetic modification and nanotechnology.

- To manage food safety risks, ensure quick distribution of information during food safety emergencies to cease the supply of contaminated food from one nation to another through International Network of Food Safety Authorities (INFOSAN).
- To provide independent international scientific guidance on microbiological and chemical hazards for the evolution of international food standards by Codex.
- To reduce the risks of transmission of zoonoses in the food chain through the consumption of animal products by providing guidance to the public.
- The five keys to safer food according to WHO for both developed and developing nations are keep clean, separate raw and cooked food, cook thoroughly, keep food at appropriate temperature, use safe raw material and water (Marusic, 2011).

Food and Agriculture Organization

The FAO is a specialized agency of the United Nations with 194 states members over 130 countries around the world. It was established in 1945 by the International Institute of Agriculture (IIA) in Quebec City, Canada, and its headquarters is in Rome, Italy. The main goals of FAO are to wipe out hunger, food uncertainty, and malnutrition, and the sustainable handling of natural resources and genetic resources for the well-being of future generations. The following are the main activities of FAO (Forsythe, 2002; FAO, 2002a,b, 2005; http://www.fao.org):

- To ensure that people have easy access to the best quality food by encouraging policies, political commitments, and up-to-date information about hunger and malnutrition challenges and solutions that support food security and good nutrition worldwide.
- To form agriculture and fisheries extra productive by serving as a knowledge network and using the expertise of agronomists, foresters, fisheries, livestock specialists, and other professionals to collect, analyze, and disseminate data that aid development.
- To reduce rural poverty by improving farm productivity and increasing off-farm employment opportunities through social protection and finding better ways for rural populations to manage and cope with risks in their environments.
- To develop inclusive, efficient agricultural and food systems by increasing the participation of smallholder farmers and agricultural producers in developing countries to achieve the goal of a world without hunger.
- To increase the prospect of livelihoods to threats, crises, and support them in preparing and responding to disasters.
- To put information within reach and supporting the transition to sustainable agriculture by serving as a knowledge network and use the expertise to collect, analyze, and disseminate data that aid development.
- To strengthen political will and sharing policy expertise to achieve rural development and hunger alleviation goals.
- To support public–private collaboration to improve smallholder agriculture by providing services to farmers and facilitate greater public and private investments in strengthening the food sector.
- To eliminate hunger, food insecurity, and malnutrition by developing mechanisms to monitor and warn about multihazard risks and threats to agriculture, food, and nutrition.

- To support countries to prevent and reduce risks by informing them on successful risk mitigation measures that can be included in agriculture policies.
- To develop a series of technical tools to provide guidance on food safety emergencies, reinforce preparedness, and bring knowledge to the field through projects.

FAO/WHO FOOD CONTROL ASSESSMENT TOOL:

The FAO/WHO started the Food Control Assessment Tool to support plans and monitor food control for the developing countries. The assessment depends on appropriate internationally recognized food control systems and Codex provisions (FAO/WHO, 2003). The systematic and evidence-based assessment by FAO/WHO Food Control Tool promotes improved responsibility in the area of capacity development, in government services, between donors and implementers, and between technical assistance providers and beneficiary nations. It also accelerates greater integrity in capacity development even by building complementarity between different involvements (FAO, 2005).

FAOLEX:

It is the world's largest, comprehensive, and up-to-date database on electronic collection of national laws and regulations on food, agriculture, and renewable natural resources. FAOLEX is constantly updated with an average of 8000 new entries every year. Presently, it has legal and policy documents drawn from over 200 countries, territories, and regional economic integration organizations and arises in more than 40 languages. It is administered by the Development Law Service of the FAO Legal Office, and it complements FAO's core function of advising its members on legal and institutional means to promote and regulate national and international cooperation in the area of food and agriculture sector. Primarily, its mandate is to collect, analyze, interpret, and circulate information related to nutrition, food, and agriculture. The terms and conditions that are applied to the use of the FAO website are applicable to use of the FAOLEX database (http://www.fao.org).

International Scientific Committees

The FAO/WHO facilitates the implementation of risk assessment in food safety that is based on scientific guidance and evidence provided by the authorities. The risk assessment according to the CAC is a scientifically based process with hazard identification and characterization, exposure assessment, and risk characterization (FAO/WHO, 2003). It provides an estimate of the possibility and severity of food-borne illnesses that is useful in determining hazards and their reduction to acceptable levels. The risk assessments and safety evaluations are dependent on available best scientific information, systemizing inputs from many authorities and publications (WHO, 2015; FAO/WHO, 2006a,b, 2015; FAO, 2016). At present, there are the following international scientific committees that provide advice to Codex, governments, industries, and researchers worldwide:

- Joint FAO/WHO Expert Committee on Food Additives (JECFA)
- Joint FAO/WHO Meeting on Pesticide Residues (JMPR)
- Joint FAO/WHO Expert Committee on Microbiological Risk Assessment (JEMRA)

JOINT FAO/WHO EXPERT COMMITTEE ON FOOD ADDITIVES:

The JECFA is an international expert scientific committee that is organized jointly by FAO/WHO and was established in 1956, primarily to assess the security of food additives. Now its function also includes assessment of food safety, naturally occurring toxicants, contaminants, and residues of veterinary drugs in foodstuff (FAO/WHO, 2003; FAO, 2005). The committee establishes acceptable daily intakes (ADIs), prepares specifications on purity of food additives, recommends maximum residue limits (MRLs), and decides norms for detecting and quantifying residues in foods. JECFA has evaluated more than 1300 food additives with 17 maximum levels (MLs) for contaminants over 4037 MLs covering 303 food additives, and MRLs for residues of veterinary drugs is 610, covering 75 veterinary drugs in foods (FAO/WTO, 2017). As on today, it has evaluated more than 2500 food additives, 40 contaminants, and residues of 90 veterinary drugs (http://www.fao.org/food/food-safety-quality/scientific-advice/jecfa/en/). It serves as a scientific advisory body to FAO, WHO, CAC, and some nations use its particulars in framing their own regulations and standards. Mainly information to the Codex is furnished by the Codex Committee on Food Additives and Contaminants (CCFAC) and the Codex Committee on Residues of Veterinary Drugs in Foods (CCRVDF) (Forsythe, 2002; FAO, 2005; FAO/WHO, 2015).

JOINT FAO/WHO MEETING ON PESTICIDE RESIDUES:

The JMPR consists of the joint assembly of the FAO/WHO committee of specialists on pesticide residues in foodstuff and in the ecosystems. It accomplishes toxicological assessment of pesticide residues, estimates the ADIs and recommends maximum residues limits (MRLs) for individual pesticides residues in or on particular food products (FAO/WHO, 2003). The MRLs for pesticides are defined according to good agricultural practices based on estimated residue levels in administered field trials. When the ADIs are exceeded, more clarified intake estimations are carried out using national food consumption data and guidance from pesticide residues monitoring programs (FAO, 2005). The JMPR manifest chemical security norms are based on a review of toxicological studies in the test animal species. As of today, JMPR has 4846 MRLs for pesticide residue covering 294 pesticides in foods (FAO/WHO, 2015; FAO/WTO, 2017).

JOINT FAO/WHO EXPERT COMMITTEE ON MICROBIOLOGICAL RISK ASSESSMENT:

The FAO/WHO have started a series of joint specialist negotiations to evaluate risk associated with microbiological contamination of eatables with the aim of providing an unambiguous review of scientific advice on the microbiological risk assessment (MRA) and to develop the quantitative risk assessments of specific pathogen−commodity combinations followed by the CAC principles and guidelines. Its work includes an assessment of existing and current risk evaluation methodologies and emphasizes their strengths and weaknesses (FAO/WHO, 2003). The aim of advice is to assist the risk assessor and manager to understand the concepts and information associated in the risk assessment. The MRA are confirmed in association with the Codex Committee on food hygiene and remove the risk evaluation of *Salmonella* spp. in broilers, *Salmonella enteriditis* in eggs, *L. monocytogenes* in ready-to-eat foods, *Campylobacter* in broiler chickens, and *Vibrio* spp. in seafood (FAO/WHO, 2006a,b; FAO/WHO, 2015).

GENETICALLY MODIFIED FOOD RISK ASSESSMENT:

The utilization of biotechnology for the genetic improvement of plants, animals, and microorganisms for the production of foods create an extra concern to certain consumer groups. The WHO and FAO perceive that modern biotechnology has promised to increase agricultural productivity and the nutritional value of foods. However, they also acknowledge that there are possible risks to human, animals, and ecosystems that require an individual assessment (FAO/WHO, 2001; FAO, 2009). FAO and WHO mutually planned a series of specialist's discussions to consider overall safety and nutritional aspects of foods derived by genetic modifications. The negotiations addressed Strategies for Assessing the Safety of Foods Produced by Biotechnology in 1990, Biotechnology and Food Safety in 1996, and Safety Aspects of Genetically Modified Foods of Plant Origin in 2000 and 2001. The latter negotiations specifically addressed survey on security that were elevated by a Codex intergovernmental ad hoc task force on foods derived from biotechnology and to examine the benchmarks essential for the risk evaluation of food and its ingredients produced with the assistance of viable or nonviable GM microorganisms (FAO/WHO, 2002; FAO/WHO, 2003).

Codex Alimentarius Commission

The Codex Alimentarius or "food code" is a collection of international codes, standards, and guidelines that have been developed and adopted by the CAC. The CAC is the central component of the joint FAO/WHO food standards program and was established by FAO and WHO in 1963. The principal objectives of CAC are to protect the health of consumers, promote fair trade practices in the food trade, and ensure coordination of all food standards work undertaken by international organizations (FAO/WHO, 2006a,b). The publication of the Codex Alimentarius (i.e., food code) is intended to supervise and encourage the refinement of definitions and requirements for foods to aid in their harmonization and facilitate transcontinental trade (Randell and Whitehead, 1997; FAO/WHO, 2015).

WHAT IS CODEX?

The Codex encompasses codes or norms for all the principal foods, whether processed, semiprocessed, or raw, meant for dispersal to the consumers. Materials for further processing into foods should be included as and when deemed inevitable to achieve the purpose of the Codex Alimentarius. It has provisions with respect to food hygiene, food additives, harmful residues of agrochemicals and veterinary drugs, contaminants, labeling and presentation, methods of sampling and analysis, and export and import inspections and certifications. It also includes a voluntary code of ethics for international trade in food (Forsythe, 2002; Gauthier and Mahabir, 2012). As of today, there are 188 member countries of the CAC. Codex norms are developed through the work of various committees with an eight-step process from proposal to adoption. The worldwide authority may also participate in the commission as observers. However, they have no voting rights, unlike countries members. A Codex standard for any food should be evaluated in accordance with the format for Codex commodity standards and contain all the relevant sections listed therein. The CAC and its associated bodies are committed to revise Codex standards and related texts if required to ensure that these become consistent with and reflect latest scientific guidance. A standard may be revised or modified in accordance with the procedures of Codex standards when required for elaboration. Every member of the CAC is responsible for identifying any new scientific advice, which may warrant amendment of any existing Codex standards and presenting to the appropriate committee for desired improvements (Randell and Whitehead, 1997; FAO/WHO, 2010).

The process for preparing and formulating standards is very well defined, open, and clear. A national authority or auxiliary committee of the Commission generally presents the proposal for a standard to be developed, and then they prepare a document for discussion that outlines what the proposed standard is expected to achieve and then a project document that indicates the time frame for the work to be executed and its priority. For the preparation of standard proposed, initially the Commission considers and reviews the project document and makes a decision whether the proposed standard should be developed, and then the Commission Secretariat may arrange and circulate the proposed draft standard to member governments, observer organizations, and other Codex committees for two rounds of comments and specific advice. It may take several years to develop standards. Once the developed standard is adopted by the commission, it is listed as a Codex standard in the Codex Alimentarius and is published on its website (Masson-Matthee, 2007; FAO/WHO, 2015, http://www.codexalimentarius.org).

Codex guidelines and codes are recommended to nations to execute them as legislation or regulations and can be accessed by the public as they are freely available on the Codex website. The MLs of contaminants and natural toxicants in food and feed according to Codex standard that are safe for goods are subject to global trade. The Codex database on food additives, MRLs for pesticides, and residues of veterinary drugs in food include the conditions and maximum limits that can be used in all foodstuffs. The number of Codex standards and guidelines by July 2016 after the decisions of the 39th Session of the CAC were 191 goods standards and 76 guidelines (Masson-Matthee, 2007; FAO/WTO, 2017).

SOME POPULAR CODEX STANDARDS:

These include guidelines on nutrition labeling, general standards for food additives, list of Codex specifications for food additives, MRLs and risk management recommendations for residues of veterinary drugs in foods. In addition, regional code of hygienic practices for street and vended foods, guidelines related to performance criteria for methods of analysis for the determination of pesticide residues in food and feed, guidelines and principles for monitoring the performance of national food control systems, code of practice for the prevention and reduction of arsenic contamination in food grains are also included in Codex (FAO/WHO, 2015; http://www.codexalimentarius.org).

ROLE OF CODEX STANDARDS IN INTERNATIONAL TRADE:

The role of the Codex has been significantly enhanced under the World Trade Organization (WTO) trading regime through the agreement on the application of SPS measures and the agreement on Technical Barriers to Trade (TBT). The WTO members are required to use international standard, and the agreement also states that the risk assessment procedure should be developed by an international organization and become part of national food regulations on food safety. The TBT agreement includes all types of standards covering quality requirements for foods and put emphasis on international standards. It also includes all measures designed to protect the consumers against fraud and deception (Forsythe, 2002; FAO/WHO, 2002; Masson-Matthee, 2007). Because of WTO, TBT and SPS agreement the Codex Alimentarius standards have become the baseline references for consumer protection vis-à-vis safe food production. Hence, the CAC has become the important baseline for the international as well as national food safety requirements. The General Agreement on Tariffs and Trade (GATT) decision on SPS states that no member countries are prevented from adopting and enforcing measures required to protect humans, animals, and plant health (GATT, 1994).

The World Trade Organization

The WTO was established in 1995 by Uruguay Round Negotiations and is located in Geneva, Switzerland, now with 164 memberships (http://www.wto.org). It is the global agency that deals with the guidelines of trade internationally with the aim to aid producers of goods and services, exporters and importers running their business freely, and solve the trade issues they face with each other. WTO members negotiate and supervise the implementation of rules under trade agreements and examine the trade plans of its members (Acharya, 2016; WTO, 2010). The goal of WTO is to assist trade flow as smoothly as possible, which leads to economic development and wholeness of the people globally. The starting of national merchandise to worldwide trade with requisite flexibilities render to sustainable development, poverty mitigation, and the upgradation of living standards (WTO, 2014). At the time of the GATT, the WTO was on ordinary customs duties (tariffs); today, it also includes all other measures that affect trade internationally. A principal development in this consideration was the emergence of the WTO, SPS, and TBT agreements (WTO, 2017; FAO/WTO, 2017). The following are the major function of WTO (http://www.wto.org):

- To supervise WTO trade conventions.
- To act as assembly for merchandise consultation.
- To settle trade conflicts.
- To monitor governmental trade strategies.
- To provide technical support and training for underdeveloped nations.
- To cooperate with other world organizations.

The Sanitary and Phytosanitary and Technical Barriers to Trade Agreements

The SPS and TBT agreements maintain fairness between members' privilege to regulate food security and ensuring that such regulations do not become obstacle to trade. Both agreements promote harmonization of international standards and provide unquestionable protection from legal objections (WTO, 2010).

SANITARY AND PHYTOSANITARY AGREEMENT:

SPS agreements are precautions and safeguards to be observed to protect humans, animals, and plants from diseases, pests, or contaminants. The SPS Agreement is one of the important documents approved at the Uruguay Round of the Multilateral Trade Negotiations in 1994. It is applicable to all sanitary (relating to animals) and phytosanitary (relating to plants) measures that may directly or indirectly impact the international trade.

The SPS agreement has 14 articles covering the rights and obligations. It includes a number of understandings (trade disciplines) on how SPS measures will be implemented to the member states, when they establish, revise, or apply their own domestic laws and regulations. The WTO agreement is a key factor in developing and executing new hygiene measures for the international trade in food (Forsythe, 2002; FAO/WHO, 2003; FAO/WTO, 2017).

Role of SPS Agreement in international food safety

The SPS agreement incorporates the following regime for food safety and animal and plant health (http://www.wto.org/sps):

- It accepts the privilege of nations to acquire and execute regulations necessary to protect humans, animals, and plant health.
- It ensures food safety and the goal of restricting the irrelevant effects of such regulations on worldwide trade.

The SPS measures include all types of regulations to fulfill these commitments, whether these are requisite for final products, processing, inspection, certification, packaging, and labeling that are directly associated with food security.

SCOPE OF SANITARY AND PHYTOSANITARY AGREEMENT:

The scope of the SPS agreement is defined by the goals of the agreement. The main targets of SPS agreement are taken to protect (FAO, 2005, http://www.wto.org/sps):

- risks arising from additives, contaminants, toxins in food and feed;
- plant- and animal-carried diseases (zoonoses) and disease-causing organisms; and
- destruction caused by the pests.

AREAS OF SANITARY AND PHYTOSANITARY MEASURES:

The three main areas in which SPS measures are functional include (FAO, 2005, http://www.wto.org/sps):

- The FAO/WHO CAC, for food safety standards
- The World Organization for Animal Health (OIE), for animal health
- The IPPC, for plant health standards

TECHNICAL BARRIERS TO TRADE AGREEMENT:

The SPS agreement is mostly applied to health-related risks; however, the TBT agreement covers a variety of product standards and regulations acquired by the nation to achieve public policy targets such as protecting human health and the environment, providing buyer information, and determining product quality. It was negotiated in the Tokyo Round of Multilateral Trade Negotiations in 1979 with the provisions for settling trade disputes arising from the use of food security and other technical limitations. The aim of the TBT agreement is to ensure that countries' laws, standards, testing, and licensing methods do not create unnecessary barriers to global trade. It covers trade in all commodities and implements three categories of measures, which are technical regulations, standards, and conformity assessment procedures (FAO/WHO, 2003; FAO, 2005; FAO/WTO, 2017).

The Agreement manages technical regulations and standards regarding the use of terminology, symbols, packaging, marketing, and labeling requirements. The TBT agreement also follows Codex standards as the global benchmark, not specifically referenced. If a trade dispute arises, the WTO can approve trade penalties against a nation that cannot justify a stringent and trade restrictive requirement specified in the Codex (Gauthier and Mahabir, 2012; WTO, 2014).

SCOPE OF TECHNICAL BARRIERS TO TRADE AGREEMENT:

The scope of the TBT agreement is defined by the objective of the measures with respect to technical regulations, standards, and conformity assessment procedures as follows (http://www.wto.org/tbt):

- Technical regulations involve product characteristics, production methods, and also deal with symbols, packaging, marking, and labeling requirements.
- Standards are agreed by an authority responsible for framing rules, guidelines, and attributes for products.
- Conformity assessment procedures are used to determine the applicable requirements in technical regulations. They include procedures for sampling, testing and inspection, evaluation, verification, and assurance of conformity, and registration, consignment, and compliance.

International Plant Protection Convention

The IPPC is an international plant health agreement, established in 1952, which expects to protect cultivated and wild plant resources by preventing the introduction and spread of pests through international trade and travel. Pest introductions and subsequent outbreaks cost farmers, governments, and consumers billions of dollars every year. Once the exotic pest species are established in any nation, their control is often very difficult, and eradicating them is costly. The IPPC allows countries to use science-based knowledge to protect their wild and cultivated plant resources. The Convention helps to protect farmers from economically destructive pests and disease outbreaks, the environment from loss of species diversity as a result of pest invasions, and industries and consumers from the costs of pest management.

It provides an international framework for plant protection, which includes the development of International Standards for Phytosanitary Measures (ISPMs) to protect plant resources. The ISPMs was developed in 2009 and includes standards for (https://www.ippc.int/en/):

- pest surveillance, survey, and monitoring;
- pest risk analysis and pest management;
- conformity procedures and phytosanitary inspection techniques;
- postentry quarantine and foreign pest emergency response; and
- export certification and import regulations.

The IPPC also has provisions for exchange of information related to export and import requirements and pest status provided by each member state. Developing nations also receive technical help for its implementation. The primary focus of IPPC is on plants and plant products in worldwide trade. It also covers research materials, biological control organisms, germplasm banks, and vector for the spread of plant pests (e.g., packaging materials, soil, vessels, containers machinery and vehicles). The Convention provides assistance to developing nations to improve the effectiveness of their National Plant Protection Organizations (NPPOs) and to aid regional plant protection organizations to understand the welfare of safe trade (Forsythe, 2002; Lupien, 2002).

World Organization for Animal Health

The OIE is also known as the World Animal Health Organization. It is the intergovernmental agency responsible for improvement of animal health all over the world. It was established in 1924 and its headquarters is in Paris, France. It is recognized as a reference organization by the WTO with 182 member states. The OIE has relations with 45 other international and regional organizations. It provides comprehensive, verified, and transparent information on voluntary sustainability standards and other similar initiatives covering issues such as safety and quality of food. The main focus of the program is on the capacity building of producers, exporters, policymakers, and buyers for participation in more sustainable food production and trade. The main purposes of OIE are ensuring transparency in the worldwide animal disease control, publishing health standards for global trade in animals and animal products, improving the veterinary services of nations, providing assurance of animal origin food, and upgrading animal welfare through a science-based approach (FAO/WHO, 2003, http://www.oie.int).

International Organization for Standardization

The International Organization for Standardization (ISO) was established in 1946 to facilitate the international coordination and integration of industrial standards. It is an independent, nongovernmental organization with a membership of 162 standard bodies based on one member per country, and its headquarters is in Geneva, Switzerland (http://www.iso.org).

WHAT IS ISO INTERNATIONAL STANDARD?

ISO International Standard is a document containing specifications for products, services, and systems to ensure quality, safety, and efficiency and facilitating international trade. ISO has published 22,371 International Standards and related documents, covering almost every industry, technology, food safety, agriculture, and health care with a very wide impact. ISO standards help:

- To make products compatible, so they fit and work well with each other.
- To identify safety issues of products and services.
- To share good ideas and solutions for best management practices.

KEY PRINCIPLES IN DEVELOPMENT OF ISO STANDARDS (HTTP://WWW.ISO.ORG)

- ISO standards respond to a need in the market from industry or other consumer groups; they first communicate to its national member and then to ISO.

- ISO standards are based on global expert opinion from all over the world, which are part of larger groups called technical committees, and experts negotiate all aspects of the standard, including its scope, key definitions, and content.
- ISO standards are developed through a multistakeholder process.
- ISO standards are based on a consensus-based approach, and comments from all stakeholders are taken into account.

Popular ISO standards: The most popular ISO standards include management system standards, ISO 9001 quality management, ISO 22000 food safety management and ISO 14000 environmental management (http://www.iso.org).

ISO 22000 FOOD SAFETY MANAGEMENT:

ISO 22000 is an international food safety standard developed in 2005. It is a derivative of ISO 9000 standard and involves the interactive communication and prerequisite programs for a food safety management system. ISO 22000 specifies the requirements for implementing food safety management systems in all types of organizations along the food chain, ranging from production transport and storage to producers of equipment, packaging material, and additives. It has been placed with ISO 9001 in order to increase the conformity of the two standards. It combines the concepts of the hazard analysis and critical control points (HACCP) and the CAC (FAO/WHO, 2003). ISO is also developing additional standards that are related to ISO 22000. These standards are known as the ISO 22000 family of standards. The new standards are developed by authorities from the food industries along with specialized international agencies and in cooperation with Codex, systemize relevant national and international food laws, and integrate HACCP principles (Gauthier and Mahabir, 2012, https://www.iso.org/iso-22000-food-safety-management.html).

Benefits of International Standards

International food laws facilitate trade by reducing unnecessary trade restrictions and encouraging economies for producers. It provides a best scientific and technical basis for objectives related to food safety and aids nations in developing SPS measures to ensure animal and plant health. International standards also provide the basis for testing, inspection, or certification that governments use to ensure the requirements for safety and dissemination of technology.

CONCLUSION

Food-borne diseases have emerged as major public health and economic concerns all over the world. In spite of significant advances in food science and technology, food-borne diseases and illness are the rising cause of mortality and morbidity in many countries. Although a number of food-borne pathogens from different groups have been identified, still there is lack of accurate data on the full extent and cost of food-borne diseases that has been a major obstacle to address food safety issues. In order to fill this gap, different local and international regulatory bodies, standards, and laws are available that comply with safety and regulatory requirements related to food products. These provide the framework for uniformity of quality and safety of food products and thus help in improving the efficiency of production and reduction in the cost and also open borders for the free transport of food products around the globe. A number of challenges including changes in environmental conditions that lead to food contamination, changes in food production and supply, changes in consumer preference and habits, and emergence of newer pathogens, toxins, and antibiotic resistance still exist that need to be taken care of. The challenges for food regulators are to maintain a food regulatory system that delivers safe food for the people and also to maintain public confidence in the food regulations.

References

Abraham, M., Pai, M., Kang, G., Asokan, G.V., Magesh, S.R., Bhattacharji, S., et al., 1997. An outbreak of food poisoning in Tamil Nadu associated with Yersinia Enterocolitica. Indian J. Med. Res. 106, 465–468.

Acha, P.N., Szyfres, B., 2001. Zoonoses and communicable diseases common to man and animals, third ed. Bacterioses and Mycoses, Volume I. Pan American Health Organization; Salmonellosis, Washington, DC, pp. 233–246.

Acharya, R., 2016. Regional Trade Agreements and the Multilateral Trading System, first ed. Cambridge University Press, United Kingdom.

Adam, R.D., 2001. Biology of *Giardia lamblia*. Clin. Microbiol. Rev. 14, 447–475.

Adams, M.R., Moss, M.O., 2003. Significance of food borne diseases, second ed. Food Microbiology, 2. RSC Publishing, Cambridge, pp. 160–164 (163).

Adams, M.R., Moss, M.O., 2008. Food Microbiology., third ed. The Royal Society of Chemistry, Cambridge, pp. 158–309.

Addis, M., Sisay, D., 2015. A review on major food borne bacterial illnesses. J. Trop. Dis. 3, 176.

Agency for Toxic Substances and Disease Registry (ATSDR), 2008. US Department of Health and Human Services. Public Health Service. Toxicological Profile for Chromium, Atlanta, GA.

Aggarwal, P., Singh, S.M., Bhattacharaya, M.M., 1985. An outbreak of food poisoning in a family due to *S. weltevreden* at Delhi. J. Diarrhoeal Dis. Res. 3 (4), 224–225.

Anderson, K., Pritchard, D., 2008. An Update on *Staphylococcus aureus* Mastitis, 4. Benjamin/Cummings Publishing Company, San Francisco, CA, p. 38.

Antony, B., Scaria, B., Dias, M., Pinto, H., 2009a. *Salmonella* wein from gastroenteritis cases encountered in Mangalore, India: a report of 10 cases and review of the literature. Indian J. Med. Sci. 63 (5), 195–197.

Antony, B., Dias, M., Shetty, A.K., Rekha, B., 2009b. Food poisoning due to *Salmonella enterica* serotype weltevreden in Mangalore. Indian J. Med. Sci. 27 (3), 257–258.

Argudín, M.A., Mendoza, M.C., Rodicio, M.R., 2010. Food poisoning and *Staphylococcus aureus* enterotoxins. Toxins. 2 (7), 1751–1773.

Asano, N., Kuroi, H., Ikeda, K., Kizu, H., Kameda, Y., Kato, A., et al., 1999. New polyhydroxylated pyrrolizidine alkaloids from *Muscari armeniacum*: structural determination and biological activity. Tetrahedron: Asymmetry. 11 (2000), 1–8.

Ayeh-Kumi, P.F., Quarcoo, S., Kwakye-Nuako, G., Kretchy, J.P., Osafo-Kantanka, A., Mortu, S., 2009. Prevalence of intestinal parasitic infections among food vendors in Accra, Ghana. J. Trop. Med. Parasitol. 32 (1), 1.

Azian, M.Y.N., San, Y.M., Gan, C.C., Yusri, M.Y., Nurulsyamzawaty, Y., Zuhaizam, A.H., et al., 2007. Prevalence of intestinal protozoa in an aborigine community in Pahang, Malaysia. Trop. Biomed. 24 (1), 55–62.

Balaban, N., Rasooly, A., 2000. Staphylococcal enterotoxins. Int. J. Food Microbiol. 61 (1), 1–10.

Banwell, J.G., Boldt, D.H., Meyers, J., Weber Jr., F.L., 1983. Phytohemagglutinin derived from red kidney bean (*Phaseolus vulgaris*): a cause for intestinal malabsorption associated with bacterial overgrowth in the rat. Gastroenterology 84, 506–515.

Bean, N.H., Griffin, P.M., 1990. Foodborne disease outbreaks in United States, 1973–1987: pathogens, vehicles, and trends. J. Food Prot. 53, 804–817.

Benedict, K., Chiller, T.M., Mody, R.K., 2016. Invasive fungal infections acquired from contaminated food or nutritional supplements: a review of the literature. Foodborne Pathog. Dis. 13 (7), 343–349.

Bennett, J.W., Klich, M., 2003. Mycotoxins. Clin. Microbiol. Rev. 16 (3), 497–516.

Berhe, B., Bugssa, G., Bayisa, S., Alemu, M., 2018. Foodborne intestinal protozoan infection and associated factors among patients with watery diarrhea in Northern Ethiopia; a cross-sectional study. J. Health Popul. Nutr. 37 (1), 5.

Bezuidenhout, S.C., Gelderblom, W.C.A., Gorst-Allman, C.P., Horak, R.M., Marasas, W.F.O., Spiteller, G., et al., 1988. Structure elucidation of the fumonisins, mycotoxins from *Fusarium moniliforme*. J. Chem. Soc. Chem. Commun. 743–745.

Bhan, A., Sarkar, N.N., 2005. Mercury in the environment: effects on health and reproduction. Rev. Environ. Health. 20 (1), 39–56.

Bhat, R.V., Krishnamachari, K.A., 1977. Follow-up study of aflatoxic hepatitis in parts of western India. Indian J. Med. Res. 66 (1), 55–58.

Bhat, R.V., Rao, N.R., 1987. Foodborne diseases in India. Indian J. Pediatr. 54 (4), 553–562.

Bonazzi, M., Lecuit, M., Cossart, P., 2009. *Listeria monocytogenes* internalin and E-cadherin: from structure to pathogenesis. Cell Microbiol. 11 (5), 693–702.

Booth, M., Bundy, D.A.P., 1992. Comparative prevalences of *Ascaris lumbricoides, Trichuris trichiura* and hookworm infections and the prospects for combined control. Parasitology 105 (1), 151–157.

Brundtland, G.H., 2001. 24th Session of the Codex Alimentarius Commission, Geneva, 2 July 2001. Available from: <https://www.who.int/fsf>.

Bryan, F.L., Mckiley, T.W., Mixon, B., 1971. Use of time temperature evaluations in detecting the responsible vehicles and contributing factors of food born disease outbreaks. J. Milk Food Technol. 34, 576–582.

Buchanan, R.L., Doyle, M.P., 1997. Food born disease significance of *E. coli* O157:H7 and other enterohemorrhagic *E. coli*. Food Technol. 5, 69–76.

Buchanan, R.L., Gorris, L.G.M., Hayman, M.M., Jackson, T.C., Whiting, R.C., 2017. A review of *Listeria monocytogenes*: an update on outbreaks, virulence, dose-response, ecology, and risk assessments. Food Control 75, 1–13.

Buckley, M., 2008. The Fungal Kingdom: Diverse and Essential Roles in Earth's Ecosystem. American Academy of Microbiology, Washington, DC, p. 48, AAM Colloquium The Fungal Kingdom: Diverse and Essential Roles in Earth's Ecosystem, Tucson, AZ, November 2–4, 2007.

Buzby, J.C., Roberts, T., 1997. Economic costs and trade implications of microbial foodborne illness. World Health Stat. Quart. 50 (1–2), 57–66.

Cartwright, E.J., Jackson, K.A., Johnson, S.D., Graves, L.M., Silk, B.J., Mahon, B.E., 2013. Listeriosis outbreaks and associated food vehicles, United States, 1998–2008. Emerg. Infect Dis. 19, 1–9.

Cassels, F.J., Wolf, M.K., 1995. Colonization factors of diarrheagenic *E. coli* and their intestinal receptors. J. Ind. Microbiol. 15, 214–226.

Centeno, J.A., Tchounwou, P.B., Patlolla, A.K., Mullick, F.G., Murakat, L., Meza, E., et al., 2005. Environmental pathology and health effects of arsenic poisoning: a critical review. In: Naidu, R., Smith, E., Smith, J., Bhattacharya, P. (Eds.), Managing Arsenic in the Environment: From Soil to Human Health. CSIRO Publishing Corp, Adelaide, Australia.

Chaudhry, R., Dhawan, B., Kumar, D., Bhatia, R., Gandhi, J.C., Patil, R.K., et al., 1998. Outbreak of suspected *Clostridium butyricum* botulism in India. Emerg. Infect. Diseases. 4 (3), 506–507.

Cheeke, P.R., 1998. Mycotoxins in cereal grains and supplements. In: Cheeke, P.R. (Ed.), Natural Toxicants in Feeds, Forages, and Poisonous Plants. Interstate Publishers, Inc, Danville, IL, pp. 87–136.

Chimelli, L., Lovalho, A.F., Takayanagui, O.M., 1998. Neurocysticercosis: contribution of autopsies in the consolidation of mandatory notification in Ribeirao Preto-SP, Brazil. Arq. Neuropsiquiatr. 56 (3B), 577–584.

Chugh, T.D., 2008. Emerging and re-emerging bacterial diseases in India. J. Biosci. 33 (4), 549–555.

Clarkson, T.W., Magos, L., Myers, G.J., 2003. The toxicology of mercury-current exposures and clinical manifestations. N. Engl. J. Med. 349 (18), 1731–1737.

Conway, D.J., Hall, A., Anwar, K.S., Rahman, M.L., Bundy, D.A.P., 1995. Household aggregation of *Strongyloides stercoralis* infection in Bangladesh. Trans. R. Soc. Trop. Med. Hyg. 89 (3), 258–261.

Crespo-Lopez, M.R., Macedo, G.L., Pereira, S.I.D., Arrifano, G.P.F., Picano-Dinc, D.L.W., doNascimento, J.L.M., et al., 2009. Mercury and human genotoxicity: critical considerations and possible molecular mechanisms. Pharmacol. Res. 60 (4), 212–220.

Crompton, D.W., 1999. How much human helminthiasis is there in the world? J. Parasitol. 85 (3), 397–403.

Crutchfield, S.R., Buzby, J.C., Roberts, T., Ollinger, M., 1999. Assessing the costs and benefits of pathogen reduction. Food Review 22 (2), 6–9.

Cruz, M.E., Schantz, P.M., Cruz, I., Espinosa, P., Preux, P.M., Cruz, A., et al., 1999. Epilepsy and neurocysticercosis in an Andean community. Int. J. Epidemiol. 28 (4), 799–803.

Dalvi, R.R., Bowie, W.C., 1983. Toxicology of solanine: an overview. Vet. Hum. Toxicol. 25 (1), 13–15.

Deshpande, S.S., 2002. Toxicants and antinutrients in plant foods. Handbook of Food Toxicology. Marcel Dekker, New York, pp. 331–372.

Dikid, T., Hpalya, S.S., Thakur, J.S., Taneja, N., Thapar, R., Kumar, R., 2009. Salmonella food poisoning outbreak in Karar town of Punjab. Indian J. Public Health. 53 (4), 265.

Ding, X., Lichti, K., Staudiner, J.L., 2006. The mycoestrogen zearalenone induces CYP3A through activation of the pregnene X receptor. Toxicol. Sci. 91 (2), 448–455.

Dolan, L.C., Matulka, R.A., Burdock, G.A., 2010. Review: naturally occurring food toxins. Toxins. 2, 2289–2332.

Dopp, E., Hartmann, L.M., Florea, A.M., Rettenmier, A.W., Hirner, A.V., 2004. Environmental distribution, analysis, and toxicity of organometal (loid) compounds. Crit. Rev. Toxicol. 34 (3), 301–333.

Duffus, J.H., 2002. Heavy metals—a meaningless term? Pure. Appl. Chem. 74 (5), 793–807.

Edwards, G., Krishna, S., 2004. Pharmacokinetic and pharmacodynamic issues in the treatment of parasitic infections. Eur. J. Clin. Microbiol. Infect. Dis. 23 (4), 233–242.

Espinosa-Cantellano, M., Martinez-Palomo, A., 2000. Pathogenesis of intestinal amoebiasis: from molecules to disease. Clin. Microbiol. Rev. 13 (2), 318–331.

European Food Safety Authority (EFSA), 2008. Scientific Opinion of the Panel on Plant Protection Products and their Residues (PPR Panel) on a request from the EFSA to evaluate the suitability of existing methodologies and, if appropriate, the identification of new approaches to assess cumulative and synergistic risks from pesticides to human health with a view to set MRLs for those pesticides in the frame of Regulation (EC) 396/2005. EFSA J. 704, 1–85. Available from: http://www.efsa.europa.eu/en/efsajournal/doc/705.pdf.

European Food Safety Authority (EFSA), 2015. The European Union summary report on trends and sources of zoonoses, zoonotic agents and food-borne outbreaks in 2013. EFSA J. 13 (1), 1–165. 3991.

FAO, 2002a. Codex proposed draft principles for the risk analysis of food derived from modern biotechnology, Rome. Available from: <https://www.fao.org/input/download/standards/10007/CXG_044e.pdf>.

FAO, 2002b. Law and sustainable development since Rio: Legal trends in agriculture and natural resource management, Vapnek, J., Mekouar, A., Legislative study no. 73, Rome.

FAO, 2005. Perspective and guidelines on food legislation with a new model food law. Vapnek, J., Spreij, M., Legislative study no. 87, Rome.

FAO, 2009. GM food safety assessment tools for trainers, Rome. Available from: <https://www.fao.org/3/a-i0110e.pdf>.

FAO, 2016. Risk based imported food control manual, Rome. Available from: <https://www.fao.org/3/a-i5381e.pdf>.

FAO/WHO, 2001. FAO/WHO joint consultation on food derived from biotechnology, Rome. Available from: <https://www.fao.org/ag/agn/food/pdf/bi03al.pdf>.

FAO/WHO, 2002. Report of the evaluation of the Codex Alimentarius and other FAO and WHO food standards work. Available from: <https://www.who.int/foodsafety/codex/en/codex_eval_report_en.pdf>.

FAO/WHO, 2003. Assuring food safety and quality: Guideline for strengthening national food control system. FAO Food and Nutrition paper 76.

FAO/WHO, 2006a. Food safety risk analysis: A guide for national food safety authorities, Rome. Available from: <http://www.fao.org/docrep/014/ba0092e/ba0092e00.pdf>.

FAO/WHO, 2006b. Understanding the Codex Alimentarius (third ed.) Rome. Available from: <ftp://ftp.fao.org/codex/Publications/understanding/Understanding_EN.pdf>.

FAO/WHO, 2010. Statutes of the Codex Alimentarius Commission, Codex Alimentarius Commission-Procedural Manual, 19th Ed. Available from: <http://www.fao.org/tempref/codex/Publications/ProcManuals/Manual_09e.pdf>.

FAO/WHO, 2015. Codex Alimentarius Commission-Procedural manual, 24th Ed., Rome. Available from: <http://www.fao.org/3/a-i5079e.pdf>.

FAO/WTO, 2017. Trade and food standards, Rome. Available from: <https://www.wto.org/english/res_e/booksp_e/tradefoodfao17_e.pdf>.

Fergusson, J.E., 1990. The Heavy Elements: Chemistry, Environmental Impact and Health Effects. Pergamon Press, Oxford.

Forsythe, S.J., 2002. The Microbiological Risk Assessment of Food. Blackwell Science Ltd, Oxford.

Fotedar, R., Stark, D., Beebe, N., Marriott, D., Ellis, J., Harkness, J., 2007. PCR detection of *Entamoeba histolytica, Entamoeba dispar*, and *Entamoeba moshkovskii* in stool samples from Sydney, Australia. J. Clin. Microbiol. 45 (3), 1035–1037.

Frenzen, P.D., Riggs, T.L., Buzby, J.C., Breuer, T., Roberts, T., Voetsch, D., et al., 1999. Salmonella cost estimate updated using FoodNet data. Food Review 22, 10–15.

FuÈrst, T., Duthaler, U., Sripa, B., Utzinger, J., Keiser, J., 2012. Trematode infections: liver and lung flukes. Infect. Dis. Clin. North Am. 26 (2), 399–419.

Fule, R.P., Ingole, K.V., Jalgaonkar, S.V., Bharthi, U., 1996. Outbreak of food poisoning due to *Salmonella paratyphi* A var durazzo (2,12:a: -) in Yavatmal (Maharashtra) in May 1995. Indian J. Med. Res. 103, 74–76.

Fürst, T., Duthaler, U., Sripa, B., Utzinger, J., Keiser, J., 2012. Trematode infections liver and lung flukes. Infect. Dis. Clin. N. Am. 26 (2012), 399–419.

García, H.H., Gonzalez, A.E., Evans, C.A.W., Gilman, R.H., 2003. *Taenia solium* cysticercosis. Lancet. 362 (9383), 547–556.

GATT, 1994. The Applications of Uruguay Round of Multilateral Trade Negotiations: The Legal Texts. World Trade Organization, Geneva92-870-1121-4.

Gauthier, A., Mahabir, M., 2012. International Food Standards. Library of Parliament, Ottawa, Canada.

Glavin, M.O.'K., 2003. A single microbial sea: food safety as a global concern. SAIS Rev. Int. Aff. 23 (1), 203–220.

Gunnell, D., Eddleston, M., 2003. Suicide by intentional ingestion of pesticides: a continuing tragedy in developing countries. Int. J. Epidemiol. 32 (6), 902–909.

Gunnell, D., Eddleston, M., Phillips, M.R., Konradsen, F., 2007. The global distribution of fatal pesticide self-poisoning: systematic review. BMC Public Health. 7, 357.

Gupta, R.C., Mostrom, M.S., Evans, T.J., 2018. Veterinary toxicology, Basic and Clinical Principles, third ed. Academic Press, pp. 1055–1063, Chapter 76.

Hall, J.D., McCroskey, L.M., Pincomb, B.J., Hatheway, C.L., 1985. Isolation of an organism resembling *Clostridium baratii* which produces type F botulinal toxin from an infant with botulism. J. Clin. Microbiol. 21, 654–655.

Harold, W.B., 1975. *Entamoeba histolytica* in Basic Clinical Parasitology, 4. Appleton-Century-Croft, New York, pp. 23–27.

Hawksworth, D.L., 2001. The magnitude of fungal diversity: the 1.5 million species estimate revisited. Mycol. Res. 105, 1422–1432.

He, Z.L., Yang, X.E., Stoffella, P.J., 2005. Trace elements in agroecosystems and impacts on the environment. J. Trace Elem. Med. Biol. 19 (2–3), 125–140.

Health Information of India, 2004. Central Bureau of Health Intelligence, Directorate General of Health Services, Ministry of Health and Family Welfare. Government of India, New Delhi.

Hernández, A.F., Parrónb, T., Tsatsakis, A.M., Requena, M., Alarcónc, R., Guarnidoa, O.L., 2013. Toxic effects of pesticide mixtures at a molecular level: their relevance to human health. Toxicology 307, 136–145.

Hirsh, D.C., Maclachian, J., Walker, R.L., 2004. Botulism, Veterinary Microbiology, second ed. Blackwell Publishing Professional, Ames, IA.

Hoelzer, K., Chen, Y., Dennis, S., Evans, P., Pouillot, R., Silk, B.J., et al., 2013. New data, strategies, and insights for *Listeria monocytogenes* dose-response models: summary of an interagency workshop. Risk Anal. 33, 1568–1581.

Hotez, P.J., Brown, S.A., 2009. Neglected tropical disease vaccines. Biologicals. 37 (3), 160–164.

Hotez, P.J., Brindley, P.J., Bethony, J.M., King, C.H., Pearce, E.J., Jacobson, J., 2008. Helminth infections: the great neglected tropical diseases. J. Clin. Invest. 118 (4), 1311–1321.

ICMSF, 1996a. International commission on microbiological specifications for food: update. Food Control 7, 99–101.

International Commission on Microbiological Specifications for Food (ICMSF), 1996b. Microorganisms in foods, Characteristics of Microbial Pathogens., vol. 5. Blackie Academic & Professional, London.

Iturriza-Gomara, M., O'Brien, S.J., 2010. Foodborne viral infections. Curr. Opin. Infect. Dis. 2016 29 (5), 495–501.

Jacobs, J.A., Testa, S.M., 2005. Overview of chromium (VI) in the environment: background and history. In: Guertin, J., Jacobs, J.A., Avakian, C.P. (Eds.), Chromium (VI) Handbook. CRC Press, Boca Raton, FL, pp. 1–22.
Janakiraman, V., 2008. Listeriosis in pregnancy: diagnosis, treatment and prevention. Rev. Obstet. Gynecol. 1 (4), 179–185.
Jensen, S., Johnel, S.G., Olsson, M., Otterlind, G., 1972. DDT and PCB in herring and cod from the Baltic, the Kattegat and the Skagerra. Atnbio. Spec. Rep. 1 (1), 71–85.
Jay, J.M., 2000. Modern food Microbiology, sixth ed. Aspen Publications, Gaithersburg, Maryland.
Jiang, Y., Jolly, P.E., Ellis, W.O., Wang, J.S., Phillips, T.D., Williams, J.H., 2005. Aflatoxin B1 albumin adduct levels and cellular immune status in Ghanaians. Int. Immunol. 17 (6), 807–814.
Joffe, A.Z., 1978. *Fusanum poae* and *F. sporotncbioides* as principal causal agents of alimentary toxic aleukia. In: Wylhe, T.D., Morehouse, L.G. (Eds.), Mycotoxic Fungi, Mycotoxins, Mycotoxicoses: An Encyclopedic Handbook, vol. 3. Marcel Dekker, New York, pp. 21–86.
Jones, J.M.J., 1995. Food Safety. Eagan Press, St. Paul, MN, p. 71, 77, 84, 87.
Käferstein, F., Abdussalam, M., 1999. Food safety in the 21st century. Bull. World Health Org. 77 (4), 347–351.
Kalpelmecher, K., 1993. The role of salmonella in food borne diseases. Microbiological Quality of Foods. Academic press, New York.
Kaper, J.B., Nataro, J.P., Mobley, H.L.T., 2004. Pathogenic *Escherichia coli*. Nat. Rev. Microbiol 2, 123–140.
Kappus, K.D., Lundgren, R.G., Juranek, D.D., Roberts, J.M., Spencer, H.C., 1994. Intestinal parasitism in the United States: update on a continuing problem. Am. J. Trop. Med. Hyg. 50 (6), 705–713.
Kenny, B., Ellis, S., Leard, A.D., Warawa, J., Mellor, H., Jepson, M.A., 2002. Co-ordinate regulation of distinct host cell signalling pathways by multifunctional enteropathogenic *Escherichia coli* effector molecules. Mol. Microbiol. 44, 1095–1107.
Khan, N.T., 2017. Review: facts concerning *Entamoeba histolytica* induced amoebiasis. Biotechnol. Ind, J. 13 (3), 137–143.
Kirk, M.D., Pires, S.M., Black, R.E., Caipo, M., Crump, J.A., Devleesschauwer, B., et al., 2015. World Health Organization estimates of the global and regional disease burden of 22 foodborne bacterial, protozoal, and viral diseases, 2010: a Data Synthesis. PLoS Med. 12 (12), e1001940.
Kohli, C., Garg, S., 2015. Food safety in India: an unfinished agenda. MAMC J. Med. Sci. 1 (3), 131–135.
Krishnamachari, K.A., Bhat, R.V., Nagarajan, V., Tilak, T.B., 1975. Hepatitis due to aflatoxicosis. An outbreak in western India. Lancet 1 (7915), 1061–1063.
Kuo, C.Y., Su, L.H., Perera, J., Carlos, C., Tan, B.H., Kumarasinghe, G., 2008. Antimicrobial susceptibility of *Shigella* isolates in eight Asian countries, 2001–2004. J. Microbiol. Immunol. Infect. 41, 107–111.
Labbe, R.G., Nolan, L.L., 1981. Stimulation of *Clostridium perfringens* enterotoxin formation by caffeine and theobromine. Infect. Immun. 34, 50–54.
Laetz, C.A., Baldwin, D.H., Collier, T.K., Hebert, V., Stark, J.D., Scholz, N.L., 2009. The synergistic toxicity of pesticide mixtures: implications for risk assessment and the conservation of endangered Pacific salmon. Environ. Health Perspect. 117, 348–353.
Lalitha, M.K., Walter, N.M., Jesudason, M., Mathan, V.I., 1983. An abstract of gastroenteritis due to *Vibrio parahaemolyticus* in Vellore. Indian J Med Res. 78, 611–615.
Langard, S., Vigander, T., 1983. Occurrence of lung cancer in workers producing chromium pigments. Br. J. Ind. Med. 40 (1), 71–74.
Lasztity, R., 2009. Food Quality and Standard, vo. I. EOLSS, UNESCO.
le Loir, Y., Baron, F., Gautier, M., 2003. *Staphylococcus aureus* and food poisoning. Genet. Mol. Res. 2 (1), 63–76.
Le Minor, L., Popoff, M.P., 1987. Designation of salmonella enteric species, nom, rev., as the type and only species of the genus salmonella. Int. J. Syst. Evol. Micr. 37, 465–468.
Leeson, S., Dias, G.J., Summers, J.D., 1995. Tricothecenes. In: Poultry Metabolic Disorders. Guelph, Ontario, Canada, pp. 190–226.
Lindsay, J.A., 1997. Chronic sequelae of foodborne disease. Emerg. Infect. Dis. 3 (4), 443–452.
Lupien, J.R., 2002. Quality and safety assessment of food derived by modern biotechnology and their international regulation. Af. J. Food Nutr. Sci. 2 (2), 31–38.
Mandokhot, U.V., Garg, S.R., Chandiramani, N.K., 1987. Epidemiological investigation of a food poisoning outbreak. Indian J. Public Health. 31 (2), 113–119.

Marasas, W.F.O., Kellerman, T.S., Gelderblom, W.C.A., Coetzer, J.A.W., Thiel, P.G., van der Lugt, J.J., 1988. Leukoencephalomalacia in a horse induced by fumonisin B1 isolated from Fusarium moniliforme. Onderstepoort J. Vet. Res. 55 (4), 197–203.

Marasas, W.F.O., Riley, R.T., Hendricks, K.A., Stevens, V.L., Sadler, T.W., Gelineau-van Waes, J., et al., 2004. Fumonisins disrupt sphingolipid metabolism, folate transport, and neural tube development in embryo culture and in vivo: a potential risk factor for human neural tube defects among populations consuming fumonisin contaminated maize. J. Nutr. 134 (4), 711–716.

Marcos, L.A., Terashima, A., Gotuzzo, E., 2008. Update on hepatobiliary flukes: fascioliasis, opisthorchiasis and clonorchiasis. Curr. Opin. Infect. Dis. 21 (5), 523–530.

Marroquín-Cardona, A.G., Johnson, N.M., Phillips, T.D., Hayes, A.,W., 2014. Mycotoxins in a changing global environment—a review. Food Chem. Toxicol. 69, 220–230.

Marusic, A., 2011. Food safety and security: what were favourite topics for research in the last decade? J. Glob. Health. 1 (1), 72–78.

Mas-Coma, S., Valero, M.A., Bargues, M.D., 2014. Fascioliasis. Adv. Exp. Med. Biol. 766, 77–114.

Masson-Matthee, M.D., 2007. The Codex Alimentarius Commission and its standards. An Examination of the Legal Aspects of the Codex Alimentarius Commission. Asser Press, The Netherlands.

McQuillen, D., McQuillen, M., 2005. Poliomyelitis. In: Jones Royden, H. (Ed.), Netter's Neurology. Elsevier Saunders, Philadelphia, PA, pp. 597–601.

Mehndiratta, M.M., Mehndiratta, P., Pande, R., 2014. Poliomyelitis: historical facts, epidemiology, and current challenges in eradication. Neurohospitalist. 4 (4), 223–229.

Melton-Celsa, A.R., O'Brien, A.D., 1998. In: Kaper, J.B., O'Brien, A.D. (Eds.), In *Escherichia coli* O157:H7 and Other Shiga Toxin-Producing *E. coli* Strains. ASM Press, Washington, DC, pp. 121–128.

Melton-Celsa, 2014. Shiga Toxin (Stx) classification, structure, and function. Microbiol. Spectr. 2 (2), 1–13.

Miles, J.R.W., Harris, C.R., 1971. Insecticide residues in a stream and a controlled drainage system in agricultural areas of south western Ontario. Pestic. Monit. J. 5, 289.

Miller, R.F., Symeonidou, C., Shaw, P.J., 2005. Pneumonia complicating *Shigella sonnei* dysentery in an HIV-infected adult male. Int. J. STD AIDS. 16, 763–765.

Moeller Jr., R.B., Puschner, B., Walker, R.L., Rocke, T., Galey, F.D., et al., 2003. Determination of the median toxic dose of type C botulinum toxin in lactating dairy cows. J. Vet. Diagn. Invest. 15, 523–526.

Moreau, C., 1979. Moulds, toxins and food. John Wiley & Sons Ltd., Chichester, UK, p. 477.

Murray, R.J., 2005. Recognition and management of *Staphylococcus aureus* toxin-mediated disease. Intern. Med. J. 35 (2), S106–S119.

Murray, C.J., Lopez, A.D., 1997. Global mortality, disability, and the contribution of risk factors: global burden of disease study. Lancet. 349 (9063), 1436–1442.

Nataro, J.P., Kaper, J.B., 1998. Diarrheagenic *Escherichia coli*. Clin. Microbiol. Rev. 11, 142–201.

National Center for Emerging and Zoonotic Infectious Disease (CDC), 2018. National Enteric Disease Surveillance: Shigella Annual Report, 2016. pp 1–18.

National Research Council (NRC), 2001. Arsenic in Drinking Water. 2001 Update. 2001. Online at: <http://www.nap.edu/books/0309076293/html/>.

Nayar, K.R., Krishnan, S.K., Gorthi, S.P., 1993. Epidemiological investigation of an outbreak of food poisoning. Med. J. Armed Forced India. 49, 257–260.

Ndimubanzi, P.C., Carabin, H., Budke, C.M., Nguyen, H., Qian, Y.-J., Rainwater, E., et al., 2010. A systematic review of the frequency of neurocysticercosis with a focus on people with epilepsy. PLoS. Negl. Trop. Dis. 4 (11), 870–888.

Nema, V., Agrawal, R., Kamboi, D.V., Goel, A.K., Singh, L., 2007. Isolation and characterization of heat resistant enterotoxigenic *Staphylococcus aureus* from a food poisoning outbreak in Indian subcontinent. Int. J. Food Microbiol. 117 (1), 29–35.

NRC (National Research Council), 1998. Ensuring Safe Food Production to Consumption. National Academy Press, Washington, DC.

Odeyemi, O.A., Sani, N.A., 2016. Antibiotic resistance and burden of foodborne diseases in developing countries. Future Sci. OA. 2 (1), 139–142.

Ohnishi, K., Kato, Y., Imamura, A., Fukayama, M., Tsunoda, T., Sakaue, Y., et al., 2004. Present characteristics of symptomatic *Entamoeba histolytica* infection in the big cities of Japan. Epidemiol. Infect. 132 (1), 57–60.

Omaye, S.T., 2004. Toxicity of nutrients. Food and Nutritional Toxicology. CRC Press, Boca Raton, FL, pp. 205–213.

Parvathy, S., Gayathri, G., Jyothilakshmi, K., 2005. Safety and microbial quality of street foods. National Seminar on Food Safety and Quality Control in India.

Pitt, J.I., 2000. Toxigenic fungi and mycotoxins. Br. Med. Bull. 56 (1), 184–192.

Pouillot, R., Klontz, K.C., Chen, Y., Burall, L.S., Macarisin, D., Doyle, M., et al., 2016. Infectious dose of *Listeria monocytogenes* in outbreak linked to ice cream United States. Emerg. Infect. Dis. 12, 2113–2119.

Prakash, A.S., Pereira, T.N., Reilly, P.E.B., Seawright, A.A., 1999. Pyrrolizidine alkaloids in human diet. Mutat. 443, 53–67.

Quinn, P.J., Markey, B.K., Carter, M.E., Demnelly, W.J., Leonard, F.C., 2001. Veterinary Microbiology and Microbial Disease, eighth ed. Blackwell publishing, Oxford, UK.

Radostits, O.M., Gay, C.C., Hinchlif, K.W., Constable, P.D., 2007. Veterinary Medicine Text Book of Disease of Cattle, Horses, Sheep, Pig and Goats, tenth ed. Saunders, Philadelphia, PA.

Randell, A.W., Whitehead, A.J., 1997. Codex Alimentarius: food quality and safety standards for international trade. Re. Sci. Tech. 16 (2), 313–321.

Rao, N.R., Sudhakar, P., Bhat, R.V., Gupta, C.P., 1989. A study of recorded cases of foodborne diseases at Hyderabad during 1984 and 1985. J. Trop. Med. Hyg. 92 (5), 320–324.

Ravdin, J.L., Petri, W.A., 1995. *Entamoeba histolytica* (Amoebiases). Principles and practice of infectious disease, fourth ed.

Ricci, K., Girosi, F., Tarr, P., Lim, Y., Mason, C., Miller, M., et al., 2006. Reducing stunting among children: the potential contribution of diagnostics. Nature 444 (1), 29–38.

Richmond, M., 1990. The microbiological safety of food part II. Report of the Committee on the Microbiological Safety of Food. HMSO, London, p. 220.

Rijksinstituut voor Volksgezondheid en Milieu (Dutch), report, 2011. *Clostridium perfringens* Associated Food Borne Disease. Netherlands National Institute for Public Health and the Environment, pp. 1–53.

Riley, R.T., Wang, E., Schroeder, J.J., Smith, E.R., Plattner, R.D., Abbas, H., et al., 1996. Evidence for disruption of sphingohpid metabolism as a contributing factor in the toxicity and carcmogemcity of fumonisins. Nat. Toxins 4 (1), 3–15.

Roberts, J.A., 1996. Economic Evaluation of Surveillance. Department of Public Health and Policy, London.

Robinson, R.K., Batt, C.A., Patel, P.D., 2000. Encyclopedia of Food Microbiology, fifth ed. Academic Press, San Diego, CA.

Rutenber, E., Katzin, B.J., Ernst, S., Collins, E.J., Mlsna, D., Ready, M.P., et al., 1991. Crystallographic refinement of ricin to 2.5 A. Proteins. 10 (3), 240–250.

Sanfeliu, C., Sebastia, J., Cristofol, R., Rodriquez-Farre, E., 2003. Neurotoxicity of organomercurial compounds. Neurotox. Res. 5 (4), 283–305.

Scallan, E., Hoekstra, R.M., Angulo, F.J., Tauxe, R.V., Widdowson, M.A., Roy, S.L., et al., 2011. Foodborne illness acquired in the United States—major pathogens. Emerg. Infect. Diseases. 17 (1), 7–15.

Schwarzer, K., 2009. Harmful effects of mycotoxins on animal physiology. In: 17th Annual ASAIM SEA Feed Technology and Nutrition Workshop, Hue, Vietnam.

Scott, P.M., 1977. Penicillium mycotoxins. In: Wyllie, T.D., Morehouse, L.G. (Eds.), Mycotoxic Fungi, Mycotoxins, Mycotoxicoses, an Encyclopedic Handbook. Vol. 1. Mycotoxigemic Fungi. Marcel Dekker, New York, pp. 283–356.

Shank, R.C., Bourgeois, C.H., Keschamras, N., Chandavimol, P., 1971. Aflatoxins in autopsy specimens from Thai children with an acute disease of unknown aetiology. Food Cosmet. Toxicol. 9 (4), 501–507.

Shibamoto, T., Bjeldanes, L.F., 1993. Natural toxins in plant foodstuffs. Introduction to Food Toxicology. Academic Press, San Diego, CA, pp. 78–79, 82–84.

Singh, M., Kalghatgi, A.T., Narayanan, K., Rao, K.S., Nagendra, A., 1998. Outbreak of Salmonella food poisoning at high attitude. Med. J. Armed Forces India. 54 (2), 96–98.

Smith, J.E., Moss, M.O., 1985. Mycotoxins: Formation, Analysis and Significance. Wiley, Chichester, UK, p. 148.

Smith, M.A., Takeuchi, K., Anderson, G., Ware, G.O., McClure, H.M., Raybourne, R.B., et al., 2008. Dose-response model for *Listeria monocytogenes*-induced stillbirths in nonhuman primates. Infect. Immun. 76 (2), 726–731.

Spencer, P.S., 1999. Food toxins, AMPA receptors, and motor neuron disease. Drug. Metab. Rev. 31 (3), 561–587.

Stepek, G., Buttle, D.J., Duce, I.R., Behnke, J.M., 2006. Human gastrointestinal nematode infections: are new control methods required? Int. J. Exp. Path. 87, 325–341.

Stoloff, L., 1977. Aflatoxins—an overview. In: Rodncks, J.V., Hesseltine, C.W., Mehlman, M.A. (Eds.), Mycotoxins in Human and Animal Health Park Forest South. Pathotox, IL, pp. 7–28.

Sur, D., Ramamurthy, T., Deen, J., Bhattacharya, S.K., 2004. Shigellosis: challenges & management issues. Indian J. Med. Res. 120, 454–462.

Tamarapu, S., McKillip, J.L., Drake, M., 2001. Development of a multiplex polymerase chain reaction assay for detection and differentiation of *Staphylococcus aureus* in dairy products. J. Food Prot. 64 (5), 664–668.

Taneja, N., Mewara, A., 2016. Review article, shigellosis: epidemiology in India. Indian J. Med. Res. 143, 565–576.

Tanyuksel, M., Petri, W.A., 2003. Laboratory diagnosis of amoebiasis. Clin. Microbiol. Rev. 16 (4), 713–729.

Tauxe, R.V., Doyle, M.P., Kuchenmüller, T., Schlundt, J., Stein, C.E., 2010. Evolving public health approaches to the global challenge of foodborne infections. Int. J. Food Microbiol. 139 (Suppl. 1), S16–S28.

Tchounwou, P.B., Wilson, B., Ishaque, A., 1999. Important considerations in the development of public health advisories for arsenic and arsenic-containing compounds in drinking water. Rev. Environ. Health. 14 (4), 211–229.

Tchounwou, P.B., Patlolla, A.K., Centeno, J.A., 2003. Carcinogenic and systemic health effects associated with arsenic exposure—a critical review. Toxicol. Pathol. 31 (6), 575–588.

Tchounwou, P.B., Centeno, J.A., Patlolla, A.K., 2004. Arsenic toxicity, mutagenesis and carcinogenesis – a health risk assessment and management approach. Mol. Cell Biochem. 255 (1–2), 47–55.

Thekdi, R.J., Lakhani, A.G., Rale, V.B., Panre, M.V., 1990. An outbreak of food poisoning suspected to be caused by vibrio fluvialis. J. Diarrhoeal Dis. Res. 8 (4), 163–165.

Thompson, R.C.A., Ash, A., 2016. Molecular epidemiology of Giardia and Cryptosporidium infections. Infect. Genet. Evol. 40, 315–323.

Tice R., α-Chaconine [20562-03-2] and α-solanine [20562-02-1], Review of toxicological literature. National Institute of Environmental Health Sciences, February, 1998. Available online: http://ntp.niehs.nih.gov/ntp/htdocs/Chem_Background/ExSumPdf/ChaconineSolanine.pdf.

Tieng, V., Le Bouguénec, C., du Merle, L., Bertheau, P., Desreumaux, P., Janin, A., et al., 2002. Binding of *Escherichia coli* adhesin AfaE to CD55 triggers cell-surface expression of the MHC class I-related molecule MICA. Proc. Natl Acad. Sci. U.S.A. 99, 2977–2982.

Tirumalai, P.S., 2013. Listeriosis and *Listeria monocytogenes* in India. Wudpecker J. Food Technol. 1 (6), 98–103.

Todd, E.C.D., 1989. Preliminary estimates of costs of foodborne disease in Canada and costs to reduce salmonellosis. J. Food Prot. 52 (8), 586–594.

Turner, P.C., Moore, S.E., Hall, A.J., Prentice, A.M., Wild, C.P., 2003. Modification of immune function through exposure to dietary aflatoxin in Gambian children. Environ. Health Perspect. 111 (2), 217–220.

United States Food and Drug Administration (FDA). Bad Bug Book. Foodborne Pathogenic Microorganisms and Natural Toxins Handbook. Phytohaemagglutinin, May 14, 2009. Available from: <http://www.fda.gov/food/foodsafety/foodborneillness/foodborneillnessfoodbornepathogens naturaltoxins/badbugbook/ucm071092.htm>.

Van der Merwe, K.J., Steyn, P.S., Foune, L., Scott, D.B., Theron, J.J., 1965. Ochratoxin A, a toxic metabolite produced by *Aspergillus ochraceus* Wilh. Nature. 205 (976), 1112–1113.

Van Walbeek, W., Scott, P.M., Harwig, J., Lawrence, J.W., 1969. Pemcillium inndicatum. Westhng: a new source of ochratoxin A. Can. J. Microbiol 15, 1281–1285.

Varga, J., Kevei, E., Rinyu, E., Teren, J., Kozakiewicz, Z., 1996. Ochratoxin production by *Aspergillus* species. Appl. Environ. Mtcrobiol. 62 (12), 4461–4464.

Varma, A., Gaur, K.J., 2002. The clinical spectrum of neurocysticercosis in the Uttaranchal region. J. Assoc. Physicians India. 50, 1398–1400.

Vemula, S.R., Kumar, R.N., Polasa, K., 2012. Foodborne diseases in India—a review. Brit. Food J. 114 (5), 661–680.

Walderhaug, M., 2007. Food borne pathogenic microorganisms and natural toxins. FDA, Center Food Saf. Appl. Nutr. 28, 48–65.

Wang, S., Shi, X., 2001. Molecular mechanisms of metal toxicity and carcinogenesis. Mol. Cell Biochem. 222 (1–2), 3–9.

WHO, 2005. International health regulations, second ed. Switzerland. Available from: <https://www.afro.who.int/sites/default/.../international_health_regulations_2005.pdf>.

World Health Organization (WHO), 2006. Fact Sheet -237: Food Safety and Food Borne Illness. Available from: <http://www.who.int/mediacentre/factsheets/fs237/edindex.html>.

World Health Organization, 2008. Foodborne disease outbreaks. Guidelines for investigation and control. Tech. Rep., Geneva, Switzerland.

WHO, 2015. WHO estimates of the global burden of foodborne diseases: foodborne disease burden epidemiology reference group 2007–2015. WHO Library Cataloguing-in-Publication Data, pp. 1–252. Available from http://apps.who.int/iris/bitstream/10665/199350/1/9789241565165.

WTO, 2010. The WTO agreement series -sanitary and phytosanitary measures. Available from: <https://www.wto.org/english/res_e/booksp_e/agrmntseries4_sps_e.pdf>.

WTO, 2014. The WTO Agreement Series—Technical Barriers to Trade. Available from: <https://www.wto.org/english/res_e/publications_e/tbttotrade_e.pdf>.

WTO, 2017. Decisions and recommendations adopted by the WTO Committee on Technical Barriers to Trade since 1st January 1995. (G/TBT/1/Rev.13).

Wu, F., 2013. Aflatoxin exposure and chronic human diseases: estimates of burden of disease. In: Unnevehr, L., Grace, D. (Eds.), Aflatoxins: Finding Solutions for Improved Food Safety. Int. Food Policy Res. Inst, Washington, DC, Focus 20, Brief 3.

Yadav, H., Mahna, R., Rekhi, T.K., 2015. A review on food safety in India with focus on food catering organizations in India. I.J.I.R.S.T. 1 (10), 214–219.

Zalups, R.K., Koropatnick, J., 2000. Temporal changes in metallothionein gene transcription in rat kidney and liver: relationship to content of mercury and metallothionein protein. J. Pharmacol. Exp. Ther. 295 (1), 74–82.

Zeliger, H.I., 2011. Human toxicology of chemical mixtures, Toxic Consequences Beyond the Impact of One-Component Product and Environmental Exposures, second ed. Elsevier, Oxford.

Zhang, W.J., 2018. Global pesticide use: profile, trend, cost / benefit and more. Proc. Int. Acad. Ecol. Environ. Sci. 8 (1), 1–27.

Further Reading

Abdul-Mutalib, N.A., Syafinaz, A.N., Sakai, K., Shirai, Y., 2015. An overview of foodborne illness and food safety in Malaysia. Int. Food Res. J. 22 (3), 896–901.

Byrd-Bredbenner, C., Berning, J., Martin-Biggers, J., Quick, V., 2013. Food safety in home kitchens: a synthesis of the literature. Int. J. Environ. Res. Public Health. 10, 4060–4085.

Cheeke, P.R., 1988. Toxicity and metabolism of pyrrolizidine alkalaoids. J. Anim. Sci. 66, 2343–2350.

Hathaway, S.C., Cook, R.L., 1997. A regulatory perspective on the potential uses of microbial risk assessment in international trade. Int. J. Food Microbiol 36 (2/3), 127–133.

Schlundt, J., Toyofuku, H., Jansen, I., Herbst, S.A., 2004. Emerging food-borne zoonoses. Re. Sci. Tech. 23 (2), 513–515.

Spencer, P.S., 1995. Lathyrism. In: de Wolff, F.A. (Ed.), Handbook of Clinical Neurology, vol. 21: Intoxications of the Nervous System, Part II. Elsevier, Amsterdam, pp. 1–20.

Tauxe, R.V., 1997. Emerging foodborne diseases: an evolving public health challenge. Emerg. Infect. Dis. 3 (4), 425–434.

Teisl, M.F., Roe, B.E., 2010. Consumer willingness to-pay to reduce the probability of retail foodborne pathogen contamination. Food Policy 35 (6), 521–530.

Vitaliano, A., Cama, D.V.M., Blaine, A., Mathison, B.S., 2015. Infections by intestinal Coccidia and *Giardia duodenalis*. Clin. Lab. Med. 35 (2), 423–444.

Web References

CAC, http://www.codexalimentarius.org.

CDC, https://www.cdc.gov/foodsafety/outbreaks/multistate-outbreaks/outbreaks-list.html.

FAO, http://www.fao.org.

FAOLEX database, http://www.fao.org/faolex/en/.

IPPC, https://www.ippc.int/en/.

ISO 2000, https://www.iso.org/iso-22000-food-safety-management.html.

ISO, http://www.iso.org.

JECFA, http://www.fao.org/food/food-safety-quality/scientific-advice/jecfa/en/.

OIE, http://www.oie.int.

SPS, https://www.wto.org/sps.
TBT, https://www.wto.org/tbt.
WHO, https://www.who.int.
WTO/TBT Agreement, https://www.wto.org/english/docs_e/legal_e/17-tbt_e.htm.
WTO, http://www.wto.org.
WTO/SPS Agreement, https://www.wto.org/english/docs_e/legal_e/15sps_01_e.htm.

Index

Note: Page numbers followed by "*f*" and "*t*" refer to figures and tables, respectively.

B

C

D

E

G

N

Y

Z

9780128163337